PETASCALE COMPUTING
ALGORITHMS AND APPLICATIONS

Chapman & Hall/CRC
Computational Science Series

SERIES EDITOR

Horst Simon

Associate Laboratory Director, Computing Sciences
Lawrence Berkeley National Laboratory
Berkeley, California, U.S.A.

AIMS AND SCOPE

This series aims to capture new developments and applications in the field of computational science through the publication of a broad range of textbooks, reference works, and handbooks. Books in this series will provide introductory as well as advanced material on mathematical, statistical, and computational methods and techniques, and will present researchers with the latest theories and experimentation. The scope of the series includes, but is not limited to, titles in the areas of scientific computing, parallel and distributed computing, high performance computing, grid computing, cluster computing, heterogeneous computing, quantum computing, and their applications in scientific disciplines such as astrophysics, aeronautics, biology, chemistry, climate modeling, combustion, cosmology, earthquake prediction, imaging, materials, neuroscience, oil exploration, and weather forecasting.

PUBLISHED TITLES

PETASCALE COMPUTING: Algorithms and Applications
Edited by David A. Bader

PETASCALE COMPUTING
ALGORITHMS AND APPLICATIONS

EDITED BY

DAVID A. BADER

Georgia Institute of Technology
Atlanta, U.S.A.

Chapman & Hall/CRC
Taylor & Francis Group
Boca Raton   London   New York

Chapman & Hall/CRC is an imprint of the
Taylor & Francis Group, an **informa** business

Chapman & Hall/CRC
Taylor & Francis Group
6000 Broken Sound Parkway NW, Suite 300
Boca Raton, FL 33487-2742

© 2008 by Taylor & Francis Group, LLC
Chapman & Hall/CRC is an imprint of Taylor & Francis Group, an Informa business

International Standard Book Number-13: 978-1-58488-909-0 (Hardcover)

Library of Congress Cataloging-in-Publication Data

Petascale computing : algorithms and applications / editor, David A. Bader.
    p. cm. -- (Computational science series)
  Includes bibliographical references and index.
  ISBN 978-1-58488-909-0 (hardback : alk. paper) 1. High performance computing. 2. Petaflops computers. 3. Parallel processing (Electronic computers) I. Bader, David A. II. Title. III. Series.

QA76.88.P475 2007
004'.35--dc22   2007044024

**Visit the Taylor & Francis Web site at
http://www.taylorandfrancis.com**

**and the CRC Press Web site at
http://www.crcpress.com**

To Sadie Rose

Chapman & Hall/CRC Computational Science Series

Computational science, the scientific investigation of physical processes through modelling and simulation on computers, has become generally accepted as the third pillar of science, complementing and extending theory and experimentation. This view was probably first expressed in the mid-1980s. It grew out of an impressive list of accomplishments in such diverse areas as astrophysics, aeronautics, chemistry, climate modelling, combustion, cosmology, earthquake prediction, imaging, materials, neuroscience, oil exploration, and weather forecasting. Today, in the middle of the first decade of the 21st century, the pace of innovation in information technology is accelerating, and consequently the opportunities for computational science and engineering abound. Computational science and engineering (CSE) today serves to advance all of science and engineering, and many areas of research in the future will be only accessible to those with access to advanced computational technology and platforms.

Progress in research using high performance computing platforms has been tightly linked to progress in computer hardware on one side and progress in software and algorithms on the other, with both sides generally acknowledged to contribute equally to the advances made by researchers using these technologies. With the arrival of highly parallel compute platforms in the mid-1990s, several subtle changes occurred that changed the face of CSE in the last decade. Because of the complexities of large-scale hardware systems and the increasing sophistication of modelling software, including multi-physics and multiscale simulation, CSE increasingly became a team science. The most successful practitioners of CSE today are multidisciplinary teams that include mathematicians and computer scientists. These teams have set up a software infrastructure, including a support infrastructure, for large codes that are well maintained and extensible beyond the set of original developers.

The importance of CSE for the future of research accomplishments and economic growth has been well established. "Computational science is now indispensable to the solution of complex problems in every sector, from traditional science and engineering domains to such key areas as national security, public health, and economic innovation," is the principal finding of the recent report of the President's Information Technology Advisory Committee (PITAC) in the U.S. (President's Information Technology Advisory Committee, *Computational Science: Ensuring America's Competitiveness* , Arlington, Virginia: National Coordination Office for Information Technology Research and Development, 2005, p. 2.)

As advances in computational science and engineering continue to grow at a rapid pace, it becomes increasingly important to present the latest research and applications to professionals working in the field. Therefore I welcomed the invitation by Chapman & Hall/CRC Press to become series editor and start this new series of books on computational science and engineering. The series aims to capture new developments and applications in the field of computational science, through the publication of a broad range of textbooks, reference works, and handbooks. By integrating mathematical, statistical, and computational methods and techniques, the titles included in the series are meant to appeal to students, researchers, and professionals, as well as interdisciplinary researchers and practitioners who are users of computing technology and practitioners of computational science. The inclusion of concrete examples and applications is highly encouraged. The scope of the series includes, but is not limited to, titles in the areas of scientific computing, parallel and distributed computing, high performance computing, grid computing, cluster computing, heterogeneous computing, quantum computing, and their application in scientific areas such as astrophysics, aeronautics, biology, chemistry, climate modelling, combustion, cosmology, earthquake prediction, imaging, materials, neuroscience, oil exploration, and weather forecasting, and others.

With this goal in mind I am very pleased to introduce the first book in the series, *Petascale Computing: Algorithms and Applications*, edited by my good colleague and friend David Bader. This book grew out of a workshop at Schloss Dagstuhl in February 2006, and is a perfect start for the series. It is probably the first book on real petascale computing. At the beginning of an exciting new phase in high performance computing, just as we are about to enter the age of petascale performance, the chapters in the book will form an ideal starting point for further investigations. They summarize the state of knowledge in algorithms and applications in 2007, just before the first petascale systems will become available. In the same way as petascale computing will open up new and unprecedented opportunities for research in computational science, I expect this current book to lead the new series to a deeper understanding and appreciation of research in computational science and engineering.

Berkeley, May 2007

Dr. Horst Simon
Series Editor
Associate Laboratory Director,
Computing Sciences
Lawrence Berkeley National Laboratory

Foreword

Over the last few decades, there have been innumerable science, engineering and societal breakthroughs enabled by the development of high performance computing (HPC) applications, algorithms, and architectures. These powerful tools have provided researchers, educators, and practitioners the ability to computationally translate and harness data gathered from around the globe into solutions for some of society's most challenging questions and problems.

An important force which has continued to drive HPC has been a community articulation of "frontier milestones," i.e., technical goals which symbolize the next level of progress within the field. In the 1990s, the HPC community sought to achieve computing at the teraflop (10^{12} floating point operations per second) level. Teraflop computing resulted in important new discoveries such as the design of new drugs to combat HIV and other diseases; simulations at unprecedented accuracy of natural phenomena such as earthquakes and hurricanes; and greater understanding of systems as large as the universe and smaller than the cell. Currently, we are about to compute on the first architectures at the petaflop (10^{15} floating point operations per second) level. Some communities are already in the early stages of thinking about what computing at the exaflop (10^{18} floating point operations per second) level will be like.

In driving towards the "<next frontier>-flop," the assumption is that achieving the next frontier in HPC architectures will provide immense new capacity and capability, will directly benefit the most resource-hungry users, and will provide longer-term benefits to many others. However, large-scale HPC users know that the ability to use frontier-level systems effectively is at least as important as their increased capacity and capability, and that considerable time and human, software, and hardware infrastructure are generally required to get the most out of these extraordinary systems.

Experience indicates that the development of scalable algorithms, models, simulations, analyses, libraries, and application components which can take full advantage of frontier system capacities and capabilities can be as challenging as building and deploying the frontier system itself.

For application codes to sustain a petaflop in the next few years, hundreds of thousands of processor cores or more will be needed, regardless of processor technology. Currently, few real existing HPC codes easily scale to this regime, and major code development efforts are critical to achieve the potential of the new petaflop systems. Scaling to a petaflop will involve improving physical models, mathematical abstractions, approximations, and other application components. Solution algorithms may need to be improved to increase accuracy of the resulting techniques. Input data sets may need to increase in resolution (generating more data points), and/or the accuracy of the input data for measured data may need to increase. Each application or algorithmic improvement poses substantial challenges in developing petascale codes, and may motivate new computer science discoveries apart from the new "domain" results of application execution.

This book presents efforts of the individuals who will likely be the first pioneers in petascale computing. Aggregating state-of-the-art efforts from some of the most mature and experienced applications teams in high performance computing and computational science, the authors herein are addressing the challenging problems of developing application codes that can take advantage of the architectural features of the new petascale systems in advance of their first deployment. Their efforts in petascale application development will require intimate knowledge of emerging petascale hardware and systems software, and considerable time to scale, test, evaluate and optimize petascale codes, libraries, algorithms, system software, all on new systems that have yet to be built.

This is an exciting period for HPC and a period which promises unprecedented discoveries "at scale" which can provide tangible benefits for both science and society. This book provides a glimpse into the challenging work of petascale's first wave of application and algorithm pioneers, and as such provides an important context for both the present and the future.

Dr. Francine Berman
San Diego Supercomputer Center, May 2007

Introduction

Science has withstood centuries of challenges by building upon the community's collective wisdom and knowledge through theory and experiment. However, in the past half-century, the research community has implicitly accepted a fundamental change to the scientific method. In addition to theory and experiment, computation is often cited as the third pillar as a means for scientific discovery. Computational science enables us to investigate phenomena where economics or constraints preclude experimentation, evaluate complex models and manage massive data volumes, model processes across interdisciplinary boundaries, and transform business and engineering practices. Increasingly, cyberinfrastructure is required to address our national and global priorities, such as sustainability of our natural environment by reducing our carbon footprint and by decreasing our dependencies on fossil fuels, improving human health and living conditions, understanding the mechanisms of life from molecules and systems to organisms and populations, preventing the spread of disease, predicting and tracking severe weather, recovering from natural and human-caused disasters, maintaining national security, and mastering nanotechnologies. Several of our most fundamental intellectual questions also require computation, such as the formation of the universe, the evolution of life, and the properties of matter.

Realizing that cyberinfrastructure is essential to research innovation and competitiveness, several nations are now in a "new arms race to build the world's mightiest computer" (John Markoff, *New York Times*, August 19, 2005). These petascale computers, expected around 2008 to 2012, will perform 10^{15} operations per second, nearly an order of magnitude faster than today's speediest supercomputer. In fact several nations are in a worldwide race to deliver high-performance computing systems that can achieve 10 petaflops or more within the next five years.

While petascale architectures certainly will be held as magnificent feats of engineering skill, the community anticipates an even harder challenge in scaling up algorithms and applications for these leadership-class supercomputing systems. This book presents a collection of twenty-four chapters from some of the international researchers who are early thought leaders in designing applications and software for petascale computing systems. The topics span several areas that are important for this task: scalable algorithm design for massive concurrency, computational science and engineering applications, petascale tools, programming methodologies, performance analyses, and scientific visualization.

The concept for this book resulted from a week-long seminar held at Schloss Dagstuhl, Germany, in February 2006, on "Architecture and Algorithms for Petascale Computing," organized by Ulrich Rüde, Horst D. Simon, and Peter Sloot. The Dagstuhl seminar focused on high end simulation as a tool for computational science and engineering applications. To be useful tools for science, such simulations must be based on accurate mathematical descriptions of the processes and thus they begin with mathematical formulations, such as partial differential equations, integral equations, graph-theoretic, or combinatorial optimization. Because of the ever-growing complexity of scientific and engineering problems, computational needs continue to increase rapidly. But most of the currently available hardware, software, systems, and algorithms are primarily focused on business applications or smaller scale scientific and engineering problems, and cannot meet the high-end computing needs of cutting-edge scientific and engineering work. This seminar primarily addressed the concerns of petascale scientific applications, which are highly compute- and data-intensive, cannot be satisfied in today's typical cluster environment, and tax even the largest available supercomputer.

This book includes a number of chapters contributed by participants of the Dagstuhl seminar, and several additional chapters were invited to span the breadth of petascale applications. Chapter 1 recognizes that petascale systems will require applications to exploit a high degree of concurrency and examines the performance characteristics of six full codes that are good potentials for the first wave of applications to run on early petascale systems. Chapter 2 discusses challenging computational science and engineering applications that are mission-critical to the United States' National Aeronautics and Space Administration (NASA). Chapters 3 and 4 focus on multiphysics simulations, using today's fastest computer at Lawrence Livermore National Laboratory (LLNL), and for the Uintah code that combines fluid-flow and material point (particle) methods using scalable adaptive mesh refinement, respectively. Chapter 5 discusses Enzo, a code for simulating cosmological evolution, from individual galaxies up to groups and clusters of galaxies, and beyond, providing a direct route to studying two of the most mysterious substances in modern physics: dark matter and dark energy. Chapter 6 describes numerical weather prediction at the mesoscale and convective scale, that captures weather events such as floods, tornados, hail, strong winds, lightning, hurricanes, and winter storms. The Community Climate System Model (CCSM), software for petascale climate science, is presented in Chapter 7. Chapter 8 moves into the area of petascale software and discusses a multiphysics application for simulating galaxies based on petascale Grid computing on distributed systems. Chapters 9, 10, and 11, discuss different aspects of performing molecular dynamics simulations on petascale systems. Chapter 9 is based on the popular NAMD code that simulates proteins and systems of molecules on highly-parallel systems using Charm++ (a petascale software framework presented in Chapter 20). Chapter 10 gives a special-purpose hardware solution called MD-GRAPE systems tailored for this problem. Chapter

11 looks at several leading candidate codes for biomolecular simulation on petascale computers, that can be used for predicting the structure, dynamics, and function of proteins on longer timescales. Large-scale combinatorial algorithms solve a number of growing computational challenges, and Chapter 12 presents several multithreaded graph-theoretic algorithms which are used in a variety of application domains. As petascale systems are likely to involve significantly more components, Chapter 13 presents a number of key areas in software and algorithms to improve the reliability of applications, such as diskless checkpointing and fault-tolerant implementation of the message passing interface (MPI) library. Chapter 14 describes a new supercomputer cluster installed at Tokyo Institute of Technology in Tokyo, Japan, called TSUB-AME (Tokyo-Tech Supercomputer and Ubiquitously Accessible Mass-storage Environment). Understanding that large-scale systems are built from collections of symmetric multiprocessor (SMP) nodes, in Chapter 15, programming methodologies are given for petascale computers, such as using MPI with OpenMP together. Naturally, benchmarks are needed to understand the performance bottlenecks in high-performance computing systems, and Chapter 16 discusses APEX-Map, which can be used to evaluate the performance and productivity of current and future systems. To fully exploit the massive capability provided by petascale systems, programmers will need to have on hand several different types of tools that scale and can provide insight on these large platforms. Chapter 17 gives a detailed summary of several types of performance analysis tools that are geared for petascale systems. Chapters 18 and 19 discuss finite elements, a popular method for solving partial differential equations on supercomputers. Chapter 18 presents a scalable multilevel finite elements solver called ParExPDE that uses expression templates to generate efficient code for petascale computing systems, while Chapter 19 describes a software framework for the high-level production of efficient finite element codes. Charm++ is a parallel programming system, overviewed in Chapter 20, that aims to enhance the programmer's productivity while producing highly scalable application codes. This chapter also illustrates several computational science and engineering applications enabled by Charm++. In Chapter 21, an annotation language is described that is embeddable in general-purpose languages and improves the performance and productivity of the scientific programmer. Managing locality in parallel programming is a critical concern as we employ petascale computing systems, and Chapter 22 discusses new productivity languages, such as Chapel, that support general parallel computation via a global-view, locality-aware, multithreaded programming model. Chapter 23 provides an historic perspective on architectural and programming issues as we move to petascale systems with the use of architectural accelerators and other technology trends. Finally, Chapter 24 discusses Cactus, an astrophysics framework for numerical relativity that simulates events from black holes to gamma ray bursts.

In addition to the contributing authors of this book, there are a few people I must mention by name who have influenced me and deserve a special thanks for their role in completing this book. First, I thank Uli Rüde, Horst Simon, and Peter Sloot, for organizing an intensive Dagstuhl seminar that included a full week of stimulating discussion surrounding the challenges of petascale computing, and for giving me the opportunity to create this book as an out come of this ground-breaking meeting. I give additional thanks to Horst, who serves as the Chapman & Hall/CRC Computational Science Series editor, not only for helping to shape this book, but also for sharing his insights and expertise in computational science with me, for guiding me professionally, and for his great friendship. I am inspired by Horst's dedication to computational science, his exuberance in high-performance computing, and his leadership in the community.

The development and production of this book would not have been possible without the able support and assistance of Randi Cohen, computer science acquisitions editor for Chapman & Hall/CRC Press. Randi brought this project from concept to production, and has been a wonderful colleague and friend throughout the process. She deserves the credit for all of the tedious work that made my job as editor appear easy. Randi's warm personality made this project fun, and her advice significantly improved the quality of this book.

I would like to express my deep appreciation to my research group at Georgia Tech: (in alphabetical order): Virat Agarwal, Aparna Chandramowlishwaran, Manisha Gajbe, Seunghwa Kang, Kamesh Madduri, and Amrita Mathuriya. They are an amazing bunch of graduate students, and have assisted me with the preparation of this book by organizing the book chapters, reading various drafts of the chapters, and proofreading the manuscript.

Finally, I thank my wife, Sara Gottlieb, and my daughter, Sadie Rose Bader-Gottlieb, for their understanding as I tried, not always successfully, to combine family time with computer time during the preparation of the book.

High-performance computing will enable breakthrough science and engineering in the 21st century. May this book inspire you to solve computational grand challenges that will help our society, protect our environment, and improve our understanding in fundamental ways, all through the efficient use of petascale computing.

Dr. David A. Bader
Georgia Institute of Technology, May 2007

About the Editor

David A. Bader is Executive Director of High-Performance Computing and a Professor in Computational Science and Engineering, a division within the College of Computing, at Georgia Institute of Technology. He received his Ph.D. in 1996 from The University of Maryland, and was awarded a National Science Foundation (NSF) Postdoctoral Research Associateship in Experimental Computer Science. He is the recipient of an NSF CAREER Award and an IBM Faculty Fellowship Award, an investigator on several NSF awards, was a distinguished speaker in the IEEE Computer Society Distinguished Visitors Program, and a member of the IBM PERCS team for the DARPA High Productivity Computing Systems program. Dr. Bader serves as Director of the Sony-Toshiba-IBM Center of Competence for the Cell Broadband Engine Processor located at Georgia Tech. He is an elected member of Internet2's Research Advisory Council, serves on the Steering Committees of the IPDPS and HiPC conferences, and has organized and chaired numerous conferences related to high-performance computing and computational science & engineering. Dr. Bader is an associate editor for several high impact publications including the IEEE Transactions on Parallel and Distributed Systems, the ACM Journal of Experimental Algorithmics, IEEE DSOnline, and Parallel Computing, a Senior Member of the IEEE Computer Society, and a Member of the ACM.

Dr. Bader has been a pioneer in the field of high-performance computing for problems in bioinformatics and computational genomics. He has co-chaired a series of meetings, the IEEE International Workshop on High-Performance Computational Biology (HiCOMB), written several book chapters, and co-edited special issues of the Journal of Parallel and Distributed Computing (JPDC) and IEEE Transactions on Parallel and Distributed Systems (TPDS) on high-performance computational biology. He has co-authored over 80 articles in peer-reviewed journals and conferences, and his main areas of research are in parallel algorithms, combinatorial optimization, and computational biology and genomics.

Contributors

Michael Aftosmis
NASA Ames Research Center
Moffett Field, California

Pratul K. Agarwal
Oak Ridge National Laboratory
Oak Ridge, Tennessee

Sadaf R. Alam
Oak Ridge National Laboratory
Oak Ridge, Tennessee

Gabrielle Allen
Louisiana State University
Baton Rouge, Louisiana

Martin Sandve Alnæs
Simula Research Laboratory and
University of Oslo, Norway

Steven F. Ashby
Lawrence Livermore National Laboratory
Livermore, California

David A. Bader
Georgia Institute of Technology
Atlanta, Georgia

Benjamin Bergen
Los Alamos National Laboratory
Los Alamos, New Mexico

Jonathan W. Berry
Sandia National Laboratories
Albuquerque, New Mexico

Martin Berzins
University of Utah
Salt Lake City, Utah

Abhinav Bhatele
University of Illinois
Urbana-Champaign, Illinois

Christian Bischof
RWTH Aachen University
Germany

Rupak Biswas
NASA Ames Research Center
Moffett Field, California

Jonathan Carter
Lawrence Berkeley National Laboratory
Berkeley, California

Eric Bohm
University of Illinois
Urbana-Champaign, Illinois

Zizhong Chen
Jacksonville State University
Jacksonville, Alabama

James Bordner
University of California, San Diego
San Diego, California

Joseph R. Crobak
Rutgers, The State University of New Jersey
Piscataway, New Jersey

George Bosilca
University of Tennessee
Knoxville, Tennessee

Roxana E. Diaconescu
Yahoo! Inc.
Burbank, California

Greg L. Bryan
Columbia University
New York, New York

Peter Diener
Louisiana State University
Baton Rouge, Louisiana

Marian Bubak
AGH University of Science and Technology
Kraków, Poland

Jack J. Dongarra
University of Tennessee, Knoxville, Oak Ridge National Laboratory, and University of Manchester

Andrew Canning
Lawrence Berkeley National Laboratory
Berkeley, California

John B. Drake
Oak Ridge National Laboratory
Oak Ridge, Tennessee

Kelvin K. Droegemeier
University of Oklahoma
Norman, Oklahoma

Tobias Gradl
Friedrich–Alexander–Universität
Erlangen, Germany

Stéphane Ethier
Princeton University
Princeton, New Jersey

William D. Gropp
Argonne National Laboratory
Argonne, Illinois

Christoph Freundl
Friedrich–Alexander–Universität
Erlangen, Germany

Robert Harkness
University of California, San Diego
San Diego, California

Karl Fürlinger
University of Tennessee
Knoxville, Tennessee

Albert Hartono
Ohio State University
Columbus, Ohio

Al Geist
Oak Ridge National Laboratory
Oak Ridge, Tennessee

Thomas C. Henderson
University of Utah
Salt Lake City, Utah

Michael Gerndt
Technische Universität München
Munich, Germany

Bruce A. Hendrickson
Sandia National Laboratories
Albuquerque, New Mexico

Tom Goodale
Louisiana State University
Baton Rouge, Louisiana

Alfons G. Hoekstra
University of Amsterdam
Amsterdam, The Netherlands

Philip W. Jones
Los Alamos National Laboratory
Los Alamos, New Mexico

Laxmikant Kalé
University of Illinois
Urbana-Champaign, Illinois

Shoaib Kamil
Lawrence Berkeley National Laboratory
Berkeley, California

Cetin Kiris
NASA Ames Research Center
Moffett Field, California

Uwe Küster
University of Stuttgart
Stuttgart, Germany

Julien Langou
University of Colorado
Denver, Colorado

Hans Petter Langtangen
Simula Research Laboratory and
University of Oslo, Norway

Michael Lijewski
Lawrence Berkeley National Laboratory
Berkeley, California

Anders Logg
Simula Research Laboratory and
University of Oslo, Norway

Justin Luitjens
University of Utah
Salt Lake City, Utah

Kamesh Madduri
Georgia Institute of Technology
Atlanta, Georgia

Kent-Andre Mardal
Simula Research Laboratory and
University of Oslo, Norway

Satoshi Matsuoka
Tokyo Institute of Technology
Tokyo, Japan

John M. May
Lawrence Livermore National Laboratory
Livermore, California

Celso L. Mendes
University of Illinois
Urbana-Champaign, Illinois

Brian O'Shea
Los Alamos National Laboratory
Los Alamos, New Mexico

Dieter an Mey
RWTH Aachen University
Germany

Christian D. Ott
University of Arizona
Tucson, Arizona

Tetsu Narumi
Keio University
Japan

James C. Phillips
University of Illinois
Urbana-Champaign, Illinois

Michael L. Norman
University of California, San Diego
San Diego, California

Simon Portegies Zwart
University of Amsterdam,
Amsterdam, The Netherlands

Boyana Norris
Argonne National Laboratory
Argonne, Illinois

Thomas Radke
Albert-Einstein-Institut
Golm, Germany

Yousuke Ohno
Institute of Physical and Chemical
Research (RIKEN)
Kanagawa, Japan

Michael Resch
University of Stuttgart
Stuttgart, Germany

Daniel Reynolds
University of California, San Diego
San Diego, California

Leonid Oliker
Lawrence Berkeley National Laboratory
Berkeley, California

Ulrich Rüde
Friedrich–Alexander–Universität
Erlangen, Germany

Ola Skavhaug
Simula Research Laboratory and
University of Oslo, Norway

Samuel Sarholz
RWTH Aachen University
Germany

Peter M.A. Sloot
University of Amsterdam
Amsterdam, The Netherlands

Erik Schnetter
Louisiana State University
Baton Rouge, Louisiana

Erich Strohmaier
Lawrence Berkeley National Laboratory
Berkeley, California

Klaus Schulten
University of Illinois
Urbana-Champaign, Illinois

Makoto Taiji
Institute of Physical and Chemical
Research (RIKEN)
Kanagawa, Japan

Edward Seidel
Louisiana State University
Baton Rouge, Louisiana

Christian Terboven
RWTH Aachen University,
Germany

John Shalf
Lawrence Berkeley National Laboratory
Berkeley, California

Mariana Vertenstein
National Center for Atmospheric Research
Boulder, Colorado

Bo-Wen Shen
NASA Goddard Space Flight Center
Greenbelt, Maryland

Rick Wagner
University of California, San Diego
San Diego, California

Daniel Weber
University of Oklahoma
Norman, Oklahoma

Hans P. Zima
Jet Propulsion Laboratory,
California Institute of Technology
Pasadena, California
and
University of Vienna, Austria

James B. White, III
Oak Ridge National Laboratory
Oak Ridge, Tennessee

Terry Wilmarth
University of Illinois
Urbana-Champaign, Illinois

Patrick H. Worley
Oak Ridge National Laboratory
Oak Ridge, Tennessee

Bryan Worthen
University of Utah
Salt Lake City, Utah

Ming Xue
University of Oklahoma
Norman, Oklahoma

Gengbin Zheng
University of Illinois
Urbana-Champaign, Illinois

Contents

8 Towards Distributed Petascale Computing 147

Alfons G. Hoekstra, Simon Portegies Zwart, Marian Bubak, and Peter M.A. Sloot

9 Biomolecular Modeling in the Era of Petascale Computing 165

Klaus Schulten, James C. Phillips, Laxmikant V. Kalé, and Abhinav Bhatele

10 Petascale Special-Purpose Computer for Molecular Dynamics Simulations 183

Makoto Taiji, Tetsu Narumi, and Yousuke Ohno

13 Disaster Survival Guide in Petascale Computing: An Algorithmic Approach   263

Jack J. Dongarra, Zizhong Chen, George Bosilca, and Julien Langou

14 The Road to TSUBAME and Beyond   289

Satoshi Matsuoka

15 Petaflops Basics - Performance from SMP Building Blocks   311

Christian Bischof, Dieter an Mey, Christian Terboven, and Samuel Sarholz

List of Tables

List of Figures

Chapter 1

Performance Characteristics of Potential Petascale Scientific Applications

John Shalf, Leonid Oliker, Michael Lijewski, Shoaib Kamil, Jonathan Carter, Andrew Canning

CRD/NERSC, Lawrence Berkeley National Laboratory, Berkeley, CA 94720

Stéphane Ethier

Princeton Plasma Physics Laboratory, Princeton University, Princeton, NJ 08453

Abstract   After a decade where HEC (high-end computing) capability was dominated by the rapid pace of improvements to CPU clock frequency, the performance of next-generation supercomputers is increasingly differentiated by varying interconnect designs and levels of integration. Understanding the trade-offs of these system designs, in the context of high-end numerical simulations, is a key step towards making effective petascale computing a reality. This work represents one of the most comprehensive performance evaluation studies to date on modern HEC systems, including the IBM Power5, AMD Opteron, IBM BG/L, and Cray X1E. A novel aspect of our study is the emphasis on full applications, with real input data at the scale desired by computational scientists in their unique domain. We examine five candidate ultra-scale applications, representing a broad range of algorithms and computational structures. Our work includes the highest concurrency experiments to date on five of our six applications, including 32K processor scalability for two of our codes and describes several successful optimization strategies on BG/L, as well as improved X1E vectorization. Overall results indicate that our evaluated codes have the potential to effectively utilize petascale resources; however, several applications will require reengineering to incorporate the additional levels of parallelism necessary to utilize the vast concurrency of upcoming ultra-scale systems.

1.1 Introduction

Computational science is at the dawn of petascale computing capability, with the potential to achieve simulation scale and numerical fidelity at hitherto unattainable levels. However, harnessing such extreme computing power will require an unprecedented degree of parallelism both within the scientific applications and at all levels of the underlying architectural platforms. Unlike a decade ago — when the trend of HEC (high-end computing) systems was clearly towards building clusters of commodity components — today one sees a much more diverse set of HEC models. Increasing concerns over power efficiency are likely to further accelerate recent trends towards architectural diversity through new interest in customization and tighter system integration. Power dissipation concerns are also driving high-performance computing (HPC) system architectures from the historical trend of geometrically increasing clock rates towards geometrically increasing core counts (multicore), leading to daunting levels of concurrency for future petascale systems. Employing an even larger number of simpler processor cores operating at a lower clock frequency, as demonstrated by BG/L, offers yet another approach to improving the power efficiency of future HEC platforms. Understanding the trade-offs of these computing paradigms, in the context of high-end numerical simulations, is a key step towards making effective petascale computing a reality. The main contribution of this work is to quantify these trade-offs by examining the effectiveness of various architectural models for HEC with respect to absolute performance and scalability across a broad range of key scientific domains.

A novel aspect of our effort is the emphasis on full applications, with real input data at the scale desired by computational scientists in their unique domain, which builds on our previous efforts [20, 21, 5, 19] and complements a number of other related studies [6, 18, 31, 11]. Our application suite includes a broad spectrum of numerical methods and data-structure representations in the areas of magnetic fusion (GTC), astrophysics (Cactus), fluid dynamics (ELBM3D), materials science (PARATEC), and AMR gas dynamics (HyperCLaw). We evaluate performance on a wide range of architectures with varying degrees of component customization, integration, and power consumption, including: the Cray X1E customized parallel vector-processor, which utilizes a tightly coupled custom interconnect; the commodity IBM Power5 and AMD Opteron processors integrated with custom fat-tree based Federation and 3D-torus based XT3 interconnects, respectively; the commodity Opteron processor integrated with the InfiniBand high-performance commodity network; and the IBM Blue Gene/L (BG/L) which utilizes a customized SOC (system on chip) based on commodity, low-power embedded cores, combined with multiple network interconnects.

This work represents one of the most comprehensive performance evaluation studies to date on modern HEC platforms. We present the highest concurrency results ever conducted for our application suite, and show that the BG/L can attain impressive scalability characteristics all the way up to 32K processors on two of our applications. We also examine several application optimizations, including BG/L processor and interconnect-mappings for the SciDAC [24] GTC code, which achieve significant performance improvements over the original superscalar version. Additionally, we implement several optimizations for the HyperCLaw AMR calculation, and show significantly improved performance and scalability on the X1E vector platform, compared with previously published studies. Overall, we believe that these comprehensive evaluation efforts lead to more efficient use of community resources in both current installations and in future designs.

1.2 Target Architectures

Our evaluation testbed uses five different production HEC systems. Table 1.1 presents several key performance parameters of *Bassi, Jacquard, BG/L, Jaguar,* and *Phoenix,* including: STREAM benchmark results [28] showing the measured EP-STREAM [13] triad bandwidth when all processors within a node simultaneously compete for main memory; the ratio of STREAM bandwidth to the peak computational rate; the measured inter-node MPI latency [7]; and the measured bidirectional MPI bandwidth per processor pair when each processor simultaneously exchanges data with a distinct processor in another node.

Bassi: Federation/Power5: The Power5-based Bassi system is located

---

‡Ratio of measured STREAM bandwidth to peak processor computational rate.

*Minimum latency for the XT3 torus. There is a nominal additional latency of $50ns$ per hop through the torus.

†Minimum latency for the BG/L torus. There is an additional latency of up to $69ns$ per hop through the torus.

TABLE 1.1: Architectural highlights of studied HEC platforms. An MSP is defined as a processor for the X1E data.

Name	Arch	Network	Network Topology	P/ Node	Clock (GHz)	Peak (GF/s/P)	STREAM BW (GB/s/P)	STREAM (B/F)‡	MPI Lat (μsec)	MPI BW (GB/s/P)
Jaguar	Opteron	XT3	3D Torus	1	2.6	5.2	2.5	0.48	5.5*	1.2
Jacquard	Opteron	InfiniBand	Fat-tree	2	2.2	4.4	2.3	0.51	5.2	0.73
Bassi	Power5	Federation	Fat-tree	8	1.9	7.6	6.8	0.85	4.7	0.69
BG/L	PPC440	Custom	3D Torus	2	0.7	2.8	0.9	0.31	2.2†	0.16
Phoenix	X1E	Custom	4D-Hcube	4	1.1	18.0	9.7	0.54	5.0	2.9

at Lawrence Berkeley National Laboratory (LBNL) and contains 122 8-way symmetric multiprocessor (SMP) nodes interconnected via a two-link network adapter to the IBM Federation HPS switch. The 1.9 GHz RISC Power5 processor contains a 64KB instruction cache, a 1.9MB on-chip L2 cache as well as a 36MB on-chip L3 cache. The IBM custom Power5 chip has two Synchronous Memory Interface (SMI) chips, aggregating four DDR 233 MHz channels for an impressive measured STREAM performance of 6.8GB/s per processor. The Power5 includes an integrated memory controller and integrates the distributed switch fabric between the memory controller and the core/caches. The experiments in this work were conducted under AIX 5.2. Several experiments were also conducted on the 1,532 node (12,256 processor) Power5-based Purple system at Lawrence Livermore National Laboratory.

Jacquard: InfiniBand/Opteron: The Opteron-based Jacquard system is also located at LBNL. Jacquard contains 320 dual-processor nodes and runs Linux 2.6.5 (PathScale 2.0 compiler). Each node contains two 2.2 GHz Opteron processors, interconnected via InfiniBand fabric in a fat-tree configuration. The Opteron uses a single-instruction multiple-data (SIMD) floating-point unit accessed via the SSE2 instruction set extensions, and can execute two double-precision floating-point operations per cycle. The processor possesses an on-chip DDR memory controller as well as tightly-integrated switch interconnection via HyperTransport. Jacquard's Infiniband network utilizes 4X single-data-rate links, configured in a 2-layer CLOS/fat-tree configuration using Mellanox switches.

Jaguar: XT3/Opteron: The Jaguar experiments were conducted on the 5,202 node XT3 system operated by the National Leadership Computing Facility (NLCF) at Oak Ridge National Laboratory running the Catamount microkernel on the compute nodes. Each node of the Cray XT3 contains a single, dual-core 2.6 GHz AMD Opteron processor. The processors are tightly integrated to the XT3 interconnect via a Cray SeaStar ASIC through a 6.4GB/s bidirectional HyperTransport interface. All the SeaStar routing chips are interconnected in a 3D-torus topology, where — similar to the BG/L system — each node has a direct link to six of its nearest neighbors on the torus with a peak bidirectional bandwidth of 7.6GB/s. Note that Jaguar is similar to Jacquard in terms of the processor architecture and peak memory bandwidth; however, in addition to having vastly different interconnect technologies, Jaguar uses dual-core technology as opposed to Jacquard's single-core processors.

BG/L: Multiple Networks/PPC440: The IBM Blue Gene/L (BG/L) system represents a unique approach to supercomputing design, which allows unprecedented levels of parallelism, with low unit cost and power consumption characteristics. Our work presents performance results on the 1024-node BG/L located at Argonne National Laboratory (ANL) running OS SLES9, as well as the 20K-node system at IBM's Watson Research Center system (BGW), currently the world's third most powerful supercomputer [17] (16K nodes available at time of testing). Each BG/L node contains two 700MHz PowerPC 440 processors, on-chip memory controller and communication logic, and only 512MB of memory. The CPU's dual FPUs (called double hummer) are capable of dispatching two MADDs per cycle, for a peak processor performance of 2.8 GFLOPS. However, the second FPU is not an independent unit, and can only be used with special SIMD instructions — thus, making it difficult for the core to perform at close to peak except for specially hand-tuned code portions. Our experiments primarily examine performance in *coprocessor mode* where one core is used for computation and the second is dedicated to communication. Additionally, several experiments were conducted using 32K processors of BGW in *virtual node mode* where both cores are used for both computation and communication.

The BG/L nodes are connected via five different networks, including a torus, collective tree, and global interrupt tree. The 3D-torus interconnect is used for general-purpose point-to-point message-passing operations using 6 independent point-to-point serial links to the 6 nearest neighbors that operate at 175MB/s per direction (bidirectional) for an aggregate bandwidth of 2.1GB/s per node. The global tree collective network is designed for high-bandwidth broadcast operations (one-to-all) using three links that operate at a peak bandwidth of 350MB/s per direction for an aggregate 2.1GB/s bidirectional bandwidth per node. Finally, the global interrupt network provides fast barriers and interrupts with a system-wide constant latency of $\approx 1.5\mu s$.

Phoenix: Custom Network/X1E: The Cray X1E is the recently released follow-on to the X1 vector platform. Vector processors expedite uniform operations on independent data sets by exploiting regularities in the computational structure. The X1E computational core, called the single-streaming processor (SSP), contains two 32-stage vector pipes running at 1.13 GHz. Each SSP contains 32 vector registers holding 64 double-precision words, and operates at 4.5 GFLOPS. The SSP also contains a two-way out-of-order superscalar processor (564 MHz) with two 16KB caches (instruction and data). Four SSPs can be combined into a logical computational unit called the multi-streaming processor (MSP), and share a 2-way set associative 2MB data Ecache, with a peak performance of 18 GFLOPS. Note that the scalar unit operates at

1/4th the peak of SSP vector performance, but offers effectively 1/16 MSP performance if a loop can neither be multi-streamed nor vectorized. The X1E interconnect is hierarchical, with subsets of 16 SMP nodes (each containing 8 MSPs) connected via a crossbar, these subsets are connected in a 4D-hypercube topology. All reported X1E experiments were performed on a 768-MSP system running UNICOS/mp 3.0.23 and operated by Oak Ridge National Laboratory.

1.3 Scientific Application Overview

Five applications from diverse areas in scientific computing were chosen to compare the performance of our suite of leading supercomputing platforms. We examine: GTC, a magnetic fusion application that uses the particle-in-cell approach to solve nonlinear gyrophase-averaged Vlasov-Poisson equations; Cactus, an astrophysics framework for high-performance computing that evolves Einstein's equations from the Theory of General Relativity; ELBM3D, a Lattice-Boltzmann code to study turbulent fluid flow; PARATEC, a first principles materials science code that solves the Kohn-Sham equations of density functional theory to obtain electronic wave functions; HyperCLaw, an adaptive mesh refinement (AMR) framework for solving the hyperbolic conservation laws of gas dynamics via a higher-order Godunov method. Table 1.2 presents an overview of the application characteristics from our evaluated simulations.

TABLE 1.2: Overview of scientific applications examined in our study.

Name	Lines	Discipline	Methods	Structure
GTC	5,000	Magnetic Fusion	Particle in Cell, Vlasov-Poisson	Particle/Grid
Cactus	84,000	Astrophysics	Einstein Theory of GR, ADM-BSSN	Grid
ELBD	3,000	Fluid Dynamics	Lattice-Boltzmann, Navier-Stokes	Grid/Lattice
PARATEC	50,000	Materials Science	Density Functional Theory, FFT	Fourier/Grid
HyperCLaw	69,000	Gas Dynamics	Hyperbolic, High-order Godunov	Grid AMR

These codes are candidate ultra-scale applications with the potential to fully utilize leadership-class computing systems, and represent a broad range of algorithms and computational structures. Communication characteristics include: nearest-neighbor and allreduce communication across the toroidal grid and poloidal grid (respectively) for the particle-in-cell GTC calculation; simple ghost boundary exchanges for the stencil-based ELBM3D and Cactus computations; all-to-all data transpositions used to implement PARATEC's

3D FFTs, and complex data movements required to create and dynamically adapt grid hierarchies in HyperCLaw. Examining these varied computational methodologies across a set of modern supercomputing platforms allows us to study the performance trade-offs of different architectural balances and topological interconnect approaches.

To study the topological connectivity of communication for each application, we utilize the IPM [27] tool, recently developed at LBNL. IPM is an application profiling layer that allows us to noninvasively gather the communication characteristics of these codes as they are run in a production environment. By recording statistics on these message exchanges we can form an undirected graph which describes the topological connectivity required by the application. We use the IPM data to create representations of the topological connectivity of communication for each code as a matrix — where each point in the graph indicates message exchange and (color coded) intensity between two given processors — highlighting the vast range of communication requirements within our application suite. IPM is also used to collect statistics on MPI utilization and the sizes of point-to-point messages, allowing us to quantify the fraction of messages that are bandwidth- or latency-bound. The dividing line between these two regimes is an aggressively designed bandwidth-delay product of 2Kb. See [14, 25] for a more detailed explanation of the application communication characteristics, data collection methodology, and bandwidth-delay product thresholding.

Experimental results show either strong scaling (where the problem size remains fixed regardless of concurrency), or weak scaling (where the problem size grows with concurrency such that the per-processor computational requirement remains fixed) — whichever is appropriate for a given application's large-scale simulation. Note that these applications have been designed and highly optimized on superscalar platforms; thus, we describe newly devised optimizations for the vector platforms where appropriate. Performance results measured on these systems, presented in GFLOPS per processor (denoted as GFLOPS/P) and percentage of peak, are used to compare the time to solution of our evaluated platforms. The GFLOPS value is computed by dividing a valid baseline flop-count by the measured wall-clock time of each platform — thus the ratio between the computational rates is the same as the ratio of runtimes across the evaluated systems. All results are shown using the fastest (optimized) available code versions.

1.4 GTC: Particle-in-Cell Magnetic Fusion

GTC is a 3D particle-in-cell code developed for studying turbulent transport in magnetic confinement fusion plasmas [15, 8]. The simulation geometry is that of a torus, which is the natural configuration of all tokamak fusion devices. As the charged particles forming the plasma move within the externally imposed magnetic field, they collectively create their own self-consistent electrostatic (and electromagnetic) field that quickly becomes turbulent under driving temperature and density gradients. The particle-in-cell (PIC) method describes this complex interaction between fields and particles by solving the 5D gyro-averaged kinetic equation coupled to the Poisson equation. In the PIC method, the interaction between particles is calculated using a grid on which the charge of each particle is deposited and then used in the Poisson equation to evaluate the field. This is the scatter phase of the PIC algorithm. Next, the force on each particle is gathered from the grid-base field and evaluated at the particle location for use in the time advance.

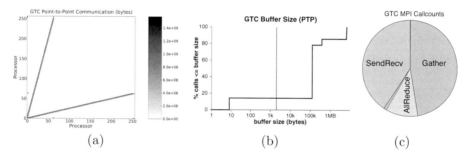

FIGURE 1.1: GTC (a) communication topology and intensity for point-to-point messaging; (b) cumulative distribution function of point-to-point message buffer sizes with the horizontal dividing line representing the border between bandwidth-bound (right of line) and latency-bound (left of line) message sizes; and (c) breakdown of MPI calls counts.

An important approximation in GTC comes from the fact that fast particle motion along the magnetic field lines leads to a quasi-two-dimensional structure in the electrostatic potential. Thus, the Poisson equation needs only to be solved on a 2D poloidal plane. GTC utilizes a simple iterative solver since the linear system is diagonally dominant [16]; note that this differs from most PIC techniques that solve a spectral system via FFTs. For 10 particles per cell per processor, the solve phase accounts only for 6% of the compute time, a very small fraction compared to the 85% spent in the scatter and gather-push phases. The primary direction of domain decomposition for GTC is in the toroidal direction in order to take advantage of the locality of the 2D Poisson solve. GTC uses a logically non-rectangular field-aligned grid, in part, to keep the number of particles per cell nearly constant. The mesh effectively twists along with the magnetic field in the toroidal direction.

GTC utilizes two levels of distributed memory parallelism. The first, based on the original GTC implementation, is a one-dimensional domain decomposition in the toroidal direction (long way around the torus) while the second is a particle distribution within each toroidal slice. The processors in charge of the same toroidal slice have a complete copy of the local grid. However, each processor within the toroidal slice holds only a fraction of the particles, and data is exchanged between slices via a local MPI sub-communicator when updating grid quantities. In the toroidal direction, a separate MPI sub-communicator supports communication between the toroidal slices to support movement of particles between the slices.

Figure 1.1 (a) shows the regular communication structure exhibited by GTC. This particle-in-cell calculation uses a one-dimensional domain decomposition across the toroidal computational grid, causing each processor to exchange data with its two neighbors as particles cross the left and right boundaries. Additionally, there is a particle decomposition within each toroidal slice as described above. Therefore, on average each MPI task communicates with 4 neighbors, so much simpler interconnect networks will be adequate for its requirements. The cumulative distribution function of message buffer sizes in Figure 1.1 (b), shows that the communication requirements are strongly bandwidth bound, where more than 80% of the message sizes are in the bandwidth-bound regime — the horizontal reference line represents the border between bandwidth-bound (right of line) and latency-bound (left of line) message sizes, (see [25] for more details). The distribution of MPI call counts shown in Figure 1.1 (c) are strongly biased towards collective calls that have very small (<128 byte) latency-bound messages sizes; at the petaflop scale level, scalable and efficient collective messaging will become an increasingly important factor for maintaining scalable performance for GTC and other PIC codes.

1.4.1   Experimental results

The benchmarks in our study were performed using 0.2 to 2 billion particles on a 2 million cell mesh (100 particles per cell per processor core). Figure 1.2 shows the results of a weak-scaling study of GTC on the platforms under comparison in both (a) raw performance and (b) percentage of peak. The size of the grid remains fixed since it is prescribed by the size of the fusion device being simulated, while the number of particles is increased in a way that keeps the same amount of work per processor for all cases.

Looking at the raw performance we see that the Phoenix platform clearly stands out with a GFLOPS/P rate up to 4.5 times higher than the second highest performer, the XT3 Jaguar. This was expected since the version of GTC used on Phoenix has been extensively optimized to take advantage of the multi-streaming vector processor [21]. In the latest improvements, the dimensions of the main arrays in the code have been reversed in order to speed up access to the memory banks, leading to higher performance. This change is not implemented in the superscalar version since it reduces cache reuse and hence slows down the code. Although still high, the performance

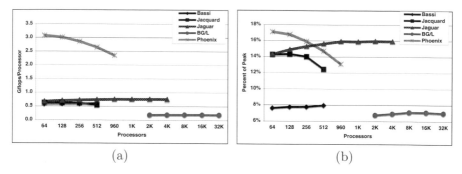

FIGURE 1.2: GTC weak-scaling performance using 100 particles per cell per processor (10 for BG/L) in (a) runtime and (b) percentage of peak. All performance data on the BG/L system were collected in virtual node mode.

per processor on the X1E decreases significantly as the number of processors, or MSPs, increases. This is probably due to the increase in intra-domain communications that arises when the number of processors per toroidal domain increases. An *allreduce* operation is required within each domain to sum up the contribution of each processor, which can lead to lower performance in certain cases. Optimizing the processor mapping is one way of improving the communications but we have not explored this avenue on Phoenix yet.

Jacquard, Bassi, and Jaguar have very similar performance in terms of GFLOPS/P although Bassi is shown to deliver only about half the percentage of peak achieved on Jaguar, which displays outstanding efficiency and scaling all the way to 5184 processors. The percentage of peak achieved by particle-in-cell codes is generally low since the gather-scatter algorithm that characterizes this method involves a large number of random accesses to memory, making the code sensitive to memory access latency. However, the AMD Opteron processor used in both Jacquard and Jaguar delivers a significantly higher percentage of peak for GTC compared to all the other superscalar processors. It even rivals the percentage of peak achieved on the vector processor of the X1E Phoenix. This higher GTC efficiency on the Opteron is due, in part, to relatively low main memory latency access. On all systems other than Phoenix, GTC exhibits near perfect scaling, including up to 5K processors on Jaguar.

The percentage of peak achieved by GTC on BG/L is the lowest of the systems under study but the scalability is very impressive, all the way to 32,768 processors! The porting of GTC to the BG/L system was straightforward but initial performance was disappointing. Several optimizations were then applied to the code, most of them having to do with using BG/L-optimized math libraries such as MASS and MASSV to accelerate the performance of transcendental functions such as `sin()`, `cos()`, and `exp()`. Whereas the original code used the default GNU implementations for the transcendentals, the

MASS and MASSV libraries include vector versions of those functions that can take advantage of improved instruction scheduling and temporal locality to achieve substantial improvements to the throughput of these operations in loops.* By replacing the default math intrinsics with their MASSV equivalents, we obtained a 30% increase in performance. Other optimizations consisted of loop unrolling and replacing calls to the Fortran `aint(x)` intrinsic function by `real(int(x))`. `aint(x)` results in a function call that is much slower than using the equivalent `real(int(x))`. These combined optimizations resulted in a performance improvement of almost 60% over original runs. We note that the results presented here are for virtual node mode where both cores on the node are used for computation. GTC maintains over 95% of its single-core (coprocessor-mode) performance when employing both cores on the BG/L node, which is quite promising as more cores will be added to upcoming processor roadmaps.

Another useful optimization performed on BGW was processor mapping. BG/L's 3D-torus interconnect topology used for point-to-point communications is ideally suited for the toroidal geometry of GTC's computational domain. Additionally, the number of toroidal-slice domains used in the GTC simulations exactly match one of the dimensions of the BG/L network torus. Thus by using an explicit mapping file that aligns GTC's point-to-point communications along these toroidal slice domains (used for moving particles between the toroidal slices) to the BG/L link topology, the performance of the code improved by an additional 30% over the default communication mapping.

1.5   ELBM3D: Lattice-Boltzmann Fluid Dynamics

Lattice-Boltzmann methods (LBM) have proven a good alternative to conventional numerical approaches for simulating fluid flows and modeling physics in fluids [29]. The basic idea is to develop a simplified kinetic model that incorporates the essential physics, and reproduces correct macroscopic averaged properties. These algorithms have been used extensively since the mid-1980s for simulating Navier-Stokes flows, and more recently extended to treat multiphase flows, reacting flows, diffusion processes, and magneto-hydrodynamics. As can be expected from explicit algorithms, LBM are prone to numerical nonlinear instabilities as one pushes to higher Reynolds numbers. These numerical instabilities arise because there are no imposed constraints to enforce the distribution functions to remain nonnegative. Entropic LBM algorithms, which do preserve the nonnegativity of the distribution functions — even in the limit of arbitrary small transport coefficients — have recently been developed for Navier-Stokes turbulence [2], and are incorporated into a recently developed code [30].

---

*The authors thank Bob Walkup for his BG/L optimization insights.

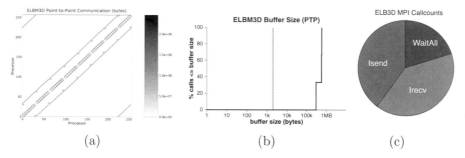

(a) (b) (c)

FIGURE 1.3: ELBM3D (a) communication topology and intensity; (b) cumulative distribution function of point-to-point message buffer sizes; and (c) breakdown of MPI calls counts.

While LBM methods lend themselves to easy implementation of difficult boundary geometries (e.g., by the use of bounce-back to simulate no slip wall conditions), here we report on 3D simulations under periodic boundary conditions, with the spatial grid and phase space velocity lattice overlaying each other. Each lattice point is associated with a set of mesoscopic variables, whose values are stored in vectors proportional to the number of streaming directions. The lattice is partitioned onto a 3-dimensional Cartesian processor grid, and MPI is used for communication — a snapshot of the communication topology is shown in Figure 1.3 (a), highlighting the relatively sparse communication pattern. As in most simulations of this nature, ghost cells are used to hold copies of the planes of data from neighboring processors.

For ELBM3D, a nonlinear equation must be solved at each time step for each grid-point, so that the collision process satisfies certain constraints. Since this equation involves taking the logarithm of each component of the distribution function the whole algorithm becomes heavily constrained by the performance of the $log()$ function.

The connectivity of ELBM3D shown in Figure 1.3 (a) is structurally similar to Cactus, but exhibits a slightly different communication pattern due to the periodic boundary conditions of the code. This topology is not quite isomorphic to a mesh or toroidal interconnect topology, but would vastly underutilize the available bisection bandwidth of a fully connected network like a fat-tree or crossbar. Figure 1.3 (c) demonstrates that like Cactus, ELBM3D is dominated by point-to-point communication while Figure 1.3 (b) shows that the point-to-point message buffer sizes are very large and therefore strongly bandwidth-bound.

1.5.1   Experimental results

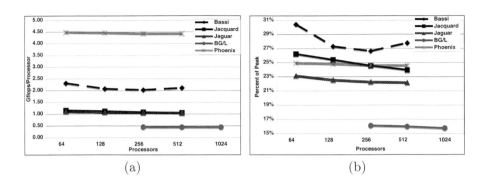

(a)   (b)

FIGURE 1.4: ELBM3D strong-scaling performance using a 512^3 grid by (a) GFLOPS and (b) percentage of peak. ALL BG/L data were collected on the ANL BG/L system in coprocessor mode.

We note that LBM codes are often compared on the basis of mega-updates-per-second as it is independent of the number of operations that the compiler generates for the LBM collision step. Our work, on the other hand, compares GFLOPS based on the flop count of a single baseline system and normalized on the basis of wall-clock time (see Section 1.3). Thus the relative results maintain consistent comparisons of delivered performance between different system architectures offered by the mega-updates-per-second.

Strong-scaling results for a system of 512^3 grid-points are shown in Figure 1.4 for both (a) raw performance and (b) percentage of peak. For each of the superscalar machines the code was restructured to take advantage of specialized `log()` functions — MASSV library for IBM and ACML for AMD — that compute values for a vector of arguments. (The benefits of these libraries are discussed in Section 1.4.) Using this approach gave ELBM3D a performance boost of between 15–30% depending on the architecture. For the X1E, the innermost grid-point loop was taken inside the nonlinear equation solver to enable full vectorization. After these optimizations, ELBM3D has a kernel of fairly high computational intensity and a percentage of peak of 15–30% on all architectures.

ELBM3D shows good scaling across all of our evaluated platforms. This is due to the dominance of nearest-neighbor point-to-point messaging, and lack of load balance issues. As expected, the parallel overhead increases as the ratio of communication to computation increases. The parallel efficiency as we move to higher concurrencies shows the least degradation on the BG/L system (although the memory requirements of the application and MPI implementation prevents running this size on fewer than 256 processors). Both Phoenix and Jaguar are very close behind, followed by Jacquard and Bassi.

Our experiments bear out the fact that the higher computational cost of the entropic algorithm, as compared to traditional LBM approaches, can be cast in a way that leads to efficient computation on commodity processors. We are thus optimistic that ELBM3D will be able to deliver exceptional performance on planned petascale platforms.

1.6 Cactus: General Relativity Astrophysics

One of the most challenging problems in astrophysics is the numerical solution of Einstein's equations following from the Theory of General Relativity (GR): a set of coupled nonlinear hyperbolic and elliptic equations containing thousands of terms when fully expanded. The BSSN-MoL application, which makes use of the Cactus Computational ToolKit [3, 10], is designed to evolve Einstein's equations stably in 3D on supercomputers to simulate astrophysical phenomena with high gravitational fluxes – such as the collision of two black holes and the gravitational waves radiating from that event. While Cactus is a modular framework supporting a wide variety of applications, this study focuses exclusively on the GR solver, which implements the ADM-BSSN (BSSN) formulation [1] with Method of Lines (MoL) integration to enable stable evolution of black holes. For parallel computation, the grid is block domain decomposed so that each processor has a section of the global grid. The standard MPI driver (PUGH) for Cactus solves the partial differential equation (PDE) on a local grid section and then updates the values at the ghost zones by exchanging data on the faces of its topological neighbors in the domain decomposition.

In the topology chart of Figure 1.5 (a), we see that the ghost-zone exchanges of Cactus result in communications with "neighboring" nodes, represented by diagonal bands. In fact, each node communicates with 6 neighbors at most due to the regular computational structure of the 3D stencil application. This topology is isomorphic to a mesh and should work well on mesh or toroidal interconnect topologies if mapped correctly. The communication is strongly bandwidth- bound and dominated by point-to-point messaging as can be seen in Figures 1.5 (b) and (c) respectively.

1.6.1 Experimental results

Figure 1.6 presents weak-scaling performance results for the Cactus BSSN-MoL application using a 60^3 per processor grid. In terms of raw performance,

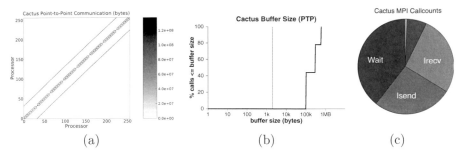

FIGURE 1.5: Cactus (a) communication topology and intensity; (b) cumulative distribution function of point-to-point message buffer sizes; and (c) breakdown of MPI calls counts.

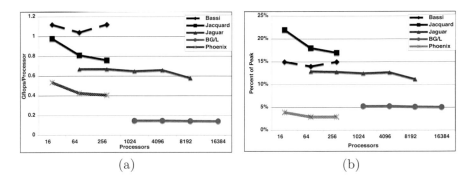

FIGURE 1.6: Cactus weak-scaling experiments on a 60^3 per processor grid in (a) runtime and (b) percentage of peak. All BG/L data were run on the BGW system. Phoenix data were collected on the Cray X1 platform.

the Power5-based Bassi clearly outperforms any other systems, especially the BG/L where the GFLOPS/P rate and the percentage of peak performance is somewhat disappointing. The lower computational efficiency of BG/L is to be expected from the simpler dual-issue in-order PPC440 processor core. However, while the per-processor performance of BG/L is somewhat limited, the scaling behavior is impressive, achieving near perfect scalability for up to 16K processors. This is (by far) the largest Cactus scaling experiment to date, and shows extremely promising results. Due to memory constraints we could not conduct virtual node mode simulations for the 60^3 data set, however further testing with a smaller 50^3 grid shows no performance degradation for up to 32K (virtual node) processors. This strongly suggests that the Cactus application will scale to extremely high-concurrency, petascale systems. All of the experiments were performed with the default processor topology mapping where the communication topology matched the torus topology of the physical interconnect. Additional investigations with alternative processor topology mappings on the BG/L showed no significant effects.

The Opteron-based Jacquard cluster shows modest scaling, which is probably due to the (relatively) loosely coupled nature of its system architecture as opposed to the tight software and hardware interconnect integration of the other platforms in our study. Separate studies of Jacquard InfiniBand interconnect also show increased messaging contention at higher concurrencies that are typical of multi tier source-routed fat-tree networks [26]. Bassi shows excellent scaling but the size of the largest concurrency was significantly smaller compared to that of the BG/L so it remains to be seen if IBM's Federation HPS interconnect will scale to extremely large systems. Likewise, Jaguar, the Cray XT3 system, shows excellent scaling up to 8192 processors. The performance per processor for Jaguar's 2.6 GHz dual-core Opteron processors is somewhat lower than for Jacquard's single-core 2.2GHz Opteron system. The effect is primarily due to the differences in the quality of code generated by the compilers (PGI v1.5.31 on Jaguar and PathScale v2.4 on Jacquard). Moving from single core to dual core on Jaguar resulted in a modest 10% change in performance that was consistently less than the performance difference attributed to the compilers.

Phoenix, the Cray X1 platform, showed the lowest computational performance of our evaluated systems. The most costly procedure for the X1 was the computation of radiation boundary conditions, which continued to drag performance down despite considerable effort to rewrite it in vectorizable form. In previous studies [5], the vectorized boundary conditions proved beneficial on a number of vector platforms including the NEC SX-8 and Earth Simulator; however the X1 continued to suffer disproportionally from small portions of unvectorized code due to the relatively large differential between vector and scalar performance, highlighting that notions of architectural balance cannot focus exclusively on bandwidth (bytes per flop) ratios.

1.7 PARATEC: First Principles Materials Science

PARATEC (PARAllel Total Energy Code [22]) performs *ab initio* quantum-mechanical total energy calculations using pseudopotentials and a plane wave basis set. The pseudopotentials are of the standard norm-conserving variety. Forces can be easily calculated and used to relax the atoms into their equilibrium positions. PARATEC uses an all-band conjugate gradient (CG) approach to solve the Kohn-Sham equations of density functional theory (DFT) and obtain the ground-state electron wave functions. DFT is the most commonly used technique in materials science, incorporating a quantum mechanical treatment of the electrons to calculate the structural and electronic properties of materials. Codes based on DFT are widely used to study properties such as strength, cohesion, growth, magnetic, optical, and transport for materials like nanostructures, complex surfaces, and doped semiconductors.

PARATEC is written in F90 and MPI and is designed primarily for massively parallel computing platforms, but can also run on serial machines. The code has run on many computer architectures and uses preprocessing to include machine specific routines such as the FFT calls. Much of the computation time (typically 60%) involves FFTs and BLAS3 routines, which run at a high percentage of peak on most platforms. For small atomic systems, PARATEC tends to be dominated by the FFTs, and becomes dominated by the BLAS3 operations for larger atomic systems. The performance of the 3D FFTs tends to suffer at high concurrencies, but it can be controlled to a limited extent by using message aggregation (described below).

In solving the Kohn-Sham equations using a plane wave basis, part of the calculation is carried out in real space and the remainder in Fourier space using parallel 3D FFTs to transform the wave functions between the two spaces. PARATEC uses its own handwritten 3D FFTs rather than library routines as the data layout in Fourier space is a sphere of points, rather than a standard square grid. The sphere is load balanced by distributing the different length columns from the sphere to different processors such that each processor holds a similar number of points in Fourier space. Effective load balancing is important, as much of the compute intensive part of the calculation is carried out in Fourier space [4].

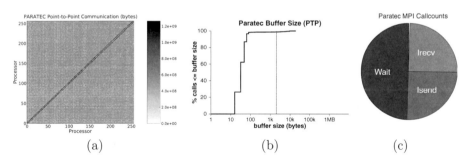

FIGURE 1.7: PARATEC (a) communication topology and intensity; (b) cumulative distribution function of point-to-point message buffer sizes; and (c) breakdown of MPI calls counts.

Figure 1.7 shows the communication characteristics of PARATEC. The global data transposes within PARATEC's FFT operations — as captured in Figure 1.7 (a) — accounting for the bulk of PARATEC's communication overhead, and can quickly become the bottleneck at high concurrencies. In general, with a fixed problem size the message sizes become smaller and increasingly latency bound at higher concurrencies, as can be seen in Figure 1.7

(b). In a single 3D FFT the size of the data packets scale as the inverse of the number of processors squared. The PARATEC code can perform all-band calculations that allow the FFT communications to be blocked together, resulting in larger message sizes and avoiding latency problems. However, the buffering required for such message aggregation consumes additional memory, which is a scarce resource on systems such as BG/L. Finally, observe in Figure 1.7 (c) that the vast majority of messaging occurs in a point-to-point fashion.

1.7.1 Experimental results

Figure 1.8 presents strong-scaling performance results for a 488 atom cdse (cadmium selenide) quantum dot (QD) system which has important technological applications due to its photoluminescent properties. Due to the use of BLAS3 and optimized one-dimensional FFT libraries, which are highly cache resident, PARATEC obtains a high percentage of peak on the different platforms studied. The results for BG/L are for a smaller system (432 atom bulk silicon) due to memory constraints.

Results show that the Power5-based Bassi system obtains the highest absolute performance of 5.49 GFLOPS/P on 64 processors with good scaling to larger processor counts. The fastest Opteron system (3.39 GFLOPS/P) was Jaguar (XT3) running on 128 processors. (Jacquard did not have enough memory to run the QD system on 128 processors.) The higher bandwidth for communications on Jaguar (see Table 1.1) allows it to scale better than Jacquard for this communication-intensive application. The BG/L system has a much lower single processor performance than the other evaluated platforms due to a relatively low peak speed of only 2.8 GFLOPS. BG/L's percent of peak drops significantly from 512 to 1024 processors, probably due to increased communication overhead when moving from a topologically packed half-plane of 512 processors to a larger configuration. The smaller system being run on the BG/L (432 atom bulk silicon) also limits the scaling to higher processor counts. Overall, Jaguar obtained the maximum aggregate performance of 4.02 TFLOPS on 2048 processors.

Looking at the vector system, results show that the Phoenix X1E achieved a lower percentage of peak than the other evaluated architectures; although in absolute terms, Phoenix performs rather well due to the high peak speed of the MSP processor.* One reason for this is the relatively slow performance of the X1E scalar unit compared to the vector processor. In consequence,

---

[†]Purple, located at LLNL, is architecturally similar to Bassi and contains 12,208 IBM Power5 processors. The authors thank Tom Spelce and Bronis de Supinksi of LLNL for conducting the Purple experiments.

[*]Results on the X1E were obtained by running the binary compiled on the X1, as running with an optimized X1E generated binary (-O3) caused the code to freeze. Cray engineers are investigating the problem.

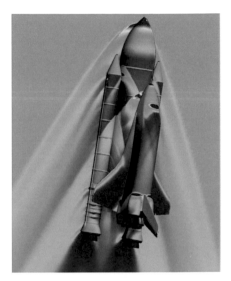

FIGURE 2.1 Full SSLV configuration including orbiter, external tank, solid rocket boosters, and fore and aft attach hardware. (a) Cartesian mesh surrounding the SSLV; colors indicate 16-way decomposition using the SFC partitioner. (b) Pressure contours for the case described in the text; the isobars are displayed at 2.6 Mach, 2.09 degrees angle-of-attack, and 0.8 degrees sideslip corresponding to flight conditions approximately 80 seconds after launch.

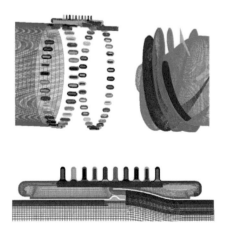

FIGURE 2.4 Liquid rocket turbopump for the SSME. (a) Surface grids for the low-pressure fuel pump inducer and flowliner. (b) Instantaneous snapshot of particle traces colored by axial velocity values.

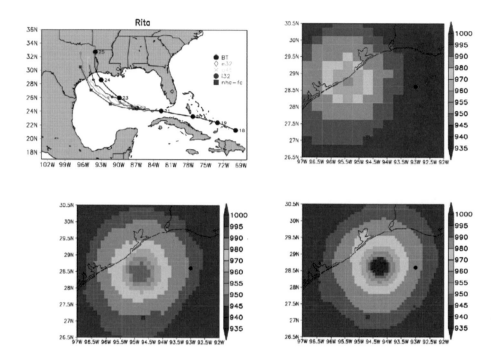

FIGURE 2.7 Four-day forecasts of Hurricane Rita initialized at 0000 UTC September 21, 2005. (a) Tracks predicted by fvGCM at 0.25° (line with diamond symbols), 0.125° (line with crosses), and 0.08° (line with circles) resolutions. The lines with hexagons and squares represent the observation and official prediction by the National Hurricane Center (NHC). (b-d) Sea level pressure (SLP) in hPa within a 4° × 5° box after 72-hour simulations ending at 0000 UTC 24 September at 0.25°, 0.125°, and 0.08° resolutions. Solid circles and squares indicate locations of the observed and official predicted hurricane centers by the NHC, respectively. The observed minimal SLP at the corresponding time is 931 hPa. In a climate model with a typical 2° × 2.5° resolution (latitude × longitude), a 4° × 5° box has only four grid points.

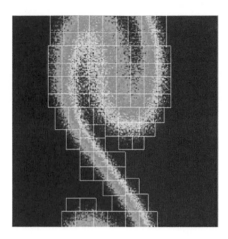

FIGURE 3.4 Simulation of a moving interface via a hybrid continuum-atomistic method. The white grid blocks show where a direct simulation Monte Carlo particle method is applied at the finest AMR grid scale to resolve the physics at interface between two fluids. A continuum-scale method is applied elsewhere in the fluid. (Adapted from Hornung et al. [13])

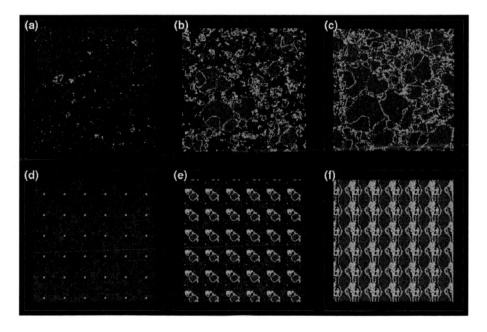

FIGURE 3.5 Three-dimensional molecular dynamics simulation of nucleation and grain growth in molten tantalum. Three snapshots in time are shown for two simulations. The top row corresponds to a simulation using 16 million atoms on the Blue Gene/L supercomputer at LLNL. This 2005 Gordon Bell Prize-winning calculation was the first to produce physically correct, size-independent results. The rich 3D detail is seen in the planar slices. The bottom row used 64,000 atoms on a smaller supercomputer. Periodic boundary conditions were used to generate the entire domain, resulting in the unphysical replicated pattern. (Image from Streitz et al. [17])

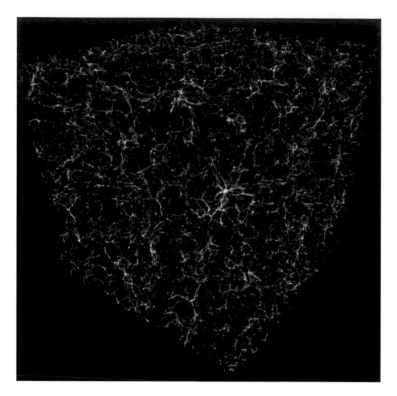

FIGURE 5.1   Enzo hydrodynamic simulation of cosmic structure in a 700 Mpc volume of the universe. Up to seven levels of adaptive mesh refinement resolve the distribution of baryons within and between galaxy clusters, for an effective resolution of $65,536^3$. Volume rendering of baryon density. Image credit M. Hall, NCSA.

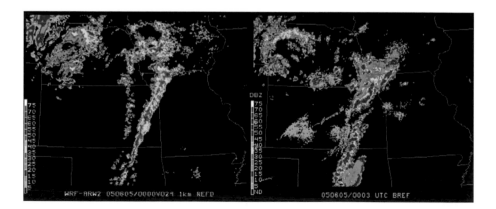

FIGURE 6.4   24-hour WRF-predicted (left) and observed (right) radar reflectivity valid at 00 UTC on June 5, 2005. Warmer colors indicate higher precipitation intensity. The WRF model utilized a horizontal grid spacing of 2 km and forecasts were produced by CAPS on the Terascale Computing System at the Pittsburgh Supercomputing Center as part of the 2005 SPC Spring Experiment.

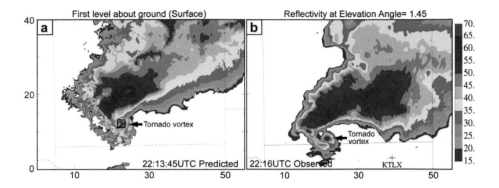

FIGURE 6.6 (a) Predicted surface reflectivity field at 13.75 minutes of the 50-m forecast valid at 2213:45 UTC and (b) observed reflectivity at the 1.45° elevation of the Oklahoma City radar observation at 2216 UTC. The domain shown is 55 km × 40 km in size, representing the portion of the 50 m grid between 20 and 75 km in the east-west direction and from 16 to 56 km in the north-south direction.

FIGURE 8.1 The Andromeda nebula, M31. A mosaic of hundreds of Earth-based telescope pointings were needed to make this image.

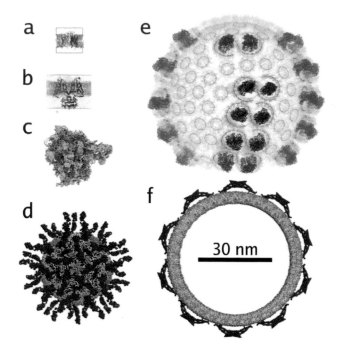

FIGURE 9.4 Example biomolecular simulations: (a) aquaporin in membrane with solvent, (b) potassium channel in membrane with solvent, (c) ribosome, (d) poliovirus with cell surface receptors, (e) photosynthetic chromatophore, (f) BAR domain vesicle cross section.

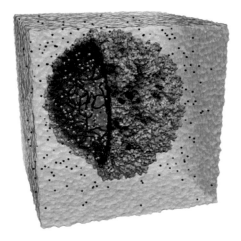

FIGURE 11.1 Biomolecular simulations of a system with >1 million atoms. This figure shows a schematic representation of of satellite tobacco mosaic virus particle. The viral particle was solvated in a water box of dimensions 220 Å × 220Å × 220 Å, consisting of about 1.06 million atoms. The protein capsid (green) is enveloping the RNA and part of the capsid is cut out to make the RNA core of the particle visible. The backbone of RNA is highlighted in red; ions were added to make the system charge neutral. Figure courtesy of Theoretical and Computational Biophysics Group. *Reprinted with permission from P.L. Freddolino et al., Structure (2006), 14, 437-449. ©Elsevier 2006.*

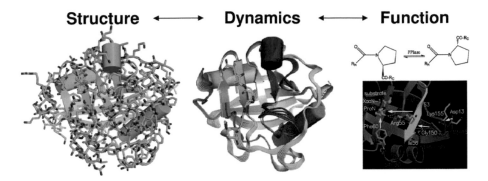

Structure ⟷ **Dynamics** ⟷ **Function**

FIGURE 11.2 The dynamic personality of proteins. An integrated view of protein structure, dynamics and function is emerging where proteins are considered dynamically active molecular machines. Biomolecular simulations spanning multiple timescales are providing new insights into the working of protein systems. Computational modeling of enzymes is leading to the discovery of a network of protein vibrations promoting enzyme catalysis in several systems including cyclophilin A, which is shown here.

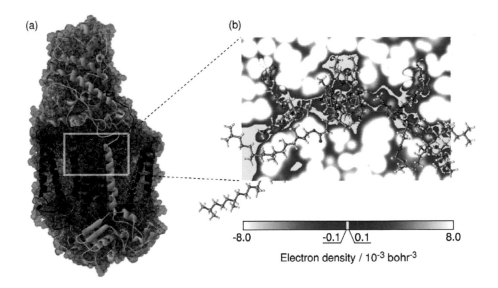

FIGURE 11.4 Full quantum calculation of a protein with 20,581 atoms. Electron densities of the photosynthetic system were computed at the quantum level (RHF/6-31G*) with the FMO method: (a) an electron density of the whole system, and (b) a differential electron density around the special pair. *Reprinted with permission from T. Ikegami et al., Proceedings of the 2005 ACM/IEEE Conference on Supercomputing 2005. ©ACM 2005.*

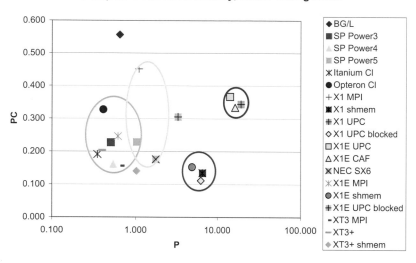

P-C Map for a Two Level Memory, Linear Timing Model

Legend:
- ◆ BG/L
- ■ SP Power3
- ▲ SP Power4
- ■ SP Power5
- ✸ Itanium CI
- ● Opteron CI
- + X1 MPI
- ▣ X1 shmem
- ✚ X1 UPC
- ◇ X1 UPC blocked
- □ X1E UPC
- △ X1E CAF
- ✕ NEC SX6
- ✳ X1E MPI
- ◉ X1E shmem
- ✚ X1E UPC blocked
- − XT3 MPI
- − XT3+
- ◆ XT3+ shmem

FIGURE 16.7 P-C Map for parallel systems based on efficiencies [accesses/cycle]. Horizontal axis is performance P in [accesses/100 cycles] and the vertical axis is complexity PC in [accesses/cycles]. Systems fall with few exceptions into 4 categories: PGAS languages (UPC, CAF), one-sided block-access (SHMEM, UPC block-mode), MPI-vector (X1-MPI and SX6), and superscalar-based MPI systems.

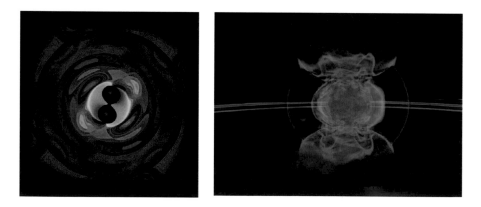

FIGURE 24.1 Left: Gravitational waves and horizons in a binary black hole in spiral simulation. Simulation by AEI/CCT collaboration, image is by W. Benger (CCT/AEI/ZIB). Right: A rotationally deformed proto-neutron star formed in the iron core collapse of an evolved massive star is pictured. Shown are a volume rendering of the rest-mass density and a 2D rendition of outgoing gravitational waves. Simulation by [28], image by R. Kähler.

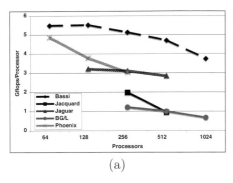
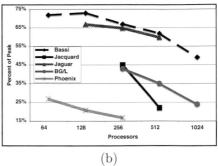

(a)   (b)

FIGURE 1.8: PARATEC strong-scaling performance on a 488 atom CdSe quantum dot. Power5 data for $P = 1024$ was run on LLNL's Purple system.[†] The BG/L data, collected on the BGW, are for a 432 atom bulk silicon due to memory constraints. Phoenix X1E data were collected with an X1 binary.

the X1E spends a smaller percentage of the total time in highly optimized 3D FFTs and BLAS3 libraries than on any of the other machines. The other code segments are handwritten F90 routines and have a lower vector operation ratio.

PARATEC results do not show any clear advantage for a torus versus a fat-tree communication network. The main limit to scaling in PARATEC are the handwritten 3D FFTs, where all-to-all communications are performed to transpose the data across the machine. In a single 3D FFT the size of the data packets scales as the inverse of the number of processors squared. PARATEC can perform an all-band calculation, allowing the FFT communications to be blocked, resulting in larger message sizes and avoiding latency problems. Overall, the scaling of the FFTs is limited to a few thousand processors. Therefore, scaling PARATEC to petascale systems with tens (or hundreds) of thousands of processors requires a second level of parallelization over the electronic band indices as is done in the QBox code [11]. This will greatly benefit the scaling and reduce per-processor memory requirements on architectures such as BG/L.

1.8   HyperCLaw: Hyperbolic AMR Gas Dynamics

Adaptive mesh refinement (AMR) is a powerful technique that reduces the computational and memory resources required to solve otherwise intractable problems in computational science. The AMR strategy solves the system of partial differential equations (PDEs) on a relatively coarse grid, and dynamically refines it in regions of scientific interest or where the coarse grid error is too high for proper numerical resolution. HyperCLaw is a hybrid C++/Fortran AMR code developed and maintained by CCSE at LBNL [9, 23] where it is frequently used to solve systems of hyperbolic conservation laws using a higher-order Godunov method.

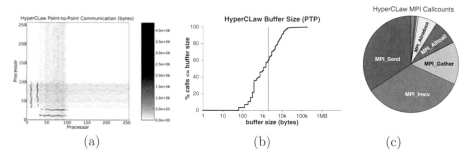

(a)   (b)   (c)

FIGURE 1.9: HyperCLaw (a) communication topology and intensity; (b) cumulative distribution function of point-to-point message buffer sizes; and (c) breakdown of MPI calls counts.

The HyperCLaw code consists of an applications layer containing the physics classes defined in terms of virtual functions. The basic idea is that data blocks are managed in C++ in which ghost cells are filled and temporary storage is dynamically allocated so that when the calls to the physics algorithms (usually finite difference methods implemented in Fortran) are made, the same stencil can be used for all points and no special treatments are required.

The HyperCLaw problem examined in this work profiles a hyperbolic shock-tube calculation, where we model the interaction of a Mach 1.25 shock in air hitting a spherical bubble of helium. This case is analogous to one of the experiments described by Haas and Sturtevant [12]. The difference between the density of the helium and the surrounding air causes the shock to accelerate into and then dramatically deform the bubble. The base computational grids for the problems studied are $512 \times 64 \times 32$. These grids were adaptively refined by an initial factor of 2 and then a further factor of 4, leading to effective sizes of $2048 \times 256 \times 256$ and $4096 \times 512 \times 256$, respectively.

HyperCLaw is also dominated by point-to-point MPI messages with minimal collective communication requirements as shown in Figure 1.9 (c). Most of the communication overhead occurs in the FillPatch operation. FillPatch starts with the computational grid at a given level, adds ghost cells five layers thick around each grid, and then fills those cells either by copying from other grids at that level, or by interpolating from lower level (coarser) cells that are covered by those ghost cells. FillPatch has a very complicated sparse nonlinear communication pattern as shown in Figure 1.9 (a). Once it completes, all the ghost cells for each grid are filled with valid data, a higher-order Godunov solver is applied to each resulting grid.

Although the communication topology appears to be a many-to-many pattern that would require a high bisection bandwidth, the message sizes shown in Figure 1.9 (b) are surprisingly small. Such small messages will have only ephemeral contention events and primarily be dominated by the interconnect latency and overhead of sending the message. Further analysis shows that if we focus exclusively on the messages that dominate the communication time, the number of *important* communicating partners drops significantly. Therefore, despite the complexity of the communication topology and large aggregate number of communicating partners, bisection bandwidth does not appear to be a critical factor in supporting efficient scaling of this AMR application. Architectural support for petascale AMR codes therefore depend more on reducing the computational overhead of sending small messages. Understanding the evolving communication requirements of AMR simulations will be the focus of future work.

1.8.1 Experimental results

The HyperCLaw problem examined in this work profiles a hyperbolic shock-tube calculation, where we model the interaction of a Mach 1.25 shock in air hitting a spherical bubble of helium. This case is analogous to one of the experiments described by Haas and Sturtevant [12]. The difference between the density of the helium and the surrounding air causes the shock to accelerate into and then dramatically deform the bubble. The base computational grids for the problems studied are $512 \times 64 \times 32$. These grids were adaptively refined by an initial factor of 2 and then a further factor of 4, leading to an effective resolution of $4096 \times 512 \times 256$.

Figure 1.10 presents the absolute runtime and percentage of peak for the weak-scaling HyperCLaw experiments. In terms of absolute runtime (at P=128), Bassi achieves the highest performance followed by Jacquard, Jaguar, Phoenix, and finally BG/L (the Phoenix and Jacquard experiments crash at P≥256; system consultants are investigating the problems). Observe that all of the platforms achieve a low percentage of peak; for example at 128 processors, Jacquard, Bassi, Jaguar, BG/L, and Phoenix achieve 4.8%, 3.8%, 3.5%, 2.5%, and 0.8% respectively. Achieving peak performance on BG/L requires double hummer mode (as described in Section 1.2), which has operand alignment restrictions that make it very difficult for the compiler to schedule efficiency for this application. Therefore BG/L delivered performance is most likely to be only half of the stated theoretical peak because of its inability to exploit the double-hummer. With this in mind, the BG/L would achieve a sustained performance of around 5%, commensurate with the other platforms in our

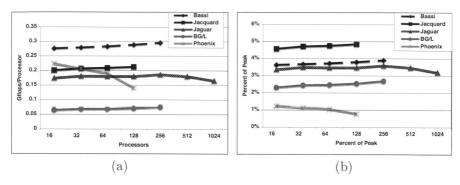

FIGURE 1.10: HyperCLaw weak-scaling performance on a base computational grid of 512 x 64 x 32 in (a) runtime and (b) percentage of peak. All BG/L data were collected on an ANL system (the BG/L system was unable to run jobs over 256 processors due to memory constraints).

study. Note that although these are weak-scaling experiments in the numbers of grids, the volume of work increases with higher concurrencies due to increased volume of computation along the communication boundaries; thus, the percentage of peak generally increases with processor count.

Although Phoenix performs relatively poorly for this application, especially in terms of its attained percentage of peak, it is important to point out that two effective X1E optimizations were undertaken since our initial study into AMR vector performance [32]. Our preliminary study showed that the *knapsack* and *regridding* phases of HyperCLaw were largely to blame for limited X1E scalability, cumulatively consuming almost 60% of the runtime for large concurrency experiments. The original knapsack algorithm — responsible for allocating boxes of work equitably across the processors — suffered from a memory inefficiency. The updated version copies pointers to box lists during the swapping phase (instead of copying the lists themselves), and results in knapsack performance on Phoenix that is almost cost-free, even on hundreds of thousands of boxes.

The function of the regrid algorithm is to replace an existing grid hierarchy with a new hierarchy in order to maintain numerical accuracy, as important solution features develop and move through the computational domain. This process includes tagging coarse cells for refinement and buffering them to ensure that neighboring cells are also refined. The regridding phase requires the computations of box list intersection, which was originally implemented in an $\mathcal{O}(N^2)$ straightforward fashion. The updated version utilizes a hashing scheme based on the position in space of the bottom corners of the boxes, resulting in a vastly improved $\mathcal{O}(NlogN)$ algorithm. This significantly reduced the cost of the regrid algorithm on Phoenix, resulting in improved performance. Nonetheless, Phoenix performance still remains low due to the non-vectorizable and short-vector-length operations necessary to maintain and regrid the hierarchical data structures.

Overall, our HyperCLaw results highlight the relatively low execution efficiency of the AMR approach as measured by the flop rate. This is due (in part) to the irregular nature of the AMR components necessary to maintain and regrid the hierarchical meshes, combined with complex communication requirements. Additionally, the numerical Godunov solver, although computationally intensive, requires substantial data movement that can degrade cache reuse. Nevertheless, the algorithmic efficiency gains associated with AMR and high-resolution discretizations more than compensate for the low sustained rate of execution as measured by flop rates. It points to the danger of measuring efficiency of an algorithm in terms of flop throughput alone rather than comparing performance on the basis of time to solution. Other key results of our study are the knapsack and regridding optimizations, which significantly improved HyperCLaw scalability [32]. These improvements in scaling behavior suggest that, in spite of the low execution efficiency, the AMR methodology is a suitable candidate for petascale systems.

1.9   Summary and Conclusions

The purpose of any HEC system is to run full-scale scientific codes, and performance on such applications is the final arbiter of a platform's utility; comparative performance evaluation data must therefore be readily available to the HEC community at large. However, evaluating large-scale scientific applications using realistic problem sizes on leading supercomputing platforms is an extremely complex process, requiring coordination of application scientists in highly disparate areas. Our work presents one of the most extensive comparative performance results on modern supercomputers available in the literature.

Figure 1.11 shows a summary of results using the largest comparable concurrencies for all five studied applications and five state-of-the-art parallel platforms, in relative performance (normalized to the fastest system) and percentage of peak. Results show that the Power5-based Bassi system achieves the highest raw performance for three of our five applications, thanks to dramatically improved memory bandwidth (compared to its predecessors), and increased attention to latency hiding through advanced prefetch features. The Phoenix system achieved impressive raw performance on GTC and ELBM3D, however, applications with non-vectorizable portions suffer greatly on this architecture due the imbalance between the scalar and vector processors. Comparing the two Opteron systems, Jacquard and Jaguar, we see that, in general, sustained performance is similar between the two platforms. However, for some applications such as GTC and PARATEC, the tight integration

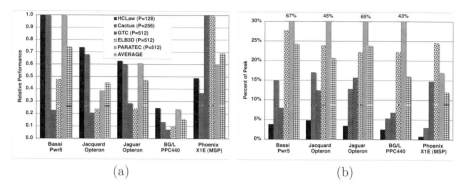

(a)   (b)

FIGURE 1.11: Summary of results for the largest comparable concurrencies; (a) relative runtime performance normalized to the fastest system; and (b) sustained percentage of peak. Cactus Phoenix results are on the X1 system. BG/L results are shown for P=1024 on Cactus and GTC, as smaller BG/L concurrencies are not available.

of Jaguar's XT3 interconnect results in significantly better scalability at high concurrency compared with Jacquard's commodity-based InfiniBand network. The BG/L platform attained the lowest raw and sustained performance on our suite of applications, however, results at very high concurrencies show impressive scalability characteristics and potential for attaining petascale performance.

Results also indicate that our evaluated codes have the potential to effectively utilize petascale resources. However, some applications, such as PARATEC, will require significant reengineering to incorporate the additional levels of parallelism necessary to utilize vast numbers of processors. Other applications, including the Lattice-Boltzmann ELBM3D and the dynamically adapting HyperCLaw simulation, are already showing scaling behavior with promising prospects to achieve ultra-scale. Finally, two of our tested codes, Cactus and GTC, have successfully demonstrated impressive scalability up to 32K processors on the BGW system. A full GTC production simulation was also performed on 32,768 processors and showed a perfect load balance from beginning to end. This, combined with its high efficiency on a multi-core processor, clearly puts GTC as a primary candidate to effectively utilize petascale resources.

Overall, these extensive performance evaluations are an important step toward conducting simulations at the petascale level, by providing computational scientists and system designers with critical information on how well numerical methods perform across state-of-the-art parallel systems. Future work will explore a wider set of computational methods, with a focus on irregular and unstructured algorithms, while investigating a broader set of HEC platforms, including the latest generation of multicore technologies.

1.10 Acknowledgments

The authors would like to gratefully thank Bob Walkup for optimizing GTC on the BG/L as well as Tom Spelce and Bronis de Supinksi of LLNL for conducting the Purple experiments. We also thank Tom Goodale for his assistance with the Cactus code, Costin Iancu for his many thoughtful suggestions, and Arin Fishkin for her graphic contribution. The authors are grateful to IBM Watson Research Center for allowing BG/L access via the BGW Consortium Day. All LBNL authors were supported by the Office of Advanced Scientific Computing Research in the Department of Energy Office of Science under contract number DE-AC02-05CH11231. Dr. Ethier was supported by the Department of Energy under contract number DE-AC020-76-CH-03073.

References

[1] M. Alcubierre, G. Allen, B. Brügmann, E. Seidel, and W.M. Suen. Towards an understanding of the stability properties of the 3+1 evolution equations in general relativity. *Phys. Rev. D*, (gr-qc/9908079), 2000.

[2] S. Ansumali and I. V. Karlin. Stabilization of the Lattice-Boltzmann method by the H theorem: A numerical test. *Phys. Rev.*, E62:7999–8003, 2000.

[3] Cactus Code Server. http://www.cactuscode.org.

[4] A. Canning, L. W. Wang, A. Williamson, and A. Zunger. Parallel empirical pseudopotential electronic structure calculations for million atom systems. *J. Comp. Phys.*, 160, 2000.

[5] J. Carter, L. Oliker, and J. Shalf. Performance evaluation of scientific applications on modern parallel vector systems. In *VECPAR: High Performance Computing for Computational Science*, Rio de Janeiro, Brazil, July 2006.

[6] T. H. Dunigan Jr., J. S. Vetter, J. B. White III, and P. H. Worley. Performance evaluation of the Cray X1 distributed shared-memory architecture. *IEEE Micro*, 25(1):30–40, Jan/Feb 2005.

[7] ORNL Cray X1 Evaluation. http://www.csm.ornl.gov/~dunigan/cray.

[8] S. Ethier, W. M. Tang, and Z. Lin. Gyrokinetic particle-in-cell simulations of plasma microturbulence on advanced computing platforms. *J. Phys.: Conf. Series*, 16, 2005.

[9] Center for Computational Sciences and Engineering, Lawrence Berkeley National Laboratory. http://seesar.lbl.gov/CCSE.

[10] T. Goodale, G. Allen, G. Lanfermann, J. Masso, T. Radke, E. Seidel, and J. Shalf. The Cactus framework and toolkit: Design and applications. In *VECPAR: 5th International Conference, Lecture Notes in Computer Science*, Berlin, 2003. Springer.

[11] F. Gygi, E. W. Draeger, B. R. de Supinski, J. Gunnels, V. Austel, J. Sexton, F. Franchetti, S. Kral, C. Ueberhuber, and J. Lorenz. Large-scale first-principles molecular dynamics simulations on the Blue Gene/L platform using the Qbox code. In *Proc. SC05: International Conference for High Performance Computing, Networking, Storage and Analysis*, Seattle, WA, Nov 12-18, 2005.

[12] J.-F. Haas and B. Sturtevant. Interaction of weak shock waves with cylindrical and spherical gas inhomogeneities. *Journal of Fluid Mechanics*, 181:41–76, 1987.

[13] HPC Challenge Benchmark. http://icl.cs.utk.edu/hpcc/index.html.

[14] S. Kamil, L. Oliker, J. Shalf, and D. Skinner. Understanding ultra-scale application communication requirements. In *IEEE International Symposium on Workload Characterization*, 2005.

[15] Z. Lin, T. S. Hahm, W. W. Lee, W. M. Tang, and R. B. White. Turbulent transport reduction by zonal flows: Massively parallel simulations. *Science*, 281(5384):1835–1837, Sep 1998.

[16] Z. Lin and W. W. Lee. Method for solving the gyrokinetic Poisson equation in general geometry. *Phys. Rev. E.*, 52:5646–5652, 1995.

[17] H.W. Meuer, E. Strohmaier, J.J. Dongarra, and H.D. Simon. TOP500 Supercomputer Sites. http://www.top500.org.

[18] K. Nakajima. Three-level hybrid vs. flat MPI on the Earth Simulator: Parallel iterative solvers for finite-element method. In *Proc. 6th IMACS Symposium Iterative Methods in Scientific Computing*, volume 6, Denver, CO, Mar 27-30, 2003.

[19] L. Oliker, A. Canning, J. Carter, C. Iancu, M. Lijewski, S. Kamil, J. Shalf, H. Shan, E. Strohmaier, S. Ethier, and T. Goodale. Scientific application performance on candidate petascale platforms. In *Proc. IEEE International Parallel & Distributed Processing Symposium (IPDPS)*, Long Beach, CA, Mar 26-30, 2007.

[20] L. Oliker, A. Canning, J. Carter, J. Shalf, and S. Ethier. Scientific computations on modern parallel vector systems. In *Proc. SC04: International Conference for High Performance Computing, Networking, Storage and Analysis*, Pittsburgh, PA, November 2004.

[21] L. Oliker, J. Carter, M. Wehner, A. Canning, S. Ethier, A Mirin, G. Bala, D. Parks, P. Worley, S. Kitawaki, and Y. Tsuda. Leading computational methods on scalar and vector HEC platforms. In *Proc. SC05: International Conference for High Performance Computing, Networking, Storage and Analysis*, Seattle, WA, Nov 12-18, 2005.

[22] PARAllel Total Energy Code. http://www.nersc.gov/projects/paratec.

[23] C. A. Rendleman, V. E. Beckner, M. Lijewski, W. Y. Crutchfield, and J. B. Bell. Parallelization of structured, hierarchical adaptive mesh refinement algorithms. *Computing and Visualization in Science*, 3(3):147–157, 2000.

[24] SciDAC: Scientific Discovery through Advanced Computing. http://www.scidac.gov/.

[25] J. Shalf, S. Kamil, L. Oliker, and D. Skinner. Analyzing ultra-scale application communication requirements for a reconfigurable hybrid interconnect. In *Proc. SC05: International Conference for High Performance Computing, Networking, Storage and Analysis*, Seattle, WA, November 2005.

[26] H. Shan, E. Strohmaier, J. Qiang, D. Bailey, and K. Yelick. Performance modeling and optimization of a high energy colliding beam simulation code. In *Proc. SC06: International Conference for High Performance Computing, Networking, Storage and Analysis*, 2006.

[27] D. Skinner. Integrated performance monitoring: A portable profiling infrastructure for parallel applications. In *Proc. ISC2005: International Supercomputing Conference*, Heidelberg, Germany, 2005.

[28] STREAM: Sustainable Memory Bandwidth in High Performance Computers. http://www.cs.virginia.edu/stream.

[29] S. Succi. The lattice Boltzmann equation for fluids and beyond. *Oxford Science Publ.*, 2001.

[30] G. Vahala, J. Yepez, L. Vahala, M. Soe, and J. Carter. 3D entropic Lattice-Boltzmann simulations of 3D Navier-Stokes turbulence. In *Proc. of 47th Annual Meeting of the APS Division of Plasma Physics*, Denver, CO, 2005.

[31] J. Vetter, S. Alam, T. Dunigan, Jr., M. Fahey, P. Roth, and P. Worley. Early evaluation of the Cray XT3. In *Proc. IEEE International Parallel*

& Distributed Processing Symposium (IPDPS), Rhodes Island, Greece, April 25-29, 2006.

[32] M. Welcome, C. Rendleman, L. Oliker, and R. Biswas. Performance characteristics of an adaptive mesh refinement calculation on scalar and vector platforms. In *CF '06: Proceedings of the 3rd Conference on Computing Frontiers*, May 2006.

Chapter 2

Petascale Computing: Impact on Future NASA Missions

Rupak Biswas

NASA Ames Research Center

Michael Aftosmis

NASA Ames Research Center

Cetin Kiris

NASA Ames Research Center

Bo-Wen Shen

NASA Goddard Space Flight Center

2.1 Introduction

To support its diverse mission-critical requirements, the National Aeronautics and Space Administration (NASA) solves some of the most unique, computationally challenging problems in the world [8]. To facilitate rapid yet accurate solutions for these demanding applications, the U.S. space agency procured a 10,240-CPU supercomputer in October 2004, dubbed Columbia. Housed in the NASA Advanced Supercomputing facility at NASA Ames Research Center, Columbia is comprised of twenty 512-processor nodes (representing three generations of SGI Altix technology: 3700, 3700-BX2, and 4700), with a combined peak processing capability of 63.2 teraflops per second (TFLOPS). However, for many applications, even this high-powered computational workhorse, currently ranked as one of the fastest in the world, does not have enough computing capacity, memory size, and bandwidth rates needed to meet all of NASA's diverse and demanding future mission requirements.

In this chapter, we consider three important NASA application areas: aerospace analysis and design, propulsion subsystems analysis, and hurricane prediction, as a representative set of these challenges. We show how state-of-the-art methodologies for each application area are currently performing on Columbia, and explore projected achievements with the availability of petascale computing. We conclude by describing some of the architecture and algorithm obstacles that must first be overcome for these applications to take full advantage of such petascale computing capability.

2.2   The Columbia Supercomputer

Columbia is configured as a cluster of 20 SGI Altix nodes, each with 512 Intel Itanium 2 processors and one terabyte (TB) of global shared-access memory, that are interconnected via InfiniBand fabric communications technology. Of these 20 nodes, 12 are model 3700, seven are model 3700-BX2, and one is the newest-generation architecture, a 4700. The 3700-BX2 is a double-density incarnation of the 3700, while the 4700 is a dual-core version of the 3700-BX2. Each node acts as a shared-memory, single-system-image environment running a Linux-based operating system, and utilizes SGI's scalable, shared-memory NUMAflex architecture, which stresses modularity.

Four of Columbia's BX2 nodes are tightly linked to form a 2,048-processor 4 TB shared-memory environment and use NUMAlink4 among themselves, which allows access to all data directly and efficiently, without having to move them through I/O or networking bottlenecks. Each processor in the 2,048-CPU subsystem runs at 1.6 GHz, has 9 MB of level-3 cache (the Madison 9M processor), and a peak performance of 6.4 gigaflops per second (GFLOPS). One other BX2 node is equipped with these same processors. (These five BX2 nodes are denoted as BX2b in this chapter.) The remaining fourteen nodes: two BX2 (referred to as BX2a here) and twelve 3700, all have processors running at 1.5 GHz, with 6 MB of level-3 cache, and a peak performance of 6.0 GFLOPS. All nodes have 2 GB of shared memory per processor.

The 4700 node of Columbia is the latest generation in SGI's Altix product line, and consists of 8 racks with a total of 256 dual-core Itanium 2 (Montecito) processors (1.6 GHz, 18 MB of on-chip level-3 cache) and 2 GB of memory per core (1 TB total). Each core also contains 16 KB instruction and data caches. The current configuration uses only one socket per node, leaving the other socket unused (also known as the bandwidth configuration). Detailed performance characteristics of the Columbia supercomputer using micro-benchmarks, compact kernel benchmarks, and full-scale applications can be found in other articles [7, 10].

2.3   Aerospace Analysis and Design

High-fidelity computational fluid dynamics (CFD) tools and techniques are developed and applied to many aerospace analysis and design problems throughout NASA. These include the full Space Shuttle Launch Vehicle (SSLV) configuration and future spacecraft such as the Crew Exploration Vehicle (CEV). One of the more commonly used high-performance aerodynamics simulation packages used on Columbia to assist with aerospace vehicle analysis and design is Cart3D. This software package enables high-fidelity characterization of aerospace vehicle design performance over the entire flight envelope.

Cart3D is a simulation package targeted at conceptual and preliminary design of aerospace vehicles with complex geometry. It solves the Euler equations governing inviscid flow of a compressible fluid on an automatically generated Cartesian mesh surrounding a vehicle. Since the package is inviscid, boundary layers and viscous phenomena are not present in the simulations, facilitating fully automated Cartesian mesh generation in support of inviscid analysis. Moreover, solutions to the governing equations can typically be obtained for about 2–5% of the cost of a full Reynolds-averaged Navier-Stokes (RANS) simulation. The combination of automatic mesh generation and high-quality, yet inexpensive, flow solutions makes the package ideally suited for rapid design studies and exploring what-if scenarios. Cart3D offers a drop-in replacement for less scalable and lower-fidelity engineering methods, and is frequently used to generate entire aerodynamic performance databases for new vehicles [1, 3].

2.3.1   Methodology

Cart3D's solver module uses a second-order cell-centered, finite-volume upwind spatial discretization combined with a multigrid-accelerated Runge-Kutta scheme for advance to steady-state [1]. As shown in Figure 2.1 (a), the package uses adaptively refined, hierarchically generated Cartesian cells to discretize the flow field and resolve the geometry. In the field, cells are simple Cartesian hexahedra, and the solver capitalizes on regularity of the mesh for both speed and accuracy. At the wall, these cells are cut arbitrarily and require more extensive data structures and elaborate mathematics. Nevertheless, the set of cut-cells is lower-dimensional, and the net cost remains low. Automation and insensitivity to geometric complexity are key ingredients in enabling rapid parameter sweeps over a variety of configurations.

Cart3D utilizes domain decomposition to achieve high efficiency on parallel machines [3, 6]. To assess performance of Cart3D's solver module on realistic problems, extensive experiments have been conducted on several large applications. The case considered here is that of the SSLV. A mesh containing approximately 4.7 million cells around this geometry is shown in Figure 2.1 (a) and was built using 14 levels of adaptive subdivision. The grid is painted to

FIGURE 2.1: (See color insert following page 18.) Full SSLV configuration including orbiter, external tank, solid rocket boosters, and fore and aft attach hardware. (a) Cartesian mesh surrounding the SSLV; colors indicate 16-way decomposition using the SFC partitioner. (b) Pressure contours for the case described in the text; the isobars are displayed at 2.6 Mach, 2.09 degrees angle-of-attack, and 0.8 degrees sideslip corresponding to flight conditions approximately 80 seconds after launch.

indicate partitioning into 16 sub-domains using the Peano-Hilbert space-filling curve (SFC) [3]. The partitions are all predominantly rectangular, which is characteristic of sub-domains generated with SFC-based partitioners, indicating favorable compute/communicate ratios.

Cart3D's simulation module solves five partial differential equations for each cell in the domain, giving this example close to 25 million degrees-of-freedom. Figure 2.1 (b) illustrates a typical result from these simulations by showing pressure contours in the discrete solution. Surface triangulation describing the geometry contains about 1.7 million elements. For parallel performance experiments on Columbia, the mesh in Figure 2.1 (a) was refined so that the test case contained approximately 125 million degrees-of-freedom. These investigations included comparisons between the OpenMP and MPI parallel programming paradigms, analyzing the impact of multigrid on scalability, and understanding the effects of the NUMAlink4 and InfiniBand communication fabrics.

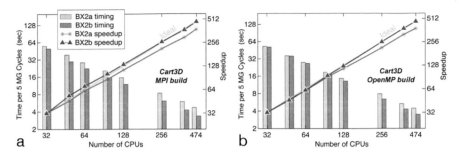

FIGURE 2.2: Comparison of execution time and parallel speedup of the Cart3D solver module on single BX2a and BX2b Columbia nodes using (a) MPI and (b) OpenMP parallelization strategies.

2.3.2 Results

The domain-decomposition parallelization strategy in the Cart3D flow simulation package has previously demonstrated excellent scalability on large numbers of processors with both MPI and OpenMP libraries [6]. This behavior makes it a suitable candidate for comparing performance of Columbia's BX2a and BX2b nodes using varying processor counts on a complete application. Figure 2.2 shows charts of parallel speedup and execution timings for Cart3D using MPI and OpenMP. Line graphs in the figure show parallel speedup between 32 and 474 CPUs; the corresponding execution time for five multigrid cycles is shown via bar charts. This comparison clearly demonstrates excellent scalability for both node types, indicating that this particular example did not exceed the bandwidth capabilities of either system. The bar charts in Figure 2.2 contain wall-clock timing data and show comparisons consistent with the slightly faster clock-speeds in the BX2b nodes.

The memory on each of Columbia's 512-CPU nodes is globally sharable by any process within the node, but cache-coherency is not maintained between nodes. Thus, all multi-node examples with Cart3D are run using the MPI communication back-end. Numerical experiments focus on the effects of increasing the number of multigrid levels in the solution algorithm. These experiments were carried out on four of Columbia's BX2b nodes. Figure 2.3 (a) displays parallel speedup for the system, comparing the baseline four-level multigrid solution algorithm with that on a single grid, using the NUMA-link4 interconnect. The plot shows nearly ideal speedup for the single grid runs on the full 2,048-CPU system. The multigrid algorithm requires solution and residual transfer to coarser mesh levels, and therefore places substantially greater demands on interconnect bandwidth. In this case, a slight degradation in performance is apparent above 1,024 processors, but the algorithm still posts parallel speedups of about 1,585 on 2,016 CPUs.

Figure 2.3 (b) shows that the InfiniBand runs do not extend beyond 1,536 processors due to a limitation on the number of connections. Notice that

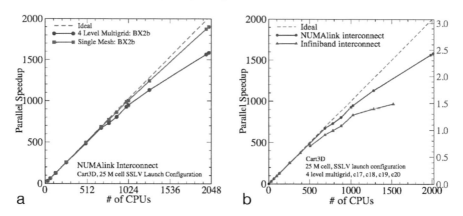

FIGURE 2.3: (a) Parallel speedup of the Cart3D solver module using one and four levels of mesh in the multigrid hierarchy with a NUMAlink4 interconnect. (b) Comparison of parallel speedup and TFLOPS with four levels of multigrids using NUMAlink4 and InfiniBand.

the InfiniBand performance consistently lags that of the NUMAlink4, and that there is an increased penalty as the number of nodes goes up. Using the standard NCSA FLOP counting procedures, net performance of the multigrid solution algorithm is 2.5 TFLOPS on 2,016 CPUs. Additional performance details are provided in [2].

2.3.3   Benefits of petascale computing to NASA

In spite of several decades of continuous improvements in both algorithms and hardware, and despite the widespread acceptance and use of CFD as an indispensable tool in the aerospace vehicle design process, computational methods are still employed in a very limited fashion. In some important flight regimes, current computational methods for aerodynamic analysis are only reliable within a narrow range of flight conditions — where no significant flow separation occurs. This is due, in part, to the extreme accuracy requirements of the aerodynamic design problem, where, for example, changes of less than one percent in the drag coefficient of a flight vehicle can determine commercial success or failure. As a result, computational analyses are currently used in conjunction with experimental methods only over a restricted range of the flight envelope, where they have been essentially calibrated.

Improvements in accuracy achieved by CFD methods for aerodynamic applications will require, among other things, dramatic increases in grid resolution and simulation degrees-of-freedom (over what is generally considered practical in the current environment). Manipulating simulations with such high resolutions demands substantially more potent computing platforms. This drive toward finer scales and higher fidelity is obviously motivated by a dawning understanding of error analysis in CFD simulations. Recent research in uncertainty analysis and formal error bounds indicates that hard quantification of simulation error is possible, but the analysis can be many times more expensive than the CFD simulation itself. While this field is just opening up for fluid dynamic simulations, it is already clear that certifiably accurate CFD simulations require many times the computing power currently available. Petascale computing would open the door to affordable, routine, error quantification for simulation data.

Once optimal designs can be constructed, they must be validated throughout the entire flight envelope, which includes hundreds of thousands of simulations of both the full range of aerodynamic flight conditions, and parameter spaces of all possible control surface deflections and power settings. Generating this comprehensive database will not only provide all details of vehicle performance, but it will also open the door to new possibilities for engineers. For example, when coupled with a six-degree-of-freedom integrator, the vehicle can be flown through the database by guidance and control (G&C) system designers to explore issues of stability and control. Digital Flight initiatives undertake the complete time-accurate simulation of a maneuvering vehicle and include structural deformation and G&C feedback. Ultimately, the vehicle's suitability for various mission profiles or other trajectories can be evaluated by full end-to-end mission simulations, and optimization studies can consider the full mission profile. This is another area where a petascale computing capability can significantly benefit NASA missions.

2.4 Propulsion Subsystem Analysis

High-fidelity unsteady flow simulation techniques for design and analysis of propulsion systems play a key role in supporting NASA missions, including analysis of the liquid rocket engine flowliner for the Space Shuttle Main Engine (SSME). The INS3D software package [11] is one such code developed to handle computations for unsteady flow through a full-scale, low- and high-pressure rocket pump. Liquid rocket turbopumps operate under severe conditions and at very high rotational speeds. The low-pressure fuel turbopump creates transient flow features such as reverse flows, tip clearance effects, secondary flows, vortex shedding, junction flows, and cavitation effects. The reverse flow originating at the tip of an inducer blade travels upstream and interacts with the bellows cavity. This flow unsteadiness is considered to be one of the major contributors to high-frequency cyclic loading that results in cycle fatigue.

2.4.1   Methodology

To resolve the complex geometry in relative motion, an overset grid approach [9] is employed where the problem domain is decomposed into a number of simple grid components. Connectivity between neighboring grids is established by interpolation at the grid outer boundaries. Addition of new components to the system and simulation of arbitrary relative motion between multiple bodies are achieved by establishing new connectivity without disturbing existing grids.

The computational grid used for the experiments reported in this chapter consisted of 66 million grid points and 267 blocks or zones. Details of the grid system are shown in Figure 2.4. The INS3D code solves the incompressible Navier-Stokes equations for both steady-state and unsteady flows. The numerical solution requires special attention to satisfy the divergence-free constraint on the velocity field. The incompressible formulation does not explicitly yield the pressure field from an equation of state or the continuity equation. One way to avoid the difficulty of the elliptic nature of the equations is to use an artificial compressibility method that introduces a time-derivative of the pressure term into the continuity equation. This transforms the elliptic-parabolic partial differential equations into the hyperbolic-parabolic type. To obtain time-accurate solutions, the equations are iterated to convergence in pseudo-time for each physical time step until divergence of the velocity field has been reduced below a specified tolerance value. The total number of required sub-iterations varies depending on the problem, time step size, and artificial compressibility parameter. Typically, the number ranges from 10–30 sub-iterations. The matrix equation is solved iteratively by using a non-factored Gauss-Seidel-type line-relaxation scheme, which maintains stability and allows a large pseudo-time step to be taken. More detailed information about the application can be found elsewhere [11, 12].

2.4.2   Results

Computations were performed to compare scalability between the multi-level parallelism (MLP) [17] and MPI+OpenMP hybrid (using a point-to-point communication protocol) versions of INS3D on one of Columbia's BX2b nodes. Both implementations combine coarse- and fine-grain parallelism. Coarse-grain parallelism is achieved through a UNIX fork in MLP, and through explicit message-passing in MPI+OpenMP. Fine-grain parallelism is obtained using OpenMP compiler directives in both versions. The MLP code utilizes a global shared-memory data structure for overset connectivity arrays, while the MPI+OpenMP code uses local copies.

Initial computations using one group and one thread were used to establish the baseline runtime for one physical time step, where 720 such time steps are required to complete one inducer rotation. Figure 2.5 displays the time per iteration (in minutes) versus the number of CPUs, and the speedup factor for

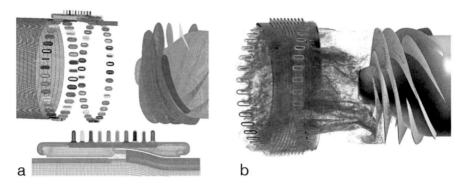

FIGURE 2.4: (See color insert following page 18.) Liquid rocket turbopump for the SSME. (a) Surface grids for the low-pressure fuel pump inducer and flowliner. (b) Instantaneous snapshot of particle traces colored by axial velocity values.

both codes. Here, 36 groups have been chosen to maintain good load balance for both versions. Then, the runtime per physical time step is obtained using various numbers of OpenMP threads (1, 2, 4, 8, and 14). It includes the I/O time required to write the time-accurate solution to disk at each time step. The scalability for a fixed number of MLP and MPI groups and varying OpenMP threads is good, but begins to decay as the number of OpenMP threads becomes large. Further scaling can be accomplished by fixing the number of OpenMP threads and increasing the number of MLP/MPI groups until the load balancing begins to fail. Unlike varying the OpenMP threads, which does not affect the convergence rate of INS3D, varying the number of groups may deteriorate it. This will lead to more iterations even though faster runtime per iteration is achieved. The results show that the MLP and MPI+OpenMP codes perform almost equivalently for one OpenMP thread, but that the latter begins to perform slightly better as the number of threads is increased. This advantage can be attributed to having local copies of the connectivity arrays in the MPI+OpenMP hybrid implementation. Having the MPI+OpenMP version of INS3D as scalable as the MLP code is promising since this implementation is easily portable to other platforms.

We also compare performance of the INS3D MPI+OpenMP code on multiple BX2b nodes against single node results. This includes running the MPI+OpenMP version using two different communication paradigms: master-worker and point-to-point. The runtime per physical time step is recorded using 36 MPI groups, and 1, 4, 8, and 14 OpenMP threads on one, two, and four BX2b nodes. Communication between nodes is achieved using the InfiniBand and NUMAlink4 interconnects, denoted as IB and XPM, respectively.

Figure 2.6 (a) contains results using the MPI point-to-point communication paradigm. When comparing performance of using multiple nodes with that of

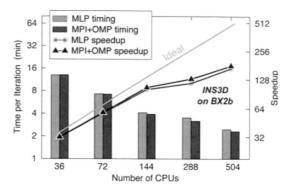

FIGURE 2.5: Comparison of INS3D performance on a BX2b node using two different hybrid programming paradigms.

a single node, we observe that scalability of the multi-node runs with NUMAlink4 is similar to the single-node runs (which also use NUMAlink4 internally). However, when using InfiniBand, execution time per iteration increases by 10–29% on two- and four-node runs. The difference between the two- and four-node runs decreases as the number of CPUs increases. Figure 2.6 (b) displays the results using the master-worker communication paradigm. Note that the time per iteration is much higher using this protocol compared to the point-to-point communication. We also see a significant deterioration in scalability for both single- and multi-node runs. With NUMAlink4, we observe a 5–10% increase in runtime per iteration from one to two nodes, and an 8–16% increase using four nodes. This is because the master resides on one node, and all workers on the other nodes must communicate with the master. However, when using point-to-point communication, many of the messages remain within the node from which they are sent. An additional 14–27% increase in runtime is observed when using InfiniBand instead of NUMAlink4, independent of the communication paradigm.

2.4.3   Benefits of petascale computing to NASA

The benefits of high-fidelity modeling of full-scale multi-component, multiphysics propulsion systems to NASA's current mission goals are numerous and have the most significant impact in the areas of: crew safety (new safety protocols for the propulsion system); design efficiency (provide the ability to make design changes that can improve the efficiency and reduce the cost of space flight); and technology advancement (in propulsion technology for manned space flights to Mars).

With petascale computing, fidelity of the current propulsion subsystem analysis could be increased to full-scale, multi-component, multi-disciplinary

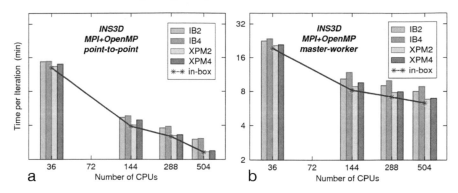

FIGURE 2.6: Performance of INS3D across multiple BX2b nodes via NU-MAlink4 and InfiniBand using MPI (a) point-to-point, and (b) master-worker communication.

propulsion applications. Multi-disciplinary computations are critical for modeling propulsion systems of new and existing launch vehicles to attain flight rationale. To ensure proper coupling between the fluid, structure, and dynamics codes, the number of computed iterations will dramatically increase, demanding large, parallel computing resources for efficient solution turnaround time. Spacecraft propulsion systems contain multi-component/multi-phase fluids (such as turbulent combustion in solid rocket boosters and cavitating hydrodynamic pumps in the SSME) where phase change cannot be neglected when trying to obtain accurate and reliable results.

2.5   Hurricane Prediction

Accurate hurricane track and intensity predictions help provide early warning to people in the path of a storm, saving both life and property. Over the past several decades, hurricane track forecasts have steadily improved, but progress on intensity forecasts and understanding of hurricane formation/genesis has been slow. Major limiting factors include insufficient model resolutions and uncertainties of cumulus parameterizations (CP). A CP is required to emulate the statistical effects of unresolved cloud motions in coarse resolution simulations, but its validity becomes questionable at high resolutions. Facilitated by Columbia, the ultra-high resolution finite-volume General Circulation Model (fvGCM) [4] has been deployed and run in real-time to study the impacts of increasing resolutions and disabling CPs on hurricane forecasts. The fvGCM code is a unified numerical weather prediction (NWP) and climate model that runs on daily, monthly, decadal, and century timescales,

TABLE 2.1:   Changes in fvGCM resolution as a function of time (available computing resources). Note that in the vertical direction, the model could be running with 32, 48, 55, or 64 stretched levels.

Resolution (lat×long)	Grid Points (y×x)	Total 2D Grid Cells	Major Application	Implementation Date
2° × 2.5°	91 × 144	13,104	Climate	1990s
1° × 1.25°	181 × 288	52,128	Climate	Jan. 2000
0.5° × 0.625°	361 × 576	207,936	Climate/Weather	Feb. 2002
0.25° × 0.36°	721 × 1000	721,000	Weather	July 2004
0.125° × 0.125°	1441 × 2880	4,150,080	Weather	Mar. 2005
0.08° × 0.08°	2251 × 4500	11,479,500	Weather	July 2005

and is currently the only operational global NWP model with finite-volume dynamics. While doubling the resolution of such a model requires an 8-16X increase in computational resources, the unprecedented computing capability provided by Columbia enables us to rapidly increase resolutions of fvGCM to 0.25°, 0.125°, and 0.08°, as shown in Table 2.1.

While NASA launches many high-resolution satellites, the mesoscale-resolving fvGCM is one of only a few global models with comparable resolution to satellite data (QuikSCAT, for example), providing a mechanism for direct comparisons between model results and satellite observations. During the active 2004 and 2005 hurricane seasons, the high-resolution fvGCM produced promising forecasts of intense hurricanes such as Frances, Ivan, Jeanne, and Karl in 2004; and Emily, Dennis, Katrina, and Rita in 2005 [4, 14, 15, 16]. To illustrate the capabilities of fvGCM, coupled with the computational power of Columbia, we discuss the numerical forecasts of Hurricanes Katrina and Rita in this chapter.

2.5.1   Methodology

The fvGCM code, resulting from a development effort of more than ten years, has the following three major components: (1) finite-volume dynamics, (2) NCAR Community Climate Model (CCM3) physics, and (3) NCAR Community Land Model (CLM). Dynamical initial conditions and sea surface temperature (SST) were obtained from the global forecast system (GFS) analysis data and one-degree optimum interpolation SST of National Centers for Environmental Prediction.

The unique features of the finite-volume dynamical core [13] include a genuinely conservative flux-form semi-Lagrangian transport algorithm which is Gibbs oscillation-free with the optional monotonicity constraint; a terrain-following Lagrangian control-volume vertical coordinate system; a finite-volume integration method for computing pressure gradients in general terrain following coordinates; and a mass, momentum, and total energy conserving algorithm for remapping the state variables from the Lagrangian control-volume to an Eulerian terrain-following coordinate. The vorticity-preserving horizontal dynamics enhance the simulation of atmospheric oscillations and vortices. Physical processes such as CP and gravity-wave drag are largely enhanced with emphasis for high-resolution simulations; they are also modified for consistent application with the innovative finite-volume dynamics.

From a computational perspective, a crucial aspect of the fvGCM development is its high computational efficiency on a variety of high-performance supercomputers including distributed-memory, shared-memory, and hybrid architectures. The parallel implementation is hybrid: coarse-grain parallelism with MPI/MLP/SHMEM and fine-grain parallelism with OpenMP. The model's dynamical part has a 1-D MPI/MLP/SHMEM parallelism in the y-direction, and uses OpenMP multi-threading in the z-direction. One of the prominent features in the implementation is the permission of multi-threaded data communications. The physical part inherits 1-D parallelism in the y-direction from the dynamical part, and further applies OpenMP loop-level parallelism in this decomposed latitude. CLM is also implemented with MPI and OpenMP parallelism, and its grid cells are distributed among processors.

All of the aforementioned features make it possible to advance the state-of-the-art of hurricane prediction to a new frontier. To date, Hurricanes Katrina and Rita are the sixth and fourth most intense hurricanes in the Atlantic, respectively. They devastated New Orleans, southwestern Louisiana, and the surrounding Gulf Coast region, resulting in losses in excess of 90 billion U.S. dollars. Here, we limit our discussion to the simulations initialized from 1200 UTC August 25, and 0000 UTC September 21, for Katrina and Rita, and show improvement of the track and intensity forecasts by increasing resolutions to 0.125° and 0.08°, and by disabling CPs.

2.5.2 Results

The impacts of increased computing power and thus enhanced resolution (in this case, to 0.125°), and disabling CPs on the forecasts of Hurricane Katrina have been documented [15]. They simulated comparable track predictions at different resolutions, but better intensity forecasts at finer resolutions. The predicted mean sea level pressures (MSLPs) in the 0.25°, 0.125°, and 0.125° (with no CPs) runs are 951.8, 895.7, and 906.5 hectopascals (hPa) with respect to the observed 902 hPa. Consistent improvement as a result of using a higher resolution was illustrated from the six 5-day forecasts with the 0.125° fvGCM, showing small errors in center pressure of only ±12 hPa. The notable improvement in Katrina's intensity forecasts was attributed to the sufficient fine resolution used for resolving hurricane near-eye structures. As the hurricane's internal structure has convective-scale variations, it was shown that the 0.125° run with disabled CPs could lead to further improvement on Katrina's intensity and structure (asymmetry).

Earlier forecasts of Hurricane Rita by the National Hurricane Center (represented by the line in Figure 2.7 (a) with the square symbols) had a bias toward the left side of the best track (the line with circles, also in Figure 2.7 (a)), predicting the storm hitting Houston, Texas. The real-time 0.25° forecast (represented by the line in Figure 2.7 (a) with the diamonds) initialized at 0000 UTC 21 September showed a similar bias. Encouraged by the successful Katrina forecasts, we conducted two experiments at 0.125° and 0.08° resolutions with disabled CPs, and compared the results with the 0.25° model. From Figures 2.7 (b-d), it is clear that a higher resolution run produces a better track forecast, predicting a larger shift in landfall toward the Texas-Louisiana border. Just before making landfall, Rita was still a Category 3 hurricane with an MSLP of 931 hPa at 0000 UTC 24. Looking at Figures 2.7 (b-d) that show the predicted minimal MSLPs (957.8, 945.5, and 936.5 hPa at 0.25°, 0.125°, and 0.08° resolutions, respectively), we can conclude that a higher-resolution run produces a more realistic intensity. Although these results are promising, note that the early rapid intensification of Rita was not fully simulated in either of the above simulations, indicating the importance of further model improvement and better understanding of hurricane (internal) dynamics.

2.5.3   Benefits of petascale computing to NASA

With the availability of petascale computing resources, we could extend our approach from short-range weather/hurricane forecasts to reliable longer-duration seasonal and annual predictions. This would enable us to study hurricane climatology [5] in present-day or global-warming climate, and improve our understanding of interannual variations of hurricanes. Petascale computing could also make the development of a multi-scale, multi-component Earth system model feasible, including a non-hydrostatic cloud-resolving atmospheric model, an eddy-resolving ocean model, and an ultra-high-resolution land model [18]. Furthermore, this model system could be coupled with chemical and biological components. In addition to the model's improvement, an advanced high-resolution data assimilation system is desired to represent initial hurricanes, thereby further improving predictive skill.

2.6   Bottlenecks

Taking full advantage of petascale computing to achieve the research challenges outlined in Sections 2.3–2.5 will first require clearing a number of computational hurdles. Known bottlenecks include parallel scalability issues, development of better numerical algorithms, and the need for dramatically higher bandwidths.

Recent studies on the Columbia supercomputer underline major parallel

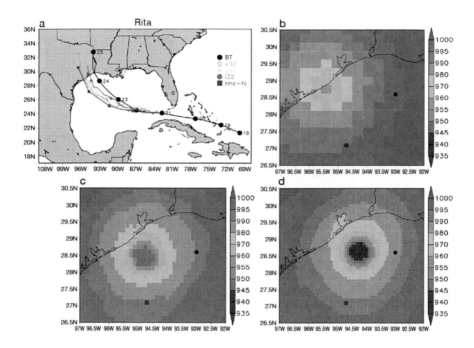

FIGURE 2.7: (See color insert following page 18.) Four-day forecasts of Hurricane Rita initialized at 0000 UTC September 21, 2005. (a) Tracks predicted by fvGCM at 0.25° (line with diamond symbols), 0.125° (line with crosses), and 0.08° (line with circles) resolutions. The lines with hexagons and squares represent the observation and official prediction by the National Hurricane Center (NHC). (b-d) Sea-level pressure (SLP) in hPa within a 4° × 5° box after 72-hour simulations ending at 0000 UTC 24 September at 0.25°, 0.125°, and 0.08° resolutions. Solid circles and squares indicate locations of the observed and official predicted hurricane centers by the NHC, respectively. The observed minimal SLP at the corresponding time is 931 hPa. In a climate model with a typical 2° × 2.5° resolution (latitude × longitude), a 4° × 5° box has only four grid-points.

scalability issues with several of NASA's current mainline Reynolds-averaged Navier-Stokes solvers when scaling to just a few hundred processors, requiring communication among multiple nodes). While sorting algorithms can be used to minimize internode communication, scaling to tens or hundreds of thousands of processors will require heavy investment in scalable solution techniques to replace NASA's current block tri- and penta-diagonal solvers.

More efficient numerical algorithms are being developed (to handle the increased number of physical time steps) which focus on scalability while increasing accuracy and preserving robustness and convergence. This means computing systems with hundreds of thousands of parallel processors (or cores) are not only desirable, but are required to solve these problems when including all of the relevant physics. In addition, unlike some Earth and space science simulations, current high-fidelity CFD codes are processor speed-bound. Runs utilizing many hundreds of processors rarely use more than a very small fraction of the available memory, and yet still take hours or days to run. As a result, algorithms which trade-off this surplus memory for greater speed are clearly of interest.

Bandwidth to memory is the biggest problem facing CFD solvers today. While we can compensate for latency with more buffer memory, bandwidth comes into play whenever a calculation must be synchronized over large numbers of processors. Systems 10x larger than today's supercomputers will require at least 20x more bandwidth, since current results show insufficient bandwidth for CFD applications on even the best available hardware.

2.7   Summary

High performance computing has always played a major role in meeting the modeling and simulation needs of various NASA missions. With NASA's 63.2 TFLOPS Columbia supercomputer, high-end computing is having an even greater impact within the agency and beyond. Significant cutting-edge science and engineering simulations in the areas of space exploration, shuttle operations, Earth sciences, and aeronautics research, are continuously occurring on Columbia, demonstrating its ability to accelerate NASA's exploration vision. In this chapter, we discussed its role in the areas of aerospace analysis and design, propulsion subsystems analysis, and hurricane prediction, as a representative set of these challenges.

But for many NASA applications, even this current capability is insufficient to meet all of the diverse and demanding future requirements in terms of computing capacity, memory size, and bandwidth rates. A petaflops-scale computing power would greatly alter the types of applications solved and approaches taken as compared with those in use today. We outlined potential benefits of petascale computing to NASA, and described some of the architecture and algorithm bottlenecks that must be overcome to achieve its full potential.

References

[1] M.J. Aftosmis, M.J. Berger, and G.D. Adomavicius. A parallel multilevel method for adaptively refined Cartesian grids with embedded boundaries. In *Proc. 38th AIAA Aerospace Sciences Meeting & Exhibit*, Reno, NV, Jan. 2000. AIAA-00-0808.

[2] M.J. Aftosmis, M.J. Berger, R. Biswas, M.J. Djomehri, R. Hood, H. Jin, and C. Kiris. A detailed performance characterization of Columbia using aeronautics benchmarks and applications. In *Proc. 44th AIAA Aerospace Sciences Meeting & Exhibit*, Reno, NV, Jan. 2006. AIAA-06-0084.

[3] M.J. Aftosmis, M.J. Berger, and S.M. Murman. Applications of space-filling-curves to Cartesian methods in CFD. In *Proc. 42nd AIAA Aerospace Sciences Meeting & Exhibit*, Reno, NV, Jan. 2004. AIAA-04-1232.

[4] R. Atlas, O. Reale, B.-W. Shen, S.-J. Lin, J.-D. Chern, W. Putman, T. Lee, K.-S. Yeh, M. Bosilovich, and J. Radakovich. Hurricane forecasting with the high-resolution NASA finite-volume General Circulation Model. *Geophysical Research Letters*, 32:L03801, doi:10.1029/2004GL021513, 2005.

[5] L. Bengtsson, K.I. Hodges, and M. Esch. Hurricane type vortices in a high-resolution global model: Comparison with observations and reanalyses. *Tellus*, submitted.

[6] M.J. Berger, M.J. Aftosmis, D.D. Marshall, and S.M. Murman. Performance of a new CFD flow solver using a hybrid programming paradigm. *Journal of Parallel and Distributed Computing*, 65(4):414–423, 2005.

[7] R. Biswas, M.J. Djomehri, R. Hood, H. Jin, C. Kiris, and S. Saini. An application-based performance characterization of the Columbia supercluster. In *Proc. SC|05*, Seattle, WA, Nov. 2005.

[8] R. Biswas, E.L. Tu, and W.R. Van Dalsem. Role of high-end computing in meeting NASA's science and engineering challenges. In H. Deconinck and E. Dick, editors, *Computational Fluid Dynamics 2006*. Springer, to appear.

[9] P.G. Buning, D.C. Jespersen, T.H. Pulliam, W.M. Chan, J.P. Slotnick, S.E. Krist, and K.J. Renze. Overflow User's Manual, Version 1.8g. Technical report, NASA Langley Research Center, Hampton, VA, 1999.

[10] R. Hood, R. Biswas, J. Chang, M.J. Djomehri, and H. Jin. Benchmarking the Columbia supercluster. *International Journal of High Performance Computing Applications*, to appear.

[11] C. Kiris, D. Kwak, and W.M. Chan. Parallel unsteady turbopump simulations for liquid rocket engines. In *Proc. SC2000*, Dallas, TX, Nov. 2000.

[12] C. Kiris, D. Kwak, and S. Rogers. Incompressible Navier-Stokes solvers in primitive variables and their applications to steady and unsteady flow simulations. In M. Hafez, editor, *Numerical Simulations of Incompressible Flows*, pages 3–24. World Scientific, 2003.

[13] S.-J. Lin. A vertically Lagrangian finite-volume dynamical core for global models. *Monthly Weather Review*, 132:2293–2307, 2004.

[14] B.-W. Shen, R. Atlas, J.-D. Chern, O. Reale, S.-J. Lin, T. Lee, and J. Chang. The 0.125-degree finite volume General Circulation Model on the NASA Columbia supercomputer: Preliminary simulations of mesoscale vortices. *Geophysical Research Letters*, 33:L05801, doi:10.1029/2005GL024594, 2006.

[15] B.-W. Shen, R. Atlas, O. Reale, S.-J. Lin, J.-D. Chern, J. Chang, C. Henze, and J.-L. Li. Hurricane forecasts with a global mesoscale-resolving model: Preliminary results with Hurricane Katrina (2005). *Geophysical Research Letters*, 33:L13813, doi:10.1029/2006GL026143, 2006.

[16] B.-W. Shen, W.-K. Tao, R. Atlas, T. Lee, O. Reale, J.-D. Chern, S.-J. Lin, J. Chang, C. Henze, and J.-L. Li. Hurricane forecasts with a global mesoscale-resolving model on the NASA Columbia supercomputer. In *Proc. AGU 2006 Fall Meeting*, San Francisco, CA, Dec. 2006.

[17] J.R. Taft. Achieving 60 Gflops/s on the production CFD code OVERFLOW-MLP. *Parallel Computing*, 27(4):521–536, 2001.

[18] W.M. Washington. The computational future for climate change research. *Journal of Physics*, pages 317–324, doi:10.1088/1742-6569/16/044, 2005.

Chapter 3

Multiphysics Simulations and Petascale Computing

Steven F. Ashby

Lawrence Livermore National Laboratory

John M. May

Lawrence Livermore National Laboratory

3.1 Introduction

The scientific computing and computational science communities have experienced a remarkable renaissance in recent years with the advent of increasingly powerful supercomputers and scalable application codes. This combination enables high fidelity simulations of physical phenomena via increased spatial-temporal resolution and the inclusion of more accurate physics. For example, today's codes and computers are able to simulate supernovae in three dimensions with detailed radiation transport models. Such simulations facilitate greater insight and advance scientific discovery.

Scientists are increasingly interested in using codes that include multiple physical models and span multiple scales. In materials science, for example, computational scientists wish to simulate extreme materials properties across scales ranging from the atomistic (via *ab initio* molecular dynamics) through the microscale (dislocation dynamics) and mesoscale (aggregate materials models) to the continuum (finite elements).

Terascale supercomputers have been used to perform detailed (i.e., high-resolution) simulations within a scale regime, but they lack the computational horsepower to simulate accurately across scales. Meeting this challenge demands petascale computing power. The supercomputing industry is addressing this need through massive parallelism, first by increasing the number of computing nodes and now by increasing the number of processing cores per chip (see Section 3.2).

This increased computing power will need to be used in new ways. In past simulation advances, computational scientists used greater computing power to resolve a process more finely in space and time, but today's scientist wishes to use the computing power to integrate across spatial and temporal scales and different physics regimes. In other words, instead of using $8n$ processors to double the resolution of a 3D simulation previously done on n processors, one might instead use $2n$ processors on each of four scales (e.g., atomistic through continuum).

The convergence of these two trends — multiphysics applications and massively parallel petascale computers — requires a fundamentally new approach to large-scale scientific simulation. In the past, most simulations were data parallel and used all of the machine's processors to perform similar operations on distributed data. Multiscale, multiphysics simulations are inherently task parallel and call for a compatible programming paradigm (Section 3.3). These applications also demand scalable numerical algorithms suited to the multiscale and multiphysics nature of the problem being solved (Section 3.4).

3.2   The Next Generation of Supercomputers

In the early 1990s, the architecture of supercomputers began to evolve into the design that is familiar today: a large number of general-purpose processors connected by a high-performance network. The processors are often identical to those used in workstations or enterprise servers. Using off-the-shelf components allows system designers to exploit the rapid improvements in processors developed for high-volume markets and thereby avoid the cost of developing custom chips for supercomputers. A few vendors, such as Cray and NEC, continue to offer supercomputers based on vector processors rather than commodity processors, but vector-based systems account for a small part of the current supercomputer market: Only seven of the world's 500 fastest computers in November 2006 were vector machines [15].

During the past 15 years, improvements in peak processing power have come both from faster clock rates in individual processors and from larger processor counts. To illustrate the evolution of these systems, Table 3.1 compares two machines from Top 500 lists in 1993 and 2006. Each is the highest-ranked commodity-processor machine from its list. In this example, increasing the processor count contributed almost as much to the overall performance gain as the higher clock speeds. The two factors together account for a 3110-fold

TABLE 3.1: Characteristics of two supercomputers based on commodity processors. The Intel Delta was ranked number 8 in June 1993, and the Cray XT3 was number 2 in November 2006. The CPU total for the Opteron system counts two cores per processor. (Data from Top 500 Supercomputer Sites Web site [15].)

System	Processor	CPUs	Clock rate	Peak FLOPS
Intel Delta	i860	512	40 MHz	20.48×10^9
Cray XT3	Opteron	26,544	2.4 GHz	127.4×10^{12}
Ratio of increase		51.8	60	6,220

increase in peak performance. (The remaining factor-of-two increase comes from the use of floating point units that can complete a multiply-add instruction every clock cycle.)

Faster clock rates give users more performance with essentially no development effort, but higher processor counts can require significant development effort to exploit the added parallelism. Application developers must ensure that their algorithms scale and that their problems express enough parallelism to keep all the processors busy.

Unfortunately, the performance of individual processor cores is likely to increase much more slowly in coming years than it has over the past two decades. The most obvious method for increasing performance is raising the clock rate, which typically involves shrinking the components on a chip. However, reducing the spacing between components increases the amount of current that leaks between them, which in turn increases the power consumption for a given clock rate. Moreover, a smaller processor must dissipate the heat it produces over a smaller area of silicon. The resulting concentration of heat degrades reliability and further increases leakage current. These problems have made it difficult to develop commodity processors that run faster than about 4 GHz. Other architectural techniques that increase a core's computational performance for a given clock rate are also producing diminishing benefits.

Chip designers have turned instead to multicore designs to increase performance [10]. Many current processor designs now incorporate two processing cores on a single chip. Keeping the clock rate constant while doubling the number of cores allows the chip to use less power than doubling the clock rate for a single-core chip. Of course, for the dual-core chip to compute at the same rate as the higher-frequency single-core chip, there must be work available for the two cores to do in parallel. Meanwhile, major chipmakers such as Intel, AMD, and IBM already offer or have announced quad-core processors. The trend toward higher levels of internal parallelism seems clear.*

If processor clock rates remain essentially fixed, performance increases for supercomputers based on commodity chips can only come from increased parallelism. IBM's Blue Gene [9] line of computers is a forerunner of this trend. The largest of these systems, Blue Gene/L, has 65,536 processors, each with two IBM PowerPC 440 cores. These relatively simple processors run at only 700 MHz, limiting both component costs and power usage. However, this massive parallelism presents a new challenge to application developers who until recently had to deal with only 10,000-way parallelism. Section 3.5 shows how some users have risen to this challenge. As the trend toward 100,000-core or even million-core machines continues, exploiting this level of parallelism will need to become a mainstream activity.

3.3   Programming Models for Massively Parallel Machines

Most applications written for the current generation of parallel computers use a data parallel model of computation, also known as single program, multiple data (SPMD). In this model, all tasks execute the same algorithm on different parts of the data at the same time. (The tasks need not execute precisely the same instruction at a given instant, since some regions of the problem domain may require additional iterations or other special handling.) In contrast to data parallelism, task parallelism assigns different types of work to different threads or processes in an application. One way to implement task parallelism is by giving multiple threads in a process separate tasks to carry out on a common data set. In practice, however, scientific applications most often use threading as an extended form of data parallelism, such as running independent loop iterations in parallel. (Other types of applications, such as those that do commercial data processing, may have a genuinely task parallel architecture, but these are outside the scope of this chapter.)

One difficulty in writing an efficient data parallel program is keeping all the cores busy all of the time. Forcing some tasks to wait while others finish larger units of work can drastically reduce overall performance. Keeping a program well load-balanced becomes more difficult with increasing parallelism: There may not be a natural way to divide a problem into 100,000 or more equal parts that can be worked on concurrently. Even if partitioning the problem is easy, there may be so little work for each core to do that the cost of communicating data between the large number of tasks overwhelms the gains from increased

---

*In the rest of this chapter, the term "processor" will refer to a discrete chip, and "core" will refer to one or more CPUs on those chips. A "node" is a unit of the computer with one or more processors that share direct access to a pool of memory.

parallelism. The remainder of this section describes several approaches to using massively parallel machines efficiently.

3.3.1   New parallel languages

Several research groups are developing new languages for parallel programming that are intended to facilitate application development. Their goals include improving programmer productivity, using parallel resources more effectively, and supporting higher levels of abstraction. Three of these new languages are being developed for the U.S. Defense Advanced Research Projects Agency (DARPA) program on High Performance Computer Systems (HPCS). They are Chapel [5] (being developed by Cray), Fortress [1] (Sun), and X10 [6] (IBM). All three languages feature a global view of the program's address space, though in Chapel and X10 this global address space has partitions visible to the programmer. The languages also include numerous methods for expressing parallelism in loops or between blocks of code. Although these languages support task-level parallelism in addition to data parallelism, the developer still writes a single application that coordinates all activity.

While new languages may simplify writing new parallel applications from scratch, they are not an attractive means to extract further parallelism from existing large parallel codes. At the Lawrence Livermore National Laboratory (LLNL), for example, several important simulation codes have been under continuous development for more than ten years, and each has a million or more source lines. Rewriting even one of these applications in a new language, however expressive and efficient it may be, is not feasible.

3.3.2   MPI-2

Another approach to writing a task parallel application is to use the job creation and one-sided communication features in the MPI-2 standard [11]. MPI-2 allows a code to partition itself into several parts that run different executables. Processes can communicate using either the standard MPI message passing calls or the "one-sided" calls that let processes store data in another process or retrieve it remotely.

While MPI-2 has the basic capabilities necessary to implement task parallel applications, it is a low-level programming model that has not yet been adopted as widely as the original MPI standard. One obstacle to broad adoption of MPI-2 has been the slow arrival of full implementations of the standard. Although at least one complete implementation appeared shortly after MPI-2 was finalized in 1997, MPI-1 implementations remain more widely available.

3.3.3   Cooperative parallelism

To offer a different way forward, LLNL is developing an alternative parallel programming model, called cooperative parallelism [14], that existing applications can adopt incrementally.

Cooperative parallelism is a task parallel, or multiple program, multiple data (MPMD) programming model that complements existing data parallel models. The basic unit of computation is a *symponent* (short for "simulation component"), which consists of one or more processes executing on one or more nodes. In other words, a symponent can be a single-process program or a data parallel program. A symponent has an object-oriented interface, with defined methods for carrying out tasks. When an application starts, it allocates a fixed set of nodes and processors on a parallel machine, and the runtime system (called Co-op) launches a symponent on a subset of these nodes. Any thread or process within this initial symponent can launch additional symponents on other processors in the pool, and these new symponents can issue further launch requests. Each symponent can run a different executable, and these executables can be written entirely separately from each other.

When a thread launches a symponent, it receives an identifying handle. This handle can be passed to other threads or processes, and even to other symponents. Any thread with this handle can issue a remote method invocation (RMI) on the referenced symponent. Remote methods can be invoked as a blocking, nonblocking, or one-way call. (One-way RMI calls never return data or synchronize with the caller.)

Except for their RMI interfaces, symponents are opaque to one another. They do not share memory or exchange messages. Issuing an RMI to a symponent requires no previous setup, other than obtaining the handle of the target symponent, and no state information regarding the RMI persists in the caller or the remote method after the call is completed. This design avoids potential scaling issues that would arise if each of many thousands of symponents maintained information on all the symponents it had ever called or might potentially call in the future.

Symponents can terminate other symponents, either when they detect an error or for other reasons, and a parent symponent can be notified when its child terminates. Co-op is implemented using Babel middleware [7], which supplies the RMI functionality and also allows functions written in different languages to call each other seamlessly. This means that symponents written in C, C++, and Fortran can call each other without any knowledge of each other's language.

3.3.4   Example uses of cooperative parallelism

Cooperative parallelism enables applications to exploit several different kinds of parallelism at the same time. Uses for this flexibility include factoring out load imbalance and enabling federated computations.

Factoring out load imbalance. Some simulations incur load imbalance because extra calculations are necessary in certain regions of the problem domain. If the application synchronizes after each time step, then all the

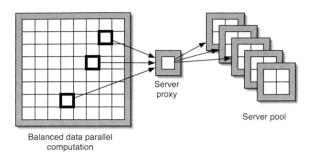

Balanced data parallel
computation

FIGURE 3.1: Cooperative parallelism can improve the performance of unbalanced computations. The boxes on the left represent a group of processors running a data parallel simulation. The highlighted boxes are tasks that have additional work to do, and they sublet this work to a pool of servers on the right. Each server is itself a parallel job. A server proxy keeps track of which servers are free and assigns work accordingly. This arrangement improves load balance by helping the busier tasks to finish their work more quickly.

tasks may wait while a subset of them complete these extra calculations. To reduce the time that the less-busy tasks wait for the busy one, the developer can assign the extra calculations to a pool of server symponents, as shown in Figure 3.1. Here, a large symponent running on many cores executes a normal data parallel computation. When any task determines that it needs additional computations, it sends a request to a *server proxy*, which forwards the request to an available server in a pool that has been allocated for this purpose. To the main simulation, this request looks like a simple function call, except that the caller can proceed with other work while it waits for the result. It can even submit additional nonblocking requests while the first one is executing. The servers themselves may run as parallel symponents, adding another level of parallelism. The server's ability to apply several processors to the extra calculation, combined with the caller's ability to invoke several of these computations concurrently, allows the caller to finish its extra work sooner. This reduces the time that less-busy tasks need to wait, so the overall load balance improves. We expect this approach to be helpful in materials science applications and a number of other fields.

Federated computations. A more ambitious use of cooperative parallelism is to build a large application from existing data parallel codes. For example, a multiphysics model of airflow around a wing could combine a fluid dynamics model with a structural mechanics model; or a climate application could combine models of the atmosphere, the ocean, and sea ice. In either case, a separate symponent would represent each independent element of the system. Periodically, the symponents would issue RMI calls to update either a central database or a group of independent symponents with data about their current state. A fully general solution would employ a collective form of RMI, in which multiple callers could send a request to multiple recipients; however, cooperative parallelism does not yet support this model. Although federated simulations can be written with current technology, they may encounter difficult load balance problems and other complications. Cooperative parallelism, when fully realized, should provide a simpler way to build multiphysics simulations as a federation of existing single-physics codes.

3.4 Multiscale Algorithms

Petascale simulations of multiphysics, multiscale phenomena require scalable algorithms and application codes that can effectively harness the power of the computers described above. Specifically, each symponent within a federated simulation needs to be scalable within its processor space. Scalability across symponents may be achieved via cooperative parallelism using an MPMD paradigm. This section discusses the importance of scalable numerical algorithms within a single physics regime, and it previews some novel approaches to scaling across multiple physics regimes.

In the scientific applications of interest here, one typically approximates the solution of a system of partial differential equations (PDEs) describing some physical phenomenon on a discretized spatial mesh. The choice of discretization scheme and numerical PDE solver are closely entwined and together determine the accuracy and scalability of the resulting simulation. The overall application code — consisting of the discretization and solver — must be scalable in both its underlying numerical algorithms and its parallel implementation. This would result in an application whose time to solution is constant as the problem size increases proportionally with the machine size (i.e, weak scaling). In practice, the scalability burden typically falls on the solver, which is the subject of the next section.

3.4.1 Parallel multigrid methods

Multigrid methods were first introduced in the 1970s and are provably optimal solvers for various classes of PDEs. Here optimality means that the algorithm has complexity $O(n)$, where n is the number of grid points on the discretized spatial mesh. In contrast, other solution techniques are $O(n^2)$ or worse. The optimality of the multigrid solver translates into mathematical scalability: The amount of work required for solution is linearly proportional to the problem (mesh) size. If this numerical method can be efficiently implemented in parallel — for example, by overlapping communication and computation — then the overall algorithm (and that portion of the application

TABLE 3.2: Execution times (in seconds) for a parallel multigrid method. The table compares two coarsening operators (C-old and C-new) for each of two data querying techniques (global and assumed partition).

Processors	Unknowns	Data query via global partition		Data query via assumed partition	
		C-old	C-new	C-old	C-new
4,096	110.6M	12.42	3.06	12.32	2.86
64,000	1.73B	**67.19**	10.45	19.85	**4.23**

code) is scalable.

A detailed discussion of multigrid methods is beyond the scope of this chapter, but the key idea is this: Multigrid methods solve successively smaller problems on a hierarchy of grids to accelerate the solution of the original fine-grid problem. Specifically, one must properly define coarsening and prolongation operators that map the intermediate approximations from one grid to another. (Coarsening restricts the approximate solution to a coarser grid; prolongation interpolates the approximate solution onto a finer grid.) These operators are highly problem-dependent. Most recent mathematical multigrid research has concentrated on defining operators in real applications so that the resulting algorithm retains the hallmark optimality (and mathematical scalability). Toward this end, considerable work has been done over the past decade in the area of algebraic multigrid methods. AMG methods, as they are known, do not presume that the underlying mesh is structured. Instead they rely on inferred algebraic properties of the underlying system of discretized equations. This information is used to define the coarsening and prolongation operators. Researchers successfully have applied AMG methods to challenging PDEs discretized on unstructured meshes for a variety of applications, including computational astrophysics, structural mechanics, and fluid dynamics.

In the past decade, considerable effort has been focused on improving the parallel scalability of algebraic multigrid methods. In serial AMG methods, key aspects of the calculation are inherently sequential and do not parallelize well on massively parallel machines. In particular, although computational complexity is optimal, storage and communication costs increase significantly on parallel computers. This problem has not been solved, but recent advances in coarsening strategies have ameliorated the complexity and setup costs by halving the storage requirements and reducing the execution times by an order of magnitude [8].

The parallel scalability issues become even more pronounced on massively parallel computers like Blue Gene/L. Consider the kernel operation of answering a global data distribution query. In a traditional parallel implementation (such as one that uses MPI's `MPI_Allgatherv` collective operation), this requires $O(p)$ storage and communication, where p is the number of processors. On a machine like Blue Gene/L with more than 100,000 processors — and future petascale machines with upwards of one million processors — storing $O(p)$ data is impractical if not impossible. A novel "assumed partition" algorithm [3] employs a rendezvous algorithm to answer queries with $O(1)$ storage and $O(\log p)$ computational costs.

To illustrate the power of efficiently implemented parallel multigrid methods, consider Table 3.2. It demonstrates the scalability of the LLNL AMG solver on the Blue Gene/L supercomputer: A problem with nearly two billion grid points (30^3 unknowns per processor) is solved in just over four seconds. The 16-fold improvement over the previous algorithm is a combination of an improved coarsening algorithm and the faster communication routine mentioned above. (It should be noted that the underlying algorithm also is mathematically scalable in terms of number of iterations required for convergence.)

3.4.2　ALE-AMR discretization

The preceding discussion touched on the importance of the underlying spatial mesh to the mathematical and parallel performance of the PDE solver. It is easier to define and implement optimally performing multigrid methods on structured Eulerian meshes (the regular communication patterns facilitate efficient parallel implementation), but such meshes may not adequately represent important problem features, such as complex moving parts in an engineering application.

To overcome this deficiency, computational scientists have typically turned to one of two competing discretization methodologies: adaptive mesh refinement (AMR) or arbitrary Lagrangian-Eulerian (ALE) meshing. In AMR, one refines the mesh during runtime based on various estimates of the solution error. This allows one to obtain the accuracy of a much finer mesh at a fraction of the storage and computational costs. The underlying grid still has a fixed topology, however. In an ALE approach, the mesh moves in response to evolving problem dynamics. This allows one to track complex physics more accurately, but the mesh often becomes so tangled that it must be remapped periodically. Constructing robust mesh motion algorithms for ALE schemes remains a central challenge.

An interesting recent idea is to combine the best features of ALE and AMR into a new discretization approach that is better suited to petascale simulation of multiphysics applications. The method, called ALE-AMR [2], is illustrated in Figure 3.2. Standard ALE is illustrated in the left graphic: One starts with an Eulerian mesh, which deforms over time in response to problem dynamics, and eventually it must be remapped. Intermediate meshes may possess highly skewed elements which present numerical difficulties. On the other hand, it nicely resolves complex features, such as shock fronts. A typical AMR grid

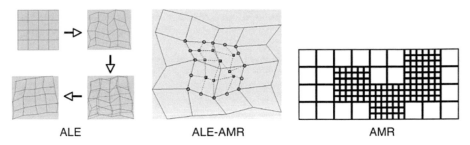

FIGURE 3.2: ALE-AMR (center) combines the moving mesh feature of ALE (left) with the adaptive refinement of AMR (right) to yield a cost-effective mesh discretization technique that accurately resolves evolving physical phenomena such as shock fronts.

hierarchy is shown on the right. The meshing is simpler with uniform elements, but the shock resolution is inferior per grid point compared to the ALE scheme. The ALE-AMR approach is shown in the center of Figure 3.2. The essential idea here is that one refines a portion of the mesh as in AMR through the dynamic insertion and deletion of grid points rather than allow it to deform too much, thereby combining the advantages of both types of adaptivity in one method. This helps to avoid many of the undesirable numerical properties associated with the highly skewed elements that arise in ALE schemes. It also allows one to leverage much of the efficient computer science machinery associated with AMR grid hierarchy management.

The combination of ALE and AMR technology — each challenging in itself — presents many numerical and software challenges that are still being researched. While some fundamentals of ALE-AMR algorithms have been established, the incorporation of more specialized physics capabilities such as sliding surfaces, globally coupled diffusion physics, and multi-material treatments continue to pose research challenges.

3.4.3   Hybrid atomistic-continuum algorithms

The discussion so far has focused on continuum methods, that is, numerical methods for approximating a solution on a spatial mesh. In multiphysics applications, however, one needs a range of models to simulate the underlying physical phenomena accurately. These applications may include some combination of continuum and atomistic (e.g., particle) methods. For instance, consider shock-induced turbulent mixing of two fluids, as shown in Figure 3.3. Continuum computational fluid dynamics (CFD) methods (e.g., Euler and Navier-Stokes) adequately describe fluid motion away from the interface, but they are limited by the smallest scales in the computational mesh. On the other hand, atomistic methods (e.g., direct simulation Monte

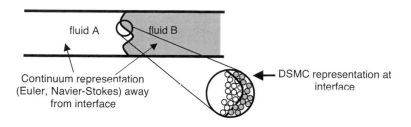

FIGURE 3.3: An illustrative multiphysics simulation of a shock propagating through two fluids. Continuum methods (e.g., based on Navier-Stokes) accurately describe the fluid motion away from the interface, but one needs an atomistic method (e.g., direct simulation Monte Carlo, DSMC) to simulate behavior at the interface. Since the atomistic method is too expensive to use throughout the domain, the use of a hybrid algorithm is attractive.

Carlo) adequately resolve the shock fronts, but they are too expensive to use throughout the problem domain.

Several researchers have recently investigated hybrid continuum-atomistic methods via adaptive mesh and algorithmic refinement (AMAR) [18]. As noted in the preceding section, traditional AMR allows one to refine a continuum calculation around dynamically moving and growing interfaces. Specifically, AMR refines the mesh around a feature of interest, say a shock front, and applies the same continuum method within the refined mesh. In a hybrid algorithm such as AMAR, one instead switches to a discrete atomistic method at the finest grid scale. This allows one to use an appropriate (but expensive) method only where it is needed. This is illustrated in Figure 3.4, where one can see the particles of an atomistic method embedded within an AMR grid hierarchy.

The implementation of hybrid methods like AMAR could be facilitated by a MPMD programming paradigm like cooperative parallelism. For example, one could easily allocate additional processors dynamically to the finer meshes or to the direct simulation Monte Carlo (DSMC) method. Although this can be done in a data parallel context, a properly implemented MPMD programming paradigm should make it easier to implement — but this remains to be demonstrated.

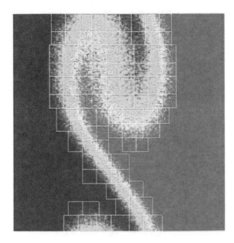

FIGURE 3.4: (See color insert following page 18.) Simulation of a moving interface via a hybrid continuum-atomistic method. The white grid blocks show where a direct simulation Monte Carlo particle method is applied at the finest AMR grid scale to resolve the physics at the interface between two fluids. A continuum-scale method is applied elsewhere in the fluid. (Adapted from Hornung *et al.* [13])

3.5 Applications Present and Future

The arrival of terascale supercomputers such as Blue Gene/L ushers in a new era of computational science, in which scientific simulation will emerge as a true peer to theory and experiment in the process of scientific discovery. In the past, simulations were viewed largely as an extension of theory. Moreover, these simulations often were lacking in some fundamental way, for example, insufficient spatial resolution due to limited computational resources. Today's supercomputers finally possess sufficient computing power (and memory) to enable unprecedented simulations of physical phenomena — simulations that often suggest new theories or guide future experiments. This section examines the state of the art in terascale simulation through two illustrative applications. It also offers a glimpse of the future of petascale simulation by describing the use of cooperative parallelism to enable a multiphysics simulation.

3.5.1 State of the art in terascale simulation

The LLNL Blue Gene/L supercomputer described in Section 3.2 represents a milestone in both scientific computing and computational science. In the former area, Blue Gene/L is the first computer to exceed 100 TFLOPS on the LINPACK benchmark [15]. It achieved 280.6 TFLOPS, 76% of the machine's theoretical peak. It is the first computer to employ more than 100,000 cores, thus redefining the term "massively parallel." This is important because it challenges the scientific computing community to think anew about parallelism and scalability. Many observers view Blue Gene/L as a stepping stone to the petascale.

The computational science milestone is even more impressive and important: In November 2005, Blue Gene/L ran the first meaningful scientific simulation to sustain more than 100 TFLOPS. Less than a year later, two additional application codes exceeded 100 TFLOPS on Blue Gene/L, including one that has since sustained 207 TFLOPS. In fact, the last two winners of the Gordon Bell Prize were LLNL's ddcMD and Qbox codes. Both codes ran molecular dynamics (MD) simulations of material properties under extreme conditions on Blue Gene/L. MD codes are particularly well suited to this machine, but other codes have been ported with excellent results.

World's first 100 TFLOPS sustained calculation (ddcMD). The 2005 Gordon Bell Prize was awarded to an LLNL-IBM team led by computational physicist Fred Streitz for the world's first 100 TFLOPS sustained calculation [17]. They simulated the solidification of tantalum and uranium via classical molecular dynamics using pseudopotentials. They ran a number of simulations, including some with more than 500 million atoms. These simulations begin to span the atomistic to mesoscopic scales.

The code, called ddcMD, sustained 102 TFLOPS over a seven-hour run, thus achieving a remarkable 28% of theoretical peak performance. It was fine-tuned by IBM performance specialists, and it demonstrated exemplary strong and weak scaling across 131,072 cores for several different simulations.

The simulation results are scientifically important in their own right. For the first time, scientists had both sufficient computing power (in Blue Gene/L) and a scalable application code (ddcMD) to fully resolve the physics of interest. Specifically, their 16 million atom simulation of grain formation was the first to produce physically correct, size-independent results. In contrast, previous simulations on smaller machines were able to use at most 64,000 atoms. In order to simulate the physical system of interest, periodic boundary conditions were imposed, resulting in an unphysical periodicity in the simulation results. The striking difference between these two simulations is readily seen in Figure 3.5, where three snapshots in time are shown. In the top row, one can see the rich 3D detail in the planar slices, whereas in the bottom row one sees the replicated pattern resulting from the under-resolved simulation. Moreover, the team proved that for this problem, no more than 16 million atoms are needed.

Current world record of 207 TFLOPS sustained (Qbox). The 2006 Gordon Bell Prize was awarded to an LLNL-UC Davis team led by computational scientist François Gygi for their 207 TFLOPS simulation of molybdenum [12]. This result is the current world record for performance of a scientific

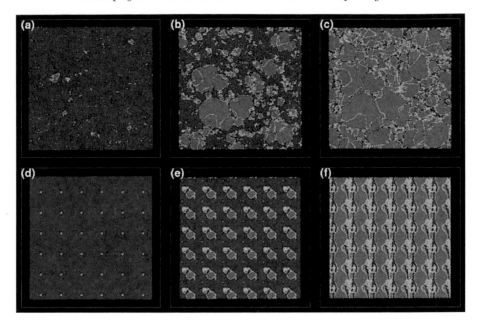

FIGURE 3.5: (See color insert following page 18.) Three-dimensional molecular dynamics simulation of nucleation and grain growth in molten tantalum. Three snapshots in time are shown for two simulations. The top row corresponds to a simulation using 16 million atoms on the Blue Gene/L supercomputer at LLNL. This 2005 Gordon Bell Prize-winning calculation was the first to produce physically correct, size-independent results. The rich 3D detail is seen in the planar slices. The bottom row used 64,000 atoms on a smaller supercomputer. Periodic boundary conditions were used to generate the entire domain, resulting in the unphysical replicated pattern. (Image from Streitz *et al.* [17])

application code on a supercomputer. The code, called Qbox, simulates material properties via quantum molecular dynamics. It is a first principles approach based on Schrödinger's equation using density functional theory (with a plane-wave basis) and pseudopotentials. This versatile code has been used to simulate condensed matter subject to extreme conditions (such as high pressure and temperature) in a variety of applications.

The code is written in C++ and MPI and is parallelized over several physics parameters (plane waves, electronic states, and k-points). It employs optimized ScaLAPACK and BLACS linear algebra routines, as well as the FFTW

Fast Fourier Transform library. The test problem involved 1000 molybdenum atoms.[†] In November 2005, this code sustained 64 TFLOPS on Blue Gene/L. Less than one year later, after considerable effort and tuning, the code achieved 207 TFLOPS on the problem above. It should be mentioned that this code's performance is highly dependent on the distribution of tasks across processors. The best performance was achieved using a quadpartite distribution across a $64 \times 32 \times 32$ processor grid.

The heroic effort behind the simulations described above should not be understated: In each case, teams of dedicated computational and computer scientists toiled to tune their codes to perform well on a radically new architecture. These pioneers deserve accolades, but future success will be measured by how routine and easy similar simulations become. The next section hints at one approach to realizing this vision.

3.5.2   Multiphysics simulation via cooperative parallelism

The two preceding examples illustrate the state of the art in large-scale simulation using single-physics codes on terascale supercomputers. In many scientific applications of interest, the desire is to integrate across multiple scales and physics regimes rather than resolve further a single regime. For example, a computational scientist might wish to federate several extant simulation codes (each scalable in its own right) into a single multiphysics simulation program. As discussed earlier, such applications require petascale computing power. The question is how to harness this computing power in an application developer-friendly way.

One approach to building such multiphysics application codes is cooperative parallelism, an MPMD programming model discussed in Section 3.3. Although this programming model is still in its early development, some work is already being done to show how it can facilitate the development of multiphysics applications. Specifically, a team has modified a material modeling code so that the coarse-scale material response computation uses parameters computed from fine-scale polycrystal simulations [4]. If the fine-scale parameters were recomputed each time they were needed, these computations would overwhelm the execution time of the simulation and make it infeasible to complete in a reasonable period of time. However, using a technique known as adaptive sampling (derived from *in situ* adaptive tabulation [16]), the simulation can eliminate many of these expensive calculations. It does so by storing results from early fine-scale computations in a database. Subsequent fine-scale computations can interpolate or extrapolate values based on previously stored results, if the estimated error is not too large. When it does become necessary

---

[†]Quantum MD codes are much more expensive than classical MD codes, so Qbox cannot model as many atoms as can ddcMD. On the other hand, Qbox is much more accurate for certain classes of problems. These are but two codes in a family of MD codes being used at LLNL for various scientific and programmatic applications.

to perform a full fine-scale computation, the simulation assigns the work to a server, as illustrated in Figure 3.1. So far, this application has run on just over 4000 cores, but the combined effect of the reduced need for fine-scale computations (because of adaptive sampling) and the improved load balance (because of the server pool model) has demonstrated performance gains ranging from one to two orders of magnitude, depending on the problem and the desired accuracy. Moreover, the application shows how MPMD programming can exploit several kinds of parallelism concurrently.

3.6 Looking Ahead

The convergence of multiscale, multiphysics applications with rapidly increasing parallelism requires computational scientists to rethink their approach to application development and execution. The largest current machine, Blue Gene/L, has challenged developers to utilize effectively more than 100,000 cores. Several teams have achieved notable successes, but accomplishments that now lead the field will need to become commonplace if the scientific computing community is to fully exploit the power of the next generation of supercomputers. Heroic efforts in application development and performance tuning must give way to standardized scalable algorithms and programming methodologies that allow computational scientists to focus on their science rather than underlying computer science issues.

Work is now underway at LLNL and other research institutions to find the best ways to harness petascale computers, but these efforts will not yield near-term results. When large-scale distributed memory computers began to dominate scientific computing in the early 1990s, there was a period of uncertainty as developers of hardware, systems software, middleware, and applications worked to determine which architectures and programming models offered the best combination of performance, development efficiency, and portability. Eventually, a more-or-less standard model emerged that embraced data parallel programming on commodity processors using MPI communication and high-speed interconnection networks.

The next generation of applications and hardware will bring about another period of uncertainty. Multiphysics simulations need programming models that more naturally lend themselves to MPMD applications. In light of this, the authors believe that the dominant paradigm for computational science applications running on petascale computers will be task parallel programming that combines multiple scalable components into higher-level programs.

3.7 Acknowledgments

We thank the following colleagues for generously contributing to this chapter: Bob Anderson, Nathan Barton, Erik Draeger, Rob Falgout, Rich Hornung, David Jefferson, David Keyes, Gary Kumfert, James Lcck, Frcd Streitz, and John Tannahill. We also acknowledge the U.S. National Nuclear Security Administration's Advanced Simulation and Computing program, the Department of Energy Office of Science's Scientific Discovery through Advanced Computing program, and LLNL's Laboratory Directed Research and Development program for their support.

This work was performed under the auspices of the U.S. Department of Energy by University of California, Lawrence Livermore National Laboratory under Contract W-7405-Eng-48. UCRL-BOOK-227348.

References

[1] E. Allen, D. Chase, J. Hallett, V. Luchangco, J.-W. Maessen, S. Ryu, G. L. Steel Jr., and S. Tobin-Hochstadt. *The Fortress Language Specification*. Sun Microsystems, Inc., 1.0 alpha edition, 2006.

[2] R. W. Anderson, N. S. Elliott, and R. B. Pember. An arbitrary Lagrangian-Eulerian method with adaptive mesh refinement for the solution of the Euler equations. *Journal of Computational Physics*, 199(2):598–617, 20 September 2004.

[3] A. H. Baker, R. D. Falgout, and U. M. Yang. An assumed partition algorithm for determining processor inter-communication. *Parallel Computing*, 31:319–414, 2006.

[4] N. R. Barton, J. Knap, A. Arsenlis, R. Becker, R. D. Hornung, and D. R. Jefferson. Embedded polycrystal plasticity and *in situ* adaptive tabulation. *International Journal of Plasticity*, 2007. In press.

[5] D. Callahan, B. L. Chamberlain, and H. P. Zima. The Cascade high productivity language. In *Proceedings of the Ninth International Workshop on High-Level Parallel Programming Models and Supportive Environments*, pages 52–60, 2004.

[6] P. Charles, C. Grothoff, V. Saraswat, C. Donawa, A. Kielstra, K. Ebcioglu, C. von Praun, and V. Sarkar. X10: An object-oriented

approach to non-uniform cluster computing. In *OOPSLA '05: Proceedings of the 20th annual ACM SIGPLAN Conference on Object Oriented Programming, Systems, Languages, And Applications*, pages 519–538, New York, NY, 2005. ACM Press.

[7] T. Dahlgren, T. Epperly, G. Kumfert, and J. Leek. *Babel User's Guide*. CASC, Lawrence Livermore National Laboratory, Livermore, CA, babel-1.0 edition, 2006.

[8] H. De Sterck, U. M Yang, and J. J. Heys. Reducing complexity in parallel algebraic multigrid preconditioners. *SIAM Journal on Matrix Analysis and Applications*, 27:1019–1039, 2006.

[9] A. Gara, M. A. Blumrich, D. Chen, G. L.-T. Chiu, P. Coteus, M. E. Giampapa, R. A. Haring, P. Heidelberger, D. Hoenicke, G. V. Kopcsay, T. A. Liebsch, M. Ohmacht, B. D. Steinmacher-Burow, T. Takken, and P. Vranas. Overview of the Blue Gene/L system architecture. *IBM Journal of Research and Development*, 49(2/3), 2005.

[10] D. Geer. Industry trends: Chip makers turn to multicore processors. *Computer*, 38(5):11–13, 2005.

[11] W. Gropp, S. Huss-Lederman, A. Lumsdaine, E. Lusk, B. Nitzberg, W. Saphir, and M. Snir. *MPI - The Complete Reference: Volume 2, The MPI-2 Extensions*. MIT Press, Cambridge, MA, 1998.

[12] F. Gygi, E. W. Draeger, M. Schulz, B. R. de Supinksi, J. A. Gunnels, V. Austel, J. C. Sexton, F. Franchetti, S. Kral, C. W. Ueberhuber, and J. Lorenz. Large-scale electronic structure calculations of high-Z metals on the Blue Gene/L platform. In *Proceedings of the IEEE/ACM SC06 Conference*, 2006.

[13] R. D. Hornung, A. M. Wissink, and S. R. Kohn. Managing complex data and geometry in parallel structured AMR applications. *Engineering with Computers*, 22(3–4):181–195, December 2006.

[14] D. R. Jefferson, N. R. Barton, R. Becker, R. D. Hornung, J. Knap, G. Kumfert, J. R. Leek, J. May, P. J. Miller, and J. Tannahill. Overview of the cooperative parallel programming model. In preparation, 2007.

[15] H.W. Meuer, E. Strohmaier, J.J. Dongarra, and H.D. Simon. TOP500 Supercomputer Sites. http://www.top500.org.

[16] S. B. Pope. Computationally efficient implementation of combustion chemistry using *in situ* adaptive tabulation. *Combustion Theory Modelling*, 1(1):41–63, January 1997.

[17] F. H. Streitz, J. N. Glosli, M. V. Patel, B. Chan, R. K. Yates, and B. R. de Supinksi. 100+ TFlop solidification simulations on Blue Gene/L. In *Proceedings of the ACM/IEEE SC\05 Conference*, 2005.

[18] H. S. Wijesinghe, R. D. Hornung, A. L. Garcia, and N. G. Hadjiconstantinou. Three-dimensional hybrid continuum-atomistic simulations for multiscale hydrodynamics. *Journal of Fluids Engineering*, 126:768–777, 2004.

Chapter 4

Scalable Parallel AMR for the Uintah Multi-Physics Code

Justin Luitjens

SCI Institute, University of Utah

Bryan Worthen

SCI Institute, University of Utah

Martin Berzins

SCI Institute, University of Utah

Thomas C. Henderson

School of Computing, University of Utah

4.1 Introduction

Large-scale multi-physics computational simulations often provide insight into complex problems that both complement experiments and help define future physical and computational experiments [13]. A good example of a production-strength code is Uintah [13, 12]. The code is designed to solve reacting fluid-structure problems involving large deformations and fragmentation. The underlying methods inside Uintah are a combination of standard fluid-flow methods and material point (particle) methods. In the case of codes, like Uintah, which solve large systems of partial differential equations on a mesh, refining the mesh increases the accuracy of the simulation. Unfortunately refining a mesh by a factor of two increases the work by a factor of 2^d, where d is the dimensionality of the problem. This rapid increase in computational effort severely limits the accuracy attainable on a particular parallel machine.

Adaptive mesh refinement (AMR) attempts to reduce the work a simulation must perform by concentrating mesh refinement on areas that have high error [3, 2] and coarsening the mesh in areas in which the error is small. One standard parallel AMR method divides the domain into rectangular regions called patches. Typically each patch contains Cartesian mesh cells of the same size. Each processor runs the simulation on a subset of the patches while communicating with neighboring patches. By using a component-based framework [13, 12, 8, 10] simulation scientists can solve their problems without having to focus on the intricacies of parallelism. For example, inside Uintah, parallelism is completely hidden from the simulation scientists [13], by means of a sophisticated task compilation mechanism.

In order to achieve good parallel performance with AMR, the component-based frameworks must be enhanced. The framework must drive refinement by locating regions that require refinement and creating patches on those regions. We refer to the process of generating patches as regridding. The patches must be load balanced onto processors in such a way that each processor is performing approximately the same amount of work while minimizing the overall communication. This can be achieved in the case of AMR calculations through the use of space-filling curves. Load balancers based on space-filling curves can create partitions quickly that keep communication between processors low while also keeping the work imbalance low [1, 16, 18, 6]. Recent work has shown that space-filling curves can be generated quickly and scalably in parallel [11]. In addition, the Uintah framework must also schedule the communication between the patches. As the simulation runs, regions needing refinement change, and as the computational mesh changes, load balancing and scheduling need to be recomputed. The regridding can occur often throughout the computation requiring each of these processes to be quick and to scale well in parallel. Poor scalability in any of these components can significantly impact overall performance [19, 20]. The Berger-Rigoutsos algorithm, which is commonly used for regridding [19, 4], creates patch sets with low numbers of patches that cover the regions needing refinement. In addition, the Berger-Rigoutsos algorithm can be parallelized using a task-based parallelization scheme [19, 20].

Present state-of-the-art AMR calculations have been shown by Wissink and colleagues [19, 20] to scale to many thousands of processors in the sense of the distribution of computational work. The outstanding issues with regard to AMR are surveyed by Freitag Daichin et al. [7], and in great depth by Steensland and colleagues [17]. In this paper we will show how scalability needs to reevaluated in terms of accuracy and consider how different components of Uintah may be made scalable.

4.2   Adaptive Mesh Refinement

Traditionally in parallel computing, scalability is concerned with maintaining the relative efficiency as the problem size grows. What is often missing is any attempt to ensure that the computational effort is, in comparative terms, well spent. The following argument was made by Keyes at the Dagstuhl workshop where his work was presented:

Consider a fixed mesh partial differential equation (PDE) calculation in three space dimensions with mesh sizes δx, δy, δz defining a regularly refined box mesh M_1 and a time step δt. Then it is possible to write the error obtained by using that mesh in some suitable norm, defined here by $||E(M_1)||$ as

$$||E(M_1)|| = C_x(\delta x)^k + C_y(\delta y)^k + C_z(\delta z)^k + C_t(\delta t)^q \tag{4.1}$$

In many cases of practical interest p and q may be less than three. The computational time, T_{cpu}, associated with this mesh is (at best) a linear function of the number of unknowns and the number of time steps and hence may be written as:

$$T_{cpu}(M_1) = C_{cpu}\frac{1}{\delta x \delta y \delta z \delta t} \tag{4.2}$$

where C_{cpu} is an appropriate constant.

In order to simplify the discussion from hereon the time dependent nature of the calculation is neglected. In the case when a new mesh, M_2 is defined by uniformly refining the original mesh by a factor of 2 in each dimension

$$\delta x = \frac{\delta x}{2} \tag{4.3}$$

and similarly for δy and δz, then the computational work increases by

$$T_{cpu}(M_2) = 8T_{cpu}(M_1) \tag{4.4}$$

while the error changes by

$$||E(M_2)|| = \frac{1}{2^k}||E(M_1)|| \tag{4.5}$$

Hence for first and second order methods $k = 1, 2$ the increase in work is greater than the decrease in error. The situation becomes worse if the work has a greater than linear dependency on the number of unknowns. Observations have spurred much work on high-order methods; see for example comparisons in the work of Ray and Steensland and others [14, 18].

We now define a mesh accuracy ratio, denoted by M_{ar} by:

$$M_{ar} = \frac{||E(M_2)||}{||E(M_1)||} \tag{4.6}$$

Assuming that the calculation on M_1 uses P_1 parallel processors and the calculation on mesh M_2 uses P_2 parallel processors, and that both solutions are delivered in the same time, then the Mesh parallel efficiency as denoted by M_{pe}, may be defined as

$$M_{pe} = M_{ar} \frac{P_2}{P_1} \qquad (4.7)$$

For the simple fixed mesh refinement steady state case above then,

$$M_{pe} = \frac{8}{2^k} \qquad (4.8)$$

It is worth remarking that if the calculation (inevitably) does not scale perfectly as far as computational work, or the different numbers of processors take different amounts of time, then the above expression may be modified to take this into account:

$$M_{pe} = M_{ar} \frac{P_2 T_2}{P_1 T_1} \qquad (4.9)$$

where T_1 and T_2 are the compute times using P_1 and P_2 processors, respectively.

As a simple example consider the case of the solution of Laplace equation on the unit cube using a Jacobi method. Suppose that an evenly spaced $N x N x N$ mesh is decomposed into p sub-cubes each of size $\frac{N^3}{p}$ on p processors. A standard second order method gives an accuracy of $\frac{C_2}{N^2}$ while use of a fourth order method [9], gives an accuracy of $\frac{C_4}{N^4}$ and where C_2 and C_4 are both known constants. The cost of the fourth order method is twice as many operations per point with a communications message length that is twice as long as in the second order case, and with perhaps r_I as many iterations, thus with an overall cost of $2r_I$ of that of the second-order method.

For the same accuracy, and $M_{ar} = 1$, it follows that

$$N_2 = \sqrt{\frac{C_4}{C_2}} N_4^2 \qquad (4.10)$$

where N_2 is the second order mesh size and N_4 is the fourth order mesh size. In order to achieve $M_{pe} = 1$ with each run having the same execution time, as estimated by a simple cost model based on $\frac{N^3}{p}$, the number of processors used by the second order mesh, P_2, must be related to the number used by the fourth order mesh, P_4 by

$$\frac{(N_2)^3}{P_2} \approx 2r_I \frac{(N_4)^3}{P_4} \qquad (4.11)$$

Hence the lower order method needs approximately the square of the number of processors of the higher order method:

$$P_2 \approx \frac{1}{2r_I} \left(\frac{C_4}{C_2} \right)^{\frac{3}{2}} (P_4)^2 \qquad (4.12)$$

The mesh parallel efficiency is also far from one:

$$M_{pe} = \frac{C_4}{C_2} \frac{2r_I}{N_2 N_4} \tag{4.13}$$

Unless the fourth order method has a significantly greater number of iterations per point than the second order method, the implications of this estimate for a petaflop machine with possibly $O(10^6)$ processors are clear.

In considering mesh refinement it is possible to start with a simple one-dimensional case. Consider a uniform mesh of $N_c\ \delta x_f$ cells. Next consider a nonuniform mesh which starts with a cell of width δx_f at its left side and then doubles with each subsequent cell. Suppose that there are q of these cells, then:

$$\delta x_f (1 + 2 + 4 + ...2^q) = \delta x_f N_c \tag{4.14}$$

Hence after summing the left side of this

$$\delta x_f (2^{q+1} - 1) = \delta x_f N_c \tag{4.15}$$

or

$$q = log_2(N_c + 1) - 1 \tag{4.16}$$

It is worth remarking that it is possible to modify the above expression to account for a mesh that increases more gradually. For example, an adaptive mesh in which two cells have the same size before the size changes gives:

$$q = log_2(N_c + 2) - 2 \tag{4.17}$$

With this simple logarithmic model of mesh changes in mind consider three cases in which mesh refinement is applied to a three-dimensional box around either one vertex, one edge or one bounding plane. Suppose that the box is discretized by using N_c^3 regular cells. While there are many refinement possibilities, typically nonuniform refinement takes place on a lower dimensional manifold than the original mesh. For example:

Refinement at a Point. In this case the mesh can increase in all three dimensions as in the simple one-dimensional example and so q^3 cells are used. In this case we assume that it is possible to increase the accuracy by only refining m cells close to the point. Hence the new mesh has $q^3 + m$ cells and

$$M_{pe} = \frac{q^3 + m}{q^3} \frac{1}{2^k} \tag{4.18}$$

Refinement on a Line. In this case the mesh can increase in only two dimensions as in the simple one-dimensional example and so $N_c q^2$ cells are used. In this case we assume that it is possible to increase the accuracy by only refining m cells close to the line. Hence the new mesh has $N_c q^2 + mN_c$ cells and

$$M_{pe} = \frac{N_c q^2 + mN_c}{N_c q^2} \frac{1}{2^k} \tag{4.19}$$

Refinement on a Plane. In this case the mesh can increase in only one dimension as in the simple one-dimensional example and so $N_c^2 q$ cells are used. In this case we assume that it is possible to increase the accuracy by only refining $N_c^2 m$ cells close to the plane. Hence the new mesh has $q^3 + m$ cells and

$$M_{pe} = \frac{N_c^2 q + m N_c^2}{N_c^2 q} \frac{1}{2^k} \tag{4.20}$$

In all three cases a mesh efficiency close to 1 requires:

$$\frac{m}{q^j} \le 2^k \tag{4.21}$$

where $j = 1, 2$ or 3 depending on the case above. Even in the case $k = 1$ and $j = 1$ (refinement of a plane), this simple analysis shows that as long as mesh refinement needs to be used on less than 50% of the existing cells in order to reduce the accuracy by half then the increase in accuracy is matched by the increase in work.

These studies show that if either high order methods or adaptive mesh approaches are used, then computational resources are used wisely with respect to the accuracy achieved. This has already been recognized for high order methods [14], but is not so widely understood for adaptive methods, such as those discussed below.

4.3   Uintah Framework Background

Uintah is a framework consisting of components such as a simulation component, the load balancer, the scheduler, and the regridder. The regridder component will be described in detail below.

4.3.1   Simulation components

The Uintah simulation components implement different algorithms and operate together or independently [13, 12, 8]. Uintah's main simulation components are based on the implicit compressible Eulerian algorithm (ICE), material point method (MPM), and Arches [13]. The simulation component will create tasks, and pass them to the scheduler, which is described below, instructing it as to what data relative to a patch that task will need. The scheduler will then execute the simulation component's tasks, one patch at a time, thus creating a parallel environment, and enabling the applications scientist to concentrate on the science issues.

4.3.2   Load balancer

The load balancer is responsible for determining which patches will be owned by each processor. There are two main load balancers in Uintah: the simple load balancer, and the dynamic load balancer. The simple load balancer simply determines the average number of patches per processor, and assigns that number of consecutive patches to each processor. This suffices for simple static problems that are easily load balanced. The dynamic load balancer attempts to achieve balance for more complicated problems. First, it orders the patches according to a space-filling curve; and second, it computes a weight for each patch, based on its size and the number of particles. The curve and the weights are then used to distribute the patches according to the average work per processor. The patches are assigned in the order of the space-filling curve placing patches that are close together in space on the same processor. This reduces the overall amount of communication that must occur. The patches are assigned so that each processor has approximately the same amount of work.

The Hilbert [15] space-filling curve is used in Uintah because it may be generated quickly, [1, 16, 18, 6, 11], in parallel [11], and provides good locality. The curve is formed over patches by using the centroid of the patches to represent them. The space-filling curve provides a linear ordering of the patches such that patches that are close together in the linear ordering are also closer together in the higher dimensional space. The curve is then broken into curve segments based on the weights of the patches. This provides approximately equally sized partitions that are clustered locally. Figure 4.1 shows an adaptive mesh partitioned using the Hilbert curve.

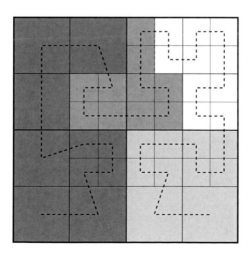

FIGURE 4.1: An example of how a space-filling curve is used in partitioning a mesh.

The space-filling curve can be generated quickly in parallel. Figure 4.2 shows how the generation performance varies for large numbers of patches on up to 2048 processors. The load balancer is also responsible to create a

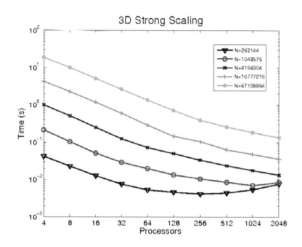

FIGURE 4.2: Scalability of the generating space-filling curves.

processor's neighborhood, which in essence is every patch on that processor along with every patch on any other processor that will communicate with a patch on that processor.

4.3.3   Scheduler

The scheduler orders the simulation component's tasks in a parallel fashion, and determines the corresponding MPI communication patterns. Its work is divided into two phases: compilation and execution. The compilation phase determines what data is required by each patch from its neighboring patches for each task. This is determined from the basic communication requirements that are provided by the simulation component. It accomplishes this by determining which patches are neighbors, and then computes the range of data the neighboring patch will need to provide [13, 12, 8]. On each processor, this algorithm is executed for each patch in the processor's neighborhood, which is on the order of the number of patches per processor, thus giving a complexity of nearly $O(\frac{N}{P} \log \frac{N}{P}^2)$, where N is the number of patches and P is the number of processors. This phase is executed only once for problems without AMR or dynamic load balancing, hence for fixed meshes its performance is not an issue. During the execution phase, each task will receive any data it requires from a neighboring patch's processor, run the simulation code for the task, and then send any data it computes to requiring tasks on other processors.

4.4   Regridder

The regridder's duty is to create a finer level, which is a set of patches with the same cell spacing, based on the refinement flags, which are created by the simulation component. The regridder determines the region of space on which to create a finer level, and then divides that space into patches with a finer resolution. It is important that the regridder considers what type of patches to produce. Producing patches that are too large can result in large load imbalances and prevent scalability. However, producing patches that are too small can cause significant overhead in other components.

The Uintah framework has constraints which require the regridder to produce certain types of patch sets. The first constraint is a minimum patch size. Each edge of a patch must be at least 4 cells in length. In addition, patch boundaries can either be coarse or fine but not a combination of the two. That is, every patch boundary must be completely filled with neighboring patches or have no neighbors at all. For the rest of this chapter we refer to the location on a boundary that moves from coarse to fine as a mixed boundary. Figure 4.3 shows two patch sets that cover the same area. The left patch set is invalid because it contains a mixed boundary; the second patch set does not contain a mixed boundary and is valid.

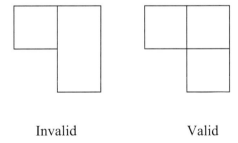

Invalid   Valid

FIGURE 4.3: Valid and invalid patch sets within Uintah. The left patch set is invalid because it contains a mixed boundary.

Regridding is commonly accomplished through the Berger-Rigoutsos algorithm [19, 4]. The algorithm starts by placing a bounding box around all of the refinement flags. A histogram of the refinement flags is then created in each dimension. This histogram is then used to determine a good location to split the bounding box in half. The process then recursively repeats on both halves of the bounding box. By having different processors evaluate different sections of the recursion this process can be made parallel. A full description of the parallel algorithm can be found in [19, 20].

The Berger-Rigoutsos algorithm produces patch sets with low numbers of patches. However, the constraints within Uintah prevent the use of the Berger-Rigoutsos algorithm. Berger-Rigoutsos produces patch sets that contain mixed boundaries. Mixed boundaries can be eliminated by splitting patches at the point where the boundary changes. However, splitting patches produced by Berger-Rigoutsos can lead to patches that violate the minimum patch size requirement. Due to the constraints within Uintah we initially used a tiled regridder. A grid was placed across the domain creating square patches. Each patch was searched for refinement flags. If a patch did not contain any refinement flags then the patch was thrown away. This produces square patches that cannot contain mixed boundaries and are larger than the minimum patch size. In addition, this regridder simplified the load balancer because all patches had the same number of cells allowing us to load balance by using simpler algorithms. Figure 4.4 shows a set of flags and a corresponding patch set produced by the tiled regridder.

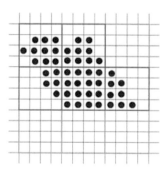

FIGURE 4.4: A patch set produced using the tiled regridder. Patches that do not contain flags are removed from the computational mesh.

As mentioned above, Uintah cannot use the original algorithm directly. However, a modified version of the Berger-Rigoutsos algorithm was devised that creates patches that satisfy Uintah's constraints. The first modification is to coarsen the refinement flags by the minimum patch size. To coarsen the refinement flags, tiles equal to the minimum patch size are laid across the domain. Each tile represents a single coarse flag. A new flag set is generated from these coarse flags. This modification guarantees that any patch created by any regridding algorithm used on these flags is at least the size of the minimum patch size. The Berger-Rigoutsos algorithm is then run on the coarse flag set producing a coarse patch set.

Next a fix-up phase is run on the coarse patch set to guarantee the boundary constraint. Each patch is searched for mixed boundaries. When a mixed boundary is found the patch is split at the point where the boundary changes, eliminating the mixed boundary. Performing this search along each boundary of each patch guarantees that the boundary condition is met. This search can easily be performed in parallel by having each processor search a subset of patches.

The modifications have both advantages and disadvantages over the original Berger-Rigoutsos algorithm. The coarsening of the flags allows the Berger-Rigoutsos algorithm to run on a coarser level speeding up the computation of the patch set. In addition, the minimum patch size can be set larger to prevent tiny or narrow patches. The disadvantage to coarsening the flags is that the final patch set will, in most cases, contain more area than it would with original flags and at best will contain the same area. In addition, the fix-up phase causes the number of patches to increase. However, this increase is small and this method is much better than the tiled algorithm. A comparison of patch sets produced by the two regridders can be found in Figure 4.5. The coarsened Berger-Rigoutsos regridder produces significantly fewer patches than the tiled regridder.

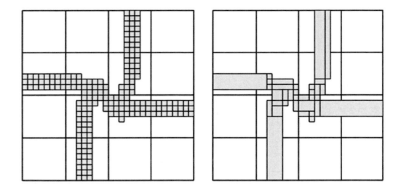

FIGURE 4.5: Two patch sets from Uintah. The left patch set is using the tiled regridder while the right is using the coarsened Berger-Rigoutsos regridder.

Finally, in order to facilitate a better load balance we have implemented one additional modification to the Berger-Rigoutsos algorithm. After the fix-up phase, patches may be subdivided further. The weights for each patch are calculated and any patches that are greater than the average amount of work per processor are split in half along the longest dimension. Patches that are larger than the average amount of work are too big and will result in large load imbalances. In addition, the dynamic load balancer can further split patches in order to load balance them more efficiently.

Changing the grid is an expensive process. Whenever the grid changes the computation must be load balanced, data must be migrated to its new location, and the task graph must be recompiled. When the grid is changing often this can greatly hinder the performance of the entire simulation. By using a process refered to as dilation we can ensure that regridding does not occur too often. Dilation expands the refinement flags outward in all directions by applying a stencil to the refinement flags creating a second set of refinement flags refered to as the dilated flags. At every timestep the refinement flags are compared to the current grid. If the refinement flags are contained within the grid then no regridding is necessary. When regridding occurs the dilated flags are used to create the new grid ensuring the grid is larger than what is dictated necessary by the refinement flags. This allows the grid to be used for multiple timesteps before a regrid is necessary.

4.4.1   Extending Uintah's components to enable AMR

The simulation, scheduling, and load balancing components all need to be extended for AMR. Any Uintah simulation component that wants to use AMR must provide a set of functions to: compute refinement flags (so the regridder can use them to create finer levels); refine the coarse–fine interface, which interpolates coarse-level data to the fine level along the boundaries of the fine level; coarsen, which interpolates the computed fine-level data to the corresponding space on the coarse level; and refine, which interpolates coarse-level data to a newly generated fine patch. These operations will increase communication costs as each patch no longer only communicates along its boundary, but must also communicate with the patches that are coarser and finer in the same region of space.

4.5   Performance Improvements

In order to analyze the performance of Uintah's AMR infrastructure we ran a 3D two-level spherically expanding blast wave problem using 128 processors. This problem is near the worst case for AMR. It has relatively low computation per cell and requires lots of regridding. Performing analysis on this problem provides good insight into where AMR overheads are coming from. This problem was ran on Zeus, which is a Linux cluster located at Lawrence Livermore National Laboratory with 288 nodes each with eight 2.4 GHz AMD Opteron processors. Each node has 16 GB of memory. Nodes are connected with an InfiniBand switch.

Figure 4.6 shows the runtime of the dominant components using the tiled regridder. This graph shows that with the original tiled regridder communication and regridding time was a major overhead. By using the new load balancer the communication dropped significantly because the code could greater exploit intra-node communication. This made the regridder the most time-consuming portion of the overhead. By switching to the Berger-Rigoutsos regridder the number of patches was significantly lowered and as a result the time for regridding and recompiling the task graph was also lowered. However, a significant amount of overhead was still due to regridding and the following recompile. By introducing dilation the number of times the grid was changed was reduced and an improvement in both regridding and recompile time was observed. These changes in total have reduced the AMR overhead by around 65% for this problem.

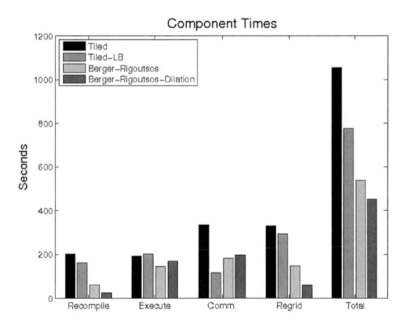

FIGURE 4.6: The component times for a 3D blast wave in Uintah on 128 processors on Zeus.

4.6   Future Work

There is still much room for improvements in the infrastructure that could decrease the overhead of AMR which will lead to good scalability. Wissink and Steensland and their colleagues have recently shown that it is possible but challenging to get specific codes to scale well. The challenge of applying and extending such ideas to a more general purpose code such as Uintah is considerable. The central issue is to ensure that the very general task compiler and task mapping components scale. Clearly if the substantial overhead of AMR does not scale then the code as a whole will not scale. One possible solution is to move toward incremental algorithms, for which AMR is an ideal problem. Often during execution only the finest level is changing. Incremental algorithms could exploit this by not recalculating on the coarser levels except when needed. In addition, when level changes are typically small, a few patches may be added and a few may be removed but the overall patch structure remains unchanged. Incremental algorithms could take advantage of this, reducing the AMR overhead considerably. For instance, the task graph compiler would only have to compile small subsets of the entire task graph and the load balancer could take the current placement of patches into consideration when deciding the placement of new patches. Ideally the entire framework would be incremental reducing the overhead associated with the framework and making the dominant costs the task computation and communication.

The communication is still a dominant portion of the runtime. We believe this is due to synchronization and are working on modifying the infrastructure to work in a more asynchronous fashion. In addition we are working on ways to reduce the overall communication needed. The infrastructure of Uintah can be made quite complex in order to perform communication as efficiently as possible while keeping the interface for simulation component developers simple. This provides an ideal scheme for having general purpose simulations that use highly complex parallel algorithms and at the same time allows simulation component developers to implement their algorithms without being hindered by the parallel complexities. Finally, given that redistributing data is expensive after load balancing it may also be appropriate to take into account the relative merits of the redistribution cost against computing with a small imbalance, see [5].

4.7   Acknowledgments

This work was supported by the University of Utah's Center for the Simulation of Accidental Fires and Explosions (C-SAFE) funded by the Department of Energy, under subcontract No. B524196.

We would like to thank the C-SAFE team for all their hard work on Uintah and would also like to thank Lawrence Livermore National Laboratory who graciously gave us access to their computing facilities where we were able to test Uintah on large numbers of processors.

References

[1] S. Aluru and F. Sevligen. Parallel domain decomposition and load balancing using space-filling curves. In *Proceedings of the 4th International Conference on High-Performance Computing*, pages 230–235, Bangalore, India, 1997.

[2] M. J. Berger and J. Oliger. Adaptive mesh refinement for hyperbolic partial differential equations. *Journal of Computat. Phys.*, 53:484–512, 1984.

[3] M.J. Berger and P. Colella. Local adaptive mesh refinement for shack hydrodynamics. *Journal of Computat. Phys.*, 82:65–84, 1989.

[4] M.J. Berger and I. Rigoutsos. An algorithm for point clustering and grid generation. *IEEE Transactions on Systems, Man and Cybernetics*, 21(5):1278–1286, 1991.

[5] M. Berzins. A new metric for dynamic load balancing. *Applied Math. Modell.*, 25:141–151, 2000.

[6] K.D. Devine, E.G. Boman, R.T. Heaply, B.A. Hendrickson, J.D. Teresco, J. Faik, J.E. Flaherty, and L.G. Gervasio. New challenges in dynamic load balancing. *Applied Numerical Mathematics*, 52(2-3):133–152, 2005.

[7] L. Freitag Daichin, R. Hornung, P. Plassman, and A. Wissink. A parallel adaptive mesh refinement. In M. Heroux, P. Raghavan, and H. Simon, editors, *Parallel Processing for Scientific Computing*, pages 143–162. SIAM, 2005.

[8] J.D. Germain, J. McCorquodale, S.G. Parker, and C.R. Johnson. *A Massively Parallel Problem Solving Environment*. IEEE Computer Society, Washington, DC, 2000.

[9] F. Gibou and R. Fedkiw. A fourth order accurate discretization for the Laplace and heat equations on arbitrary domains, with applications to the Stefan problem. *Journal of Computat. Phys.*, 202(2):577–601, 2005.

[10] R.D. Hornung and S.R. Kohn. Managing application complexity in the SAMRAI object-oriented framework. *Concurrency and Computation: Practice and Experience*, 14:347–368, 2002.

[11] J. Luitjens, M. Berzins, and T. Henderson. Parallel space-filling curve generation through sorting. *Concurrency and Computation: Practice and Experience*, 19(10):1387–1402, 2007.

[12] S.G. Parker. A component-based architecture for parallel multi-physics PDE simulation. *Future Generation Comput. Sys.*, 22(1):204–216, 2006.

[13] S.G. Parker, J. Guilkey, and T. Harman. A component-based parallel infrastructure for the simulation of fluid-structure interaction. *Eng. with Comput.*, 22(1):277–292, 2006.

[14] J. Ray, C. A. Kennedy, S. Lefantzi, and H.N. Najm. Using high-order methods on adaptively refined block-structured meshes i - derivatives, interpolations, and filters. *SIAM Journal on Scientific Computing*, 2006.

[15] H. Sagan. *Space-Filling Curves*. Springer-Verlag, Berlin, 1994.

[16] M. Shee, S. Bhavsar, and M. Parashar. Characterizing the performance of dynamic distribution and load-balancing techniques for adaptive grid hierarchies. In *Proc. of the IASTED Int. Conf., Parallel and Distributed Computing and Systems*, Cambridge, MA, November 1999.

[17] J. Steensland and J. Ray. A partitioner-centric model for structured adaptive mesh refinement partitioning trade-off optimization. *Part I. International Journal of High Performance Computing Applications*, 19(4):409–422, 2005.

[18] J. Steensland, S. Söderberg, and M. Thuné. A comparison of partitioning schemes for blockwise parallel SAMR algorithms. In *PARA '00: Proc. of the 5th Int. Workshop on Appl. Parallel Comput., New Paradigms for HPC in Industry and Academia*, pages 160–169, London, 2001. Springer-Verlag.

[19] A.M. Wissink, R.D. Hornung, S.R. Kohn, S.S. Smith, and N. Elliott. Large scale parallel structured AMR calculations using the SAMRAI framework. In *Supercomputing '01: Proc. of the 2001 ACM/IEEE Conference on Supercomputing*, page 6, New York, 2001. ACM Press.

[20] A.M. Wissink, D. Hysom, and D.R. Hornung. Enhancing scalability of parallel structured AMR calculations. In *ICS '03: Proc. of the 17th Ann. Int. Conf. on Supercomputing*, pages 336–347, New York, 2003. ACM Press.

Chapter 5

Simulating Cosmological Evolution with Enzo

Michael L. Norman, James Bordner, Daniel Reynolds, Rick Wagner

Laboratory for Computational Astrophysics, Center for Astrophysics and Space Sciences, University of California San Diego, 9500 Gilman Dr., La Jolla, CA 92093, (mlnorman, jobordner, drreynolds, rpwagner)@ucsd.edu

Greg L. Bryan

Astronomy Department, Columbia University, 550 West 120th St., New York, NY 10027, gbryan@astro.columbia.edu

Robert Harkness

San Diego Supercomputer Center, University of California San Diego, 9500 Gilman Drive, La Jolla, CA 92093, harkness@sdsc.edu

Brian O'Shea

Theoretical Astrophysics (T-6), Los Alamos National Laboratory, P.O. Box 1663, Los Alamos, NM 87545, bwoshea@lanl.gov

5.1 Cosmological Structure Formation

The universe is homogeneous and isotropic on scales exceeding half a billion light years, but on smaller scales it is clumpy, exhibiting a hierarchy of structures ranging from individual galaxies up to groups and clusters of galaxies, and on the largest scales, the galaxies are aligned in a cosmic web of filaments and voids. In between the galaxies is a diffuse plasma which is the reservoir of matter out of which galaxies form. Understanding the origin and evolution of these structures is the goal of *cosmological structure formation* (CSF).

It is now understood that CSF is driven by the gravitational clustering of dark matter, the dominant mass constituent of the universe. Slight inhomogeneities in the dark matter distribution laid down in the early universe seed CSF. The rate at which structure develops in the universe depends upon the power spectrum of these perturbations, now well measured on scales greater than one million light years [18]. It also depends on the cosmic expansion rate, which in turn is influenced by dark energy, a form of energy which exhibits negative pressure and whose existence was revealed only in 1998 by the discovery that the expansion rate is accelerating. The quantitative study of CSF is thus a direct route to studying two of the most mysterious substances in modern physics: dark matter and dark energy.

The part of the universe that astronomers can see directly is made up of ordinary "baryonic" matter. Thus, in order to make contact with observations, we must simulate the detailed dynamics and thermodynamics of the cosmic plasma — mostly hydrogen and helium — under the influence of dark matter and dark energy, as well as ionizing radiation backgrounds. Such simulations are called hydrodynamic cosmological simulations, and are what we consider in this chapter. CSF is inherently nonlinear, multidimensional, and involves a variety of physical processes operating on a range of length- and timescales. Large scale numerical simulation is the primary means we have of studying it in detail.

To give a feeling for the range of scales involved, the large scale distribution of galaxies in the present universe traces out a web-like pattern on typical scales of 100 million light years. Individual galaxies are 10^3 to 10^4 times smaller in linear scale. Resolving the smallest galaxies with ten resolution elements per radius yields a range of scales of 2×10^5 throughout the large scale volume. A uniform grid of this size in 3D is out of the question, now and in the near future. However, such a high dynamic range is not needed everywhere, but only where the galaxies are located. Using adaptive mesh refinement (AMR) techniques we are close to achieving this dynamic range running on today's terascale computers including a simple decription of the baryonic fluid (Figure 5.1). With petascale platforms, even higher dynamic ranges will be achievable including the complex baryonic physics that governs the formation and evolution of galaxies.

In this paper we describe Enzo [8] a multiphysics, parallel, AMR application for simulating CSF developed at University of California, San Diego, and Columbia. We describe its physics, numerical algorithms, implementation, and performance on current terascale platforms. We also discuss our future plans and some of the challenges we face as we move to the petascale.

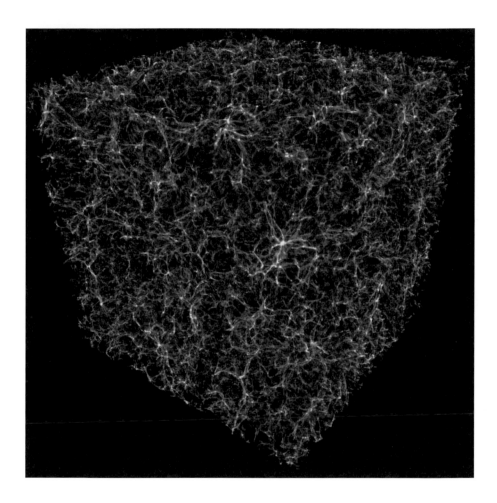

FIGURE 5.1: (See color insert following page 18.) **Enzo** hydrodynamic simulation of cosmic structure in a 700 Mpc volume of the universe. Up to seven levels of adaptive mesh refinement resolve the distribution of baryons within and between galaxy clusters, for an effective resolution of $65,536^3$. Shown is a volume rendering of baryon density. Image credit: M. Hall, NCSA.

5.2 The Enzo Code

5.2.1 Physical model and numerical algorithms

Matter in the universe is of two basic types: baryonic matter composed of atoms and molecules out of which stars and galaxies are made, and non-baryonic "dark" matter of unknown composition, which is nevertheless known to be the dominant mass constituent in the universe on galaxy scales and larger. Enzo self-consistently simulates both components, which evolve according to different physical laws and therefore require different numerical algorithms.

Baryonic matter is evolved using a finite volume discretization of the Euler equations of gas dynamics cast in a frame which expands with the universe. Energy source and sink terms due to radiative heating and cooling processes are included, as well as changes in the ionization state of the gas [1]. We use the piecewise parabolic method (PPM), which is a higher-order Godunov scheme developed by Colella and Woodward for ideal gas dynamics calculations [7]. The species abundances for H, H+, He, He+, He++, and e- (and optionally H_2, HD and related species) are solved out of equilibrium by integrating the rate equations including radiative and collisional processes [1]. Radiation fields are modeled as evolving but spatially homogeneous backgrounds using published prescriptions.

Dark matter is assumed to behave as a collisionless phase fluid, obeying the Vlasov-Poisson equation. Its evolution is solved using particle-mesh algorithms for collisionless N-body dynamics [11]. In particular, we use the spatially second-order-accurate cloud-in-cell (CIC) formulation, together with leapfrog time integration, which is formally second-order-accurate in time. Dark matter and baryonic matter interact only through their self-consistent gravitational field. The gravitational potential is computed by solving the Poisson equation on the uniform or adaptive grid hierarchy using fast Fourier transform (FFT) and multigrid techniques. In generic terms, Enzo is a 3D hybrid code consisting of a multispecies hydrodynamic solver for the baryons coupled to a particle-mesh solver for the dark matter via a Poisson solver.

Matter evolution is computed in a cubic domain of length $L = a(t)X$, where X is the domain size in co-moving coordinates, and $a(t)$ is the homogenous and isotropic scale factor of the universe which is an analytic or numerical solution of the Friedmann equation, a first order ODE. For sufficiently large X compared to the structures of interest, any chunk of the universe is statistically equivalent to any other, justifying the use of periodic boundary conditions. The speed of FFT algorithms and the fact that they are ideally suited to periodic problems make them the Poisson solver of choice given the large grids employed — $1,024^3$ or larger.

CSF simulations require very large grids and particle numbers due to two

competing demands: large boxes are needed for a fair statistical sample of
the universe; and high mass and spatial resolutions are needed to adequately
resolve the scale lengths of the structures which form. For example, in order
to simultaneously describe galaxies' large scale distribution in space (large
scale structure) and adequately resolve their internal structure, a dynamic
range of 10^5 per spatial dimension and 10^9 in mass is needed *at a minimum*,
as discussed above.

5.2.2   Adaptive mesh refinement

The need for higher resolution than afforded by uniform grids motivated
the development of `Enzo`. `Enzo` uses structured adaptive mesh refinement
(SAMR, [3, 6]) to achieve high resolution in gravitational condensations. The
central idea behind SAMR is simple to describe but challenging to implement
efficiently on parallel computers. The idea is this: while solving the desired set
of equations on a coarse uniform grid, monitor the quality of the solution and
when necessary, add an additional, finer mesh over the region that requires
enhanced resolution. This finer (child) mesh obtains its boundary conditions
from the coarser (parent) grid or from other neighboring (sibling) grids with
the same mesh spacing. The finer grid is also used to improve the solution on
its parent. In order to simplify the bookkeeping, refined patches are required
to have a cell spacing which is an integer number divided by the parent's
spacing. In addition, refined patches must begin and end on a parent cell
boundary. As the evolution continues, it may be necessary to move, resize or
even remove the finer mesh. Refined patches themselves may require further
refinement, producing a tree structure that can continue to any depth. We
denote the level of a patch by the number of times it has been refined compared
to the root grid. If the cell spacing of the root grid (level 0) is Δx, then the
cell spacing of a mesh at level l is $\Delta x / r^l$ where r is the integer refinement
factor (typically 2).

To advance our system of coupled equations in time on this grid hierarchy,
we use a recursive algorithm. The `EvolveLevel` routine is passed the level of
the hierarchy it is to work on and the new time. Its job is to march the grids
on that level from the old time to the new time.

Inside the loop which advances the grids on this level, there is a recursive call
so that all the levels with finer subgrids are advanced as well. The resulting
order of time steps is like the multigrid W-cycle.

Before we update the hyperbolic gas dynamics equations and solve the elliptic Poisson equation, we must set the boundary conditions on the grids. This
is done by first interpolating from a grid's parent and then copying from sibling grids, where available. Once the boundary values have been set, we solve
the Poisson equation using the procedure `PrepareDensityField` and evolve
the hydrodynamic field equations using the procedure `SolveHydroEquations`.
The multispecies kinetic equations are integrated by the procedure `SolveRate`
`Equations`, followed by an update to the gas energy equation due to radiative

```
EvolveLevel(level, ParentTime)
begin
    SetBoundaryValues(all grids)
    while (Time < ParentTime)
    begin
        dt = ComputeTimeStep(all grids)
        PrepareDensityField(all grids, dt)
        SolveHydroEquations(all grids, dt)
        SolveRateEquations(all grids, dt)
        SolveRadiativeCooling(all grids, dt)
        Time += dt
        SetBoundaryValues(all grids)
        EvolveLevel(level+1, Time)
        RebuildHierarchy(level+1)
    end
end
```

FIGURE 5.2: Enzo AMR algorithm.

cooling by the procedure `SolveRadiative Cooling`. The final task of the
`EvolveLevel` routine is to modify the grid hierarchy to reflect the changing
solution. This is accomplished via the `RebuildHierarchy` procedure, which
takes a level as an argument and modifies the grids on that level and all finer
levels. This involves three steps: First, a refinement test is applied to all
the grids on that level to determine which cells need to be refined. Second,
rectangular regions are chosen which cover all of the refined regions, while
attempting to minimize the number of unnecessarily refined points. Third,
the new grids are created and their values are copied from the old grids (which
are deleted) or interpolated from parent grids. This process is repeated on
the next refined level until the grid hierarchy has been entirely rebuilt.

5.2.3   Implementation

Enzo is written in a mixture of C++ and Fortran. High-level functions
and data structures are implemented in C++ and computationally intensive
lower-level functions are implemented in Fortran. As described in more detail
below, Enzo is parallelized using the MPI message-passing library [10] and uses
the HDF5 data format [16] to write out data and restart files in a platform-
independent format. The code is quite portable and has been run on numerous
parallel shared and distributed memory systems, including the IBM Power N
systems, SGI Altix, Cray XT3, IBM BG/L, and numerous Beowulf-style Linux
clusters.

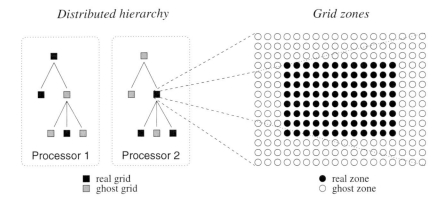

FIGURE 5.3: Real and ghost grids in a hierarchy; real and ghost zones in a grid.

The AMR grid patches are the primary data structure in Enzo. Each individual patch is treated as a separate object, and can contain both field variables and particle data. Individual patches are organized into a dynamic distributed AMR mesh hierarchy using two different methods: the first is a hierarchical tree and the second is a level-based array of linked lists. This allows grids to be quickly accessed either by level or depth in the hierarchy. The tree structure of a small illustrative 2D AMR hierachy — six total grids in a three-level hierarchy distributed across two processors — is shown on the left in Figure 5.3.

Each data field within a grid is an array of zones of 1,2 or 3 dimensions (typically 3D in cosmological structure formation). Zones are partitioned into a core block of *real zones* and a surrounding layer of *ghost zones*. Real zones are used to store the data field values, and ghost zones are used to temporarily store neighboring grid values when required for updating real zones. The ghost zone layer is three zones deep in order to accomodate the computational stencil of the hydrodynamics solver, as indicated in the right panel in Figure 5.3. These ghost zones can lead to significant computational and storage overhead, especially for the smaller grid patches that are typically found in the deeper levels of an AMR grid hierarchy.

5.2.4   Parallelization

Parallelization is achieved by distributing the patches, with each grid object locally resident in memory. Communication between grids on different processors is carried out using message-passing. The structure of the hierarchy, in the form of the set of linked lists described earlier, is stored redundantly on every processor to facilitate communication. However, if we kept all the grid data on all processors that would obviously consume too much memory, so instead we have developed the concept of a *ghost grid*, which is a nearly empty grid structure used to represent those grids that reside on other processors. For every *real grid*, which contains all grid data and resides on a given processor, there are $p - 1$ *ghost grids* which represent that grid on other processors (assuming p processors). This is feasible because *ghost grids* consume orders of magnitude less memory than *real grids* (but see below for a discussion of how this must change for very large numbers of processors). This structure is shown graphically in Figure 5.3.

Since child grids depend on parent grids, parallelization can only proceed over the grids on a given level (different levels must be computed in a serial fashion). Therefore, all grids on a given level are distributed over processors separately, starting with the root grid. The root grid is simply tiled and split into a number of patches which is at least as large as the number of processors. Then, as grids are added, each grid is placed by default on the same processor as its parent, minimizing communication. Once the rebuild of the hierarchy has completed on a given level, the load balancing ratio between processors is computed and grids are transfered between processors in an attempt to even the load. Because grids are discrete objects, this cannot in general be done in an optimal fashion (although we have experimented with grid splitting, see [15]). For the large problems typical of CSF, there are many grids per processor so this is not typically a problem.

Communication is overlapped with computation by precomputing communication pathways and starting non-blocking MPI calls. Care is taken to generate these as soon as possible so that the data will be ready when required by another processor. This is one of the reasons why it is important to have the entire hierarchy in memory on each processor, so that all grid overlaps can be found and data transfer initiated early in the compute cycle.

5.2.5   Fast sibling grid search

As the code has been expanded and run on more and more processors with more and more grids, a number of performance bottlenecks have been identified and eliminated. We describe one such example here. Many binary operations between grids, such as copying boundary values, require first identifying which grids overlap. In early versions of the code, this was done using a simple double-loop to perform a comparison between each grid and all other grids. This is an $O(N_{\text{grid}}^2)$ operation but because the number of operations per grid comparison is very small, this bottleneck did not appear until we ran simulations with 100s of processors, generating more than 10,000 grids.

The problem was solved by carrying out a chaining-mesh search to identify neighboring grids. First, a coarse mesh is constructed over the entire domain (with 4^3 times fewer cells than the root grid), and a linked list is begun for each chaining-mesh cell. Then we loop over all the grids and find which chaining-mesh cell(s) that grid belongs to, adding that grid to the appropriate linked list(s). In this way we generate a coarse localization of the grids, so that when we need to find the list of neighbors for a given grid, we simply need to check the grids in the same chaining-mesh (or multiple chaining-meshes if the grid covers more than one) This reduces the number of comparisons by a factor of about N_{chain}^{-3}, and since the number of chaining mesh cells, N_{chain}, scales with the problem size, the procedure reduces the operation count back to $O(N_{\text{grid}})$ scaling.

5.2.6 Enzo I/O

All input/output (I/O) operations in Enzo are performed using the Hierarchical Data Format 5 library (HDF5). HDF5, specifically hdf5 v.1.6.5, has a number of very attractive features for management of large, complex data sets and the associated metadata. HDF5 has an excellent logical design and is available on every major computing platform in the NSF/DOE arena. One outstanding advantage of HDF5 is that one can easily restart a model on a different computational platform without having to worry about differences in endian-ness or internal floating-point format.

The basic data model in Enzo is that a given MPI task "owns" all of the I/O required by the set of top level grids and subgrids present in that task. A single method is used to provide checkpointing and restart capability and for science dumps at specified intervals.

All I/O is designed for moving the largest possible amount of data per operation (subject to chunking constraints described below), and all memory references are stride 1 for maximum efficiency. Each task performs all of its I/O independently of the other tasks and there is no logical need for concurrency, i.e, there is actually no advantage to using the parallel cooperative interface available in HDF5. Although synchronization is not required in principle, it is convenient to synchronize all processes after an entire dump has been written so that it is safe to hand off to asynchronous processes which may move the data across a network or to archival storage.

The basic object in HDF5 is a file containing other HDF5 objects such as data sets and associated metadata attributes. HDF5 also supports a truly hierarchical organization through the group concept. Enzo uses each of these features. The group concept, in particular, allows Enzo to pack potentially huge numbers of logically separate groups of data sets (i.e., one such set for each subgrid) into a single Unix file resulting in a correspondingly large reduction in the number of individual files and making the management of the output at the operating system level more convenient and far more efficient. In an Enzo AMR application running on N processors with G subgrids per MPI task with each subgrid having D individual baryon fields and/or particle lists this results in only N HDF5 files per data dump instead of N*G files without grouping or N*G*D files in a simplistic case with no groups or data sets and a separate file for each physical variable.

The packed-AMR scheme necessarily involves many seek operations and small data transfers when the hierarchy is deep. A vital optimization in Enzo is the use of in-core buffering of the assembly of the packed-AMR HDF5 files. This is very simply achieved using the HDF5 routine `H5Pset_fapl_core` to set the in-core buffering properties for the file. `Enzo` uses a default buffer size of 1 MByte per file. At present, in HDF5 v.1.6.5 this is only available on output where increases in performance by $> 120\times$ have been observed with *Lustre* file systems. The lack of input buffering implies that reading restart files can be relatively expensive compared to writing such files but in a typical batch run there may be dozens of write operations for a single read operation in the restart process. When input buffering becomes available in a future version of HDF5 the cost of I/O operations will be a negligible fraction of the runtime even for extremely deep hierarchies.

For very large uniform grid runs (e.g., 2048^3 cells and particles) we encounter different problems. Here individual data sets are so large it is necessary to using chunking so that any individual read/write operation does not exceed certain internal limits in some operating systems. The simplest strategy is to ensure that access to such data sets is by means of HDF5 hyperslabs corresponding to a plane or several planes of the 3D data. For problem sizes beyond 1280^3 it is necessary to use 64-bit integers to count dark matter particles or compute top grid data volumes. `Enzo` uses 64-bit integers throughout so it is necessary to handle MPI and HDF5 integer arguments by explicitly casting integers back to 32 bits.

5.3   Performance and Scaling on Terascale Platforms

`Enzo` has two primary usage modes: *unigrid*, in which a nonadaptive uniform Cartesian grid is used, and *AMR*, in which adaptive mesh refinement is enabled in part or all of the computational volume. In actuality, unigrid mode is obtained by setting the maximum level of refinement to zero in an AMR simulation, and precomputing the relationships between the root grid tiles. Otherwise, both calculations use the same machinery in `Enzo` and are updated according to Figure 5.2. The scaling and performance of `Enzo` is very different in these two modes, as are the memory and communication behaviors. Therefore we present both in this section.

5.3.1   Unigrid application

Figure 5.4(a) shows the results of a recent unigrid simulation carried out on 512 processors of NERSC *Bassi*, and IBM Power5 system. The simulation was performed on a grid of 1024^3 cells and the same number of dark matter particles. The simulation tracked 6 ionization states of hydrogen and helium including nonequilibrium photoionization and radiative cooling. Figure 5.4(a) plots the total wall-clock time per time step versus time step as well as the cost of major code regions corresponding to the procedure calls in Figure 5.2. Unigrid performance is quite predictable, with the cost per time step for the different code regions being roughly constant. In this example hydrodynamics dominates the cost of a time step (45%). The remaining cost breaks down as follows: radiative cooling (15%), boundary update 2 (11%), self-gravity (7%), boundary update 1 (7%), rate equations (3%). An additional 12% is spent in message-passing. Figure 5.4(b) shows the CPU time per processor for the job. We find that the master processor in each 8-processor SMP node is doing 9% more work than the other 7, otherwise the workload is nearly uniform across processors.

5.3.2 AMR application

Figure 5.5(a) plots the wall-clock time per root grid time step versus the time step for an AMR simulation with identical cosmology and physics as our unigrid example, only here the base grid has dimensions 512^3, and up to 4 levels of refinement are permitted in dense regions. Prior to the onset of mesh refinement at time step 103, the AMR simulation behaves identically to a unigrid simulation, with hydrodynamics dominating the cost per time step. By time step 156, the cost of a root grid time step has increased a hundred fold. There are several reasons for this. First, we use hierarchical time stepping in Enzo, which means that for each root grid timestep, a subgrid of level ℓ will be time-stepped roughly 2^ℓ times. The total number of substeps per root grid time step for a fully refined region is $\sum_{\ell=1}^{\ell_{max}} 2^\ell$, which is 30 for $\ell_{max}=4$. By time step 156, only a small fraction of the total volume is refined, and the average number of substeps per root grid time step is 8, far less than the factor 100 we observe. The dominant cause for the upturn is the cost of procedure boundary_update_2, which exchanges ghost zone data between every subgrid at every level and substep. Secondary causes for the upturn are the procedures rate_equations and radiative_cooling. These routines are both subcycled on a chemical timescale that is short compared to the hydrodynamic time step. The separation between these two timescales increases as the gas density increases. AMR allows the gas to become very dense, and thus the ionization of cooling calculations grows to dominate the hydrodynamics calculation.

Figure 5.5(b) shows the cumulative CPU time as a function of processor number. There is considerably more spread between the largest and smallest time (30,000/10,000=3) compared with the unigrid run (5,800/4,800=1.21). This is due to true load imbalances arising from mesh refinement, chemistry/-cooling subcycling, and communication loads. At present our dynamic load-balancing algorithm in Enzo only trys to equalize the zone-time steps among processors. It does not attempt to balance communications and subcycling loads. This is an obvious area for improvement.

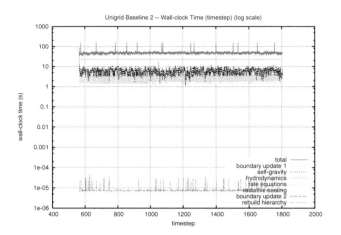

(a)

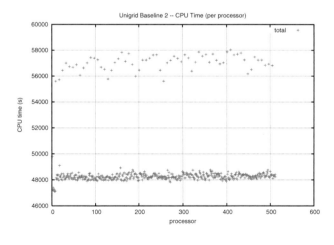

(b)

FIGURE 5.4: (a) Wall-clock time per time step versus time step broken down according to major code region for the 1024^3 unigrid simulation. Spikes in total time correspond to I/O events. (b) CPU time per processor versus processor. Simulations were carried out on 512 CPUs on the NERSC IBM Power5 system *Bassi*.

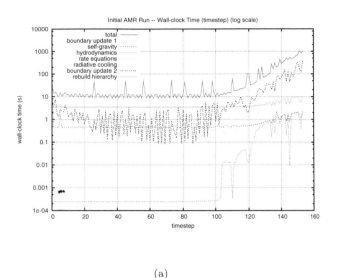

(a)

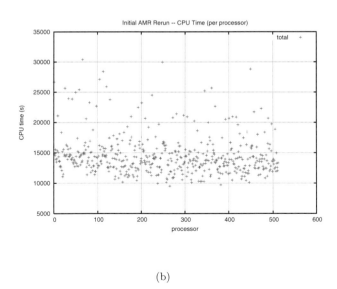

(b)

FIGURE 5.5: (a) Wall-clock time per root grid time step versus root grid time step broken down according to major code region for the 512^3 4-level AMR simulation. (b) CPU time per processor versus processor. Simulations were carried out on 512 CPUs on the NERSC IBM Power5 system *Bassi*.

5.3.3   Parallel scaling

Because the cost per time step of a unigrid simulation is roughly constant, varying by only a factor of a few as physical conditions change, parallel scaling tests are easy to perform for even the largest grids and particle counts. The same is not true of AMR simulations since the load is constantly changing in a problem-dependent way. Time-to-solution is the only robust metric. Because large AMR simulations are so costly, it is not possible to run many cases to find the optimal machine configuration for the job. Consequently we do not have parallel-scaling results for AMR simulations of the size which require terascale platforms (for smaller-scaling tests, see [4].) In practice, processor counts are driven by memory considerations, not performance. We expect this situation to change in the petascale era where memory and processors will be in abundance. Here, we present our unigrid parallel-scaling results for Enzo running on a variety of NSF terascale systems available to us.

Figure 5.6 shows the results of strong-scaling tests of Enzo unigrid simulations of the type described above for grids of size $256^3, 512^3, 1024^3$ and 2048^3 and an equal number of dark matter particles. We plot cell updates/sec/CPU versus processor count for the following machines: *Lemieux*, a Compaq DEC alpha cluster at Pittsburgh Supercomputing Center (PSC), *Mercury*, an IBM Itanium2 cluster at National Computational Science Alliance (NCSA), and *DataStar*, an IBM Power4 cluster at San Diego Supercomputer Center (SDSC). Ideal parallel scaling would be horizontal lines for a given archictecture, differentiated only by their single processor speeds. We see near-ideal scaling for the 256^3 test on *DataStar* and *Lemieux* up to 32 processors, followed by a gradual rollover in parallel performance to $\sim 50\%$ at 256 processors. Non-ideality sets in at 16 processors on *Mercury*, presumably due to its slower communications fabric. As we increase the problem size on any architecture, parallel efficiency increases at fixed NP, and the rollover moves to higher NP. Empirically, we find that parallel efficiency suffers if grid blocks assigned to individual processors are smaller than about 64^3 cells. Using blocks of size $128^2 \times 256$, we find that the cell update rate for our 2048^3 simulation on 2048 *DataStar* processors is $\sim 80\%$ the cell update rate of our 256^3 on 4 processors. Enzo in unigrid mode is very scalable.

5.4   Toward Petascale Enzo

5.4.1   New AMR data structures

For moderate numbers of processors, the current data structure used for storing the AMR grid hierarchy is adequate. Even though the hierarchy topology is stored redundantly on each processor, because the data fields are vastly

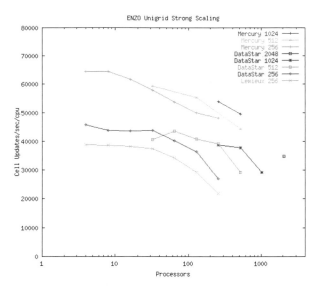

FIGURE 5.6: **Enzo** strong-scaling results for unigrid cosmology runs of size $256^3, 512^3, 1024^3$ and 2048^3 on NSF terascale platforms.

larger than **Enzo**'s individual C++ **grid** objects, the extra memory overhead involved is insignificant. However, as the number of processors increases, this memory overhead increases as well. For the processor counts required for petascale-level computing, the storage overhead would overwhelmingly dominate, to the point where the largest computation would actually not be limited by the number of processors, but rather by the amount of memory available for each processor. Thus, for **Enzo** to scale to the petascale level, the memory overhead for storing the AMR hierarchy must be reduced.

The memory required for storing **Enzo**'s current AMR data structure can be approximated as $|F| + np|G|$, where the first term $|F|$ is the field variable data, and the second term $np|G|$ is the overhead for storing the grid hierarchy data structure. Here n is the number of grid patches, p is the number of processors, and $|G|$ is the storage required by a C++ **grid** object.

One approach to reducing the size of the overhead term $np|G|$ is to reduce the size of the **grid** object by removing unnecessary member variables from the **grid** class. Since some member variables are constant for a hierarchy level, and some are constant for the entire hierarchy, the amount of memory required would be reduced. This refactoring is indicated by the first transition in Figure 5.7. Although this would indeed save some memory, preliminary estimates indicate that the savings would only be about 13.5%.

A second approach to reducing the size of the overhead term would be to split the **grid** class into two subclasses **grid_local** and **grid_remote**, and use **grid_local** objects for local grids that contain field data, and **grid_remote**

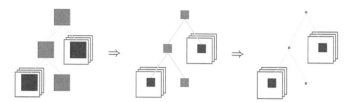

FIGURE 5.7: Reducing AMR data structure storage overhead. Solid dark squares represent local `grid` objects, solid lighter squares represent remote `grid` objects, and open boxes represent data fields. The first transition illustrates reducing the size of all `grid` objects, and the second transition illustrates splitting the single `grid` class into two `grid_local` and `grid_remote` subclasses.

objects for grid patches whose data fields reside on another processor. This refactoring is indicated by the second transition in Figure 5.7. This modification would change the overhead storage term from $np|G|$ to $n((p-1)|G_r| + |G_l|)$, where $|G_r|$ is the size of the `grid_remote` class and $|G_l|$ is the size of the `grid_local` class. The advantage of doing this is that the `grid_remote` class could be made much smaller, since most of the variables in the `grid` class are only required for local grids. Also, a vast majority of the grid classes are these much smaller `grid_remote` objects. Thus the memory savings for this modification would be quite large — a factor of roughly 14 over the already slightly reduced `grid` size from the first approach.

5.4.2   Hybrid parallelism

Another improvement would of course be not to store the entire grid hierarchy on each processor. The easiest approach would be to store one copy per shared memory node, which would decrease the memory storage further by a factor equal to the number of processors per node. This could be implemented with a hybrid parallel programming model in which instead of assigning one MPI task per processor which serially executes its root grid tile and all its subgrids, we assign one MPI task per node. This heavy weight node would execute a somewhat larger root grid tile and its more numerous subgrids, using shared memory parallelism wherever possible. For example, every subgrid at a given level of refinement would be processed concurrently using OpenMP threads [17].

While the above modifications to the AMR data structure should allow `Enzo` to run on machines with on the order of 10^4 to 10^5 processors, extending to 10^6 processors would require reducing the overhead even further. The ultimate improvement memory-wise would be to store a single copy of the grid hierarchy, though depending on the node interconnect that would cause a communication bottleneck. A refinement on this approach would be to store one copy of the hierarchy for every M processor, where M is some machine-dependent number chosen to balance the trade-off between memory and communication overhead.

5.4.3 Implicitly coupled radiation hydrodynamics

We are working to incorporate radiation transfer processes within the `Enzo` framework to improve the physical realism of cosmological modeling of self-regulated star formation and predictions on the epoch of cosmic re-ionization. These efforts are unique within the computational astrophysics community, because unlike traditional approaches to such multiphysics couplings, we are coupling the radiation transfer implicitly with both the fluid energy and chemical kinetics processes. This approach promises to provide a fully consistent coupling between the physical processes, while enabling the use of highly scalable solvers for the coupled solution.

Through implicitly coupling the radiation–hydrodynamics–chemical kinetics processes together, we have the benefits of numerical stability in time (regardless of step size) and the ability to use high-order time discretization methods. On the other hand, this coupling results in a nonlinear system of partial differential equations that must be solved at each time step. For this coupled solution we will use *inexact Newton methods*, which in recent years have been shown to provide an efficient and scalable approach to solving very large systems of nonlinear equations [14, 13]. This scalability arises due to a number of factors, notably the fact that for many problems Newton's method exhibits a convergence rate that is independent of spatial resolution, so long as the inner linear solver scales appropriately [19]. As radiation diffusion processes are elliptic in nature, and radiative couplings to fluid energy and chemical kinetics occur only locally, we plan to achieve such optimal scalability in the linear solver through the use of a Schur complement formulation ([5]) to reduce the coupled Newton systems to scalar diffusion problems, which will then be solved through optimally scalable multilevel methods, provided by the state-of-the-art HYPRE linear solver package [9, 12]. This solver library has been shown to scale up to massively parallel architectures ([2]), and as the computational heart of the inexact Newton approach is the inner linear solver, we are confident that such an implicit formulation and solution will enable radiation–hydrodynamics–chemical kinetics simulations in `Enzo` to the petascale and beyond.

5.4.4 Inline analysis tools

Petascale cosmology simulations will provide significant data analysis challenges, primarily due to the size of simulation data sets. For example, an `Enzo` Lyman alpha forest calculation on a 1024^3 grid, the current state of the art, requires approximately 110 GB of disk space per simulation output. Tens or hundreds of these outputs are required per simulation. If scaled to a 8192^3

grid, a reasonable size for a petascale computation, this would result in 56 TB of data per simulation output, and multiple petabytes of data written to disk in total. Given these data set sizes, analysis can become extremely time-consuming: doing even the simplest *post facto* analysis may require reading petabytes of data off of disk. For more complex analysis, such as spectrum or N-point correlation function generation, the data analysis may be comparable in computational cost to the simulation itself. Analysis tools for petascale data sets will by necessity be massively parallel, and new forms of data analysis that will be performed hundreds or thousands of times during the course of a calculation may make it exceedingly cumbersome to store all of the data required. Furthermore, it will become extremely useful to be able to monitor the status and progress of petascale calculations as they are in progress, in an analogous way to more conventional experiments.

For these reasons, among others, doing data analysis while the simulation itself is running will become necessary for petascale-level calculations. This will allow the user to save disk space and time doing I/O, greatly enhance the speed with which simulations are analyzed (and hence improve simulation throughput), and have access to data analysis strategies which are otherwise impractical or impossible. Analysis that can be done during cosmological simulations include the calculation of structure functions and other global snapshots of gas properties; dark matter power spectra, halo and substructure finding, and the generation of merger trees; radial profile generation of baryon properties in cosmological halos; production of projections, slices, and volume-rendered data for the creation of movies; the generation of synthetic observations, or "light cones"; the calculation of galaxy population information such as spectral energy distributions and color-magnitude relations; global and halo-specific star formation rates and population statistics; ray tracing for strong and weak lensing calculations and for Lyman alpha forest or DLA spectrum generation; and essentially any other type of analysis that can be massively parallelized. Analysis of simulations in this way, particularly if done with greater frequency in simulation time, may enhance serendipitous discovery of new transient phenomena.

A major constraint to inline analysis of cosmological simulations is that it requires very careful and possibly time-consuming planning on the part of the scientists designing the calculation, and is limited to analysis techniques that can be heavily parallelized. As a result, inline analysis techniques will not completely negate the necessity of writing out significant amounts of data for follow-up analysis and data exploration. Furthermore, inline analysis will require careful integration of analysis tools with the simulation code itself. For example, the simulation and analysis machinery will, for the sake of efficiency, require shared code, data structures, and parallelization strategies (i.e., domain-decomposed analysis if the simulations are similarly decomposed). If these relatively minors hurdles can be surmounted, however, this sort of analysis will result in gains far beyond the additional computational power used.

5.5 Acknowlegements

Enzo has been developed with the support of the National Science Foundation via grants ASC-9318185, AST-9803137, AST-0307690, and AST-0507717; the National Center for Supercomputing Applications via PACI subaward ACI-9619019; and the San Diego Supercomputer Center via the Strategic Applications Partners program. Greg Bryan acknowledges support from NSF grants AST-05-07161, AST-05-47823 and AST-06-06959.

References

[1] P. Anninos, Y. Zhang, T. Abel, and M. L. Norman. Cosmological hydrodynamics with multi-species chemistry and nonequilibrium ionization and cooling. *New Astronomy*, 2:209–224, August 1997.

[2] A. H. Baker, R. D. Falgout, and U. M. Yang. An assumed partition algorithm for determining processor inter-communication. *Parallel Computing*, 32:319–414, 2006.

[3] M. J. Berger and P. Colella. Local adaptive mesh refinement for shock hydrodynamics. *J. Comp. Phys.*, 82:64–84, 1989.

[4] J. Bordner, A. Kritsuk, M. Norman, and R. Weaver. Comparing parallel CAMR and SAMR hydrodynamic applications. Technical report, Laboratory for Computational Astrophysics, Center for Astrophysics and Space Sciences, University of California, San Diego, 2007.

[5] P. N. Brown and C. S. Woodward. Preconditioning strategies for fully implicit radiation diffusion with material-energy coupling. *SIAM J. Sci. Comput.*, 23:499–516, 2001.

[6] G. L. Bryan. Fluids in the universe: Adaptive mesh in cosmology. *Computing in Science and Engineering*, 1(2):46, 1999.

[7] G.L. Bryan, M.L. Norman, J.M. Stone, R. Cen, and J.P. Ostriker. A piecewise parabolic method for cosmological hydrodynamics. *Computer Physics Communication*, 89:149–168, 1995.

[8] Enzo Web page. http://lca.ucsd.edu/portal/software/enzo.

[9] R. D. Falgout and U. M. Yang. *Hypre: a Library of High Performance Preconditioners, in Computational Science – ICCS 2002 Part III*. Springer-Verlag, 2002.

[10] W. Gropp, S. Huss-Lederman, A. Lumsdaine, E. Lusk, B. Nitsberg, W. Saphir, and M. Snir. *MPI—The Complete Reference: Volume 2, The MPI Extensions.* Scientific and Engineering Computation Series. MIT Press, Cambridge, MA, September 1998.

[11] R. W. Hockney and J. W. Eastwood. *Computer Simulation using Particles.* McGraw-Hill, New York, January 1988.

[12] Hypre Web page. `http://www.llnl.gov/casc/linear_solvers`.

[13] D. E. Keyes, D. R. Reynolds, and C. S. Woodward. Implicit solvers for large-scale nonlinear problems. *J. Phys.: Conf. Ser.*, 46:433–442, 2006.

[14] D. A. Knoll and D. E. Keyes. Jacobian-free Newton–Krylov methods: A survey of approaches and applications. *J. Comp. Phys*, 193:357–397, 2004.

[15] Z. Lan, V. Taylor, and G. Bryan. Dynamic load balancing for structured adaptive mesh refinement applications. In *Proc. of 30th Int. Conference on Parallel Processing*, Valencia, Spain, 2001.

[16] National Center for Supercomputing Applications. NCSA Hierarchical Data Format (HDF). `hdf.ncsa.uiuc.edu`.

[17] OpenMP Web page. `http://www.openmp.org`.

[18] D. N. Spergel and *et al.* Wilkinson microwave anisotropy probe (WMAP) three year results: Implications for cosmology. eprint arXiv astro-ph0603449, 2006.

[19] M. Weiser, A. Schiela, and P. Deuflhard. Asymptotic mesh independence of Newton's method revisited. *SIAM J. Numer. Anal.*, 42:1830–1845, 2005.

Chapter 6

Numerical Prediction of High-Impact Local Weather: A Driver for Petascale Computing

Ming Xue, Kelvin K. Droegemeier

Center for Analysis and Prediction of Storms and School of Meteorology, University of Oklahoma, Norman, OK

Daniel Weber

Center for Analysis and Prediction of Storms, University of Oklahoma, Norman, OK

6.1 Introduction

The so-called mesoscale and convective scale weather events, including floods, tornadoes, hail, strong winds, lightning, hurricanes and winter storms, cause hundreds of deaths and average annual economic losses greater than $13 billion in the United States each year [16, 5]. Although the benefit of mitigating the impacts of such events on the economy and society is obvious, our ability to do so is seriously constrained by the available computational resources which are currently far from sufficient to allow for explicit real-time numerical prediction of these hazardous weather events at sufficiently high spatial resolutions or small enough grid spacings.

Operational computer-based weather forecasting, or numerical weather prediction (NWP), began in the late 1910s with a visionary treatise by L.F. Richardson [18]. The first practical experiments, carried out on the ENIAC (electronic numerical integrator and computer) some three decades later, established the basis for what continues to be the foundation of weather forecasting. A NWP model solves numerically the equations governing relevant atmospheric processes including fluid dynamics for air motion, thermodynamic processes, and thermal energy, moisture and related phase changes. Physical processes related to long-wave and short-wave radiation, heat and momentum exchanges with the land and ocean surfaces, cloud processes and their interaction with radiation, and turbulence processes often cannot be explicitly represented because of insufficient spatial and temporal resolution. In such cases, these processes are "parameterized," i.e., treated in a simplified form by making them dependent upon those quantities which the model *can* explicitly resolve. Parameterization schemes are, however, often empirical or semiempirical; they are the largest source of uncertainty and error in NWP models, in addition to resolution-related truncation errors.

Operational NWP has always been constrained by available computing power. The European Center for Medium-Range Weather Forecasting (ECMWF) operates the world's highest-resolution global NWP model with an effective horizontal resolution* of approximately 25 km. The model is based on a spherical harmonic representation of the governing equations, triangularly truncated at total wave number of 799 in the longitudinal direction with 91 levels in the vertical. The daily forecasts extend to 10 days and are initialized using the four-dimensional variational (4DVAR) data assimilation method [17]. At the same time, a 51-member Ensemble Prediction System (EPS) is also run daily at a reduced 40-km grid spacing. The EPS provides probabilistic information that seeks to quantify how uncertainty in model initial conditions can lead to differing solutions. Such information is very important for decision making.

The global deterministic (single high-resolution) prediction model operated by the U.S. National Weather Service currently has 382 spectral wave numbers and 64 vertical levels, while its probabilistic ensemble prediction system contains 14 members and operates at the spectral truncation of 126. Forecasts are produced four times a day. To obtain higher resolution over North America, regional deterministic and ensemble forecasts also are produced at 12- and roughly 40-km horizontal grid spacings, respectively.

Even at 12-km horizontal grid spacing, important weather systems that are directly responsible for meteorological hazards including thunderstorms, heavy precipitation and tornadoes cannot be directly resolved because of their small sizes. For individual storm cells, horizontal grid resolutions of at least 1-km grid are generally believed to be necessary (e.g., [34]), while even higher resolutions are needed to resolve less organized storms and the internal circulations within the storm cells. Recent studies have also shown that to resolve the inner wall structure of hurricanes and to capture hurricane eye

---

*The term "resolution" is used here in a general manner to indicate the ability of a model to resolve atmospheric features of certain spatial scales. In grid-point models, the term "grid spacing" is more appropriate whereas in spectral or Galerkin models, "spectral truncation" more appropriately describes the intended meaning.

wall replacement cycles that are important for intensity forecasts, 1- to 2-km resolution is necessary [3, 8]. Furthermore, because the smallest scales in unstable convective flows tend to grow the fastest, the resolution of convective structures will always benefit from increased spatial resolutions (e.g., [1]), though the extent to which superfine resolution is necessary for non-tornadic storms remains to be established. To predict one of nature's most violent phenomena, the tornado, resolutions of a few tens of meters are required [31].

Because most NWP models use explicit time-integration schemes with a time step size limited by the CFL (Courant-Friedrichs-Lewy) linear stability criterion, the allowable time step size is proportional to the effective grid spacing. Thus, a doubling in 3D of the spatial resolution, and the requisite halving of the time step, requires a factor of 2^4 =16 increase in processing power and a factor of 8 increases in the memory. Data I/O volume is proportional to memory usage or grid size, and the frequency of desired model output tends to increase with the number of time steps needed to complete the forecast.

In practice, the vertical resolution does not need to be increased as much because it is already relatively high compared to the horizontal resolution. The time step size is currently constrained more by the vertical grid spacing than the horizontal one because of the relatively small vertical grid spacing therefore the time step size often does not have to decreased by a factor of two when the horizontal resolution doubles. However, physics parameterizations usually increase in complexity as the resolution increases. Taking these factors into account, a factor of 8 increases in the processing power when the horizontal resolution doubles is a good estimate. Therefore, to increase the operational North American model from its current 12-km resolution to 1-km resolution would require a factor of 12^3 =1728 or nearly a factor of a two thousand increase in raw computing power. It is estimated that a 30-hour continental-U.S.-scale severe thunderstorm forecast using a state-of-the-art prediction model and 1-km grid spacing would require a total of 3×10^{11} floating point calculations. Using 3,000 of today's processors, this forecast will take 70 hours to complete [25] while for operational forecasts, this needs to be done within about one hour. This assumes that the code runs at 10% of peak performance of the supercomputer and there are 10,000 floating-point calculations per grid point per time step. To operate a global 1-km resolution model will be an even greater challenge that is beyond the petascale, but such models are necessary to explicitly resolve convective storms and capture the mutual interaction of such events with short- and long-term climate. Furthermore, the need to run high-resolution ensemble forecasts will require a factor of ~100 increase in the computing power, assuming the ~100 ensemble members also have ~1-km grid spacing.

The development of new high-resolution nonhydrostatic[†] models, coupled

---

[†]In the hydrostatic equations of fluid dynamics, vertical accelerations are assumed to be small. This approximation is valid for large-scale flows, where the atmosphere is shallow compared to its lateral extent. In thunderstorms, vertical accelerations are quite large and

with continued rapid increases in computing power, are making the explicit prediction of convective systems, including individual thunderstorms, a reality. Advanced remote-sensing platforms, such as the operational U.S. WSR-88D (Weather Surveillance Radar – 1988 Doppler) weather radar network, provide 3D volumetric observations at fine scales for initializing convection-resolving models. Unfortunately, only one dimension of the wind, in the radial direction of radar electromagnetic wave radiation, is observed, along with a measure of precipitation intensity in terms of the power of the electromagnetic waves reflected by the precipitating particles (raindrops and ice particles). The latter is called radar reflectivity. From time series volumes, or radar scan volumes, of these quantities, along with other available observations and specified constraints, one must infer the complete state of the atmosphere.

In this chapter, we address the computational needs of convection-resolving NWP, which refers to predictions that capture the most energetically relevant features of storms and storm systems ranging from organized mesoscale convective systems down to individual convective cells.

6.2   Computational Methodology and Tools

With upcoming petascale computing systems, routine use of kilometer-scale resolutions covering continent-sized computational domains, with even higher-resolution nests over subdomains, will be possible in both research and operations. Accurate characterization of convective systems is important not only for storm-scale NWP, but also for properly representing scale interactions and the statistical properties of convection in long-duration climate models. However, one cannot overlook the fact that models are not perfect and thus even with advanced assimilation methods and excellent observations, model error needs to be minimized. One of the simplest methods is to increase model resolution, and for this reason, sub-kilometer grid spacing may be required during radar data assimilation cycles and for short-range convective storm forecasting.

6.2.1   Community weather prediction models

Two of the community models used most frequently for storm-scale research and experimental forecasting are the Advanced Regional Prediction System (ARPS, [30, 29, 28, 34]) and the Weather Research and Forecast (WRF) model [15, 23]. Both were designed to be scalable (e.g., [11, 4, 19, 20, 21, 15]) and have been run on numerous computing platforms. Owing to the variety

---

thus the non-hydrostatic equations must be used.

of architectures available at the time of their design, and because of their research orientation and the need to serve a large user base, these systems are not specifically optimized for any particular platform.

Both ARPS and WRF solve a fully compressible system of equations, and both utilize finite difference numerical techniques and regular computational grids. The ARPS uses a generalized terrain-following curvilinear coordinate based on geometric height, and its horizontal grid is rectangular in map projection space but horizontally nonuniform in physical Earth coordinates [29], [28]. The same is true for the WRF model except that it uses a mass-based vertical coordinate [23] that is close to the hydrostatic pressure-based sigma-coordinate used in large-scale NWP models. However, the WRF does not invoke the hydrostatic assumption in the vertical so that a prognostic equation for the vertical equation of motion is solved.

Both ARPS and WRF employ the split-explicit approach of time integration [22] in which the fast acoustic modes‡ are integrated using a small time step while the terms responsible for slower processes, including advection, diffusion, gravity wave modes and physical parameterizations, are integrated using a large time step. For most applications, the vertical grid spacing, especially that near the ground, is much smaller than the horizontal spacing. Therefore, an explicit integration of the vertical acoustic modes would impose a severe restriction on the small time step size. For this reason, both models use an implicit integration scheme in the vertical for terms responsible for vertically propagating acoustic waves. A solver for tri-diagonal systems of equations is used by the implicit scheme.

6.2.2 Memory and performance issues associated with peta-scale systems

The upcoming petascale computing systems are expected to be comprised of hundreds of thousands of processor cores. These systems will rely on multicore technology and in most cases will contain cache sizes similar to existing technology ($< 10Mb$). Most of today's supercomputers make use of scalar-based processor technology in a massively parallel configuration to achieve terascale performance. Such individual processors are capable of billions of floating point calculations per second but most applications, in particular weather prediction models and large CFD (computational fluid dynamics) codes, cannot fully realize the hardware potential, largely due to the fact that the memory storage hierarchy is not tuned to the processor clock rate. For example, the 3.8 GHz Pentium 4 CPU has a theoretical rating of 7.6 GFLOPS. To achieve this performance, the processor needs to be fed with data at a rate

---

‡Acoustic modes have no physical importance in the atmosphere, except possibly in the most intense tornadoes where the Mach number could approach 0.5. They are contained in the compressible systems of equations, however, and affect the stability of explicit time integration schemes.

of 182.4 GB per second assuming no reuse of data in cache. The current RAM and memory bus (e.g., the 800 MHz front side bus) can only move data at a theoretical peak of 6.4 GB per second, a factor of 28.5 slower than what is needed to keep the processor fully fed. This memory access-data supply mismatch will intensify with the proliferation of multicore processors. To avoid severe penalties due to a slow memory bus, the efficiency of the fast cache utilization has to be significantly improved, and to do so usually requires significant efforts at the level of application software design.

Relatively few supercomputers of today and of the near future use vector processors with fast memory technology due to economic reasons. An evaluation of the ARPS on scalar and vector-based computers indicates that the nested do-loop structures were able to realize a significant percentage of the peak performance of vector platforms, but on commodity-processor-based platforms the utilization efficiency is typically only 5–15% (Figure 6.1) [25]. The primary difference lies with the memory and memory access speeds. Since weather forecast models are largely memory bound, they contain far more loads/stores than computations and, as currently written, do not reuse in-cache data efficiently. One approach, called *supernoding* [10] or tiling, can reduce memory access overhead by structuring the computations so as to reuse data from the high level cache. It holds promise for increasing the scalar processing efficiency for weather models up to 30–40%. Tiling involves the further splitting of the original decomposed sub-domains on each processor into smaller regions so that all of their data used in a series of calculations fit into the level 2 and/or level 3 cache. This approach allows for the reuse of data residing in the much faster cache memory and reduces processor wait states while accessing the main memory. Tiling usually requires changing loop bounds in order to perform more calculations on a sub-domain and it is possible to specify the tiling parameters at runtime to match the cache size. Tiling has been implemented into a research version of the ARPS and shows approximately a 20% improvement in performance for the small time step solver [25]. Recognizing that the scalability of software on a massively parallel computer is largely tied to the parallel processing capability, the tiling concept must be used in conjunction with efficient message passing techniques to realize as much of the peak performance as possible.

6.2.3   Distributed-memory parallelization and message passing

Because of the use of explicit finite difference schemes in the horizontal, it is relatively straightforward to use 2D domain decomposition for the ARPS and WRF on distributed memory platforms. No domain decomposition is done in the vertical direction because of the implicit solver, and because a number of physical parameterization schemes, including those involving radiation and cumulus convection, are column based, i.e., their algorithms have vertical dependencies. MPI is the standard strategy for inter-domain communication, while mixed distributed and shared memory parallelization with MPI and OpenMP, respectively, are supported by WRF.

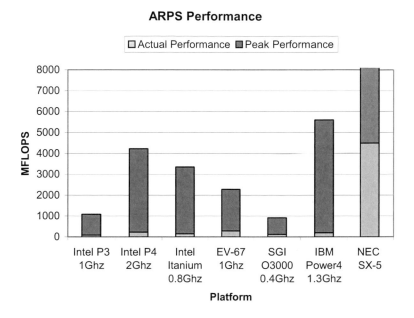

ARPS Performance

FIGURE 6.1: ARPS MFLOPS ratings on a variety of computing platforms. Mesh sizes are the same for all but the NEC-SX, which employed a larger number of grid-points and with radiation physics turned off [25].

Domain decomposition involves assigning sub-domains of the full computational grid to separate processors and solving all prognostic and/or diagnostic equations for a given sub-domain on a given processor; no global information is required at any particular grid-point and inter-processor communications are required only at the boundaries of the sub-domains (Figure 6.2). The outer border data resident in the local memories of adjacent processors are supplied by passing messages between processors. As discussed earlier, grid-point-based meteorological models usually employ two-dimensional domain decomposition because of common column-based numerics and physical parameterizations that have global dependencies in the vertical.

In the ARPS, even though the model contains fourth-order advection and diffusion options that are commonly used, only one extra zone of grid-points is defined outside the non-overlapping sub-domain. This zone contains values from the neighboring processors when this sub-domain is not at the edge of the

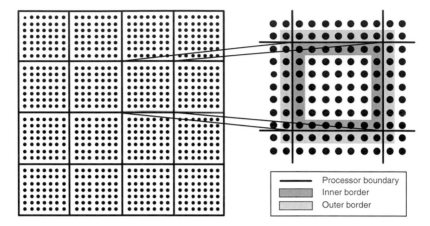

FIGURE 6.2: Two-dimensional domain decomposition of the ARPS grid. Each square in the left panel corresponds to a single processor with its own memory space. The right panel details a single processor. Grid-points having a white background are updated without communication, while those in dark stippling require information from neighboring processors. To avoid communication in the latter, data from the outer border of neighboring processors (light stippling) are stored locally.

physical boundary. Such a zone is often referred to as the "fake zone." With a leapfrog time integration scheme used in the large time steps, advection and diffusion terms are evaluated once every time step but their calculations involve 5 grid-points in each direction. This is illustrated for the fourth-order computational diffusion term [29] in the x-direction, $-K\partial^4\phi/\partial x^4$, where K is the diffusion coefficient. In standard centered difference form this term becomes

$$-K\delta_{xxxx}\phi \equiv -K\delta_{xx}\left[\delta_{xx}\phi\right]$$
$$\equiv -K\left(\phi_{i-2} - 4\phi_{i-1} + 6\phi_i - 4\phi_{i+1} + \phi_{i+2}\right)/\left(\Delta x\right)^4 \qquad (1)$$

where we define the finite difference operator

$$\delta_{xx}\phi \equiv \left(\phi_{i-1} - 2\phi_i + \phi_{i+1}\right)/\left(\Delta x\right)^2 \qquad (2)$$

and i is the grid-point index.

For calculating this term on the left boundary of the subdomain, values at $i-1$ and $i-2$, which reside on the processor to the left, are needed. However, the ARPS has only one fake zone to store the $i-1$ value. This problem is solved by breaking the calculations into two steps. In the first step, the term $\delta_{xx}\phi$ is evaluated at each interior grid-point and its value in the fake zone is then obtained from neighboring processors via MPI. In the second step, the finite difference operation is applied to δ^{ϕ}_{xx}, according to

$$-K\delta_{xxxx}\phi \equiv -K\delta_{xx}\left[\delta_{xx}\phi\right] \equiv -K\left(\left[\delta_{xx}\phi\right]_{i-1} - 2\left[\delta_{xx}\phi\right]_{i} + \left[\delta_{xx}\phi\right]_{i+1}\right)/\left(\Delta x\right)^{4}$$
(3)

After the entire diffusion term is calculated at interior points, it is used to update the prognostic variable, ϕ. Usually, the update of fake zone values of individual terms in the prognostic equation, including this diffusion term, is not necessary. The update of the prognostic variable is necessary only after completion of one time step of the integration. However, for the reason noted above, one additional set of messages from the neighboring processors is needed to complete the calculation of fourth-order horizontal diffusion and advection terms. An alternative is to define more fake zones and fill them with values from neighboring processors. In this case, the amount of data communicated is about the same, but the number of associated message passing calls is reduced. In the case of even higher-order diffusion and/or advection schemes, such as the 6th-order diffusion scheme recommended by [29] and the 5th- and 6th-order advection schemes commonly used in the WRF, more communications are needed. Furthermore, the WRF model commonly uses the 3rd-order Rouge-Kutta time integration scheme, which involves three evaluations of the advection terms during each large time step [26]. As a result, the associated MPI communications are tippled. Fortunately, this time integration scheme allows for a larger time step size.

An issue unique to the split-explicit time integration scheme is the need to exchange boundary values within small time steps, which incurs additional MPI communication costs. Most of the time, message sizes are relatively small and thus communication latency is a larger issue than bandwidth. Because boundary zone values of dependent variables cannot be communicated until the time step integration is completed, frequent message passing can incur a significant cost. One possible solution is to sacrifice memory and CPU processing in favor of fewer but larger messages. This can be done by defining a much wider fake zone for each sub-domain. For example, if five small time steps are needed in each large time step, one can define a fake zone that is five grid-points wide instead of only one. These fake zone points then can be populated with values from neighboring processors at the beginning of each large time step through a single message-passing call for each boundary. The small time step integration then will start from the larger expanded sub-domain, and decrease by one grid zone at each boundary after each small step integration. This way no additional boundary communication is required throughout the five small steps of integration. This strategy has been tested with simple codes but is yet to be attempted with the full ARPS or WRF codes. The actual benefit will depend on the relative performance of the CPU versus network, and for the network also on the bandwidth and latency performance ratio. The sub-domain size and the large–small time step ratio also influence the effectiveness of this strategy. To further reduce communication overhead, more sophisticated techniques, such as asynchronous communications and message overlap or hiding, may be exploited. Computations, performed between message passing points, are used to "hide" the network processes, through operation overlapping. To fully utilize the tiling and message-hiding techniques, it is best that the software is designed from the beginning to accommodate them; otherwise, the "retrofitting" efforts will be very significant, especially for software that contains a large number of loops with hardcoded loop bounds. Most applications, including the ARPS and WRF, will need significant restructuring to take full advantage of these techniques.

6.2.4 Load balancing

Even though all processors run the same code with the ARPS and WRF, domain decomposition is subject to load imbalances when the computational load is data-dependent. This is most often true for spatially intermittent physical processes such as condensation, radiation, and turbulence in atmospheric models. Because atmospheric processes occur nonuniformly within the computational domain, e.g., active thunderstorms may occur within only a few sub-domains of the decomposed domain, the load imbalance across processors can be significant. For sub-domains that contain active thunderstorms, some 20–30% additional computational time may be needed. Most implementations of atmospheric prediction models do not perform dynamic load balancing, however, because of the complexity of the associated algorithms and because of the communication overhead associated with moving large blocks of data across processors.

6.2.5 Timing and scalability

The scalability of the WRF on several large parallel systems is shown here for a relatively simple thunderstorm simulation case. The size of the sub-domain on each processor is held fixed at $61 \times 33 \times 51$ points as the number of processors is increased, i.e., the global domain size increases linearly with the number of processors. When 1000 processors are used, the entire model contains approximately 92-million grid-points. With the WRF and ARPS using some 150 3D arrays, the total memory usage is approximately 60 GB for this problem.

Figure 6.3 shows the timings of the WRF tests on the Pittsburgh Supercomputing Center (PSC) Cray XT3 and Terascale Computing System (TCS), and the Datastar at the San Diego Supercomputing Center (SDSC). The PSC Cray XT3 system contains 2,068 compute nodes linked by a custom-designed interconnect, and each node has two 2.6 GHz AMD Opteron processors with 2 GB of shared memory. The TCS consists of 764 Compaq ES45 AlphaServer nodes with four 1-GHz Alpha processors and 4 GB of shared memory. The nodes are interconnected using a Quadrics network. The SDSC Datastar has 272 IBM P655+ nodes with eight 1.5 or 1.7 GHz CPUs on each node. It is clear from the plot that the scaling performance[§] is reasonable for processor counts ranging from tens to a couple of hundreds, but it becomes poor for more than 256 processors for the Datastar and XT3. The scalability deterioration is particularly severe on the XT3 for more than 512 processors. Interestingly, the scalability after 512 processors is excellent on PSC TCS, with the curve remaining essentially flat, indicating perfect scaling when count the processor increases from 512 to 1024.

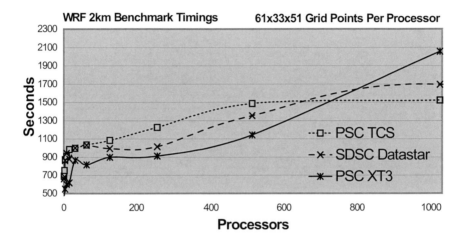

FIGURE 6.3: Performance of the WRF model on various TeraGrid computing platforms. The WRF is executed using $61 \times 33 \times 51$ grid-points per processor such that the overall grid domain increases proportionally as the number of processors increases. For a perfect system, the runtime should not increase as processors are added, resulting in a line of zero slope.

For all three systems, rather poor scaling performance occurs when the processor counts increase from one to a few tens, as indicated by the steep slopes of the timing curves for a small number of processors. In all cases, the performance is best when running a single sub-domain on the single node, for which the code has exclusive access to the CPU-memory channel and all levels of the CPU cache. When every processor on a given node is allocated a sub-domain, the memory access contention through the shared memory bus

---

[§]Ideally, the runtime should remain constant as the number of processors is increased (i.e., a horizontal line). Any deviation is due to communication/memory access overhead and load imbalances.

degrades performance, and when the program spans multiple nodes, the MPI communications cause further degradation. Except for PSC TCS, which has slower processors but a faster network, the other two systems clearly do not provide the needed scalability for larger numbers of processors. For future petascale systems having orders of magnitude larger numbers of processor cores, significantly faster network interconnects as well as aggressive algorithm optimization and/or redesign will be needed to achieve useful performance. The expected use of CPUs with dozens of processor cores that share the same path to memory will make the scaling issue even more challenging.

6.2.6   Other essential components of NWP systems

An equally important component of NWP systems is the data assimilation system, which optimally combines new observations with a previous model forecast to arrive at a best estimate of the state of the atmosphere that is also dynamically compatible with the model's equations. The four-dimensional variational (4DVAR) and ensemble-Kalman filter (EnKF) techniques are the leading methods now available but are also computationally very expensive [12]. 4DVAR obtains an initial condition for the prediction model by minimizing a cost function that measures the discrepancy between observations collected within a defined time window (called the assimilation window) and a model forecast made within the same window. The process involves setting to zero the gradient of the cost function with respect to control variables, usually the initial conditions, at the start of the assimilation window. To do so, the adjoint of the prediction model (mathematically, the adjoint is the transpose of the linear tangent version of the original nonlinear prediction model, see, [12]) is integrated "backward" in time and an optimization algorithm is used to adjust the control variables using the gradient information thus obtained. Such an iterative adjustment procedure usually has to be repeated 50 to 100 times to find the cost function minimum, and for problems with larger degrees of freedom, stronger nonlinearity and/or more observations, the iteration count is correspondingly larger. The adjoint model integration is usually 2–3 times the cost of the forward model integration, and domain decomposition strategies appropriate for the forward model usually apply. Because of the high computational cost of 4DVAR, operational implementations usually use coarser resolution for the iterative minimization and the resulting information is added to the high-resolution prior estimate to obtain an initial condition for the high-resolution forecast.

The development and maintenance of an adjoint code are extremely labor intensive though mostly straightforward. 4DVAR is known to exhibit difficulty with high nonlinearity in the prediction model and/or in the operators used to convert model-dependent variables (e.g., rain mass) to quantities that are observed (e.g., radar reflectivity). A promising alternative is the ensemble Kalman filter [6, 7] whose performance is comparable to 4DVAR. EnKF has the additional advantage of being able to explicitly evolve the background error covariance, which is used to relate the error of one quantity, such as temperature to the error of another, such as wind, and provide a natural set of initial conditions for ensemble forecasts. The probabilistic information provided by ensemble forecasting systems that include tens to hundreds of forecast members has become an essential part of operational NWP [12].

Efficient implementation of the EnKF analysis algorithms on distributed memory systems is nontrivial. With the commonly used observation-based ensemble filter algorithms, the observations need to be sorted into batches, with those in the same batch not influencing the same grid-point [13, 14]. Such implementations are rather restrictive and may not work well for a large number of processors and/or when spatially inhomogeneous data have different spatial influence ranges. A variant of the EnKF, the local ensemble transport Kalman filter (LETKF) [9], is more amenable to parallelization because of the use of independent local analysis sub-domains. This method does not, however, ensure processor independence of the results, although it appears to be the most viable approach to date for parallelization. Finally, all ensemble-based assimilation algorithms require the covariance calculations using the model state variables from all ensemble members that are within a certain radius of an observation or within a local analysis domain. This requires global transpose operations when individual ensemble members are distributed to different processors. The cost of moving very large volumes of 3D-gridded data among processors can be very high. Further, owing to the nonuniform spatial distribution of observational data, load balancing can be a major issue. For example, ground-based operational weather radar networks provide the largest volumes of weather observations yet they are only available over land and the most useful data are in precipitation regions. Orbiting weather satellites provide data beneath their flight paths, creating nonuniform spatial distributions for any given time.

Other essential components of NWP systems include data quality control and preprocessing, data post-processing and visualization, all of which need to scale well for the much larger prediction problems requiring petascale systems.

6.2.7   Additional issues

In addition to processor speed, core architecture and memory bandwidth, I/O is a major issue for storm-scale prediction using petascale system. As the simulation domains approach sizes of order 10^{12} points, the magnitude of the output produced will be enormous, requiring very-high-performance parallel file systems and distributed parallel post-processing software for analysis, mining and visualization. Parallel I/O strategies are needed where the subdomains in a simulation are stored separately instead of being gathered into full three-dimensional arrays. However, most analysis and visualization software assumes access to full three-dimensional arrays. Such software will have to be adapted to utilize the sub-domains directly. In fact, the ARPS graphics plotting software, ARPSPLT, reads ARPS output stored in sub-domains and the software itself supports MPI. In addition, very-high-resolution graphical displays will be needed for display at scales of one pixel per grid-point. Interactive visualization software with parallel computational and rendering backends also will be essential. To aid in analyzing nested-grid simulations, visualization methods for viewing simultaneously all nested grids within a simulation will be needed [27].

As petascale systems are expected to contain hundreds of thousands of processors, the likelihood of node failure increases exponentially. Although hardware and operating systems, and MPI implementations, should account for most of such failures, complete dependency on the system is idealistic and NWP models will require built-in fault tolerance. This includes enhancing model codes to take advantage of the fault tolerance in the system software as well as in MPI implementations.

6.3   Example NWP Results

6.3.1   Storm-scale weather prediction

For the purpose of demonstrating and evaluating convective storm-scale prediction capabilities in a quasi-operational setting, the Center for Analysis and Prediction of Storms at the University of Oklahoma produced daily 30-hour forecasts using the WRF model at an unprecedented 2-km resolution, over 2/3rds of the continental US, during the spring 2005 southern Great Plains storm season. The project addressed key scientific issues that are vital for forecasting severe weather, including the development and movement of thunderstorm complexes, quantitative precipitation forecasting, and convective initiation. Figure 6.4 presents the 24-hour model forecast radar reflectivity (proportional to precipitation intensity, with warm colors indicating heavier rates) at 00 UTC on June 5, 2005 (left) which compares favorably with the radar-observed reflectivity (right). The forecast required 8 hours of wall-clock time using 1100 processors on the TCS at PSC.

During the spring of 2007, the above forecast configuration was further expanded to include 10-member WRF ensemble forecasts at 4-km grid spacing [32]. Using 66 processors of the PSC Cray XT3 system, each 33-hour forecast ensemble member took 6.5 to 9.5 hours, with the differences being caused by the use of different physics options in the prediction model. A single forecast in the same domain using a horizontal grid spacing of 2 km ($1501 \times 1321 \times 51$ points) and 600 Cray XT3 processors took about 9 hours for a 33-hour forecast, including full data dumps every 5 minutes. The data dumps into a parallel Luster file system accounted for over 2 hours of the total time. For truly operational implementations, such forecasts will need to be completed within one hour.

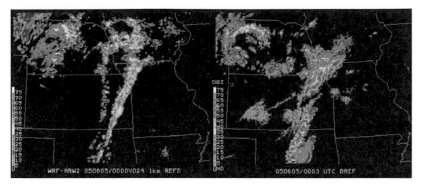

FIGURE 6.4: (See color insert following page 18.) 24-hour WRF-predicted (left) and observed (right) radar reflectivity valid at 00 UTC on June 5, 2005. Warmer colors indicate higher precipitation intensity. The WRF model utilized a horizontal grid spacing of 2 km and forecasts were produced by CAPS on the Terascale Computing System at the Pittsburgh Supercomputing Center as part of the 2005 SPC Spring Experiment.

6.3.2 Very-high-resolution tornado simulation

Using the Terascale Computing System at PSC and the ARPS model, the first author obtained the most intense tornado ever simulated within a realistic supercellular convective storm. The highest resolution simulations used 25-m horizontal grid spacing and 20-m vertical spacing near the ground. The simulation domain was $48 \times 48 \times 16$ km^3 in size and included only non-ice phase microphysics. No radiation or land surface processes were included. The use of a uniform-resolution grid large enough to contain the entire parent storm of the tornado eliminates uncertainties associated with the typical use of nested grids for this type of simulation. The maximum ground-relative surface wind speed and the maximum pressure drop in the simulated tornado were more than 120 ms^{-1} and 8×10^3 Pascals, respectively. The peak wind speed places the simulated tornado in the F5 category of the Fujita tornado intensity scale, the strongest of observed tornadoes. This set of simulations used 2048 Alpha processors and each hour of model simulation time required 15 hours of wall-clock time, producing 60 terabytes of data dumped at 1-second intervals. These output data were used to produce extremely realistic 3D visualizations of the storm clouds and the tornado funnel. Fig. 6.5 shows a 3D volume rendering of the model-simulated cloud water content in a $7.5 \times 7.5 \times 3$ km domain, with a tornado condensation funnel reaching the ground.

In a 600-second long restart simulation using 2048 Alpha processors on 512 nodes, the message-passing overhead consumed only 12% of the total simulation time. The small time step integration used more than 26%, and the sub-grid-scale turbulence parameterization used 13%. The initialization

FIGURE 6.5: Three-dimensional volume rendering of model-simulated cloud water content in a $7.5 \times 7.5 \times 3$ km domain, showing the tornado condensation funnel reaching the ground. The lowered cloud base to the left of the tornado funnel is known as the wall cloud (Rendering courtesy of Greg Foss of the Pittsburgh Supercomputing Center.)

of this restart simulation that involves the reading of 3D initial condition fields from a shared file system took over 11%. The 1-second output written to local disks of each compute node took only 8% but the copying of output from the node disks to a shared file system using a PSC-built parallel copying command at the end of the simulation took as much as 1/5th of the total simulation time. Clearly I/O performance was a major bottleneck.

6.3.3   The prediction of an observed supercell tornado

Using the ARPS and its 3DVAR¶ and cloud analysis data assimilation system, [32] obtained a successful prediction of an observed thunderstorm and embedded tornado that occurred on May 8, 2003 in the Oklahoma City area. The horizontal grid had a spacing of 50 m and was one-way nested, successively, within grids of 100-m, 1-km and 9-km horizontal spacings. The 80×60 km^2 horizontal grid of 50-m spacing had 1603×1203 grid-points and

---

¶3DVAR is a subset of 4DVAR that does not include a time integration component.

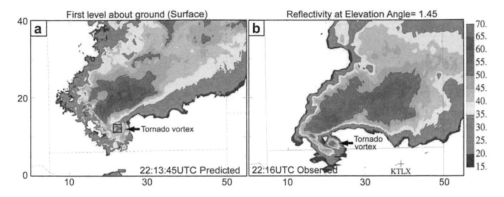

FIGURE 6.6: (See color insert following page 18.) (a) Predicted surface reflectivity field at 13.75 minutes of the 50-m forecast valid at 2213:45 UTC and (b) observed reflectivity at the 1.45° elevation of the Oklahoma City radar observation at 2216 UTC. The domain shown is 55 km × 40 km in size, representing the portion of the 50 m grid between 20 and 75 km in the east-west direction and from 16 to 56 km in the north-south direction.

the 40-minute forecast took 2 days to complete using 1600 processors on the PSC TCS. A very significant portion of the time was consumed by writing output at 15-second intervals to a slow shared file system. A single processor was gathering data from all processors and writing out to this shared file system.

During the 40-minute forecast, two successive tornadoes of F1-F2 intensity with life spans of about 5 minutes each were produced within the period of the actual tornado outbreak, and the predicted tornadoes traveled along a path about 8 km north of the observed damage track with correct orientations. A half-hour forecast lead time was achieved by the 50-m forecast nested within the 100-m grid.

The surface reflectivity at 13.75 minutes for the 50-m forecast (valid at about 2214 UTC) is plotted in Figure 6.6, together with observed reflectivity at the 1.45° elevation of the Oklahoma City operational weather radar at 2216 UTC. At this time, the predicted tornado shows a pronounced hook-shaped reflectivity pattern containing a inwardly spiraling reflectivity feature at the southwest end of the precipitation region (Figure 6.6). The observed low-level reflectivity approximately 2 minutes later also contains a similar hook echo pattern (Figure 6.6). Due to resolution limitations of the radar, the observed reflectivity shows fewer structural details than the prediction.

6.4 Numerical Weather Prediction Challenges and Requirements

A grand vision of NWP for the next 20 years is a global NWP model running at 1-km horizontal grid spacing with over 200 vertical levels. Such resolution is believed to be the minimum required to resolve individual thunderstorms, which are the main causes of hazardous local weather. To cover the roughly 510-million square kilometers of Earth's surface, about half a billion model columns will be needed, giving rise to 100-billion grid cells when 200 levels are used. With each grid cell carrying about 200 variables so as to accommodate more sophisticated physical processes, a total of 80 terabytes of memory will be needed to run a single prediction. If this model is distributed to a petascale system with 1-million processor cores, each core will have 510 model columns with which to work. This corresponds to roughly a $26 \times 20 \times 200$ sub-domain, a relatively small domain in terms of the interior domain-boundary interface ratio, yet relatively large in terms of fitting required information into the CPU cache of expected sizes. If 100 forecasts are to be run simultaneously as part of an ensemble system, each forecast can be run on a subset of the processors. In this case, each processor will get 100 times more columns to work with, i.e., the sub-domain size will be 100 times larger, at $255 \times 200 \times 200$. In this case, the ratio of MPI communication to computation is significantly reduced, and the memory access will become the dominant performance issue.

Before 100-member global ensembles of 1-km spacing can be carried out in a timely manner, regional models can be run for smaller domains, say covering North America. The ARPS and WRF models are mainly designed for such purposes. Further increases in horizontal resolution are possible and can be very desirable in order to resolve smaller, less organized convective storms. To be able to explicitly predict tornadoes, resolutions less than 100 m are essential and a better understanding of storm dynamics, coupled with better observations and parameterizations of land–atmosphere interactions, is needed. Because small-scale intense convective weather can develop rapidly and can be of short duration (e.g., 30 minutes), numerical prediction will require very rapid model execution. For example, a tornado prediction having a 30-minute lead time should be produced in a few minutes so as to provide enough time for response.

The ensemble data assimilation system [24, 33] that assimilates full volumes of weather radar data poses another grand challenge problem. The current U.S.-operational Doppler weather radar network consists of over 120 radars. Producing full volume scans every 5 minutes, roughly 600 million observations of radial wind velocity and reflectivity are collected by the entire network every 5 minutes assuming that all radars operate in precipitation mode. The planned doubling of the azimuthal resolution of reflectivity will further increase the data volume, as will the upgrade of the entire network to gain polarimetric[||] capabilities. These, together with high-resolution satellite data, pose a tremendous challenge to the data assimilation problem, and there is no limit to the need for balanced computing resources in the foreseeable future.

In addition to basic atmospheric data assimilation and prediction problems, the inclusion of pollution and chemistry processes and highly detailed microphysical processes, and the full coupling of multilevel nested atmospheric models with ocean and wave models that are important, e.g., for hurricane prediction [2], will further increase the computational challenge.

6.5 Summary

The challenges faced in numerically predicting high-impact local weather were discussed, with particular emphasis given to deep convective thunderstorms. Two community regional numerical weather prediction (NWP) systems, the ARPS and WRF, were presented and their current parallelization strategies described. Key challenges in applying such systems efficiently on petascale computing systems were discussed, along with computing requirements for other important components of NWP systems including data assimilation with 4D variational or ensemble Kalman filter methods. Several examples demonstrating today's state-of-the-art simulations and predictions were presented.

6.6 Acknowledgment

This work was mainly supported by NSF grants ATM-0530814 and ATM-0331594. Gregory Foss of PSC created the 3D tornado visualization with assistance from the first author. Kevin Thomas performed the timing tests shown in Figure 6.3.

[||]Polarimetric refers to radars that transmit an electromagnetic pulse in one polarization state (e.g., horizontal) and analyze the returned signal using another state (e.g., vertical). This allows the radar to discriminate among numerous frozen and liquid precipitation species.

References

[1] G. H. Bryan, J. C. Wyngaard, and J. M. Fritsch. Resolution requirements for the simulation of deep moist convection. *Mon. Wea. Rev.*, pages 2394–2416, 2003.

[2] S. S. Chen, J. F. Price, W. Zhao, M. A. Donelana, and E. J. Walsh. The CBLAST-Hurricane program and the next-generation fully coupled atmosphere–wave–ocean models for hurricane research and prediction. *Bull. Amer. Meteor. Soc.*, pages 311–317, 2007.

[3] S. S. Chen and J. E. Tenerelli. Simulation of hurricane lifecycle and inner-core structure using a vortex-following mesh refinement: Sensitivity to model grid resolution. *Mon. Wea. Rev.*, 2006. (Submitted).

[4] K. K. Droegemeier and coauthors. Weather prediction: A scalable storm-scale model. In G. Sabot, editor, *High Performance Computing*, pages 45–92. Addison-Wesley, 1995.

[5] K. K. Droegemeier and coauthors. Service-oriented environments for dynamically interacting with mesoscale weather. *Comput. Sci. Engineering*, pages 12–27, 2005.

[6] G. Evensen. The ensemble Kalman filter: Theoretical formulation and practical implementation. *Ocean Dynamics*, pages 343–367, 2003.

[7] G. Evensen. *Data Assimilation: The Ensemble Kalman Filter*. Springer, 2006.

[8] R. A. Houze Jr., S. S. Chen, W.-C. Lee, R.t F. Rogers, J. A. Moore, G. J. Stossmeister, M. M. Bell, J. Cetrone, W. Zhao, and S. R. Brodzik. The hurricane rainband and intensity change experiment: Observations and modeling of hurricanes Katrina, Ophelia, and Rita. *Bull. Amer. Meteor. Soc.*, pages 1503–1521, 2006.

[9] B. R. Hunt, E. J. Kostelich, and I. Szunyogh. Efficient data assimilation for spatiotemporal chaos: A local ensemble transform Kalman filter. *Physics*, page 11236H, 2005.

[10] F. Irigoin and R. Triolet. Supernode partitioning, 1989.

[11] K. W. Johnson, G. A. Riccardi, J. Bauer, K. K Droegemeier, and M. Xue. Distributed processing of a regional prediction model. *Mon. Wea. Rev.*, 1994.

[12] E. Kalnay. *Atmospheric Modeling, Data Assimilation, and Predictability*. Cambridge University Press, 2002.

[13] C. L. Keppenne. Data assimilation into a primitive-equation model with a parallel ensemble Kalman filter. *Mon. Wea. Rev.*, pages 1971–1981, 2000.

[14] C. L. Keppenne and M. M. Rienecker. Initial testing of a massively parallel ensemble Kalman filter with the Poseidon isopycnal ocean general circulation model. *Mon. Wea. Rev.*, pages 2951–2965, 2002.

[15] J. Michalakes, J. Dudhia, D. Gill, T. Henderson, J. Klemp, W. Skamarock, and W. Wang. The weather research and forecast model: Software architecture and performance. In *Proc. 11th ECMWF Workshop on the Use of High Performance Computing In Meteorology*, Reading, U.K., 2004.

[16] R. A. Pielke and R. C. Weather impacts, forecasts, and policy. *Bull. Amer. Meteor. Soc.*, pages 393–403, 2002.

[17] F. Rabier, H. Jarvinen, E. Klinker, J.-F. Mahfouf, and A. Simmons. The ECMWF operational implementation of four-dimensional variational assimilation. I: Experimental results with simplified physics. *Quart. J. Roy. Met. Soc.*, pages 1143–1170, 2000.

[18] L. F. Richardson. *Weather Prediction by Numerical Process*. Cambridge University Press. Reprinted by Dover Publications, 1965, 1922.

[19] A. Sathye, G. Bassett, K. Droegemeier, and M. Xue. *Towards Operational Severe Weather Prediction Using Massively Parallel Processing*. Tata McGraw Hill, 1995.

[20] A. Sathye, G. Bassett, K. Droegemeier, M. Xue, and K. Brewster. Experiences using high performance computing for operational storm scale weather prediction. In *Concurrency: Practice and Experience, special issue on Commercial and Industrial Applications on High Performance Computing*, pages 731–740. John Wiley & Sons, Ltd., 1996.

[21] A. Sathye, M. Xue, G. Bassett, and K. K. Droegemeier. Parallel weather modeling with the advanced regional prediction system. *Parallel Computing*, pages 2243–2256, 1997.

[22] W. Skamarock and J. B. Klemp. Efficiency and accuracy of the Klemp-Wilhelmson time-splitting technique. *Mon. Wea. Rev.*, pages 2623–2630, 1994.

[23] W. C. Skamarock and coauthors. A description of the advanced research WRF version 2, 2005.

[24] M. Tong and M. Xue. Ensemble Kalman filter assimilation of Doppler radar data with a compressible nonhydrostatic model: OSS experiments. *Mon. Wea. Rev.*, pages 1789–1807, 2005.

[25] D. B. Weber and H. Neeman. *Experiences in Optimizing a Numerical Weather Prediction Model: an Exercise in Futility?* 7th Linux Cluster Institute Conference, 2006.

[26] L. J. Wicker and W. C. Skamarock. Time-splitting methods for elastic models using forward time schemes. *Mon. Wea. Rev.*, pages 2088–2097, 2002.

[27] WRF-RAB. *Research Community Priorities for WRF-System Development.* WRF Research Applications Board Report, 2006.

[28] M. Xue and coauthors. The advanced regional prediction system (ARPS) - a multi-scale nonhydrostatic atmospheric simulation and prediction tool. Part II: Model physics and applications. *Meteor. Atmos. Phys.*, pages 143–166, 2001.

[29] M. Xue, K. K. Droegemeier, and V. Wong. The advanced regional prediction system (ARPS) - a multiscale nonhydrostatic atmospheric simulation and prediction tool. Part I: Model dynamics and verification. *Meteor. Atmos. Physics*, pages 161–193, 2000.

[30] M. Xue, K. K. Droegemeier, V. Wong, A. Shapiro, and K. Brewster. ARPS version 4.0 user's guide. http://www.caps.ou.edu/ARPS, 1995.

[31] M. Xue and M. Hu. Numerical prediction of 8 May 2003 Oklahoma City supercell tornado with ARPS and radar data assimilation. *Geophys. Res. Letters*, 2007. In review.

[32] M. Xue, F. Kong, D. Weber, K. W. Thomas, Y. Wang, K. Brewster, K. K. Droegemeier, J. S. Kain, S. J. Weiss, D. R. Bright, M. S. Wandishin, M. C. Coniglio, and J. Du. CAPS realtime storm-scale ensemble and high-resolution forecasts as part of the NOAA hazardous weather testbed 2007 spring experiment. In *18th Conference on Numerical Weather Prediction*, Park City, Utah, June 2007. Meteoro. Soc.

[33] M. Xue, M. Tong, and K. K. Droegemeier. An OSSE framework based on the ensemble square-root Kalman filter for evaluating impact of data from radar networks on thunderstorm analysis and forecast. *J. Atmos. Ocean Tech.*, pages 46–66, 2006.

[34] M. Xue, D.-H. Wang, J.-D. Gao, K. Brewster, and K. K. Droegemeier. The advanced regional prediction system (ARPS), storm-scale numerical weather prediction and data assimilation. *Meteor. Atmos. Physics*, pages 139–170, 2003.

Chapter 7

Software Design for Petascale Climate Science

John B. Drake

Oak Ridge National Laboratory

Philip W. Jones

Los Alamos National Laboratory

Mariana Vertenstein

National Center for Atmospheric Research

James B. White III

Oak Ridge National Laboratory

Patrick H. Worley

Oak Ridge National Laboratory

7.1 Introduction

Prediction of the Earth's climate is a computational grand-challenge problem. Spatial scales range from global atmospheric circulation to cloud microphysics, and time-scales important for climate range from the thousand year overturning of the deep ocean to the nearly instantaneous equilibration of solar radiation balance. Mathematical models of geophysical flows provide the theoretical structure for our understanding of weather patterns and ocean currents. These now-classical developments have been extended through numerical analysis and simulation to encompass model equations that cannot be solved except using supercomputers. That mathematics and physics lie at the heart of climate modeling is the primary basis for thinking that we can predict complex climate interactions. The degree of confidence in climate models is bounded by their ability to accurately simulate historical climate equilibrium and variation, as well as their ability to encapsulate the scientific understanding of instantaneous interactions at the process level. Observational weather data, as well as measurements of physical, chemical, and biological processes, are a constant check on the validity and fidelity of the models and point to areas that are poorly understood or represented inadequately. In large part because of the ability of ever-more-powerful computers to integrate increasingly rich data and complex models, the understanding of coupled climate feedbacks and responses has advanced rapidly.

Facilities with computers reaching a (peak) petascale capability are expected in 2009 or before. The Cray XT and IBM Blue Gene lines, with processor counts ranging from 25,000 to 250,000, are candidate architectures. Enabling climate models to use tens of thousands of processors effectively is the focus of efforts in the U.S. Department of Energy (DOE) laboratories and the National Center for Atmospheric Research (NCAR) under the auspices of the DOE Scientific Discovery through Advanced Computing (SciDAC) program. This exposition describes the software design of the Community Climate System Model (CCSM) [8, 5], one of the primary tools for climate science studies in the United States. In particular, it addresses issues and proposed solutions in readying the CCSM to use petascale computers.

The outline of the chapter is as follows. Section 7.2 describes near-term climate science goals and the corresponding computational requirements. Section 7.3 describes petascale computer architectures that will become available in the next five years, focusing in particular on how these architectures differ from current supercomputers and the implications of these differences for CCSM development. Section 7.4 is a description of the CCSM software architecture, including recent and planned functionality introduced to improve processor and problem-size scalability. Section 7.5 summarizes the promise and problems introduced by petascale architectures, and how the CCSM is being readied to exploit the new generation of petascale computers.

7.2   Climate Science

Predicting the climate for the next century is a boundary-value problem with initial conditions starting in 1870 and boundary data from the historical, observational record. Parts of the boundary data are the solar input, including solar variability, and the changing atmospheric chemical composition, for example, the level of carbon dioxide (CO_2) and other greenhouse gases. Because the future boundary conditions are not known, "scenarios" are constructed that bracket the likely input parameters. By considering a range of scenarios, the possible future climate states can be bounded.

The CCSM is a modern world-class climate code consisting of atmosphere, ocean, land, and sea-ice components coupled through exchange of mass, momentum, energy, and chemical species. The time-dependent partial differential equations for stratified liquids and gases are posed in a spherical geometry and a rotating frame. As a dynamical system, the climate is weakly forced and strongly nonlinear [41]. Increases in greenhouse gases in the atmosphere change the absorption of long-wave radiation and thus alter the energy balance of the planet. The inherent nonlinearities of the system are evident in the constantly shifting weather patterns that arise from differential heating of the Earth's surface and instabilities of the atmospheric flow. The resolution of the models, the number of grid-points used to approximate the partial differential equations representing the physical conservation laws, must adequately represent the relevant dynamical interactions. For climate simulations that include a diurnal cycle, the minimum resolution is about 300 km per gridpoint. Regional detail begins to emerge as the grid spacing is decreased to 100 km and smaller.

The dynamical algorithms of the atmosphere and ocean include explicit and implicit finite-volume methods, as well as transform methods (fast Fourier and Legendre transforms) to approximate spherical operators and to solve elliptic equations. The current physical parameterizations, approximations to subgrid phenomena, are computed for each horizontal position, or *column*, independently, in parallel. In contrast, the discrete differential operators require at least nearest-neighbor communication. The computational requirements projected for future versions of climate models are discussed in the ScaLeS report [40], where it was shown that a factor of 10^{12} increase in computing capability could be exploited by the research community. A limitation on the increase in resolution is that the allowable time-step size decreases proportionally with grid spacing. So more time-steps are required to simulate a century of climate as the grid spacing decreases. Since the time-advancement algorithms are inherently sequential, parallelism is limited, and the software structure must be designed carefully to expose as much parallelism as possible.

CCSM developers have adopted software-engineering practices intended to make the model maintainable, extensible, and portable across a wide variety of computers while maintaining computational performance [15]. However, any computational experiment, or set of experiments, has an associated throughput requirement or limitation on computer resources. Thus the configuration of the model and the specification of the computational experiment must be balanced with the computer capabilities and speeds, forcing compromises on grid resolution and accuracy of physical parameterizations. Petascale systems offer the promise of sufficient computing power to move beyond the current limitations. With this promise in mind, CCSM developers are focused on improving the fidelity of the physical climate simulation and providing new components that better account for the carbon cycle among the atmosphere, land, and ocean.

7.3   Petaflop Architectures

Over the past few decades, climate simulation has benefited from the exponential performance increases that have accompanied the progression of "Moore's Law" [30], the doubling of transistor counts in integrated circuits every two years or less. Until recently, processor performance has followed a curve similar to the one for transistor counts, primarily through increases in clock frequency. Frequency increases have stalled in recent years, however, because of limitations arising from increased power consumption and heat dissipation, while the increases in transistor counts continue. Instead of clock frequency, it is the *parallelism* in integrated-circuit chips that is growing dramatically.

A visible direction for this growth is the number of processor cores on a single chip. In the previous decade, growth in parallelism on a chip was primarily through increasing the number of functional units in a single processor and through pipelining those units to process multiple instructions at a time. This strategy increased performance without requiring explicit parallelism in the software, but it also dramatically increased the complexity of the processor architecture. The replication of cores on a chip minimizes added complexity but requires parallelism in software. Other strategies for taking advantage of the growth in transistor counts include multi-threading and vectorization, strategies that can add parallelism to a single core that is then replicated on a multicore chip.

Emerging petascale computers combine the growth in parallelism in single chips with increasing numbers of those chips united by high-performance interconnects. The growth in chip counts for individual high-end computers in recent years seems as dramatic as the growth in transistors in each chip; such computers may have 5,000 to 50,000 chips.

The path to dramatic increases in the computational rate for climate simulation is clear; the software must expose more parallelism. Of course, the details of how the software should do this depend on the specific performance characteristics of the target computers. Because CCSM combines multiple complex components, it stresses multiple aspects of computer architectures, including raw computation rate, memory bandwidth and latency, and parallel-interconnect bandwidth and latency. CCSM development has been influenced by the relative balances of these system characteristics in the available computers over time.

Most CCSM components evolved from data-parallel implementations, targeting vector or single-instruction, multiple-data (SIMD) architectures. As clusters grew in prominence, CCSM components adopted distributed-memory parallelism. The limited memory performance and interconnect performance of clusters, along with their incorporation of shared-memory multiprocessor nodes, led CCSM developers toward a variety of strategies in pursuit of portable performance. These strategies now include block-oriented computation and data structures, hybrid parallelism, and modular parallel implementation.

In *block-oriented computation*, each computational kernel is performed on a subset, or block, of the data domain at a time, not on the whole domain, nor on a single element at a time. Each phase then has an outer loop over blocks. The block size is variable and may be tuned for the specifics of the target computer. Block sizes are larger than one element to enable efficient vectorization, pipelining, and loop unrolling by compilers. Block sizes are smaller than the whole domain to enable efficient reuse of cache memory and registers.

Hybrid parallelism in general refers to the use of more than one parallel strategy at a time, typically hierarchically. For CCSM components, the hybrid parallelism specifically combines shared-memory parallelism through OpenMP [11] with distributed-memory parallelism, primarily through the Message Passing Interface (MPI) [19]. The shared-memory implementation may target the same dimension of parallelism as the distributed-memory implementation, such as over parallel blocks, or it may target a different dimension, such as parallelism within a block. This two-level strategy originally targeted clusters of shared-memory systems, where the number of shared-memory threads within a distributed-memory process is tuned based on the number of processors in each cluster node and the capability of the distributed-memory communication infrastructure.

Within this hybrid parallel implementation, *modularity* has been a necessity to deal with the multitude of rapidly changing computer architectures. This modularity allows the use of different communication protocols, different parallel algorithms, and varying process counts based on the details of a given run, such as target computer, problem size, total process count, and relative performance of each component. Performance tuning can involve runtime changes in algorithm choices, targeted use of specialized communication protocols (such as SHMEM [17] or Co-Array Fortran [32]), and load balancing across components by changing the relative process counts.

These strategies for performance portability of CCSM components are well suited for the emerging petascale architectures. Block-oriented computation will help with efficient use of cache within multicore memory hierarchies, along with efficient use of the various forms of pipelining and vectorization within each processor core. Examples of such vectorization include the Streaming SIMD Extensions (SSE) in AMD and Intel x86-64 architectures [38], Vector/SIMD Multimedia Extension Technology in IBM Power processors [39], the SIMD floating-point unit in the PowerPC processors in IBM's Blue Gene [2], and the Cray X1E [1] and successors.

Hybrid parallelism continues to be important for many high-end computers, though the largest computers today, the IBM Blue Gene and the Cray XT4, support only distributed memory. This should change, however, with the introduction of higher core counts on each chip. The initial petascale systems, successors to the current IBM Blue Gene and Cray XT4, should not only support but may require hybrid parallelism to reach the extreme levels of scalability required for a petaflop of performance.

Even with hybrid parallelism, such scaling is likely to require further tuning of the parallel algorithms and implementation. The CCSM components may have to specialize for particular interconnect topologies and communication protocols. Some phases of computation may have limited scalability, such that adding processes beyond a given point will actually decrease aggregate throughput, so these phases should use fewer processors than other phases do. The modular parallel implementation of the components will simplify these tuning tasks.

Some aspects of potential petascale computers present unresolved questions and issues for CCSM developers. The need for maximum parallelism is driving developers toward aggressive elimination of serial bottlenecks and undistributed memory, such as is found in initialization and checkpointing of components. In a related issue, parallelization of input and output (I/O) will be of growing importance.

It is unclear if memory and interconnect bandwidths will be able to keep up with the continuing increases in the aggregate computation rate, but it is clear that latencies will *not* see commensurate improvements. Tolerance of growing relative latencies will be critical to reaching the petascale. As with memory and interconnects, it is also unclear if I/O bandwidths will keep up. Parallelization of I/O may not be adequate to mitigate a growing relative cost; petascale simulation may require asynchronous I/O or in-memory data analysis and reduction at runtime.

Finally, there is uncertainty in the basic architecture of the processor, such that the current software strategies for portable performance, though demonstrably flexible, may not be flexible enough. Examples of potentially disruptive shifts in processor architecture include various forms of nonsymmetric coprocessor architectures, such as the IBM Cell Broadband Engine, the Clear-Speed accelerator, graphics-processing units, and field-programmable gate arrays. Initial petascale computers will be successors to the current Cray XT and IBM Blue Gene, with symmetric multicore processors, but scaling to sustained petaflops may require more complex processor architectures, perhaps incorporating novel coprocessors, or designed more as an integrated system, such as the "adaptive supercomputing" strategy embodied by the Cray Cascade project [9].

7.4 Community Climate System Model

7.4.1 Overview of the current CCSM

CCSM3.0 consists of a system of four parallel geophysical components (atmosphere, land, ocean, and sea ice) that run concurrently and on disjoint processor sets and periodically exchange boundary data (flux and state information) through a parallel flux coupler. The flux coupler serves to remap the boundary-exchange data in space and time. CCSM3.0 is implemented with multiple executables, where each component model is a separate executable, and all component executables run concurrently on disjoint sets of processors. Although this multiple-binary architecture permits each component model to keep its own build system and prevents name-space conflicts, it also increases the difficulty of porting and debugging. The concurrency of the CCSM components is not perfect; the atmosphere, land and sea-ice models are partially serialized in time, limiting the fraction of time when all five CCSM components are all executing simultaneously. (See Figure 7.1.)

To realize the full benefit of approaching petascale computers, each CCSM component, as well as the entire CCSM model system, must target higher-resolution configurations and use processor counts that are one to two orders of magnitude greater than present settings. These new architectures will require CCSM components and coupling mechanisms to scale in both memory and throughput to such processor counts.

7.4.2 Community Atmosphere Model

The Community Atmosphere Model (CAM) is a global atmosphere model developed at NCAR with contributions from researchers funded by DOE and by the National Aeronautics and Space Administration (NASA) [6]. CAM is also the atmosphere component of the CCSM and is the largest consumer of computing resources in typical CCSM simulations.

CAM is a hybrid parallel application, using both MPI and OpenMP protocols. It is characterized by two computational phases: the *dynamics*, which advances the evolution equations for the atmospheric flow, and the *physics*, which approximates sub-grid phenomena such as precipitation processes, clouds, long- and short-wave radiation, and turbulent mixing [7]. Control moves between the dynamics and the physics at least once during each simulation time step.

CAM is a community model that is constantly evolving to include new science. Thus it has been very important that CAM be easy to maintain and port to new systems, and that CAM performance be easy to optimize for new systems or for changes in problem specification or processor count. The software design that supports these portability and maintainability requirements

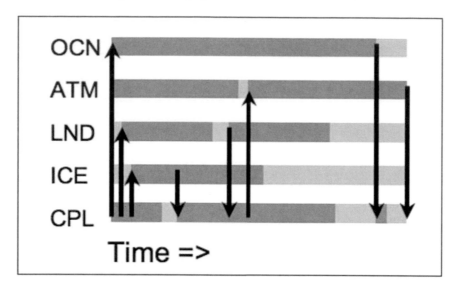

FIGURE 7.1: Example timing diagram for the CCSM. Dark grey indicates "busy," light grey indicates "idle," and arrows indicate communication between components.

should also ease the port to petascale computers, but a number of new issues will need to be addressed. This section begins with a discussion of relevant features in the existing software architecture. The design is described in more detail in [29, 33, 43].

An explicit interface exists between the dynamics and the physics, and the physics data structures and parallelization strategies are independent from those in the dynamics. A dynamics-physics coupler, internal to CAM and different from the CCSM coupler, moves data between data structures representing the dynamics state and the physics state. While some minor performance inefficiency results from the decoupled data structures, the maintainability and ability to optimize the dynamics and physics independently have proven very important. Note that the coupler is distributed, and that the dynamics, physics, and coupler run one at a time, each one in parallel across the same processors.

The explicit physics/dynamics interface enables support for multiple options for the dynamics, referred to as *dynamical cores* or *dycores*, one of which is selected at compile time. Three dycores are currently available: a spectral Eulerian (EUL) [22], a spectral semi-Lagrangian (SLD) [42], and a finite-volume semi-Lagrangian (FV) [24]. This dynamics *modularity* makes it relatively easy to develop new dycores for CAM. Some of the new dycores currently under development are expected to be better suited to petascale architectures than the current dycores, as described later.

Both the physics and dynamics use application-specific messaging layers. For the most part, these are simple wrappers for calls to MPI routines. However, by using layers, we have been able to experiment with other messaging protocols, for example, Co-Array Fortran and SHMEM. The messaging layers are also used to implement a large number of runtime options, choosing, for example, between one-sided and two-sided MPI implementations of a required interprocess communication.

CAM has numerous compile-time and runtime options for performance optimization. Primary examples are the MPI protocol options mentioned previously, static load-balancing options that trade off communication overhead with improved physics load balance, the aspect ratio of two-dimensional domain decompositions in the dynamics, the number of MPI processes and the number of OpenMP threads per process, and the granularity (block size) of parallel tasks in the physics (affecting exposed parallelism, memory requirements, and vector lengths).

Portable application-specific performance instrumentation allows relevant performance data to be collected easily when porting and tuning CAM. While not as powerful as many vendor and third-party performance-analysis tools, the imbedded instrumentation is sufficient for the majority of performance-optimization tasks.

Performance scalability in CAM is not yet adequate for petascale computation. Figure 7.2 describes strong processor scaling for a production-size problem using the spectral Eulerian dycore and for a large problem (for climate) using the finite-volume dycore. The scaling problems arise primarily from the parallel algorithms employed in the dycores. The spectral dycores use a one-dimensional decomposition of the three-dimensional computational grid, and, for example, are limited to 128 MPI processes in the production-size problem. The finite-volume dycore supports the option of a two-dimensional decomposition of the latitude-longitude-vertical computational grid, but the two-dimensional decomposition switches from latitude-longitude to latitude-vertical and back again in each time step, incurring significant communication overhead. Additionally, the number of vertical levels is small compared to the longitude dimension, limiting the available parallelism compared to a purely horizontal two-dimensional decomposition. For the problem size in Figure 7.2, at most 960 MPI processes can be used. Note that additional parallelism is available in the physics, and this can be exploited using OpenMP parallelism on a computer with shared-memory nodes.

Planned modifications to CAM will impact both performance and performance scalability. The introduction of atmospheric chemistry will significantly increase the number of fields that will be advected in the dycore, increase the computational complexity, change the nature of the load balance in the physics, and increase the volume of output. The expectation is that the size of the horizontal dimensions of the computational grid will also increase for certain science studies. Expected constraints of initial petascale computers include limited per-process memory, relatively high-cost I/O, and the need to

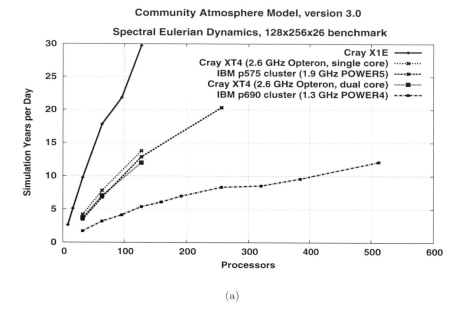

(a)

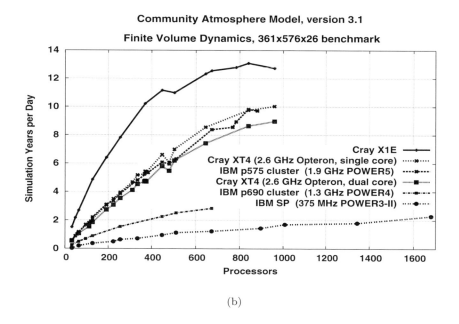

(b)

FIGURE 7.2: The graphs detail current CAM scalability (top: EUL dycore with a production problem instance; bottom: FV dycore with a large problem instance).

exploit additional distributed-memory parallelism.

Various approaches to improve the ability of CAM to exploit petascale computation are being examined. The physics can efficiently exploit more parallelism than the dynamics, but it is currently restricted to using the same number of MPI processes as the dynamics. This restriction is not necessary, and efforts are under way to allow the physics to use more MPI processes than the dynamics. More speculative research is ongoing into using a three-dimensional domain decomposition for the atmospheric chemistry in the physics and parallelizing over the fields being advected in the dynamics. In both cases, the number of active MPI processes changes as control passes through the different phases of CAM.

A number of new dycores are currently under development that are expected to exhibit improved parallel scalability compared to the current CAM dycores [27, 10, 34]. These new dycores use computational grids and numerical formulations that allow a single two-dimensional horizontal domain decomposition to be used. These dycores also attempt to minimize the need for global communications.

In addition to increasing the inherent parallelism in CAM, work is needed to exploit existing parallelism that is currently hidden within serial implementations. For example, CAM now employs a single reader/writer approach to I/O. This is not only a serial bottleneck, not taking advantage of the parallel file systems that are necessary components of petascale computers, but it also increases the per-process memory requirements (as currently implemented). Efforts are underway to evaluate and use parallel I/O libraries such as Parallel NetCDF [23] to eliminate both the I/O and memory bottlenecks.

For convenience, a number of data structures are currently replicated or not maximally decomposed. This too increases the per-process memory requirements, making porting to computers with small per-processor memory difficult, especially when increasing the problem size. An effort is ongoing to identify these data structures and determine how best to eliminate the replication.

At the minimum, a careful performance analysis and evaluation will be required as CAM runs with significantly more processors, much larger grids, and new physical processes. The current load-balancing schemes will also need to be modified to take into account the impact of, for example, atmospheric chemistry, and the exploitation of additional parallelism.

These various development activities are being funded by a number of different projects, including internal efforts within NCAR, collaborations between members of the CCSM Software Engineering Working Group, and multiple projects funded by the DOE SciDAC program.

7.4.3   Parallel Ocean Program

The Parallel Ocean Program (POP) is an ocean general circulation model that solves the incompressible Navier-Stokes equations in spherical coordinates [35]. Spatial derivatives are computed using finite-difference methods. The time integration is split into the fast vertically uniform barotropic wave mode and the full three-dimensional baroclinic modes. An implicit free-surface formulation [16] of the barotropic mode is solved using a preconditioned conjugate-gradient method, while the baroclinic equations are integrated explicitly in time using a leapfrog method. POP supports generalized orthogonal meshes in the horizontal, including displaced-pole [37] and tripole [31] grids that shift polar singularities onto continental land masses. In the vertical, a stretched Eulerian (depth) coordinate is used with higher resolution near the surface to better resolve the ocean mixed layer. A partial bottom cell option is available for a smoother representation of bottom topography. A variety of options are available for subgrid-scale mixing and parameterizations of mixed-layer processes [36], all of which can be selected at runtime.

Parallelization of POP is through domain decomposition in the horizontal. The logical domain is subdivided into two-dimensional blocks that are then distributed across processors or nodes. The optimal block size depends on both the chosen horizontal grid and the computer architecture on which POP will be run. Small blocks can be used as a form of cache-blocking for commodity microprocessors, while large blocks can be specified for vector architectures where longer vector lengths are desired. Blocks that contain only land points are eliminated from the computation, so smaller block sizes tend to result in more land-point elimination. However, because each block maintains a halo of points to reduce inter-block communication, smaller blocks lead to higher surface-to-volume ratios and a corresponding reduction in parallel efficiency.

The remaining ocean blocks can be distributed using a few different methods and can be oversubscribed to nodes, providing a mechanism for hybrid parallelism. Currently, the model supports three block distribution schemes. The Cartesian distribution simply divides blocks in a two-dimensional logically rectangular array and distributes them across a similar two-dimensional array of processors or nodes. A second load-balanced distribution starts with a Cartesian geometry and then uses a modified rake algorithm [18] in each logical direction to shift blocks from nodes with excess work to neighboring nodes with less work. A third scheme was recently introduced by Dennis [12], using a space-filling curve algorithm to create a load-balanced distribution of blocks across nodes. Once the blocks are distributed across nodes, standard message-passing (MPI) is used to communicate between nodes and threading (OpenMP) is used to distribute work across processors within a shared-memory node. Because the messaging routines are packaged in a very few (5) modules, other messaging implementations based on SHMEM and Co-Array Fortran have been used, but they are not currently part of the standard release.

Performance of POP is determined by the two major components of the model. The *baroclinic* section of the model that integrates the full three-dimensional modes is dominated by computational work per grid-point, with only two boundary updates required during each time step. Performance of this section of the model depends primarily on the performance of a given processor. Both floating-point performance and local memory bandwidth are important, for the stencil operators used in POP result in a high ratio of loads and stores to floating-point operations. Scaling of POP is determined by the *barotropic* mode solver. The preconditioned conjugate-gradient (PCG) solver used for the barotropic mode consists of a two-dimensional nine-point operator followed by a halo update and the global reductions inherent in PCG dot products. Each iteration of the solver therefore has very little computational work and frequent small messages and reductions. Performance of this section is sensitive to both load imbalance and message latency; it limits scalability because the already small amount of computational work per process decreases as the processor count increases.

As a component of CCSM, POP is currently run at a fairly coarse resolution (approximately one degree or 100 km) in order to integrate the full climate model for multiple centuries in time. At this low resolution, parallel efficiency on most architectures begins to decrease in the 128–256 processor range [21]. For example, see Fig. 7.3. Future simulations run on petascale computers are more likely to include an eddy-resolving ocean simulation. Production simulations of the ocean at one-tenth degree (10 km) resolution have been performed to resolve mesoscale eddies and simulate their impact on global circulation [28]. Such high resolution and resolved eddy dynamics are required to accurately reproduce many of the features of global ocean circulation. High-resolution simulations have been integrated only for decades due to computational expense, but an effort is currently in progress to use an eddy-resolving configuration in the full CCSM model.

At high resolution, POP scales effectively up to thousands of processors (see Figure 7.3). However, scaling of the barotropic solver can still impact performance at the yet-higher processor counts envisioned for petascale computers. Recently, scaling was improved through the better load balancing of the space-filling curve scheme described previously and by gathering active ocean points into a linear list [13] to further eliminate land points. Future versions of POP may adopt such a structure throughout the entire model to support variable-resolution unstructured grids and further eliminate land points from the domain. Such grids will enable higher resolution in regions where it is needed and will reduce the total number of ocean grid-points, improving the performance of the baroclinic part of POP.

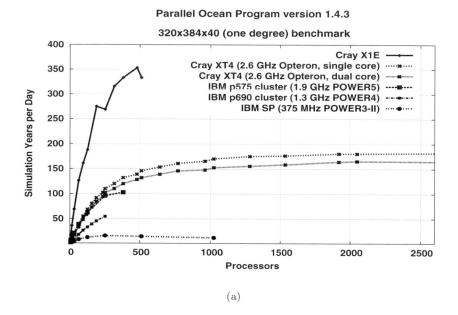

(a)

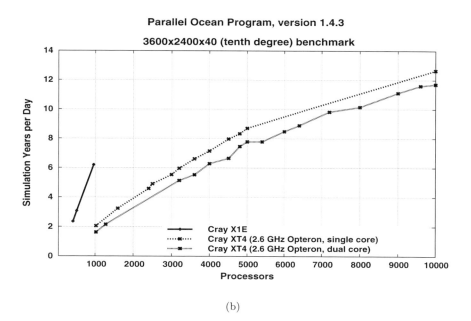

(b)

FIGURE 7.3: The graphs detail current POP scalability (top: one-degree horizontal grid resolution; bottom: tenth-degree resolution).

7.4.4 Community Land Model

The CCSM land component, the Community Land Model (CLM), is a single-column (snow-soil-vegetation) biophysical model providing surface albedos, upward long-wave radiation, heat and water vapor fluxes, and surface stresses to the atmosphere component and river runoff to the ocean component [14]. Spatial land-surface heterogeneity is represented as a nested sub-grid hierarchy in which grid cells are composed of multiple land units, snow/soil columns, and plant functional types. In CCSM3.0, CLM was constrained to run on the same grid as the CCSM atmospheric component. The current CLM has added the capability of running on an independent, higher-resolution grid, thereby including the influence of fine-resolution, sub-grid land use/land cover in the climate simulation. The capability to decouple the atmosphere and land resolutions (as well as grids) should prove beneficial for the target resolutions associated with petascale simulations. In particular, new atmosphere grids that address CAM scalability problems (see Section 7.4.2) are not well suited for land-surface modeling. Separating these grids provides the flexibility for each component to choose the grid most appropriate for both science and computational requirements.

Future CLM software will need to address several issues in order to run effectively on petascale computers. CLM is fundamentally a scalable code with no communication between grid cells other than in the computation of input fields to the river-routing module. Memory scaling, however, is very poor because of numerous non-distributed arrays and the absence of a parallel I/O implementation. Current CLM development efforts are addressing these problems with the goal that tenth-degree (10-km grid spacing) resolution stand-alone simulations will be run in 2007.

7.4.5 Community Sea Ice Model

The sea-ice component (CSIM) is based on the Los Alamos CICE model. It simulates the dynamics of sea ice using an elastic-viscous-plastic formulation [20] with an energy-conserving model of ice thermodynamics [3]. Ice is transported using an incremental remapping scheme [26]. Though the ice is mostly two-dimensional at the surface of the ocean, the ice is divided into several thickness categories to better simulate its behavior. The sea-ice component is run using the same horizontal grid as the ocean (see Section 7.4.3), and the parallel infrastructure of CICE version 4.0 is based on that of POP [25].

The sea ice poses several challenges for petascale computation. The two-dimensional nature of the sea-ice model implies fewer degrees of parallelism. In addition, ice does not cover the globe uniformly, and the ice extent has a strong seasonal variation. While part of the global domain can be eliminated for some simulations based on climatology, a substantial area must be retained in case ice extends beyond known boundaries in climate-change scenarios. To reduce computational work, the ice model is structured to check for active ice regions and gather only ice points for further calculation based on this dynamic ice mask.

It is likely that future high-resolution ice modeling may require new model formulations, since treating ice as a viscous-plastic material may be less valid at these resolutions. More discrete algorithms may be required that can track individual ice floes and their interactions.

7.4.6   Model coupling

As described eariler, CCSM currently runs as five separate concurrent executables, four component models communicating through a fifth, flux-coupling component. Recent creation of a single-executable concurrent CCSM provides a model that is simpler to port and debug, but one that still exhibits the load imbalances and associated performance penalties inherently associated with a concurrent configuration of components that runs partly serialized in time. Development efforts are further targeting alternative full sequential and sequential/hybrid designs that aim to address this load-balancing problem, especially as it relates to petascale computation. In a full sequential system, a top-level application driver replaces the current flux coupler as the coordinator of component communications and the controller of time evolution. The driver runs each *parallel* component model *sequentially*, where only one component runs at a given time and utilizes all the CCSM processors. Regridding and redistribution of boundary data between any two model components occur through coupler modules that are invoked from the top-level driver and that span the processors associated with the two model components (which in this case corresponds to the full set of CCSM processors).

The full sequential implementation has several distinct advantages over the current concurrent design. First, it improves communication efficiency by eliminating the separate coupler component. In the concurrent flux-coupler configuration, the atmosphere-to-ocean exchange involves an M-to-N communication between the atmosphere and coupler components, a regridding calculation on the coupler processors, and another N-to-P communication between the coupler and ocean components. In the sequential system, regridding and redistribution are performed in one step across all processors spanning the source and destination components. This also has the additional advantage that MPI communication is eliminated for source and destination grid cells that reside on the same processor. Secondly, the sequential configuration eliminates the component load-balancing process required to optimize concurrent throughput. Since load balancing of model components is currently a process that involves expert knowledge of the system, the sequential CCSM greatly simplifies the performance-tuning process. Finally, the sequential implementation makes it generally possible to construct a coupled model system where some data can be communicated between model components by memory copies whereas the remainder are obtained from a small number of neighboring processes. This feature is expected to prove very beneficial for running CCSM on petascale architectures.

Despite the above advantages, a sequential configuration also imposes additional scaling requirements. In a concurrent configuration, components that do not scale well can use fewer processors, and the full load can be balanced to take this limitation into account. The sequential configuration will be less forgiving of suboptimal component scaling.

As a generalization of the sequential configuration, a hybrid configuration will permit a subset of the components (e.g., the land and ice components) to run concurrently with each other, but sequentially with the other components (e.g., atmospheric and ocean components). This hybrid configuration should provide the greatest flexibility in creating a model system that will address scalability requirements while minimizing the inefficiencies inherent in the original concurrent design.

Petascale computation will require scalability improvements in more than just communication. Undistributed data structures and I/O will require parallelization. As an example, in the current CCSM, mapping weights are read in by the flux coupler on only a single processor and are subsequently scattered to the other coupler processors. As the model resolution increases, this mechanism will eventually hit local-memory limitations. Flexible, parallel I/O functionality (both binary and NetCDF [4]) will have to be implemented across the CCSM.

7.5 Conclusions

Prediction of the Earth's climate is a computational grand-challenge problem, one that will require computers at the petascale and beyond. CCSM is a modern world-class climate code, and the CCSM research community has mapped out needs that could exploit a factor of 10^{12} increase in computing capability. Over the past few decades, exponential increases in computing capability have been accessible through sequential or modestly parallel execution, but petascale computers will require significantly greater parallelism. The various CCSM components expose parallelism at multiple levels using a few common strategies, including block-oriented computation, hybrid parallelism, and modularity. These strategies are designed for flexibility and tune-ability in the face of rapidly changing computer architectures.

Beyond these shared strategies, each component has unique features and challenges for exploiting petascale computers. For example, in CAM, the computation is divided into "physics" and "dynamics" phases, and the introduction of atmospheric chemistry will affect these differently. The well-defined interface between physics and dynamics will allow the execution of these phases on different numbers of processors, as it allows dycores and physics modules to be interchanged independently. POP already scales to thousands of processors, but the barotropic solver is starting to limit scalability even at high resolutions. A new load-balancing data structure improves scalability and may be worth incorporating in the baroclinic phase as well. CLM, which uses land points coupled only through their interaction with the atmosphere, poses little algorithmic limitation on scalability. CICE, on the other hand, does pose some limitations because of the dynamic extent of its domain and its limited inherent parallelism. To enable simpler and more flexible tuning of the load among these various components, the coupler component is now under development to allow concurrent, sequential, and hybrid execution of the model components.

Challenges for all the components include elimination of most remaining incompletely decomposed data structures, parallelization of I/O, and potential changes in model formulations required at higher resolutions. Development work to meet these challenges is underway, driven by the continuing need for more computing capability.

7.6 Acknowledgments

This research used resources (Cray X1E and Cray XT4) of the National Center for Computational Sciences at Oak Ridge National Laboratory, which is supported by the Office of Science of the U.S. Department of Energy under contract no. DE-AC05-00OR22725. It used resources (IBM p690 cluster) of the Center for Computational Sciences, also at Oak Ridge National Laboratory. It used resources (IBM SP, IBM p575 cluster) of the National Energy Research Scientific Computing Center, which is supported by the Office of Science of the U.S. Department of Energy under contract no. DE-AC03-76SF00098.

The work of Drake, White, and Worley was supported by the Climate Change Research Division of the Office of Biological and Environmental Research and by the Office of Mathematical, Information, and Computational Sciences, both in the Office of Science, U.S. Department of Energy, under contract no. DE-AC05-00OR22725 with UT-Batelle, LLC.

The work of Jones was supported by the Climate Change Research Division of the Office of Biological and Environmental Research in the Office of Science, U.S. Department of Energy, under contract no. DE-AC52-06NA25396 with Los Alamos National Security, LLC.

The work of Vertenstein was supported by the National Science Foundation under contract No. NSF0001, by the U.S. Department of Energy under contract No. 2ER6338 and by NASA under contract no. NCC5623.

Accordingly, the U.S. Government retains a nonexclusive, royalty-free license to publish or reproduce the published form of this contribution, or allow others to do so, for U.S. Government purposes.

References

[1] Cray X1E. http://www.cray.com/products/x1e.

[2] L. Bachega, S. Chatterjee, K. A. Dockser, J. A. Gunnels, M. Gupta, F. G. Gustavson, C. A. Lapkowski, G. K. Liu, M. P. Mendell, C. D. Wait, and T. J. C. Ward. A high-performance SIMD floating point unit for Blue Gene/L: Architecture, compilation, and algorithm design. In *PACT '04: Proceedings of the 13th International Conference on Parallel Architectures and Compilation Techniques*, pages 85–96, Washington, DC, 2004. IEEE Computer Society.

[3] C. M. Bitz and W. H. Lipscomb. An energy-conserving thermodynamic model of sea ice. *Journal of Geophysical Research*, 104:15669–15677, 1999.

[4] S. A. Brown, M. Folk, G. Goucher, and R. Rew. Software for portable scientific data management. *Computers in Physics*, 7:304–308, 1993. http://www.unidata.ucar.edu/packages/netcdf/.

[5] W. D. Collins, C. M. Bitz, M. L. Blackmon, G. B. Bonan, C. S. Bretherton, J. A. Carton, P. Chang, S. C. Doney, J. H. Hack, T. B. Henderson, J. T. Kiehl, W. G. Large, D. S. McKenna, B. D. Santer, and R. D. Smith. The Community Climate System Model version 3 (CCSM3). *J. Climate*, 19:2122–2143, 2006.

[6] W. D. Collins, P. J. Rasch, B. A. Boville, J. J. Hack, J. R. McCaa, D. L. Williamson, B. P. Briegleb, C. M. Bitz, S.-J. Lin, and M. Zhang. The formulation and atmospheric simulation of the Community Atmosphere Model: CAM3. *Journal of Climate*, 19(11), 2006.

[7] W. D. Collins, P. J. Rasch, and *et al.* Description of the NCAR Community Atmosphere Model (CAM 3.0). Technical Report NCAR Tech Note NCAR/TN-464+STR, National Center for Atmospheric Research, Boulder, CO, 2004.

[8] Community Climate System Model. http://www.ccsm.ucar.edu/.

[9] Cray, Inc. Cray Cascade project. www.cray.com/cascade/.

[10] Cubed-sphere finite-volume dynamical core. `http://sivo.gsfc.nasa.gov/cubedsphere.html`.

[11] L. Dagum and R. Menon. OpenMP: an industry-standard API for shared-memory programming. *IEEE Computational Science & Engineering*, 5:46–55, 1998.

[12] J. M. Dennis. Inverse space-filling curve partitioning of a global ocean model. In *Proceedings International Parallel and Distributed Processing Symposium IPDPS07*, Long Beach, CA, March 2007.

[13] J. M. Dennis and E. R. Jessup. Applying Automated Memory Analysis to Improve Iterative Algorithms. Technical Report CU-CS-1012-06, Department of Computer Science, University of Colorado, 2006.

[14] R. E. Dickinson, K. W. Oleson, G. Bonan, F. Hoffman, P. Thornton, M. Vertenstein, Z.-L. Yang, and X. Zeng. The community land model and its climate statistics as a component of the climate system model. *Journal of Climate*, 19:2032–2324, 2006.

[15] J. B. Drake, P. W. Jones, and G. Carr. Overview of the software design of the Community Climate System Model. *International Journal of High Performance Computing Applications*, 19:177–186, 2005.

[16] D. K. Dukowicz and R. D. Smith. Implicit free-surface method for the Bryan-Cox-Semtner ocean model. *J. Geophys. Res.*, 99:7991–8014, 1994.

[17] K. Feind. Shared memory access (SHMEM) routines. In R. Winget and K. Winget, editors, *CUG 1995 Spring Proceedings*, pages 303–308, Eagen, MN, 1995. Cray User Group, Inc.

[18] C. Fonlupt, P. Marquet, and J.-L. Dekeyser. Analysis of synchronous dynamic load balancing algorithms. In E. H. D'Hollander, G. R. Joubert, F. J. Peters, and D. Trystram, editors, *Parallel Computing: State-of-the-Art and Perspectives, Proceedings of the Conference ParCo'95, 19-22 September 1995, Ghent, Belgium*, volume 11 of *Advances in Parallel Computing*, pages 455–462, Amsterdam, 1996. Elsevier, North-Holland.

[19] W. Gropp, M. Snir, B. Nitzberg, and E. Lusk. *MPI: The Complete Reference*. MIT Press, Boston, 2nd edition, 1998.

[20] E. C. Hunke and J. K. Dukowicz. An elastic-viscous-plastic model for sea ice dynamics. *J. Phys. Oceanogr.*, 27:1849–1867, 1997.

[21] P. W. Jones, P. H. Worley, Y. Yoshida, J. B. White III, and J. Levesque. Practical performance portability in the Parallel Ocean Program (POP). *Concurrency and Computation: Practice and Experience*, 17:1317–1327, 2005.

[22] J. T. Kiehl, J. J. Hack, G. Bonan, B. A. Boville, D. L. Williamson, and P. J. Rasch. The National Center for Atmospheric Research Community Climate Model: CCM3. *J. Climate*, 11:1131–1149, 1998.

[23] J. Li, W. Liao, A. Choudhary, R. Ross, R. Thakur, W. Gropp, R. Latham, A. Siegel, B. Gallaghar, and M. Zingale. Parallel netCDF: A high-performance scientific I/O interface. In *Proceedings of the ACM/IEEE Conference on High Performance Networking and Computing (SC03)*, pages 15–21, Los Alamitos, CA, November 2003. IEEE Computer Society Press.

[24] S.-J. Lin. A "vertically Lagrangian" finite-volume dynamical core for global models. *Mon. Wea. Rev.*, 132:2293–2307, 2004.

[25] W. H. Lipscomb. personal communication.

[26] W. H. Lipscomb and E. C. Hunke. Modeling sea ice transport using incremental remapping. *Mon. Wea. Rev.*, 132:1341–1354, 2004.

[27] R. D. Loft, S. J. Thomas, and J. M. Dennis. Terascale spectral element dynamical core for atmospheric general circulation models. In *Proceedings of the ACM/IEEE Conference on High Performance Networking and Computing (SC01)*, pages 10–16, Los Alamitos, CA, November 2001. IEEE Computer Society Press.

[28] M. E. Maltrud and J. L. McClean. An eddy resolving global 1/10 degree ocean simulation. *Ocean Modelling*, 8:31–54, 2005.

[29] A. Mirin and W. B. Sawyer. A scalable implemenation of a finite-volume dynamical core in the Community Atmosphere Model. *International Journal of High Performance Computing Applications*, 19:203–212, 2005.

[30] G. E. Moore. Cramming more components onto integrated circuits. *Electronics Magazine*, 38, 1965.

[31] R. J. Murray. Explicit generation of orthogonal grids for ocean models. *J. Comp. Phys.*, 126:251–273, 1996.

[32] R. W. Numrich and J. K. Reid. Co-Array Fortran for parallel programming. *ACM Fortran Forum*, 17:1–31, 1998.

[33] W. Putman, S. J. Lin, and B. Shen. Cross-platform performance of a portable communication module and the NASA finite volume general circulation model. *International Journal of High Performance Computing Applications*, 19:213–224, 2005.

[34] D. A. Randall, T. D. Ringler, R. P. Heikes, P. W. Jones, and J. Baumgardner. Climate modeling with spherical geodesic grids. *Computing in Science and Engg.*, 4:32–41, 2002.

[35] R. D. Smith, J. K. Dukowicz, and R. C. Malone. Parallel ocean general circulation modeling. *Phys. D*, 60:38–61, 1992.

[36] R. D. Smith and P. Gent. Reference manual for the Parallel Ocean Program (POP), ocean component of the Community Climate System Model (CCSM2.0). Technical Report LAUR-02-2484, Los Alamos National Laboratory, Los Alamos, NM, 2002.

[37] R. D. Smith, S. Kortas, and B. Meltz. Curvilinear coordinates for global ocean models. Technical Report LAUR-95-1146, Los Alamos National Laboratory, Los Alamos, NM, 1995.

[38] S. Thakkar and T. Huff. The Internet streaming SIMD extensions. *Intel Technology Journal*, Q2:1–8, 1999.

[39] J. Tyler, J. Lent, A. Mather, and H. Nguyen. AltiVec: Bringing vector technology to the PowerPC processor family. In *Proceedings of the Performance, Computing and Communications Conference*, pages 437–444, 1999. IEEE International, February 10-12.

[40] United States Department of Energy (DOE), Office of Science. *A Science-Based Case for Large-Scale Simulation*. July 30, 2003. `http://www.pnl.gov/scales/`.

[41] W. Washington and C. Parkinson. *An Introduction to Three-Dimensional Climate Modeling*. University Science Books, Sausalito, CA, 2nd edition, 2005.

[42] D. L. Williamson and J. G. Olson. Climate simulations with a semi-Lagrangian version of the NCAR Community Climate Model. *Mon. Wea. Rev.*, 122:1594–1610, 1994.

[43] P. H. Worley and J. B. Drake. Performance portability in the physical parameterizations of the Community Atmosphere Model. *International Journal of High Performance Computing Applications*, 19:187–202, 2005.

Chapter 8

Towards Distributed Petascale Computing

Alfons G. Hoekstra

Section Computational Science, University of Amsterdam, Kruislaan 403, Amsterdam, The Netherlands, alfons@science.uva.nl

Simon Portegies Zwart

Section Computational Science and Astronomical Institute "Anton Pannekoek," University of Amsterdam, Kruislaan 403, Amsterdam, The Netherlands, spz@science.uva.nl

Marian Bubak

Section Computational Science, University of Amsterdam, Kruislaan 403, Amsterdam, The Netherlands, and AGH University of Science and Technology, Kraków, Poland, bubak@science.uva.nl

Peter M.A. Sloot

Section Computational Science, University of Amsterdam, Kruislaan 403, Amsterdam, The Netherlands, sloot@science.uva.nl

8.1 Introduction

Recent advances in experimental techniques have opened up new windows into physical and biological processes on many levels of detail. The resulting data explosion requires sophisticated techniques, such as grid computing and collaborative virtual laboratories, to register, transport, store, manipulate, and share the data. The complete cascade from the individual components to the fully integrated multi-science systems crosses many orders of magnitude

147

in temporal and spatial scales. The challenge is to study not only the fundamental processes on all these separate scales, but also their mutual coupling through the scales in the overall multi-scale system, and the resulting emergent properties. These complex systems display endless signatures of order, disorder, self-organization and self-annihilation. Understanding, quantifying and handling this complexity is one of the biggest scientific challenges of our time [8].

In this chapter we will argue that studying such multi-scale multi-science systems gives rise to inherently hybrid models containing many different algorithms best serviced by different types of computing environments (ranging from massively parallel computers, via large-scale special purpose machines to clusters of PCs) whose total integrated computing capacity can easily reach the PFLOPS scale. Such hybrid models, in combination with the by now inherently distributed nature of the data on which the models "feed" suggest a distributed computing model, where parts of the multi-scale multi-science model are executed on the most suitable computing environment, and/or where the computations are carried out close to the required data (i.e., bring the computations to the data instead of the other way around).

Prototypical examples of multi-scale multi-science systems come from biomedicine, where we have data from virtually all levels between "molecule and man" and yet we have no models where we can study these processes as a whole. The complete cascade from the genome, proteome, metabolome, physiome to health constitutes multi-scale, multi-science systems, and crosses many orders of magnitude in temporal and spatial scales [19, 48]. Studying biological modules, their design principles, and their mutual interactions, through an interplay between experiments and modeling and simulations, should lead to an understanding of biological function and to a prediction of the effects of perturbations (e.g., genetic mutations or presence of drugs). [54].

A good example of the power of this approach, in combination with state-of-the-art computing environments, is provided by the study of the heart physiology, where a true multi-scale simulation, going from genes, to cardiac cells, to the biomechanics of the whole organ, is now feasible [43]. This "from genes to health" is also the vision of the Physiome project [26, 27], and the ViroLab [49, 55], where a multi-scale modeling and simulation of human physiology is the ultimate goal. The wealth of data now available from many years of clinical, epidemiological research and (medical) informatics, advances in high-throughput genomics and bioinformatics, coupled with recent developments in computational modeling and simulation, provides an excellent position to take the next steps towards understanding the physiology of the human body across the relevant 10^9 range of spatial scales (nm to m) and 10^{15} range of temporal scales, (μs to a human lifetime) and to apply this understanding to the clinic [5, 26]. Examples of multi-scale modeling are increasingly emerging (see for example, [15, 30, 32, 50]).

In Section 8.2 we will consider the grid as the obvious choice for a distributed computing framework, and we will then explore the potential of computational grids for petascale computing in Section 8.3. Section 8.4 presents the *Virtual Galaxy* as a typical example of a multi-scale multi-physics application, requiring distributed petaFLOPS computational power.

8.2 Grid Computing

The radical increase in the amount of IT-generated data from physical, living and social systems brings about new challenges related to the sheer size of data. It was this data "deluge" that originally triggered the research into grid computing [20, 24]. Grid computing is an emerging computing model that provides the ability to share data and instruments and to perform high throughput computing by taking advantage of many networked computers able to divide process execution across a distributed infrastructure.

As the grid is ever more frequently used for collaborative problem solving in research and science, the real challenge is in the development of new applications for a new kind of user through virtual organizations. Existing grid programming models are discussed in [6, 36].

Workflow is a convenient way of distribution of computations across a grid. A large group of composition languages have been studied for formal description of workflows [53] and they are used for orchestration, instantiation, and execution of workflows [37]. Collaborative applications are also supported by problem-solving environments which enable users to handle application complexity with web-accessible portals for sharing software, data, and other resources [56]. Systematic ways to building grid applications are provided through object-oriented and component technology, for instance the Common Component Architecture which combines the IDL-based distributed framework concept with requirements of scientific applications [4]. Some recent experiments with computing across grid boundaries, workflow composition of grid services with semantic description, and development of collaborative problem-solving environments are reported in [13, 42, 45]. These new computational approaches should transparently exploit the dynamic nature of the grid and virtualization of grid infrastructure. The challenges are efficient usage of knowledge for automatic composition of applications [46].

Allen et al. in [3] distinguish four main types of grid applications: (1) community-centric; (2) data-centric ; (3) computation-centric; and (4) interaction-centric. Data-centric applications are, and will continue to be the main driving force behind the grid. Community-centric applications are about bringing people or communities together, as, e.g., in the Access Grid, or in distributed collaborative engineering. Interaction-centric applications are those that require "a man in the loop", for instance in real-time computational steering of simulations or visualizations (as, e.g., demonstrated by the CrossGrid Project [45]).

In this chapter we focuss on computation-centric applications. These are the traditional high performance computing (HPC) and high throughput computing (HTC) applications which, according to Allen et al. [3] "turned to parallel computing to overcome the limitations of a single processor, and many of them will turn to Grid computing to overcome the limitations of a parallel computer." In the case of parameter sweep (i.e., HTC) applications this has already happened. Several groups have demonstrated successful parameter sweeps on a computational grid (see e.g., [51]). For tightly coupled HPC applications this is not so clear, as common wisdom is that running a tightly coupled parallel application in a computational grid (in other words, a parallel job actually running on several parallel machines that communicate with each other in a grid) is of no general use because of the large overheads that will be induced by communications between computing elements (see, e.g., [36]). However, in our opinion this certainly is a viable option, provided that the granularity of the computation is large enough to overcome the admittedly large communication latencies that exist between compute elements in a grid [25]. For PFLOPS scale computing we can assume that such required large granularity will be reached. Recently a computation-centric application running in parallel on compute elements located in Poland, Cyprus, Portugal, and the Netherlands was successfully demonstrated [23, 52].

8.3 Petascale Computing on the Grid

Execution of multi-scale multi-science models on computational grids will in general involve a diversity of computing paradigms. On the highest level functional decompositions may be performed, splitting the model in sub-models that may involve different types of physics. For instance, in a fluid-structure interaction application the functional decomposition leads to one part modeling the structural mechanics, and another part modeling the fluid flow. In this example the models are tightly coupled and exchange detailed information (typically, boundary conditions at each time step). On a lower level one may again find a functional decomposition, but at some point one encounters single-scale, single-physics sub-models, that can be considered as the basic units of the multi-scale multi-science model. For instance, in a multi-scale model for crack propagation, the basic units are continuum mechanics at the macroscale, modeled with finite elements, and molecular dynamics at the microscale [12]. Another example is provided by Plasma Enhanced Vapor Deposition where mutually coupled chemical, plasma physical and mechanical models can be distinguished [34]. In principle all basic modeling units can be executed on a single (parallel) computer, but they can also be distributed to several machines in a computational grid.

These basic model units will be large scale simulations by themselves. With an overall performance on the PFLOPS scale, it is clear that the basic units will also be running at impressive speeds. It is difficult to estimate the number of such basic model units. In the example of the fluid-structure interaction, there are two, running concurrently. However, in the case of, for instance, a multi-scale system modeled with the Heterogeneous Multiscale Method [16] there could be millions of instances of a microscopic model that in principle can execute concurrently (one on each macroscopic grid-point). So, for the basic model units we will find anything between single processor execution and massively parallel computations.

A computational grid offers many options of mapping the computations to computational resources. First, the basic model units can be mapped to the most suitable resources. So, a parallel solver may be mapped to massively parallel computers, whereas for other solvers special purpose hardware may be available, or just single PCs in a cluster. Next, a distributed simulation system is required to orchestrate the execution of the multi-scale multi-science models.

A computational grid is an appropriate environment for running functionally decomposed distributed applications. A good example of research and development in this area is the CrossGrid Project which aimed at elaboration of an unified approach to development of and running large scale interactive distributed, compute- and data-intensive applications, like biomedical simulation and visualization for vascular surgical procedures, a flooding crisis team decision support system, distributed data analysis in high energy physics, and air pollution combined with weather forecasting [45]. The following issues were of key importance in this research and will also play a pivotal role on the road towards distributed PFLOPS scale computing on the grid: porting applications to the grid environment; development of user interaction services for interactive start-up of applications, online output control, parameter study in the cascade, runtime steering, and online, interactive performance analysis based on online monitoring of grid applications. The elaborated CrossGrid architecture consists of a set of self-contained subsystems divided into layers of applications, software development tools and grid services [14].

Large-scale grid applications require online performance analysis. The application monitoring system, OCM-G, is a unique online monitoring system in which requests and response events are generated dynamically and can be toggled at runtime. This imposes much less overhead on the application and therefore can provide more accurate measurements for the performance analysis tool like G-PM, which can display (in the form of various metrics) the behavior of grid applications [7].

The High Level Architecture (HLA) fulfills many requirements of distributed interactive applications. HLA and the grid may complement each other to support distributed interactive simulations. The G-HLAM system supports for execution of legacy HLA federates on the grid without imposing major modifications of applications. To achieve efficient execution of HLA-based simulations on the grid, we introduced migration and monitoring mechanisms for such applications. This system has been applied to run two complex distributed interactive applications: N-body simulation and virtual bypass surgery [47].

In the next section we explore in some detail a prototypical application where all the aforementioned aspects need to be addressed to obtain distributed petascale computing.

8.4   The Virtual Galaxy

A grand challenge in computational astrophysics, requiring *at least* the PFLOPS scale, is the simulation of the physics of formation and evolution of large spiral galaxies like the Milky Way. This requires the development of a hybrid simulation environment to cope with the multiple timescales, the broad range of physics and the shear number of simulation operations [29, 39]. The nearby grand design spiral galaxy M31 in the constellation Andromeda, as displayed in Figure 8.1, provides an excellent bird's-eye view of how the Milky Way probably looks.

This section presents the virtual galaxy as a typical example of a multi-physics application that requires PFLOPS computational speeds, and has all the right properties to be mapped to distributed computing resources. We will introduce in some detail the relevant physics and the expected amount of computations (i.e., floating-point operations) needed to simulate a virtual galaxy. Solving Newton's equations of motion for any number of stars is a challenge by itself, but to perform this in an environment with the number of stars as in the galaxy, and over the enormous range of density contrasts and with the inclusion of additional chemical and nuclear physics, does not make the task easier. No single computer will be able to perform the resulting multitude of computations, and therefore it provides an excellent example for a hybrid simulation environment containing a wide variety of distributed hardware. We end this section with a discussion on how a virtual galaxy simulation could be mapped to a PFLOPS scale grid computing environment. We believe that the scenarios that we outline are prototypical and also apply to a multitude of other multi-science multi-scale systems, like the ones that were discussed in Sections 8.1 and 8.3.

FIGURE 8.1: (See color insert following page 18.) The Andromeda nebula, M31. A mosaic of hundreds of Earth-based telescope pointings were needed to make this image.

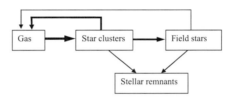

FIGURE 8.2: Schematic representation of the evolution of the gas content of the galaxy.

8.4.1 A multiphysics model of the Galaxy

The Galaxy today contains a few times 10^{11} the solar mass $(M_\odot)$ in gas and stars. The life cycle of the gas in the Galaxy is illustrated in Figure 8.2, where we show how gas transforms to star clusters, which again dissolve to individual stars. The ingredients for a self-consistent model of the Milky Way galaxy are based on these same three ingredients: the gas, the star clusters and the field stellar population. The computational cost and physical complexity for simulating each of these ingredients can be estimated based on the adopted algorithms.

8.4.1.1 How gas turns into star clusters

Stars and star clusters form from giant molecular clouds which collapse when they become dynamically unstable. The formation of stars and star clusters is coupled with the galaxy formation process. The formation of star clusters themselves has been addressed by many research teams and most of the calculations in this regard are a technical endeavor which is mainly limited by the lack of resources.

Simulations of the evolution of a molecular cloud up to the moment it forms stars are generally performed with adaptive mesh refinement and smoothed particles hydrodynamics algorithms. These simulations are complex, and some calculations include turbulent motion of the gas [10], solve the full magnetic hydrodynamic equations [57, 58], or include radiative transport [44]. All the currently performed dynamical cloud collapse simulations are computed with a relatively limited accuracy in the gravitational dynamics. We adopt the smoothed particle hydrodynamics methodology to calculate the gravitational collapse of a molecular cloud, as it is relatively simple to implement and has scalable numerical complexity. These simulation environments are generally based on the Barnes-Hut tree code [9] for resolving the self gravity between the gas or dust volume or mass elements, and have a $\mathcal{O}(n_{\mathrm{SPH}} \log n_{\mathrm{SPH}})$ time complexity [31]. Simulating the collapse of a molecular cloud requires at least $\sim 10^3$ SPH particles per star, a star cluster that eventually (after the simulation) consists of $\mathcal{O}(10^4)$ stars then requires about $n_{\mathrm{SPH}} \sim 10^7$ SPH particles. The collapse of a molecular cloud lasts for about $\tau_J \simeq 1/\sqrt{G\rho}$, which for a $10^4 \mathrm{M}_\odot$ molecular cloud with a size of $10\,\mathrm{pc}$ is about a million years. Within this time span the molecular cloud will have experienced roughly 10^4 dynamical timescales totaling the CPU requirements to about $\mathcal{O}(10^{11})$ floating-point operations for calculating the gravitational collapse of one molecular cloud.

8.4.1.2 The evolution of the individual stars

Once most of the gas is cleared from the cluster environment, an epoch of rather clean dynamical evolution mixed with the evolution of single stars and binaries starts. In general, star cluster evolution in this phase may be characterized by a competition between stellar dynamics and stellar evolution. Here we focus mainly on the nuclear evolution of the stars.

With the development of shell-based Henyé codes [17] the nuclear evolution of a single star for its entire lifetime requires about 10^9 floating-point operations [41]. Due to efficient step size refinement the performance of the algorithm is independent of the lifetime of the star; a $100\,\mathrm{M}_\odot$ star is as expensive in terms of compute time as a $1\,\mathrm{M}_\odot$ star. Adopting the mass distribution with which stars are born [33] about one in six stars requires a complete evolutionary calculation. The total compute time for evolving all the stars in the galaxy over its full lifetime then turns out to be about 10^{20} floating-point operations.

Most ($\gtrsim 99\%$) of all the stars in the Galaxy will not do much apart from burning their internal fuel. To reduce the cost of stellar evolution we can therefore parameterize the evolution of such stars. Excellent stellar evolution prescriptions at a fraction of the cost ($\lesssim 10^4$ floating-point operations) are available [18, 28], and could be used for the majority of stars (which is also what we adopted in Section 8.4.2).

8.4.1.3 Dynamical evolution

When a giant molecular cloud collapses one is left with a conglomeration of bound stars and some residual gas. The latter is blown away from the cluster by the stellar winds and supernovae of the young stars. The remaining gas depleted cluster may subsequently dissolve in the background on a time-scale of about 10^8 years.

The majority (50–90%) of star clusters which are formed in the Galaxy dissolve due to the expulsion of the residual gas [11, 22]. Recent reanalysis of the cluster population of the Large Magellanic cloud indicates that this process of *infant mortality* is independent of the mass of the cluster [35]. Star clusters that survive their infancy engage in a complicated dynamical evolution which is quite intricately coupled with the nuclear evolution of the stars [59].

The dynamical evolution of a star cluster is best simulated using direct N-body integration techniques, like NBODY4 [1, 2] or the `starlab` software environment [59].

For dense star clusters the compute time is completely dominated by the force evaluation. Since each star has a gravitational pull at all other stars this operation scales with $\mathcal{O}(N^2)$ for one dynamical time step. The good news is that the large density contrast between the cluster central regions and its outskirts can cover 9 orders of magnitude, and stars far from the cluster center are regularly moving whereas central stars have less regular orbits [21]. By applying smart time-stepping algorithms one can reduce the $\mathcal{O}(N^2)$ to $\mathcal{O}(N^{4/3})$ without loss of accuracy [40]. In fact one actually gains accuracy since taking many unnecessary small steps for a regularly integrable star suffers from numerical round-off.

The GRAPE-6, a special purpose computer for gravitational N-body simulations, performs dynamical evolution simulations at a peak speed of about 64 TFLOPS [38], and is extremely suitable for large-scale N-body simulations.

8.4.1.4 The galactic field stars

Stars that are liberated by star clusters become part of the galactic tidal field. These stars, like the Sun, orbit the galactic center in regular orbits. The average time-scale for one orbital revolution for a field star is about 250 Myr. These regularly orbiting stars can be resolved dynamically using a relatively unprecise N-body technique, we adopt here the $\mathcal{O}(N)$ integration algorithm which we introduced in Section 8.4.1.1.

In order to resolve a stellar orbit in the galactic potential about 100 integration time steps are needed. Per galactic crossing time (250 Myr) this code then requires about 10^6 operations per star, resulting in a few times $10^7 N$ floating-point operations for simulating the field population. Note that simulating the galactic field population is a trivially parallel operation, as the stars hover around in their self-generated potential

8.4.2 A performance model for simulating the galaxy

Next we describe the required computer resources as a function of lifetime of a virtual galaxy. The model is relatively simple and the embedded physics is only approximate, but it will give an indication on what type of calculation is most relevant in what state of the evolution of the galaxy.

According to the model we start the evolution of the galaxy with amorphous gas. We subsequently assume that molecular clouds are formed with power-law mass function with an index of -2 between $10^3\,M_\odot$ and $10^7\,M_\odot$, with distribution in time which is flat in $\log t$. We assume that the molecular cloud lives for between 10 Myr and 1 Gyr (with an equal probability between these moments). The star formation efficiency is 50%, and the cluster has an 80% change to dissolve within 100 Myr (irrespective of the cluster mass). The other 20% of clusters dissolve on a timescale of about $t_{\rm diss} \sim 10\sqrt{R^3 M}$ Myr. During this period they lose mass at a constant rate. The field population is enriched with the same amount of mass.

The resulting total mass in molecular clouds, star clusters and field stars is presented in Figure 8.3. At an early age, the galaxy completely consists of molecular clouds. After about 10 Myr some of these clouds collapse to form star clusters and single stars, indicated by the rapidly rising solid (field stars) and dashed (star clusters) curves. The maximum number of star clusters is reached when the galaxy is about a Gyr old. The field population continues to rise to reach a value of a few times $10^{11}\,M_\odot$ at today's age of about 10 Gyr. By that time the total mass in star clusters has dropped to several $10^9\,M_\odot$, quite comparable with the observed masses of the field population and the star cluster content.

In Figure 8.4 we show the evolution of the amount of floating-point operations required to simulate the entire galaxy, as a function of its lifetime. The flop count along the vertical axis is given in units of numbers of floating-points operations per million years in galactic evolution. For example, to evolve the galaxy's population of molecular clouds from 1000 Myr to 1001 Myr requires about 10^{16} floating-point operations.

8.4.3 Petascale simulation of a virtual galaxy

From Figure 8.4 we see that the most expensive submodels in a virtual galaxy are the star cluster simulations, the molecular cloud simulations, and the field star simulations. In the following discussion we neglect the other components. A virtual galaxy model, viewed as a multi-scale multiphysics model, can then be decomposed as in Figure 8.5.

The by far most expensive operation is the star cluster computations. We have $O(10^4)$ star clusters, each cluster can be simulated independent of the others. This means that a further decomposition is possible, down to the individual cluster level. A single star cluster simulation, containing $O(10^4)$ stars, still requires computational speeds at the TFLOPS scale (see also below). The

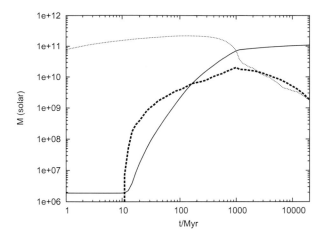

FIGURE 8.3: The evolution of the mass content in the galaxy via the simple model described in Section 8.4.2. The dotted curve gives the total mass in giant molecular clouds, the thick dashed curve in star clusters and the solid curve in field stars, which come from dissolved star clusters.

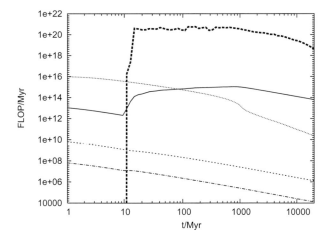

FIGURE 8.4: The number of floating-points operations expenditures per million years for the various ingredients in the performance model. The solid, thick short dashed and dotted curves are as in Figure 8.3. New in this figure are the two-dotted and dash-dotted lines near the bottom, which represent the CPU time needed for evolving the field star population (lower dotted curve) and dark matter (bottom curve).

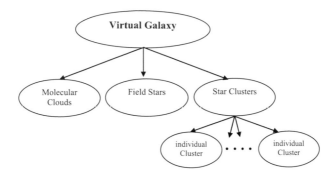

FIGURE 8.5: Functional decomposition of the virtual galaxy

cluster simulations require 10^{21} floating-point operations per simulated Myr of lifetime of the galaxy. the molecular clouds plus the field stars need, on average over the full lifetime of the Galaxy, 10^{15} floating-point operations per simulated Myr of lifetime, and can be executed on general purpose parallel machines.

A distributed petascale computing infrastructure for the virtual galaxy could consist of a single or two general-purpose parallel machines to execute the molecular clouds and fields stars at a sustained performance of 1 TFLOPS, and a distributed grid of special purpose Grapes to simulate the star clusters. We envision, for instance, 100-next generation GrapeDR systems*, each delivering 10 TFLOPS, providing a sustained 1 PFLOPS for the star cluster computations. We can now estimate the expected runtime on a virtual galaxy simulation on this infrastructure. In Table 8.1 we present the estimated wall-clock time needed for simulating the Milky Way galaxy, a smaller subset and a dwarf galaxy using the distributed petascale resource described above. Note that in the reduced galaxies the execution time goes linearly down with the reduction factor, which should be understood as a reduction of mass in the molecular clouds and a reduction of the total number of star clusters (but with the same amount of stars per star cluster).

With such a performance it will be possible to simulate the entire Milky Way galaxy for about 10 Myr which is an interesting timescale on which stars form, massive stars evolve and infant mortality of young newly born star clusters operates. Simulating the entire Milky Way galaxy on this important time-scale will enable us to study these phenomena with unprecedented detail.

At the same performance it will be possible to simulate part (1/10th) of the galaxy on a time-scale of 100 Myr. This time-scale is important for the

---

*Currently some 100 Grape6 systems, delivering an average performance of 100 GFLOPS are deployed all over the world.

TABLE 8.1: Estimated runtimes of the virtual galaxy simulation on a distributed petascale architecture as described in the main text.

Age	Milky Way Galaxy	Factor 10 reduction	Dwarf Galaxy (factor 100 reduction)
10 Myr	3 hour	17 min.	2 min.
100 Myr	3 year	104 days	10 days
1 Gyr	31 year	3 year	115 days
10 Gyr	320 year	32 year	3 year

evolution of young and dense star clusters, the major star formation mode in the galaxy.

Simulating a dwarf galaxy, like the Large Magellanic Cloud for its entire lifetime will become possible with a PFLOPS scale distributed computer. The entire physiology of this galaxy is largely not understood, as well as the intricate coupling between stellar dynamics, gas dynamics, stellar evolution and dark matter.

8.5 Discussion and Conclusions

Multi-scale multi-science modeling is the next (grand) challenge in computational science. Not only in terms of formulating the required couplings across the scales or between multi-science models, but also in terms of the sheer computational complexity of such models. The latter can easily result in requirements on the PFLOPS scale.

We have argued that simulating these models involves high-level functional decompositions, finally resulting in some collection of single-scale single-science sub-models, that by themselves could be quite large, requiring simulations on, e.g., massively parallel computers. In other words, the single-scale single-science sub-models would typically involve some form of high performance- or high throughput computing. Moreover, they may have quite different demands for computer infrastructure, ranging from supercomputers, via special purpose machines, to the single workstation. We have illustrated this by pointing to a few models from biomedicine and in more detail in the discussion on the virtual galaxy.

We believe that the grid provides the natural distributed computing environment for such functionally decomposed models. The Grid has reached a stage of maturity that in essence all the necessary ingredients needed to develop a PFLOPS scale computational grid for multi-scale multi-science simulations are available. Moreover, in a number of projects grid-enabled functionally decomposed distributed computing has been successfully demonstrated, using many of the tools that were discussed in Section 8.2.

Despite these successes the experience with computational grids is still relatively small. Therefore, a real challenge lies ahead in actually demonstrating the feasibility of grids for distributed petascale computing, and realizing grid-enabled problem-solving environments for multi-scale multi-science applications.

References

[1] S. J. Aarseth. From NBODY1 to NBODY6: The growth of an industry. *Publications of the Astronomical Society of the Pacific*, 111:1333–1346, 1999.

[2] S. J. Aarseth and M. Lecar. Computer simulations of stellar systems. *Annual Review of Astronomy and Astrophysics*, 13:1–88, 1975.

[3] G. Allen, T. Goodale, M. Russell, E. Seidel, and J. Shalf. Classifying and enabling grid applications. In F. Berman, G. Fox, , and A. J. G. Hey, editors, *Grid Computing, Making the Global Infrastructure a Reality*, chapter 23. John Wiley and Sons, 2003.

[4] R. Armstrong, G. Kumfert, L. Curfman McInnes, S. Parker, B. Allan, M. Sottile, T. Epperly, and T. Dahlgren. The CCA component model for high-performance scientific computing. *Concurr. Comput.: Pract. Exper.*, 18:215–229, 2006.

[5] N. Ayache, J.-P. Boissel, S. Brunak, G. Clapworthy, J. Fingberg, G. Lonsdale, A. Frangi, G. Deco, P. Hunter, P. Nielsen, M. Halstead, R. Hose, I. Magnin, F. Martin-Sanchez, P. Sloot, J. Kaandorp, A. Hoekstra, S. van Sint Jan, and M. Viceconti. Towards virtual physiological human: Multi-level modelling and simulation of the human anatomy and physiology, 2005. White paper.

[6] H. Bal, H. Casanova, J. Dongarra, and S. Matsuoka. Application tools. In I. Foster and C. Kesselman, editors, *The Grid2 - Blueprint for a New Computing Infrastructure*, chapter 24, pages 463–489. Morgan Kaufmann, 2004.

[7] B. Balis, M. Bubak, W. Funika, R. Wismueller, M. Radecki, T. Szepieniec, T. Arodz, and M. Kurdziel. Grid environment for on-line application monitoring and performance analysis. *Scientific Pogrammning*, 12:239–251, 2004.

[8] A. Barabasi. Taming complexity. *Nature Physics*, 1:68–70, 2005.

[9] J. Barnes and P. Hut. A hierarchical $O(N \log N)$ force-calculation algorithm. *Nature*, 324:446–449, 1986.

[10] M. R. Bate and I. A. Bonnell. The origin of the initial mass function and its dependence on the mean Jeans mass in molecular clouds. *Monthly Notices of the Royal Astronomical Society*, 356:1201–1221, 2005.

[11] C. M. Boily and P. Kroupa. The impact of mass loss on star cluster formation - II. Numerical N-body integration and further applications. *Monthly Notices of the Royal Astronomical Society*, 338:673–686, 2003.

[12] J. Q. Broughton, F. F. Abraham, N. Bernstein, and E. Kaxiras. Concurrent coupling of length scales: Methodology and application. *Phys. Rev. B*, 60:2391–2403, 1999.

[13] M. Bubak, T. Gubala, M. Kapalka, M. Malawski, and K. Rycerz. Workflow composer and service registry for grid applications. *Fut. Gen. Comp. Sys. Grid Computing*, 21:77–86, 2005.

[14] M. Bubak, M. Malawski, and K. Zając. Architecture of the grid for interactive applications. In P. M. A. Sloot et al., editor, *Proceedings of Computational Science - ICCS 2003 International Conference Melbourne, Australia and St. Petersburg, Russia, June 2003, Lecture Notes in Computer Science*, volume 2657, pages 207–213. Springer, 2003.

[15] P. F. Davies, J. A. Spaan, and R. Krams. Shear stress biology of the endothelium. *Ann Biomed Eng*, 33:1714–1718, 2005.

[16] W. E, B. Engquist, X. Li, W. Ren, and E. van den Eijnden. Heterogeneous multiscale methods: A review. *Commun. Comput. Phys.*, 2:367–450, 2007.

[17] P. Eggleton. *Evolutionary Processes in Binary and Multiple Stars*. Cambridge University Press (Cambridge, UK), 2006.

[18] P. P. Eggleton, M. J. Fitchett, and C. A. Tout. The distribution of visual binaries with two bright components. *Astrophysical Journal*, 347:998–1011, 1989.

[19] A. Finkelstein, J. Hetherington, L. Li, O. Margoninski, P. Saffrey, R. Seymour, and A. Warner. Computational challenges of system biology. *Computer*, 37:26–33, 2004.

[20] I. Foster, C. Kesselman, and S. Tuecke. The anatomy of the grid: Enabling scalable virtual organizations. *International Journal of High Performance Computing Applications*, 15:200–222, 2001.

[21] J. Gemmeke, S. Portegies Zwart, and C. Kruip. Detecting irregular orbits in gravitational N-body simulations. *ArXiv Astrophysics e-prints*, 2006.

[22] S. P. Goodwin. The initial conditions of young globular clusters in the Large Magellanic Cloud. *Monthly Notices of the Royal Astronomical Society*, 286:669–680, 1997.

[23] A. Gualandris, S. Portegies Zwart, and A. Tirado-Ramos. Performance analysis of direct N-body algorithms for astrophysical simulations on distributed systems. *Parallel Computing*, 2007. in press, (also: ArXiv Astrophysics e-prints astro-ph/0412206).

[24] A. J. G. Hey and A. E. Trefethen. The data deluge: An e-science perspective. In F. Berman, G. Fox, and A. J. G. Hey, editors, *Grid Computing, Making the Global Infrastructure a Reality*, chapter 36, pages 555–278. John Wiley and Sons, 2003.

[25] A. G. Hoekstra and P. M. A. Sloot. Introducing grid speedup gamma: A scalability metric for parallel applications on the grid. In P. M. A. Sloot, A. G. Hoekstra, T. Priol, A. Reinefeld, and M. T. Bubak, editors, *Advances in Grid Computing - EGC 2005*, volume 3470 of *Lecture Notes in Computer Science*. Springer, Berlin, Heidelberg, 2005.

[26] P. J. Hunter and T. K. Borg. Integration from proteins to organs: The physiome project. *Nature Reviews Molecular and Cell Biology*, 4:237–243, 2003.

[27] P. J. Hunter, W. W. Li, A. D. McCulloch, and D. Noble. Multiscale modeling: Physiome project standards, tools, and databases. *Computer*, 39:48–54, 2006.

[28] J. R. Hurley, O. R. Pols, and C. A. Tout. Comprehensive analytic formulae for stellar evolution as a function of mass and metallicity. *Monthly Notices of the Royal Astronomical Society*, 315:543–569, 2000.

[29] P. Hut. Dense stellar systems as laboratories for fundamental physics. In *ArXiv Astrophysics e-prints*, 2006.

[30] G. Iribe, P. Kohl, and D. Noble. Modulatory effect of calmodulin-dependent kinase II (CaMKII) on sarcoplasmic reticulum Ca2+ handling and interval-force relations: A modelling study. *Philos Transact A Math Phys Eng Sci.*, 364(2006):1107–1133, 2006.

[31] A. Kawai, J. Makino, and T. Ebisuzaki. Performance analysis of high-accuracy tree code based on the pseudoparticle multipole method. *The Astrophysical Journal Supplement Series*, 151:13–33, 2004.

[32] D. Kelly, L. Mackenzie, P. Hunter, B. Smaill, and D. A. Saint. Gene expression of stretch-activated channels and mechanoelectric feedback in the heart. *Clin Exp Pharmacol Physiol.*, 33:642–648, 2006.

[33] P. Kroupa, C. A. Tout, and G. Gilmore. The low-luminosity stellar mass function. *Monthly Notices of the Royal Astronomical Society*, 244:76–85, 1990.

[34] V. V. Krzhizhanovskaya, P. M. A. Sloot, and Y. E. Gorbachev. Grid-based simulation of industrial thin-film production. *Simulation: Transactions of the Society for Modeling and Simulation International*, 81:77–85, 2005. Special Issue on Applications of Parallel and Distributed Simulation.

[35] H. J. G. L. M. Lamers, M. Gieles, and S. F. Portegies Zwart. Disruption time scales of star clusters in different galaxies. *Astron. Astrophys.*, 429:173–179, 2005.

[36] C. Lee and D. Talia. Grid programming models: Current tools, issues, and directions. In F. Berman, G. C. Fox, and A. J. G. Hey, editors, *Grid Computing - Making a Global Infrastructure a Reality*, chapter 21, pages 555–578. John Wiley and Sons, 2003.

[37] B. Ludäscher, I. Altintas, C. Berkley, D. Higgins, E. Jaeger, M. Jones, E. A. Lee, J. Tao, and Y. Zhao. Scientific workflow management and the Kepler system. *Concurrency and Computation: Practice and Experience*, 18:1039–1065, 2006.

[38] J. Makino. Direct simulation of dense stellar systems with GRAPE-6. In S. Deiters, B. Fuchs, A. Just, R. Spurzem, and R. Wielen, editors, *ASP Conf. Ser. 228: Dynamics of Star Clusters and the Milky Way*, volume 87, 2001.

[39] J. Makino. GRAPE and Project Milkyway. *Journal of the Korean Astronomical Society*, 38:165–168, 2005.

[40] J. Makino and P. Hut. Performance analysis of direct N-body calculations. *The Astrophysical Journal Supplement Series*, 68:833–856, 1988.

[41] J. Makino and P. Hut. Bottlenecks in simulations of dense stellar systems. *Astrophysical Journal*, 365:208–218, 1990.

[42] M. Malawski, M. Bubak, M. Placzek, D. Kurzyniec, and V. Sunderam. Experiments with distributd component computing across grid boundaries. In *IEEE HPC-GECO/CompFrame*, pages 109–116, Paris, June 2006.

[43] D. Noble. Modeling the heart - from genes to cells to the whole organ. *Science*, 295:1678–1682, 2002.

[44] P. Padoan and Nordlund. The stellar initial mass function from turbulent fragmentation. *Astrophysical Journal*, 576:870–879, 2002.

[45] CrossGrid Project. http://www.crossgrid.org.

[46] K-WfGrid Project. http://www.kwfgrid.eu.

[47] K. Rycerz, M. Bubak, M. Malawski, and P. M. A. Sloot. A framework for HLA-based interactive simulations on the grid. *Simulation*, 81:67–76, 2005.

[48] P. M. A. Sloot, D. Frenkel, H. A. Van der Vorst, A. van Kampen, H. Bal, P. Klint, R. M. M. Mattheij, J. van Wijk, J. Schaye, H.-J. Langevelde, R. H. Bisseling, B. Smit, E. Valenteyn, H. Sips, J. B. T. M. Roerdink, and K. G. Langedoen. White paper on computational e-science, studying complex systems in silico, a national research initiative, December 2006.

[49] P. M. A. Sloot, A. Tirado-Ramos, I. Altintas, M. T. Bubak, and C. A. Boucher. From molecule to man: Decision support in individualized e-health. *IEEE Computer*, 39:40–46, 2006.

[50] P.M.A. Sloot, A.V. Boukhanovsky, W. Keulen, A. Tirado-Ramos, and C.A. Boucher. A Grid-based HIV expert system. *J Clin Monit Comput.*, 19:263–278, 2005.

[51] W. Sudholt, K. Baldridge, D. Abramson, C. Enticott, and S. Garic. Parameter scan of an effective group difference pseudopotential using grid computing. *New Generation Computing*, 22:125–135, 2004.

[52] A. Tirado-Ramos, A. Gualandris, and S. F. Portegies Zwart. Performance of a parallel astrophysical N-body solver on pan-European computational grids. In P. M. A. Sloot, A. G. Hoekstra, T. Priol, A. Reinefeld, and M. T. Bubak, editors, *Advances in Grid Computing - EGC 2005*, volume 3470 of *Lecture Notes in Computer Science*. Springer, Berlin, Heidelberg, 2005.

[53] W. M. P. van der Aalst and A. H. M. ter Hofstede. YAWL: Yet another workflow language. *Information Systems*, 30:245–275, 2005.

[54] B. Di Ventura, C. Lemerle, K. Michalodimitrakis, and L. Serrano. From in vivo to in silico biology and back. *Nature*, 443:527–533, 2006.

[55] Virolab Project. http:\\www.virolab.org.

[56] D.W. Walker and E. Houstis (guest editors). Complex problem-solving environments for grid computing. *Fut. Gen. Comp. Sys. Grid Computing*, 21:841–968, 2005.

[57] S. C. Whitehouse and M. R. Bate. The thermodynamics of collapsing molecular cloud cores using smoothed particle hydrodynamics with radiative transfer. *Monthly Notices of the Royal Astronomical Society*, 367:32–38, 2006.

[58] D. Zengin, E. R. Pekünlü, and E. Tigrak. Collapse of interstellar molecular clouds. In B. Uyaniker, W. Reich, and R. Wielebinski, editors, *The Magnetized Interstellar Medium*, pages 133–136, 2004.

[59] S. F. Portegies Zwart, S. L. W. McMillan, P. Hut, and J. Makino. Star cluster ecology - IV: Dissection of an open star cluster: photometry. *Monthly Notices of the Royal Astronomical Society*, 321(2):199–226, 2001.

Chapter 9

Biomolecular Modeling in the Era of Petascale Computing

Klaus Schulten

Beckman Institute, University of Illinois at Urbana-Champaign

James C. Phillips

Beckman Institute, University of Illinois at Urbana-Champaign

Laxmikant V. Kalé

Department of Computer Science, University of Illinois at Urbana-Champaign

Abhinav Bhatele

Department of Computer Science, University of Illinois at Urbana-Champaign

9.1 Introduction

The structure and function of biomolecular machines are the foundation on which living systems are built. Genetic sequences stored as DNA translate into chains of amino acids that fold spontaneously into proteins that catalyze chains of reactions in the delicate balance of activity in living cells. Interactions with water, ions, and ligands enable and disable functions with the twist of a helix or rotation of a side chain. The fine machinery of life at the molecular scale is observed clearly only when frozen in crystals, leaving the exact mechanisms in doubt. One can, however, employ molecular dynamics simulations to reveal the molecular dance of life in full detail. Unfortunately, the stage provided is small and the songs are brief. Thus, we turn to petascale parallel computers to expand these horizons.

Biomolecular simulations are challenging to parallelize. Typically, the molecular systems to be studied are not very large in relation to the available memory on computers: they contain ten thousand to a few million atoms. Since the size of basic protein and DNA molecules to be studied is fixed, this number does not increase in size significantly. However, the number of time steps to be simulated is very large. To simulate a microsecond in the life of a biomolecule, one needs to simulate a billion time steps. The challenge posed by biomolecules is that of parallelizing a relatively small amount of computation at each time step across a large number of processors, so that billions of time steps can be performed in a reasonable amount of time. In particular, an important aim for science is to effectively utilize the machines of the near future with tens of petaflops of peak performance to simulate systems with just a few million atoms. Some of these machines may have over a million processor cores, especially those designed for low power consumption. One can then imagine the parallelization challenge this scenario poses.

NAMD [15] is a highly scalable and portable molecular dynamics (MD) program used by thousands of biophysicists. We show in this chapter how NAMD's parallelization methodology is fundamentally well-suited for this challenge, and how we are extending it to achieve the goals of scaling to petaflop machines. We substantiate our claims with results on large current machines like IBM's Blue Gene/L and Cray's XT3. We also talk about a few biomolecular simulations and related research being conducted by scientists using NAMD.

9.2   NAMD Design

The design of NAMD rests on a few important pillars: a (then) novel strategy of hybrid decomposition, supported by dynamic load balancing, and adaptive overlap of communication with computation across modules, provided by the Charm++ runtime system [11].

9.2.1   Hybrid decomposition

The current version of NAMD is over ten years old. It has withstood the progress and changes in technology over these ten years very well, mainly because of its from-scratch, future-oriented, and migratable-object-based design. Prior to NAMD, most of the parallel MD programs for biomolecular simulations were extensions of (or based on) their preexisting serial versions [2, 21]. It was reasonable then to extend such programs by using a scheme such as atom decomposition (where atoms were partitioned based on their static atom numbers across processors). More advanced schemes were proposed [16, 8] that used force decomposition, where each processor was responsible for a square section of the $N \times N$ interaction matrix, where N is the total number of atoms.

In our early work on NAMD, we applied isoefficiency analysis [6] to show that such schemes were inherently unscalable: with an increasing number of processors, the proportion of communication cost to computation cost increases even if one were to solve a larger problem. For example, the communication-to-computation ratio for the force decomposition schemes of [16, 8] is of order $\sqrt{P}$, independent of N, where P is the number of processors. We showed that spatial decomposition overcomes this problem, but suffers from load balance issues.

At this point, it is useful to state the basic structure of a MD program: the forces required are those due to electrostatic and van der Waals interactions among all atoms, as well as forces due to bonds. A naïve implementation of the force calculation will lead to an $O(N^2)$ algorithm. Instead, for periodic systems, one uses an $O(N \log N)$ algorithm based on three-dimensional (3-D) fast Fourier transforms (FFTs) called the particle mesh Ewald (PME) method, in conjunction with explicit calculation of pairwise forces for atoms within a cutoff radius r_c. This suggests a spatial decomposition scheme in which atoms are partitioned into boxes of a size slightly larger than r_c. The extra margin is to allow atoms to be migrated among boxes only after multiple steps. It also facilitates storing each hydrogen atom on the same processor that owns its "mother" atom — recall that a hydrogen atom is bonded to only one other atom.

NAMD [10, 9] extends this idea of spatial decomposition, used in its early version in 1994, in two ways: first, it postulates a new category of objects called the *compute* objects. Each compute object is responsible for calculating interactions between a pair of cubical cells (actually brick-shaped cells, called *patches* in NAMD). This allows NAMD to take advantage of Newton's third law easily, and creates a large supply of work units (the compute objects) that an intelligent load balancer can assign to processors in a flexible manner. The Charm++ system, described in Chapter 20, is used for this purpose. As we will show later, it also helps to overlap communication and computation adaptively, even across multiple modules. As our 1998 paper [10] states, "the compute objects may be assigned to any processor, regardless of where the associated patches are assigned." The strategy is a hybrid between spatial and force decomposition. Recently, variations of this hybrid decomposition idea have been used by the programs Blue Matter [3] and Desmond [1] and a proposed scheme by M. Snir [19], and it has been called evocatively the "neutral territory method" [1]. Some of these methods are clever schemes that statically assign the computation of each pair of atoms to a specific processor, whereas NAMD uses a dynamic load-balancing strategy that should be superior due to its adaptive potential (see Section 9.2.2).

NAMD allows spatial decomposition of atoms into boxes smaller than the cutoff distance. In particular, it allows each dimension of a box to be $1/2$ or

1/3 of the cutoff radius plus the margin mentioned above. This allows more parallelism to be created when needed. Note that when each dimension of the cell is halved, the number of patches increases eightfold. But since each patch now must interact with patches two-away from it in each dimension (to cover the cutoff distance), a set of $5 \times 5 \times 5$ compute objects must now access its atoms. Accounting for double counting of each compute and for self-compute objects, one gets a total of $8 \times 63/14$ more work units to balance across processors. Note further that these work units are highly variable in their computation load: those corresponding to pairs of patches that share a face are the heaviest (after self-computation objects) and those corresponding to patches that are two hops away along *each* dimension have the least load, because many of their atom-pairs are beyond the cutoff distance for explicit calculation. Early versions of NAMD, in 1998, restricted us to either use full-size patches, or 1/8th-size patches (or 1/27th-size patches, which were found to be inefficient). More recent versions have allowed a more flexible approach: along each dimension, one can use a different decomposition. For example, one can have a two-away X and Y scheme, where the patch size is halved along the X and Y dimensions but kept the same (i.e., $r_c + margin$) along the Z dimension.

9.2.2   Dynamic load balancing

NAMD uses measurement-based load-balancing capabilities provided by the Charm++ runtime [23]. The runtime measures the load of each compute object and each processor during a few instrumented iterations and then assigns objects to processors based on the collected information. After the first load-balancing step, many computes are migrated to under-loaded processors because the initial assignment of computes to processors is arbitrary and as a result suboptimal. The subsequent load-balancing decisions, which use a refinement-based strategy, tend to minimize the number of migrations. This serves to keep communication volume in check and does not break the runtime's assumption of predictability of load.

On machines such as Blue Gene/L, the load balancer also uses knowledge of the three-dimensional (3-D) torus interconnect to minimize the average number of hops traveled by all communicated bytes, thus minimizing contention in the network. While doing the initial mapping of cells to processors, the runtime uses a scheme similar to orthogonal recursive bisection (ORB) [13]. The 3-D torus of processors is divided recursively until each cell can be assigned a processor and then the 3-D simulation box of cells is mapped onto the torus. In subsequent load-balancing steps, the load balancer tries to place the computes on under-loaded processors near the cells, with which this compute will interact.

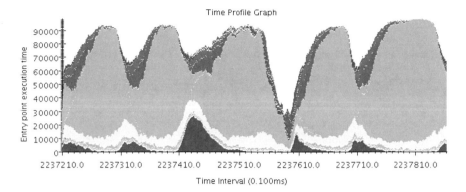

FIGURE 9.1: Time profile of NAMD running ApoA1 benchmark on 1024 processors of Blue Gene/L for five timesteps. Shades of gray show different types of calculations overlapping.

9.3 Petascale Challenges and Modifications

When NAMD was designed over ten years ago [14], million-processor machines were beyond the imagination of most people. Yet, by virtue of its parallel design, NAMD has demonstrated good scaling up to thousands of processors. As we moved to terascale machines (typically having tens of thousands of processors), NAMD faced a few challenges to maintain scalability and high efficiency.

The emergence of Blue Gene/L (which has only 256 MB of memory per processor) posed the problem of using a limited amount of memory for the initial startup (loading the molecular structure information), the actual computation, and load balancing. During startup, the molecular structure is read from a file on a single node and then replicated across all nodes. This is unnecessary and limits our simulations to about 100,000 atoms on Blue Gene/L. Making use of the fact that there are some common building blocks (amino acids, lipids, water) from which biomolecular simulations are assembled and their information need not be repeated, this scheme has been changed. Using a compression scheme, we can now run million atom simulations on the Blue Gene/L as we will see in Section 9.3.1.

The other major obstacle to scaling to large machines was the previous implementation of the particle mesh Ewald (PME) method. The PME method uses 3-D fast Fourier transforms (FFTs), which were implemented via a one-dimensional (1-D) decomposition. This limited the number of processors that could be used for this operation to a few hundred depending upon the number of planes in the grid. To overcome this limitation, a commonly used two

TABLE 9.1:   Benchmarks and their simulation sizes

Molecular System	Atom Count	Cutoff (Å)	Simulation Cell (Å³)	Time step (fs)
IAPP	5,570	12	46.70 × 40.28 × 29.18	2
ApoA1	92,224	12	108.86 × 108.86 × 77.76	1
STMV	1,066,628	12	216.83 × 216.83 × 216.83	1

dimensional (2-D) decomposition of the grid into pencils is now used. This has led to higher efficiency on large numbers of processors and has also helped to better overlap the FFT with other operations as can be seen in Figure 9.1.

This figure has been generated using the performance analysis tool in the Charm++ framework called Projections. For each 100 μs time interval (along the X-axis), the figure shows the execution time of each function added across all 1024 processors. Various shades of light gray consuming most of the graph represent the compute work. The black peaks at the bottom represent patch integration and the deep gray bordering the hills represents communication. The area in black in the valley in the center represents the PME computation, which overlaps well with other functions.

9.3.1   Current performance

The performance of NAMD on various platforms substantiates the claims made in the previous sections. NAMD has shown excellent scaling to thousands of processors on large parallel machines like the Blue Gene/L and Cray XT3. The benchmarks used for results presented are shown in Table 9.1. These molecular systems are representative of almost all sizes of the simulations interesting to biophysicists. They range from a few thousand atoms to millions of atoms.

The Blue Gene/L (BG/L) machine at IBM T. J. Watson has 20,480 nodes. Each node contains two 700 MHz PowerPC 440 cores and has 512 MB of memory shared between the two. The machine can be operated in coprocessor mode or virtual node mode. In the coprocessor mode, we use only one processor on each node for computation. In the virtual node mode, both processors on each node are used for computation. The nodes on BG/L are connected in a 3-D torus. The Cray XT3 at the Pittsburgh Supercomputing Center (called BigBen) has 2068 compute nodes, each of which has two 2.6 GHz AMD Opteron processors. The two processors on a node share 2 GB of memory. The nodes are connected into a 3-D torus by a custom C-star interconnect.

Figures 9.2(a) and 9.2(b) show NAMD scaling to 32,768 processors of the Blue Gene/L machine and to 4,000 processors of Cray XT3. Different techniques are at work together as we run on machines with large numbers of processors. At some point on the plots, depending on the atoms per processors

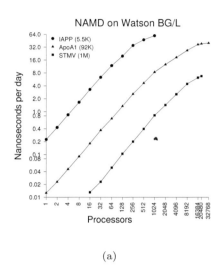

(a)

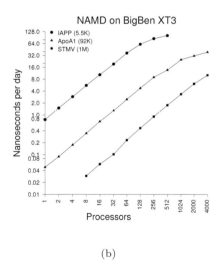

(b)

FIGURE 9.2: Performance of NAMD on Blue Gene/L and Cray XT3.

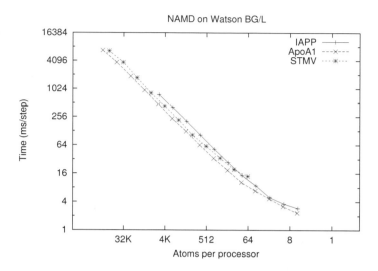

FIGURE 9.3: Time per step plotted against ratio of atoms to processors for NAMD running on BG/L.

and the machine, we switch from 1-AwayX to 2-AwayX and then 2-AwayXY decomposition of the patches. We also shift from 1-D decomposition of grids to 2-D for the PME computation when required. These decisions are automated so as to relieve the user of the burden of identifying optimal configuration parameters.

9.3.2   Performance on future petascale machines

To model running large molecular systems (many millions of atoms) on petascale machines with millions of processors, we plotted number of atoms per processor versus time step for different molecular systems. As can be seen in Figure 9.3 we get similar performance for a given ratio of number of atoms to processors for all the three benchmarks (which are quite varied in their sizes).

This plot suggests that NAMD will perform well for larger sized molecular systems on new petascale machines. For example, consider a 100-million atom molecular system which we wish to run on a hypothetical petascale machine consisting of 5-million processors. This gives us an atom-to-processor ratio of 20, which is within the regime presented in the above plot. The fraction of CPU cycles spent of FFT/PME increases as $N \log N$ as the number of atoms (N) increases, but this is a relatively small effect. We also validated these conclusions using our BigSim performance prediction framework [24] for some petascale designs.

9.3.3 Acceleration co-processors

There is currently excitement about the potential of heterogeneous clusters in which the bulk of computation is off-loaded to a specialized (or at least less-flexible) but higher performance coprocessor. Examples include the Cell Broadband Engine processor, graphics processors (GPUs), field-programmable gate arrays (FPGAs), and the special-purpose MD-GRAPE. Although these application accelerators are capable of sustaining hundreds of gigaflops for well-suited application kernels, data transfer between the CPU and the coprocessor limits performance to perhaps a factor of ten over a traditional CPU core. Since accelerated nodes are likely to outrun interconnect bandwidth, their impact will be seen most on throughput-oriented clusters, while leadership-class petascale machines will employ multicore processors for maximum code portability and a more balanced design. The parallel design of NAMD would be little-changed by the addition of acceleration coprocessors.

9.4 Biomolecular Applications

Scientifically interesting simulations of biomolecular systems currently range from ten thousand to a few million atoms, while future simulations may extend to 100,000,000 atoms. Progress will be on two fronts: supporting simulations of larger systems and increasing simulation rates on more powerful machines as they appear. For smaller systems the length of simulation that can be obtained is limited by latency and serial bottlenecks (strong scaling), while for larger simulations the size of the available machine limits performance (weak scaling). Table 9.2 summarizes expected simulation capabilities for NAMD running on the latest leadership-class hardware. Values for years 2004 and 2006 reflect capabilities already achieved. All simulations use a 12 Å cutoff, PME full electrostatics, and 1 femtosecond time steps. Achieving simulation rates higher than 100 nanoseconds/day would require significant improvement in network latencies, which is not easily foreseen in the next 5 years. Examples of biomolecular simulations in each size range are given in the table, illustrated in Figure 9.4, and described below.

9.4.1 Aquaporins

Water constitutes about 70% of the mass of most living organisms. Regulation of water flow across cell membranes is critical for maintaining fluid balance within the cell. The transportation of water in and out of a cell is mediated by a family of membrane proteins named aquaporins (AQPs), which are widely distributed in all domains of life. Through modulating water permeability of cellular membranes, AQPs play a crucial role in water

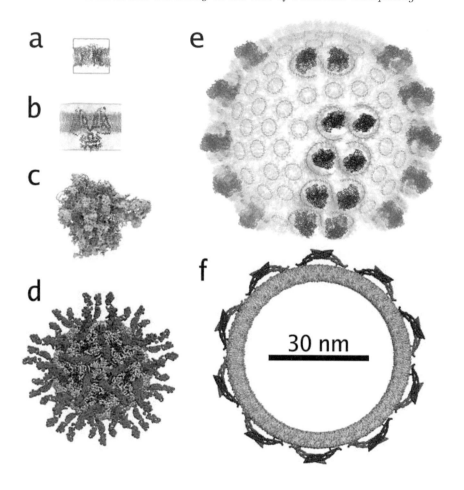

FIGURE 9.4: (See color insert following page 18.) Example biomolecular simulations: (a) aquaporin in membrane with solvent, (b) potassium channel in membrane with solvent, (c) ribosome, (d) poliovirus with cell surface receptors, (e) photosynthetic chromatophore, (f) BAR domain vesicle cross section.

homeostasis of living cells. In the human body, there are at least 11 different AQPs, whose physiological importance is reflected in the many pathophysiological situations associated with their absence/malfunction. For example, cataracts and diabetes insipidus have been linked to the impaired functions of AQP0 and AQP2, respectively. A particularly intriguing property of AQPs is their ability to block protons while allowing water to pass. In the past few

TABLE 9.2: Forecast of production simulation capabilities

Size	100 K atoms	1 M atoms	10 M atoms	100 M atoms
e.g.	aquaporin, potassium chan.	STM virus, ribosome	poliovirus	chromatophore, BAR dom. vesicle
2004	4 ns/day			
	500-core Alpha			
2006	10 ns/day	4 ns/day		
	500-core XT3	2000-core XT3		
2008	100 ns/day	10 ns/day	1 ns/day	
	5,000-core machine			
2010	100 ns/day	100 ns/day	10 ns/day	1 ns/day
	50,000-core machine			
2012	100 ns/day	100 ns/day	100 ns/day	10 ns/day
	500,000-core machine			

years, MD simulations [20] have contributed significantly to the understanding of this unique property of AQPs, and also of the molecular basis of their function and selectivity. A single solvated and membrane-embedded AQP tetramer simulation [22] comprises ~100,000 atoms.

9.4.2 Potassium channels

Ions crossing potassium channels are responsible for the generation and spread of electrical signals in the nervous system. A number of high-resolution crystal structures of potassium channels have been resolved over the last decade, recognized in part by the 2003 Nobel prize, awarded to Dr. MacKinnon for his pioneering work on structure-function relationships of these channels. However, how potassium channels dynamically transit between open, inactive, or conductive states upon application of voltage remains highly debated. Molecular dynamics simulations seem to be an ideal tool to tackle this question, but relevant gating events occur on the millisecond timescale, while current MD simulations only reach nanosecond timescales. Moreover, all-atom descriptions are required to faithfully model the behavior of the channel, e.g., its ion selectivity. A system of 350,000 atoms containing one potassium channel, a patch of membrane, water, and ions is being simulated already at 100 ns per month on 512 Cray XT3 CPUs [12]. To study channel gating, however, a tenfold improvement in simulation speed (1 μs per month) is required.

9.4.3   Viruses

Satellite Tobacco Mosaic Virus (STMV), one of the smallest and simplest viruses known, consists of a ball-shaped RNA genome enclosed in an icosahedral capsid composed of 60 identical protein units; the complete particle is roughly 17 nm in diameter. Although small for a virus, the complete simulation of STMV immersed in a water box with ions contains about one million atoms. Based on the results of the first simulation of the complete STMV particle [5], researchers were able to propose a possible assembly pathway for the virus. Further efforts on this project focus on the disassembly of STMV, which is known to be mediated by a change in pH (this holds for many other viruses), although the exact mechanism is unclear.

The poliovirus is larger and more complex than STMV (the polio capsid is about 30 nm in diameter and composed of 240 protein units), and the disassembly is believed to be triggered by contact between the capsid and host cell membrane and receptor proteins. Structures of the poliovirus itself and of the virus in complex with the receptor are available, although structure of the receptor is known only at a low resolution of $\sim$10 Å. Researchers have already developed a homology model of the poliovirus receptors and started MD simulations of a single capsid-receptor bundle ($\sim$250,000 atoms). The systems intended for further simulations, focusing on capsid disassembly, include a portion of the capsid in contact with a membrane (about 3 million atoms) and the complete poliovirus capsid ($\sim$10 million atoms). Further, a portion of the capsid in contact with a membrane will be simulated (up to more than 3 million atoms), to elucidate the role of the cellular membrane in the opening of the capsid and release of the genome. Finally, a simulation of the complete poliovirus capsid might be necessary for investigation of how the disassembly proceeds over the surface of the whole virus, which would require building a system consisting of about 10 million atoms.

9.4.4   Ribosome

The translation of genetic information into protein sequences is essential for life. At the core of the translation process lies the ribosome, a 2.5–4.5 MDa ribonucleoprotein complex where protein synthesis takes place. The ribosome is not only interesting because of its fundamental role in the cell, it is also a major target for drug discovery and design. Many antibiotics in clinical use block protein synthesis in the bacterial ribosome. With the emergence of high-resolution structures of the ribosome complexed with antibiotics, it has become clear that chemically diverse antibiotics target only a few ribosomal sites [17]. Structure-based drug design targeting these specific sites is an attractive option for discovering new antibiotics [4]. The Sanbonmatsu team at Los Alamos National Laboratory performed ground-breaking large-scale (2,640,000 atoms) all-atom MD simulations of the entire ribosome beginning in 2003. These simulations, performed with NAMD, discovered a corridor of 20 universally conserved ribosomal RNA bases interacting with the tRNA during accommodation. The study was published in 2005 and also demonstrated the feasibility of simulating conformational changes in multimillion-atom molecular machines using NAMD [18].

9.4.5   Chromatophore

One of the most fundamental processes for life on Earth is the transformation of light energy into the synthesis of ATP. This transformation is achieved through different specialized organelles, one such organelle being the chromatophore of the purple bacterium *Rhodobacter sphaeroides*. Chromatophores are sheet-like or bulb-like indentations of the bacterial plasma membrane. The chromatophore contains six types of proteins: about twenty photosynthetic reaction centers, about twenty light-harvesting complexes 1 (LH1), about 150 light harvesting complexes 2 (LH2), about ten bc1 complexes, about five cytochrome c2s, and usually one ATP synthase. These proteins are all individually structurally known, though not all from the same species. The chromatophore with its 200 proteins carries out a cardinal function in the bacterium, the absorption of sunlight by about 4,000 chlorophylls (contained in LH1 and LH2 along with 1,300 carotenoids) and the transformation of its energy into the synthesis of adenosine-triphosphate (ATP) from adenosine-diphosphate (ADP). The entire chromatophore model, an archetypal example of systems studied in structural systems biology, consists of more than 200 proteins in a $(90 \text{ nm})^3$ system containing about 70 million atoms.

9.4.6   BAR domain vesicle

Proteins containing BAR domains play an important role in essential cellular processes (such as vesicle endocytosis at synaptic nerve terminals) by inducing or sensing membrane curvature. The U.S. National Science Foundation (NSF) solicitation *Leadership-Class System Acquisition—Creating a Petascale Computing Environment for Science and Engineering* provides the following model problem, involving protein BAR domains, for the proposed machine:

> A molecular dynamics (MD) simulation of curvature-inducing protein BAR domains binding to a charged phospholipid vesicle over 10 ns simulation time under periodic boundary conditions. The vesicle, 100 nm in diameter, should consist of a mixture of dioleoylphosphatidylcholine (DOPC) and dioleoylphosphatidylserine (DOPS) at a ratio of 2:1. The entire system should consist of 100,000 lipids and 1,000 BAR domains solvated in 30 million water molecules, with NaCl also included at a concentration of 0.15 M, for a total system size of 100 million atoms. All system components should be modeled using the CHARMM27 all-atom

empirical force field. The target wall-clock time for completion
of the model problem using the NAMD MD package with the
velocity Verlet time-stepping algorithm, Langevin dynamics temperature coupling, Nose-Hoover Langevin piston pressure control,
the particle-mesh Ewald algorithm with a tolerance of 1.0e-6 for
calculation of electrostatics, a short-range (van der Waals) cut-off
of 12 Angstroms, and a time step of 0.002 ps, with 64-bit floating
point (or similar) arithmetic, is 25 hours. The positions, velocities, and forces of all the atoms should be saved to disk every 500
time steps.

The requirements for this simulation are similar to the requirements of
the chromatophore simulation described above. In both cases, systems containing hundreds of proteins and millions of atoms need to be simulated to
gain insights into biologically relevant processes. In order to accomplish such
projects, NAMD must be ported to petascale parallel computers.

9.5   Summary

Each time step in a biomolecular simulation is small, yet we need many million of them to simulate a small interval of time in the life of a biomolecule.
Therefore, one has to aggressively parallelize a small computation with high
parallel efficiency. The NAMD design is based on the concept of Charm++ migratable objects and is fundamentally adequate to scale to petascale machines—
this is indicated by the 1–2 milliseconds time per step achieved by NAMD for
some benchmarks, with ratio of atoms to processor in a similar range that
we expect to see on petascale machines. We have demonstrated scalability
to machines with tens of thousands of processors on biomolecular simulations
of scientific importance. Implementation strategies have been reworked to
eliminate obstacles to petascale through memory footprint reduction and fine
grained decomposition of the PME computation. All this has made the study
of large molecules such as the ribosome and entire viruses possible today and
will enable even larger and longer simulations on future machines.

9.6   Acknowledgments

Various students and staff of the Parallel Programming Laboratory and the
Theoretical and Computational Biophysics Group at the University of Illinois
have been of great assistance in the development of NAMD. Sameer Kumar,

at IBM, has helped significantly in improving NAMD performance on Blue Gene/L. The images of Figure 9.4 were created with the program VMD [7] by Anton Arkhipov, Jen Hsin, Fatemeh Khalili, Leonardo Trabuco, Yi Wang, and Ying Yin. Access to the Blue Gene/L machine at the IBM T. J. Watson Research Center was provided by Glenn J. Martyna and Fred Mintzer. Access to the Cray XT3 was provided by the Pittsburgh Supercomputing Center (PSC) via Large Resources Allocation Committee grant MCA93S028. We are grateful to Shawn T. Brown for helping us with the runs at PSC. This work was supported by the National Institutes of Health under grant P41-RR05969.

References

[1] K. J. Bowers, E. Chow, H. Xu, R. O. Dror, M. P. Eastwood, B. A. Gregersen, J. L. Klepeis, I. Kolossvary, M. A. Moraes, F. D. Sacerdoti, J. K. Salmon, Y. Shan, and D. E. Shaw. Scalable algorithms for molecular dynamics simulations on commodity clusters. In *SC '06: Proceedings of the 2006 ACM/IEEE Conference on Supercomputing*, New York, 2006. ACM Press.

[2] B. R. Brooks, R. E. Bruccoleri, B. D. Olafson, D. J. States, S. Swaminathan, and M. Karplus. CHARMM: A program for macromolecular energy, minimization, and dynamics calculations. *Journal of Computational Chemistry*, 4(2):187–217, 1983.

[3] B. Fitch, R. Germain, M. Mendell, J. Pitera, M. Pitman, A. Rayshubskiy, Y. Sham, F. Suits, W. Swope, T. Ward, Y. Zhestkov, and R. Zhou. Blue Matter, an application framework for molecular simulation on Blue Gene. *Journal of Parallel and Distributed Computing*, 63:759–773, 2003.

[4] F. Franceschi and E. M. Duffy. Structure-based drug design meets the ribosome. *Biochem. Pharmacol.*, 71:1016–1025, 2006.

[5] P. L. Freddolino, A. S. Arkhipov, S. B. Larson, A. McPherson, and K. Schulten. Molecular dynamics simulations of the complete satellite tobacco mosaic virus. *Structure*, 14:437–449, 2006.

[6] A. Grama, A. Gupta, and V. Kumar. Isoefficiency: Measuring the scalability of parallel algorithms and architectures. *IEEE Parallel & Distributed Technology*, 1(August), 1993.

[7] W. F. Humphrey, A. Dalke, and K. Schulten. VMD – Visual molecular dynamics. *Journal of Molecular Graphics*, 14(1):33–38, 1996.

[8] Y.-S. Hwang, R. Das, J. Saltz, M. Hodoscek, and B. Brooks. Parallelizing molecular dynamics programs for distributed memory machines. *IEEE Computational Science & Engineering*, 2(Summer):18–29, 1995.

[9] L. Kalé, R. Skeel, M. Bhandarkar, R. Brunner, A. Gursoy, N. Krawetz, J. Phillips, A. Shinozaki, K. Varadarajan, and K. Schulten. NAMD2: Greater scalability for parallel molecular dynamics. *Journal of Computational Physics*, 151:283–312, 1999.

[10] L. V. Kalé, M. Bhandarkar, and R. Brunner. Load balancing in parallel molecular dynamics. In *Fifth International Symposium on Solving Irregularly Structured Problems in Parallel*, volume 1457 of *Lecture Notes in Computer Science*, pages 251–261, 1998.

[11] L. V. Kale and S. Krishnan. Charm++: Parallel programming with message-driven objects. In G. V. Wilson and P. Lu, editors, *Parallel Programming Using C++*, pages 175–213. MIT Press, 1996.

[12] F. Khalili-Araghi, E. Tajkhorshid, and K. Schulten. Dynamics of K^+ ion conduction through Kv1.2. *Biophys. J.*, 91:L72–L74, 2006.

[13] S. Kumar, C. Huang, G. Zheng, E. Bohm, A. Bhatele, J. C. Phillips, H. Yu, and L. V. Kalé. Scalable molecular dynamics with NAMD on Blue Gene/L. *IBM Journal of Research and Development: Applications of Massively Parallel Systems*, 2007. (to appear).

[14] M. Nelson, W. Humphrey, A. Gursoy, A. Dalke, L. Kale, R. D. Skeel, and K. Schulten. NAMD—a parallel, object-oriented molecular dynamics program. *Intl. J. Supercomput. Applics. High Performance Computing*, 10(Winter):251–268, 1996.

[15] J. C. Phillips, R. Braun, W. Wang, J. Gumbart, E. Tajkhorshid, E. Villa, C. Chipot, R. D. Skeel, L. Kalé, and K. Schulten. Scalable molecular dynamics with NAMD. *Journal of Computational Chemistry*, 26(16):1781–1802, 2005.

[16] S. J. Plimpton and B. A. Hendrickson. A new parallel method for molecular-dynamics simulation of macromolecular systems. *J Comp Chem*, 17:326–337, 1996.

[17] J. Poehlsgaard and S. Douthwaite. The bacterial ribosome as a target for antibiotics. *Nat. Rev. Microbiol.*, 3:870–881, 2005.

[18] K. Y. Sanbonmatsu, S. Joseph, and C. S. Tung. Simulating movement of tRNA into the ribosome during decoding. *Proc. Natl. Acad. Sci. USA*, 102:15854–15859, 2005.

[19] M. Snir. A note on N-body computations with cutoffs. *Theory of Computing Systems*, 37:295–318, 2004.

[20] S. Törnroth-Horsefield, Y. Wang, K. Hedfalk, U. Johanson, M. Karlsson, E. Tajkhorshid, R. Neutze, and P. Kjellbom. Structural mechanism of plant aquaporin gating. *Nature*, 439:688–694, 2006.

[21] P. K. Weiner and P. A. Kollman. AMBER: Assisted model building with energy refinement a general program for modeling molecules and their interactions. *Journal of Computational Chemistry*, 2:287, 1981.

[22] J. Yu, A. J. Yool, K. Schulten, and E. Tajkhorshid. Mechanism of gating and ion conductivity of a possible tetrameric pore in aquaporin-1. *Structure*, 14:1411–1423, 2006.

[23] G. Zheng. *Achieving High Performance on Extremely Large Parallel Machines: Performance Prediction and Load Balancing*. PhD thesis, Department of Computer Science, University of Illinois at Urbana-Champaign, 2005.

[24] G. Zheng, G. Kakulapati, and L. V. Kalé. BigSim: A parallel simulator for performance prediction of extremely large parallel machines. In *18th International Parallel and Distributed Processing Symposium (IPDPS)*, page 78, Santa Fe, NM, April 2004.

Chapter 10

Petascale Special-Purpose Computer for Molecular Dynamics Simulations

Makoto Taiji

Genomic Sciences Center, RIKEN

Tetsu Narumi

Genomic Sciences Center, RIKEN

Permanent Address: *Faculty of Science and Technology, Keio University*

Yousuke Ohno

Genomic Sciences Center, RIKEN

10.1 Introduction

A universal computer is one of the greatest inventions in the modern history of technology. It enables us to utilize essentially the same machine for various applications, and thus, the hardware can be designed independently of the applications. This feature makes the development of a computer easier, since its engineering can be focused on the design of a single architecture. Along with the rapid advancements in semiconductor technologies, an enormous growth in processor speeds has been observed in the last 60 years. Their applications have also rapidly expanded — the universal computer probably has a wider range of applications than any other machine in history.

Since the invention of integrated circuits, the transistor density in microchips has been increased in accordance with Moore's law, which has resulted in an enormous performance increase. A simple floating-point arithmetic unit uses only about a few hundred thousand transistors, but nowadays a single microchip can have tens of millions of transistors. Therefore, an efficient parallel use of multiple arithmetic units is of primary importance in increasing processor performance. In recent times, multicore technology has become popular, and an advanced processor can perform 16 or more arithmetic operations in parallel. However, this is still quite low as compared with the number that can be achieved by accumulating arithmetic units with a high density. By using 45-nm technology, it is possible to accumulate thousands of double-precision floating point units (FPUs) operating around gigahertz speed in a single LSI. In conventional processors, there are two bottlenecks limiting parallel calculations — the memory bandwidth bottleneck and heat dissipation problem. These bottlenecks can be overcome by developing specialized architectures for specific applications and sacrificing some of the universality.

The GRAvity PipE (GRAPE) project is one of the most successful attempts to develop such high-performance, competitive special-purpose systems [25, 16]. The GRAPE systems are specialized for simulations of classical particles such as simulations for gravitational N-body problems or molecular dynamics (MD) simulations. In these simulations, most of the computing time is spent on the calculation of long-range forces such as gravitational, Coulomb, and van der Waals forces. Therefore, the special-purpose engine calculates only these forces, and all the other calculations are performed by a conventional host computer that is connected to the system. This style makes the hardware very simple and cost effective. This strategy of using specialized computer architectures predates the GRAPE project. Its application to MD simulations was pioneered by Bakker et al. on the Delft Molecular Dynamics Processor (DMDP) [2] and by Fine et al. on the FASTRUN processor [6]. However, neither system was able to achieve effective cost performance. The most important drawback was the architectural complexity of these machines, which demanded considerable time and money in the developmental stages. Since electronic devices technology continues to develop very rapidly, the speed of development is a crucial factor affecting the cost-to-performance ratio.

Figure 10.1 shows the advancements in the GRAPE computers. The project began at the University of Tokyo and is now run by two groups, one at the National Astronomical Observatory/University of Tokyo and the other at the RIKEN Institute. GRAPE-4 [27] was built in 1995, and it was the first machine to break the teraflops barrier in nominal peak performance. To date, eight Gordon-Bell Prizes (1995, 1996, 1999, 2000(double), 2001, 2003, and 2006) have been awarded to simulations using the GRAPE systems. The GRAPE architecture can achieve such high performances because it solves the problem of memory bandwidth bottlenecks and lessens the heat dissipation problem. In 2002, we launched a project to develop the MDGRAPE-3 system, a petaflops special-purpose computer system for MD simulations, and we completed its development in June 2006 [29]. The MDGRAPE-3 is a successor of GRAPE-2A [9], MD-GRAPE [28, 8], and MDM(MD Machine)/MDGRAPE-2 [18], which are also machines for MD simulations.

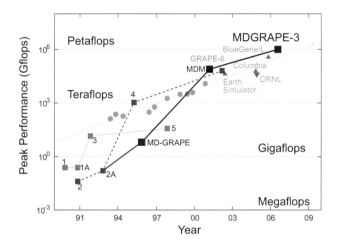

FIGURE 10.1: Advancements in the GRAPE systems. For astrophysics, there are two product lines. The even-numbered systems, GRAPE-2, 2A, 4, and 6 (dashed line), are the high-precision machines, and the systems with odd numbers, GRAPE-1, 1A, 3, and 5 (dotted line), are the low-precision machines. GRAPE-2A, MD-GRAPE, MDM/MDGRAPE-2, and MDGRAPE-3 (thick solid line) are the machines for MD simulations.

In recent years there have been very rapid advances in structural genomics, and many three-dimensional structures of proteins and other biomolecules have been solved. In Japan, for example, the national "Protein 3000" project has successfully solved the structures of 3,000 proteins between 2002 and 2007. Such MD simulations by means of high-performance, special-purpose computers will be a very powerful tool for applying the new structural knowledge to advances in bioscience and biotechnology. The main targets of MDGRAPE-3 are high-precision screening for drug design and the large-scale simulations of huge proteins/complexes [24]. In this chapter, we describe the hardware architecture of the MDGRAPE-3 system and report early results of its performance evaluation.

10.2 Hardware of MDGRAPE-3

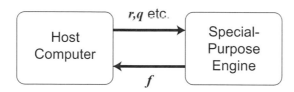

FIGURE 10.2: The basic architecture of MDGRAPE-3 and the other GRAPE systems.

First, we will describe the basic architecture of GRAPE systems, as shown in Figure 10.2. The GRAPE hardware is an accelerator for specific applications, and it is attached to general-purpose commercial computers. In the case of MDGRAPE-3, it consists of a host PC cluster with special-purpose boards for MD simulations. The host sends the coordinates and other data of particles to the special-purpose engine, which then calculates the forces between the particles and returns the results to the host computer. In the MD simulations, most of the calculation time is spent on nonbonded forces, i.e., the Coulomb and van der Waals forces. Therefore, the special-purpose engine calculates only the nonbonded forces, and all other calculations are performed by the host computer. This makes the hardware and the software quite simple. For the hardware, we have to consider only the calculation of forces, which is rather simple and has little variation. On the other hand, in the past on machines like DMDP, almost all the work required for the MD simulation was performed by the hardware; hence, the hardware was extremely complex and very time-consuming to build. In the GRAPE systems, no detailed knowledge on the hardware is required to write programs, and a user simply uses a subroutine package to perform the force calculations. All other aspects of the system, such as the operating system, compilers, etc., rely on the host computer and do not need to be specially developed.

The communication time between the host and the special-purpose engine is proportional to the number of particles, N, while the calculation time is proportional to its square, N^2, for the direct summation of the long-range forces, or is proportional to NN_c, where N_c is the average number of particles within the cutoff radius of the short-range forces. Since N_c usually exceeds a value of several hundred, the calculation cost is considerably higher than the communication cost. In the MDGRAPE-3 system, the ratio between the communication speed and calculation speed of the special-purpose engine is 0.2 Gbytes/sec·TFLOPS = 0.2 bytes for one thousand operations. This ratio is fairly smaller than that in the commercially available parallel processors. Such a low communication speed is adequate to make efficient use of the special-purpose engine.

10.3 The Calculations Performed by MDGRAPE-3

Next, we describe the calculations that are performed by the special-purpose engine of the MDGRAPE systems. It calculates two-body forces on the ith particle $\vec{F}_i$ as

$$\vec{F}_i = \sum_j a_j g(b_j r_s^2) \vec{r}_{ij} \tag{10.1}$$

where $\vec{r}_{ij} = \vec{r}_j - \vec{r}_i, r_s^2 = r_{ij}^2 + \epsilon_i^2$. The vectors $\vec{r}_i, \vec{r}_j$ are the position vectors of the i, jth particles, and ϵ_i is a softening parameter to avoid numerical divergence. For the sake of convenience, we hereafter refer to the particles on which the force is calculated as the "i-particle," and the particles that exert a force on the i-particle as the "j-particle." The function $g(\zeta)$ is an arbitrary smooth function. For example, in the case of Coulomb forces, the force is given by

$$\vec{F}_i / q_i = \sum_j \frac{q_j}{r_{ij}^3} \vec{r}_{ij} \tag{10.2}$$

where q_i, q_j are the charges of the ith and the jth particles, respectively. This can be calculated by using $a_j = q_j, b_j = 1, g(\zeta) = \zeta^{-3/2}, \epsilon_i = 0$. The multiplication by q_i is performed by the host computer.

In the case of a force due to Lennard-Jones potential, the force between the particles is given by

$$\vec{f}_{ij} = \left[\frac{A_{ij}}{r_{ij}^8} - \frac{B_{ij}}{r_{ij}^{14}} \right] \vec{r}_{ij} \tag{10.3}$$

where A_{ij} and B_{ij} are constants determined from the equilibrium position and the depth of a potential. These constants depend on the species of ith and jth particles. This force law can be evaluated by choosing $g(\zeta), a_j, b_j, \epsilon_i$ as follows:

$$g(\zeta) = \zeta^{-4} - \zeta^{-7}$$
$$a_j = A_{ij}^{7/3} B_{ij}^{-4/3} \tag{10.4}$$
$$b_j = \left(\frac{A_{ij}}{B_{ij}} \right)^{1/3}$$
$$\epsilon_i = 0$$

The other potentials, including the Born-Mayer type repulsion ($U(r) = A \exp(Br)$), can also be evaluated.

In addition to Equation (10.1), MDGRAPE can also calculate the following equations:

$$c_i = \sum_j a_j g(\vec{k}_i \cdot \vec{r}_j) \tag{10.5}$$

$$\vec{F}_i = \sum_j \vec{k}_j c_j g(\vec{k}_j \cdot \vec{r}_i + \phi_j) \tag{10.6}$$

$$\phi_i = \sum_j a_j g(b_j r_s^2) \tag{10.7}$$

Equations (10.5) and (10.6) are used to calculate discrete Fourier transformations for wave-space sums in the Ewald method [7], and the genetic algorithm-direct space method for X-ray crystallography [20]. The potential energies and isotropic virials are calculated by Equation (10.7).

Virial tensors, which are necessary for simulation under a tensile stress by using the Parrinello-Rahman method [22], can be calculated from the forces. At first, in the case of open boundary conditions, virial tensors are defined by

$$\Phi = \sum_i \vec{F}_i : \vec{r}_i \tag{10.8}$$

which we can calculate directly from forces $\vec{F}_i$. In the case of periodic boundary conditions, it has been well established that a different formula must be used [1]:

$$\Phi = \frac{1}{2} \sum_{i,j,\alpha} \vec{f}_{ij}^{\alpha} : (\vec{r}_i - \vec{r}_j^{\alpha}) \tag{10.9}$$

where α denotes neighboring cells. To calculate this, a force $\vec{F}_i$ must be divided into several forces from each cell α as

$$\vec{F}_i^{\alpha} = \sum_j \vec{f}_{ij}^{\alpha} \tag{10.10}$$

and then, the virial tensor (10.9) can be evaluated as

$$\Phi = \frac{1}{2} \sum_{i,\alpha} \vec{F}_i^{\alpha} : (\vec{r}_i - \vec{R}_{\alpha}) \tag{10.11}$$

where $\vec{R}_{\alpha}$ is a translation vector to a cell α. MDGRAPE-3 has a function for calculating the forces from each periodic cell separately, which is useful for evaluating a virial tensor.

There are several clever algorithms for the efficient calculation of long-range forces. In high-precision calculations such as those in MD simulations, these algorithms use direct summation for the near-field forces. Therefore, the GRAPE systems can accelerate these algorithms. There are also several implementations of these algorithms, for example, the tree algorithm [13], the Ewald method [7, 19], and the modified fast multipole method [10, 11].

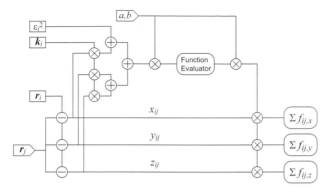

FIGURE 10.3: Block diagram of the force calculation pipeline in the MDGRAPE-3 chip.

10.4 MDGRAPE-3 Chip

In this section, we describe the MDGRAPE-3 chip, that performs force calculations in the MDGRAPE-3 system [26]. We begin with an explanation of the LSI because it is the most important part of the system.

10.4.1 Force calculation pipeline

Figure 10.3 shows the block diagram of the force calculation pipeline in the MDGRAPE-3 chip. It calculates Equations (10.1), (10.5), (10.6), and (10.7) using the specialized pipeline. The pipeline consists of three subtractor units, six adder units, eight multiplier units, and one function-evaluation unit. It can perform approximately 36 equivalent operations per cycle while calculating the Coulomb force. Here, we assume that both an inversion and a square root need 10 arithmetic operations. In such a case, the function-evaluation unit calculates $x^{-3/2}$, which is equivalent to 20 floating-point operations. The number depends on the force to be calculated. Most of the arithmetic operations are performed in a 32-bit single-precision floating-point format, with the exception of the force accumulation. The force $\vec{F}_i$ is accumulated in a 80-bit fixed-point format and it can be converted to a 64-bit double-precision floating-point format. The coordinates $\vec{r}_i$ and $\vec{r}_j$ are stored in a 40-bit fixed-point format because this makes the implementation of periodic boundary conditions easy. Since the dynamic range of the coordinates is relatively small in MD simulations, the use of the fixed-point format causes no trouble.

The function evaluator, which allows the calculation of an arbitrary smooth function, is the most important part of the pipeline. This block is almost the same as those in MD-GRAPE [28, 8]. It has a memory unit that contains a table for polynomial coefficients and exponents, and a hardwired pipeline for fourth-order polynomial evaluations. It interpolates an arbitrary smooth function $g(x)$ using segmented fourth-order polynomials by the Horner's method

$$g(x_0 + \Delta x) = (((c_4[x_0]\Delta x + c_3[x_0]) \Delta x + c_2[x_0])$$
$$\Delta x + c_1[x_0]) \Delta x + c_0[x_0] \tag{10.12}$$

where x_0 is the center of the segmented interval, $\Delta x = x - x_0$ is the displacement from the center, and $c_k[x_0]$ are the coefficients of polynomials. The coefficient table has 1,024 entries so that it can evaluate the forces by the Lennard-Jones (6–12) potential and the Coulomb force with a Gaussian kernel in a single precision.

In the case of Coulomb forces, the coefficients a_j and b_j in Equation (10.1) depend only on the species of the j-particle, that are read from the j-particle memories. Actually, the charge of the jth particle q_j corresponds to a_j, and $b_j = 1$ is constant. On the other hand, in the case of the van der Waals force, these coefficients depend on the species of both the i-particle and the j-particle, as can be seen in Equation (10.4). Thus, these coefficients must be changed when the species of the i-particle changes. Since considerable time is required to exchange them with each change in the species, the pipeline contains a table of the coefficients. The table generates the coefficients from the species of both i-particles and j-particles. Each pipeline has a table implemented by a random-access memory with a depth of 1024, and thus, the number of species is limited to 32 (since $1024 = 32^2$). We can combine four tables of four pipelines into one in order to increase the number of species to 64. In this case, the i-particle species in the combined four pipelines should be identical. In the MD simulations, we usually must exclude nonbonded interactions between bonded atoms. The MDGRAPE-3 chip has a function to exclude forces from the specified atoms for each pipeline.

10.4.2 *j*-Particle memory and control units

Figure 10.4 shows the block diagram of the MDGRAPE-3 chip. It has 20 force calculation pipelines, a j-particle memory unit, a cell-index controller, a force summation unit, and a master controller. The master controller manages the timings and inputs/outputs (I/O) of the chip. The j-particle memory unit holds the coordinates of j-particles for 32,768 bodies, and it corresponds to the "main memory" in general-purpose computers. Thus, the chip is designed by using the memory-in-the-chip architecture and no extra memory is necessary on the system board. The amount of memory is 6.6 Mbits and is constructed by a static RAM. The memory operates at half the pipeline clock speed. The same output of the memory is sent to all the pipelines simultaneously. Each pipeline calculates using the same data from the j-particle unit and individual data stored in the local memory of the pipeline. Since the two-body force

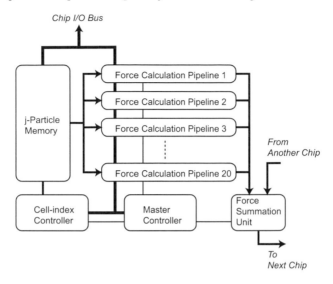

Chip I/O Bus

FIGURE 10.4: Block diagram of the MDGRAPE-3 chip.

calculation is given by

$$\vec{F}_i = \sum_j \vec{f}(\vec{r}_i, \vec{r}_j) \qquad (10.13)$$

for the parallel calculation of multiple $\vec{F}_i$, we can use the same $\vec{r}_j$. This parallelization scheme — "the broadcast memory architecture" — is one of the most important advantages of the GRAPE systems. It enables the efficient parallelization at a low bandwidth realized by a simple hardware. In addition, we can extend the same technique to the temporal axis. If a single pipeline calculates two forces, $\vec{F}_i$ and $\vec{F}_{i+1}$, in every two cycles by time-sharing, we can reuse the same $\vec{r}_j$. This means that an input $\vec{r}_j$ is required every two cycles. We refer to this parallelization scheme as a "virtual multiple pipeline" [15], because it allows a single pipeline to behave like multiple pipelines operating at half the speed. The advantage of this technique is reduction in the "virtual frequency." The requirement for the memory bandwidth is reduced by half, and it can be decreased further by increasing the number of the "virtual" pipelines. The hardware cost for the virtual pipeline consists of only the registers for the coordinates and forces, which are very small.

In the MDGRAPE-3 chip, there are two virtual pipelines per physical pipeline, and thus, the total number of logical pipelines is 40. The physical bandwidth of the j-particle unit will be 2.5 Gbytes/s, but the virtual bandwidth will reach 100 Gbytes/s. This allows a fairly efficient parallelization in the chip. The MDGRAPE-3 chip has 340 arithmetic units and 20 function-evaluator units that work simultaneously. The chip has a high performance of 180 GFLOPS at a modest speed of 250 MHz. This advantage will become even more important in the future. The number of transistors will continue to increase over the next ten years, but it will become increasingly difficult to use additional transistors to enhance performance. In fact, in a general-purpose scalar processor, the number of floating-point units in the CPU core appears to reach a ceiling at around 4–8 because of the limitation in the bandwidth of the memory, that of the register file, and the difficulties in the software. On the other hand, in the GRAPE systems, the chip can house more and more arithmetic units with little performance degradation. Thus, the performance of the chip exactly follows the performance of the semiconductor device technology, which is the product of the number of transistors and the speed. In a general-purpose processor, the performance increases roughly 10 times every 5 years, a rate slower than that of the semiconductor device, which is about 30 times every 5 years. Therefore, a special-purpose approach is expected to become increasingly more advantageous than a general-purpose approach. The demerit of the broadcast memory parallelization is, of course, its limitation with respect to applications. However, it is possible to find several applications other than particle simulations. For example, the calculation of dense matrices [21] and the dynamic programming algorithm for hidden Markov models can be accelerated by the broadcast memory parallelization.

The size of the j-particle memory, 32,768 bodies = 6.6 Mbits, is sufficient for the chip despite the remarkable calculation speed, since MD simulation is a computation-intensive application and not a memory-intensive application such as fluid dynamics simulations. Of course, although the number of particles often exceeds this limit, we can use many chips in parallel to increase the capacity. We can divide the force $\vec{F}_i$ on the ith particle as

$$\vec{F}_i = \sum_{k=1}^{n_j} \vec{F}_i^{(k)} \tag{10.14}$$

by grouping j-particles into n_j subsets. Thus, by using n_j chips, we can treat n_j times more particles at the same time. However, if a host computer collects all the partial forces $\vec{F}_i^{(k)}$ and sums them, the communication between the host and special-purpose engines increases n_j times. To solve this problem, the MDGRAPE-3 chip has a force summation unit, which calculates the summation of the partial forces. In the MDGRAPE-3 system, the 12 MDGRAPE-3 chips on the same board are connected by a unidirectional ring, as explained later. Therefore, each chip receives partial forces from the previous chip, adds the partial force calculated in the chip, and sends the results to the next chip. The force summation unit calculates the sum in the same 80-bit fixed-point format as is used for the partial force calculations in the pipelines. The result can be converted to a 64-bit floating-point format at the end of the chain. The summation of the force subsets on different boards is performed by the host computer. If the number of particles still exceeds the total memory size, the forces must be subdivided into several subsets, and the contents of the j-particle memory should be replaced for each subset. In this case the number of particles is quite huge, and hence, the overhead for communication/calculation is not as important. Also, in such cases we usually use PME or treecode or cutoff methods, where systems are divided into cells.

The j-particle memory is controlled by the cell-index controller, which generates the addresses for the memory. To calculate the short-range forces, it is not necessary to calculate two-body interactions with all particles. There exists a cutoff r_c above which forces can be ignored, i.e., when $r > r_c$. If we calculate the contributions inside the sphere $r < r_c$, the cost of force calculations decreases by $(L/r_c)^3$, where L is the size of a system. Thus, when $L \gg r_c$, the efficiency increases substantially. The standard technique for this approach is called the cell-index method [1]. In this method, a system is divided into cells with sides l. When we calculate the forces on the particles in a cell, we treat only those that belong to cells within the cutoff.

The implementation of the cell-index method is almost the same as that for the MD-GRAPE/MDGRAPE-2 systems, except for two new major functions; one is used to calculate forces from different cells separately and the other is a function to generate cell numbers automatically. These functions decrease communication with a host. The host computer divides particles into cells and sorts them by the cell indices. Thus, the particles that belong to the same cell have sequential particle numbers. Hence, the address of the particles in a cell can be generated from the start and end addresses. The cell-index table contains the start and end addresses for all cells. The controller generates the cell number involved in the interaction and looks up the table to obtain the start address and the end one. The counter then generates the sequential address for the j-particle memory. When the count reaches to the end address, the same process is repeated for the next cell. Further, the chip has a function to translate cells according to the periodic boundary condition, and a support for the virial tensor calculation. As shown in Equation (10.11), the forces separated by each mirror image must be evaluated in order to calculate a virial tensor under periodic boundary conditions. The cell-index controller controls the force calculation pipelines to calculate these partial forces.

10.4.3 Chip specifications

By combining all the above-mentioned units, the MDGRAPE-3 chip has almost all the elements of the MD-GRAPE/MDGRAPE-2 system board other than the bus interface. The chip is designed to operate at 250 MHz under the worst conditions (1.08 V, 85°C, process factor = 1.3). It has 20 pipelines, and each pipeline performs 36 equivalent operations per cycle; thus, the peak performance of the chip will reach 180 GFLOPS. This is 12.5 times faster than the previous MDGRAPE-2 chip. More than 80% of the LSI operates at 300 MHz and 216 GFLOPS. The chip was manufactured by Hitachi Device Development Center using HDL4N 0.13 μm CMOS technology. It consists of

6-M gates and 10-M bits of memory, and the chip has a size of 15.7 mm × 15.7 mm. At 300 MHz, the power dissipation of the LSI was measured as 17 W with full-speed calculations. It will be possible to design a chip having a speed of gigahertz; however, the power dissipation will become above 80 W and the number of transistors per operation will increase by at least 80%. Since the parallelization is very efficient in our architecture, the gigahertz speed will cause considerable difficulties in the power dissipation and transistor layout, but this will not result in any performance gain of the system. Therefore, we chose the modest speed of 250 MHz.

Its power per performance was less than 0.1 W/GFLOPS, which is much better than those of the conventional CPUs. For example, the thermal design power of a Pentium 4 3GHz processor produced by a similar 0.13 μm process is 82 W, while its nominal peak performance is 6 GFLOPS. Therefore, its power per performance is approximately 14 W/GFLOPS, which is a hundred times worse than that of MDGRAPE-3. There are several reasons for this high power efficiency. First, the accuracy of the arithmetic units is smaller than that of the conventional CPUs. Since the number of gates of a multiplier is roughly proportional to the square of the word length, this precision has a significant impact on the power consumption. Secondly, in the MDGRAPE-3 chip, 90% of the transistors are used for arithmetic operations, and the rest are used for control logic (the transistors for the memory are not counted). The specialization and broadcast memory architecture make the efficient usage of silicon possible. Finally, the MDGRAPE-3 chip operates at the modest speed of approximately 300 MHz. At gigahertz speed, the depth of the pipeline significantly increases, and the ratio of the pipeline registers tends to increase. Since the pipeline registers dissipate power but do not perform any calculation, there is a decrease in the power efficiency. In our applications, it is better to parallelize calculations instead of increasing the frequency. Thus, the GRAPE approach is very effective to suppress power consumption. This is another important advantage of the GRAPE. IBM Blue Gene/L also uses a similar approach with massively parallel calculations at a modest speed [31]. When the feature size of the semiconductor decreases, almost all the parameters are improved other than the power density. Therefore, the power dissipation will become important in the future, and the GRAPE architecture can alleviate the problem.

10.5   System Architecture

In this section, the system architecture of MDGRAPE-3 is explained. Figure 10.5 shows the photograph and block diagram of the MDGRAPE-3 system. The system consists of a PC cluster with special-purpose engines attached. For the host PC cluster we used Intel Xeon® processors. The system has 100 nodes with 200 CPUs. For the host network, we use Infiniband (10 Gbit/s). The network topology is two-stage fat-tree, but the bandwidth between the root and leaf switches is one-fifth of the speed in order to achieve nonblocking communication. It is implemented by using six 24-port switches.

Each node has the special-purpose engines of two boards with 12 MDGRAPE-3 chips on each board. Since the MDGRAPE-3 chip has a performance of 216 GFLOPS at 300MHz the performance of the nodes will be 5.18 TFLOPS. In detail, the system consists of 400 boards, which have the following specifications

Board with 12 chips at 300 MHz $\cdots$ 304
Board with 12 chips at 250 MHz $\cdots$ 74
Board with 11 chips at 300 MHz $\cdots$ 22

On some boards, a chip is disabled because it malfunctions. Since it is faster to use 11 chips at 300 MHz than 12 chips at 250 MHz, a chip is disabled by design for several boards regardless of its operation at 250 MHz in order to increase the operational frequency. In total, we have 4,778 chips, and among them, 3,890 chips operate at 300 MHz and 888 chips operate at 250 MHz. This corresponds to the nominal peak performance of 1.0 PFLOPS.

The MDGRAPE-3 chips are connected to the host by two PCI-X buses at 100 MHz. Figure 10.6 shows the photograph and block diagram of the MDGRAPE-3 system board. The board has twelve MDGRAPE-3 chips, and each chip is connected in series to send/receive the data. Since the memory is embedded in the MDGRAPE-3 chip, the board is extremely simple. The speed of the communication between the chips is 1.5 Gbytes/sec, which corresponds to an 80-bit word transfer at 150 MHz. GTL I/O is used for these connections. The board has a control FPGA (Xilinx XC2VP30) which performes I/O. The FPGA has I/O to the MDGRAPE-3 chips, and two high-speed serial interconnections with a speed of 10 Gbit/sec for both up and down streams. This was realized by using a bundle of four 2.5 Gbit/sec LVDS channels (Xilinx RocketIO), and an Infiniband 4X cable is used for a physical connection. Two boards are mounted in a 2U chassis, and they are connected by a daisy-chain using the high-speed serial interconnection. Two chassis with a total of 48 MDGRAPE-3 chips are connected to each host via two independent PCI-X buses. The PCI-X interface board has also been developed for the high-speed serial interconnection by using Xilinx XC2VP7 FPGA.

The MDGRAPE-3 units and PCs are mounted in 42U-height 19" racks. The system consists of 22 racks in total, including the storage units and the

(a)

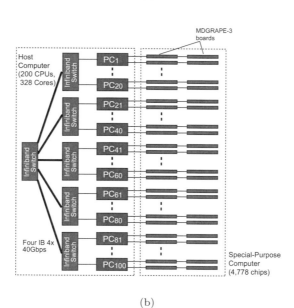

(b)

FIGURE 10.5: (a) Photograph and (b) block diagram of the MDGRAPE-3 system.

control node. The total power dissipation is about 200 kW and it occupies 80 m². Because of the high-density packaging in addition to the high performance of the chip, the installation area is fairly small even with a petaflops system. We have also developed a small board with two MDGRAPE-3 chips that can be attached directly to a PCI-X bus. The board is commercially available from a venture company.

10.6   Software for MDGRAPE-3

As mentioned above, in the GRAPE systems, a user does not need to consider the detailed architecture of the special-purpose engine, and all the necessary computation is defined in the subroutine package. In order to use MDGRAPE-3, we only have to modify the program in order to call the application programming interface (API) of this system. We have already ported AMBER 8 [4] and CHARMM [3] to run with a single board of the MDGRAPE-3 system. For large-scale parallel simulations, we use a software called MOA developed by T. Koishi, which is specially designed to operate efficiently with MDGRAPE systems. In the following discussions, we illustrate the typical procedure for using the MDGRAPE system.

The typical APIs used in the MD programs are summarized in Table 10.1. All the routines, except those in the "overlap calculation" category, are identical to the earlier MDGRAPE-2 routines [19]. APIs for the overlap calculation have been recently included to ensure an efficient use of the MDGRAPE-3 system. The MDGRAPE-3 board can perform force calculations on many *i*-particles without any control from the host computer. Therefore, the host computer can perform other calculations, while the board performs the force calculations. On the other hand, the host computer is required to replace the particle positions in the chip registers and read the calculated forces after every calculation by the previous MDGRAPE-2 board. Therefore, it was more difficult to implement parallel calculations between the host and MDGRAPE-2 since fine-grain background calculations are required. On the other hand, this can be conveniently performed by the MDGRAPE-3 system since it can perform long calculations without a host computer. The core of the MD program is shown in Table 10.2. The sequence of calculations for a time step using the MDGRAPE-3 system is as follows:

1. Set the function table in the MDGRAPE-3 chips to the van der Waals force (line 4).

2. Set all the coordinates and atom types of the j-particle memory in the MDGRAPE-3 chips (lines 5–6).

3. Activate overlap calculation mode (line 7).

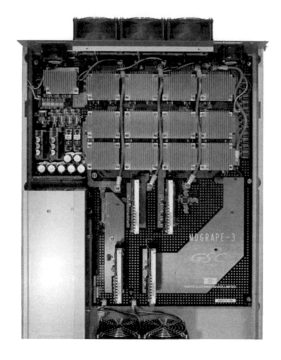

(a)

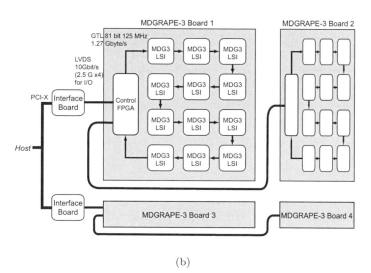

(b)

FIGURE 10.6: Photograph and block diagram of the MDGRAPE-3 board. "MDG3 LSI" denotes the MDGRAPE-3 chip.

TABLE 10.1: APIs used in a typical MD program with the MDGRAPE-3 system

	Name	Function
Initialize and Finalize		
	m3_allocate_unit	Acquire the MDGRAPE-3 board
	m3_set_function	Change the function table of the MDGRAPE-3 chips
	m3_free_unit	Release MDGRAPE-3 board
Set *j*-particle data and van der Waals parameters		
	m3_set_positions	Set positions of j-particles
	m3_set_types	Set atom types of j-particles
	m3_set_charges	Set charges of j-particles
	m3_set_rscale_matrix	Set the scaled van der Waals radii
	m3_set_charge_matrix	Set the well depths of van der Waals potential
Calculate forces or potentials		
	m3_set_cells	Set cell-index information to MDGRAPE-3
	m3_set_pipeline_types	Set atom types of i-particles
	m3_calculate_forces	Calculate forces on specified i-particles
	m3_calculate_potentials	Calculate potentials on specified i-particles
Overlap calculation		
	m3_setup_overlap	Activate overlap calculation mode
	m3_start_overlap_calculation	Start actual calculation on MDGRAPE-3
	m3_wait_overlap_calculation	Wait until MDGRAPE-3 finishes calculation

4. Loop over cells (line 8).

5. Set the cell information of 27 neighboring cells in the MDGRAPE-3 chips (line 9). Atoms in these neighboring cells are used as the *j*-particle in the subsequent force calculation routine.

6. Set coordinates and atom types of the i-particles in the MDGRAPE-3 chip registers and calculate the forces on these particles (lines 10–12).

7. The Coulomb forces are calculated in a similar manner (lines 14–20).

8. Start direct-memory-access (DMA) transfer to the MDGRAPE-3 board (line 21). Steps 1–7 prepare only the data and commands in the DMA buffer in the host memory. The commands include the "start calculation," "wait until calculation finishes," and "transfer calculated forces to the host memory."

TABLE 10.2: Typical MD program to use MDGRAPE-3 system

```
 1 unit=m3_allocate_unit("grav.tblmd3",M3_FORCE,xmin,xmax);
 2 Other initialization
 3 for(step=0;step<nstep;step++){
 4   m3_set_function("lj01.tblmd3",M3_FORCE,xmin,max);
 5   m3_set_types(unit, atypej, N);
 6   m3_set_positions(unit, posj, N);
 7   m3_setup_overlap(unit);
 8   for(icell=0;icell<num_icell;icell++){
 9     m3_set_cells(unit, neighbor_cells_lj[icell], 27);
10     m3_set_pipeline_types(unit, atypei[icell], ni_lj[icell]);
11     m3_calculate_forces(unit, posi[icell], ni_lj[icell],
12                                         force_lj[icell]);
13   }
14   m3_set_function("grav.tblmd3",M3_FORCE,xmin,xmax);
15   m3_set_charges(unit, chargej, N);
16   for(icell=0;icell<num_icell;icell++){
17     m3_set_cells(unit, neighbor_cells_cl[icell], 27);
18     m3_calculate_forces(unit, posi[icell], ni_cl[icell],
19                                         force_cl[icell]);
20   }
21   m3_start_overlap_calculation(unit);
22   Calculate bond, angle, torson (force_bond) by host
23   Calculate van der Waals and Coulomb forces (force_exclude)
24           for bonded part by host
25   m3_wait_overlap_calculation(unit);
26   Get total force by adding force_lj, force_cl and force_bond
27   Subtract force_exclude
28   if (is_potential_calculated_for_this_step?) {
29     Potential calculation with MDGRAPE-3
30   }
31   Update position of atoms
32   Communicate atom position, etc. between the nodes
33 }
34 m3_free_unit(unit);
35 Other finalization
```

9. Calculate the bonding, Coulomb, and van der Waals forces for the bonded atoms with the assistance of the host computer (lines 22–24). These calculations are performed in parallel with the MDGRAPE-3 system.

10. Wait until MDGRAPE-3 finishes the calculation (line 25).

11. Obtain the final forces by adding several forces (lines 26–27). The Coulomb and van der Waals forces between the bonded atoms must be subtracted because MDGRAPE-3 basically calculates the forces between all the pairs of i- and j-particles in the cell list.

12. The Coulomb and van der Waals potentials are calculated by using the MDGRAPE-3 system in a similar manner (lines 28–30). We can reduce the calculation time by reducing the frequency of the potential calculations; this is because their calculation times are the same as those for the force calculations using the MDGRAPE-3 system. The MD simulations in microcanonical or canonical ensembles do not require potential energy to proceed to the subsequent coordinates of atoms.

10.7   Performance of MDGRAPE-3

In this section, we briefly explain the early results of sustained performance evaluations. First, a single-node performance with AMBER-8 is described and then, the entire system performance with large-scale calculations is shown.

Figure 10.7 shows the comparison of the calculation time using MDGRAPE-3 and the host machine. Here the protein scytalone dehydratase solvated in water was treated in all simulations and the number of atoms was controlled by the number of water molecules. The sander module of AMBER-8 was used for all the simulations. The time per step was calculated from a measured wall-clock time of simulations over a thousand steps. In the case of the free boundary condition with direct force summation, we obtained enormous accelerations. Here, we use a single core Intel Xeon® 5160 processor with 2-chip and 12-chip MDGRAPE-3 boards. With the MDGRAPE-3 systems, we obtained speeds that were more than 100 times faster for systems over 5,000 particles.

Next, we show the results of the particle-mesh Ewald method. Table 10.3 shows the results of the joint-AMBER-CHARMM (JAC) benchmark. Here, again, we use the same host with a 2-chip MDGRAPE-3 board. In this case, we obtained a modest acceleration of 5.4, since our machine only accelerates direct force calculations. The direct part was accelerated nearly 30 times; however, the reciprocal part still persists. Now, we attempt to accelerate the reciprocal part by using parallel execution on the host machines. Recently, smart cutoff methods that can substitute PME have been proposed by several groups [33, 34, 5]. These methods are suitable for the MDGRAPE accelerators, since they require only the direct part for nonbonded forces. The implementation of a modified Wolf method using the 12-chip MDGRAPE-3 board was 30 times faster than the PME calculation using the single-core host [12].

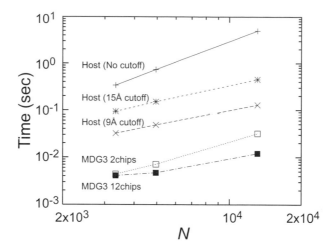

FIGURE 10.7: Calculation time per step versus number of atoms. The upper three lines are the results using only the host machine with various cutoff lengths. The bottom two lines are those using 2-chip and 12-chip MDGRAPE-3 (MDG3) systems without cutoff.

TABLE 10.3: Breakdown of calculation time for JAC benchmark.

	original		MDG3
Cutoff(Å)	9	10.1	10.1
N_{fft}	64	54	54
Direct (sec)	195.2	264.3	9.0
Reciprocal (sec)	40.9	31.4	27.8
List (sec)	21.2	25.7	0.5
Shake, etc. (sec)	10.0	10.2	12.0
Total (sec)	267.3	331.6	49.3
Acceleration	1	0.8	5.4

Next, we explain the performance of the entire system with large-scale calculations [17]. In this simulation, we investigated the amyloid-forming process of the Sup35 protein of yeast. Here, we report a result with the highest sustained performance. The simulation system contained 462 random coil peptides with seven residues (GNNQQNY) and 4,276,993 water molecules. It has no amyloid nucleus. The solute molecules in the system were enclosed in a box having a size of $400.5\text{Å} \times 445\text{Å} \times 712\text{Å}$ filled with water molecules. The TIP4P water model and a cutoff of 44.5 Åare used. The software used is the MOA. The host operating system is CentOS 4.3 for x86_64 with the Voltaire MPI library; further, the Intel C compiler for x86_64 was used. In

TABLE 10.4: Performance and parameters for the fastest simulation

Description	
Number of total particles	17,157,406
Number of particles that interact via van der Waals potential	4,326,427
Number of particles in a cell	11,915
Number of j-particles	321,701
Number of j-particles via van der Waals potential	81,121
Number of floating-point operations between two particles that interact by only the Coulomb potential	36
Number of floating-point operations between two particles that interact by the Coulomb and van der Waals potentials	46
Number of floating-point operations per time step	2.02×10^{14}
Number of time steps	10,000
Total number of floating-point operations in the simulation	2.02×10^{18}
Total time for the simulation (sec)	5,464
Peak speed for the simulation system (TFLOPS)	822
Calculation speed (TFLOPS)	370
Effective speed (TFLOPS)	185
Efficiency (%)	45

this run, all the system boards are set to 250 MHz, and the nominal peak performance corresponds to 860 TFLOPS.

Table 10.4 summarizes the obtained performance and system parameters for the simulation, wherein we achieved a performance of 185 TFLOPS. We counted the number of floating-point operations only for the Coulomb and van der Waals force calculations performed on the MDGRAPE-3 board because other calculations performed by the host computer are negligible. We obtained the effective speed from the following three steps:

1. Count the number of pairwise force calculations for the Coulomb and van der Waals forces (lines 2–6 in Table 10.4).

2. Estimate the equivalent floating-point count for a pairwise force calculation using the MDGRAPE-3 system (lines 7–8 in Table 10.4).

3. Eliminate duplicated calculations on the MDGRAPE-3 system (lines 8 and 14 in Table 10.4).

In the simulation, we truncated the forces at a finite distance by using the cell-index method. Each atom interacts with the atoms in its neighboring 3^3 cells. Thus, the number of j-particles N_j is approximated as $27N_{cell}$, where N_{cell} denotes the average number of atoms in a cell. It should be noted that the number of particles that interact with the van der Waals forces — N_{vdw} — is roughly one-fourth of the total number of atoms because only oxygen atoms in a water molecule have van der Waals interaction in the TIP4P model. The effective number of floating-point operations in a pairwise force calculation using the MDGRAPE-3 system should be converted to a reasonable value that is consistent with that of the general-purpose computers for each force. Since the pipelines in the MDGRAPE-3 system modify the potential functions by updating the coefficient table of segmented polynomials, the speed in evaluating a pairwise interaction is independent of the potentials. We count both division and square-root operations as ten floating-point operations. Then, the floating-point operation count for the Coulomb and van der Waals forces are 36 and 32, respectively. Some pairs of atoms interact with both Coulomb and van der Waals forces. Since the calculation of an inverse (square) distance between a pair is commonly performed in these force calculations, we used a conversion count of 46 for the pairs interacting with both Coulomb and van der Waals forces. It should be noted that when the MDGRAPE-3 system has to perform the force calculation twice, it performs more calculations than required. By assuming all the abovementioned estimations, the effective calculation speed is reduced to 370 TFLOPS. Furthermore, the use of action-reaction symmetry can reduce the calculation cost to half its value. Hence, from these assumptions, we obtained a final effective speed of 185 TFLOPS. The action-reaction symmetry is often used in simulations with general-purpose machines; however, its hardware implementation inhibits the use of the broadcast memory architecture, and the hardware costs increase to a considerable extent. Therefore, the MDGRAPE-3 system calculates the same pairwise interaction twice for each pair of atoms. By eliminating all these duplicated operations in the MDGRAPE-3 system, the effective number of floating-point operations per cycle per pipeline is 16.7 for the current simulation. By using this value, the effective peak performance of the system for the simulation — the performance at 100% efficiency of the MDGRAPE-3 system — is calculated to be 411 TFLOPS. The efficiency of the simulations should be examined on the basis of this performance and not the nominal peak performance of 860 TFLOPS for the Coulomb force calculations. By taking this into account, the obtained sustained performance of 185 TFLOPS corresponds to an efficiency of 45%, which is fairly appropriate.

10.8   Summary and Future Directions

We have successfully built the MDGRAPE-3, a special-purpose computer system for MD simulations with a petaflops nominal peak speed. It consists of a host cluster with special-purpose engines having 4,778 dedicated LSIs. The custom ASIC MDGRAPE-3 chip has a peak performance of 216 GFLOPS at 300 MHz and 17W. The total cost was about 8.6 million U.S.

dollars including labor costs and no cost for the host computer. We were able to achieve sustained performance of 185 TFLOPS during the practical simulation. Currently, we are using the system for various types of simulations — protein-folding simulations, MD-based drug designs, studies on protein dynamics, and virus simulations.

Such an accelerator approach will become more popular in the future. Currently, several projects using heterogeneous processors are running. For example, the GRAPE-DR project, which is operated by the National Astronomical Observatory, University of Tokyo, and RIKEN aims to develop a PC cluster with quasi-general-purpose accelerators of 2 PFLOPS (single precision) [14]. They have successfully developed its processor with the nominal peak performance of 512 GFLOPS, which is essentially considered as a SIMD accelerator with the broadcast memory architecture. The TSUBAME system in the Tokyo Institute of Technology introduced Clearspeed advance accelerator boards having 35 TFLOPS in total [32]. The Roadrunner was begun by Los Alamos National Laboratory and IBM as a project to develop a heterogeneous machine with a sustained performance of 1 PFLOPS by using Cell Broadband Engine® and AMD Opteron® processors. Since it will become more difficult to utilize an increasing number of transistors in future microprocessors, these accelerator approaches will become more popular. Another prospective architecture for future high-performance processors is a tile processor. Several projects already exist, for example, the MIT RAW microprocessor [30] or University of Texas Austin TRIPS [23]. Since the tile processor can be also considered as reconfigurable pipelines, it is closely related to the special-purpose approach in the GRAPE architecture. Such a pipelined operation may be one of its advantages in comparison with SIMD accelerators. The combination of a tile processor and dedicated units, a function evaluator or a force pipeline itself, for example, is also promising. Such dedicated units can be easily inserted into a pipeline implemented in a tile processor, and will enhance performance considerably for a small number of transistors. For the next-generation system, MDGRAPE-4, we are currently investigating the possibility of a tile processor with dedicated units. Such processors will realize high performance, low power, and reasonable flexibility at the same time.

10.9   Acknowledgments

The authors would like to express their sincere gratitude to their coworkers in the team, Dr. Noriaki Okimoto, Dr. Atsushi Suenaga, Dr. Noriyuki Futatsugi, Dr. Naoki Takada, and Ms. Ryoko Yanai. The authors thank Professor Takahiro Koishi for providing MOA software. The simulations of Sup35 were performed under collaboration with the above-listed people and

Dr. Ryutaro Himeno, Dr. Shigenori Fujikawa and Dr. Mitsutu Ikei. The authors thank Dr. Yoshiyuki Sakaki, Dr. Tomoya Ogawa, Professor Shigeyuki Yokoyama, and Dr. Toshikazu Ebisuzaki for providing their support to this project. The authors acknowledge all people in the GRAPE projects, especially Emeritus Professor Daiichiro Sugimoto, Professor Junichiro Makino, Professor Tomoyoshi Ito, Dr. Toshuiyuki Fukushige, Dr. Atsushi Kawai, and Dr. Ryutaro Susukita. The authors also thank Japan SGI Inc. for system support. In particular, we would like to thank Masahiro Yamaguchi, Hiroshige Tatsu, and Junya Shinohara for their assistance in the system installation. The authors thank the Digital Enterprise Group, Intel Corporation, for providing their early processor and its system support. This work was partially supported by Research Revolution 2002 "Protein 3000 Project," operated by the Ministry of Education, Culture, Sports, Science and Technology, Japan.

References

[1] M. P. Allen and D. J. Tildesley. *Computer Simulation of Liquids*. Oxford University Press, Oxford, U.K., 1987.

[2] A. F. Bakker and C. Bruin. Design and implementation of the Delft molecular-dynamics processor. In Berni J. Alder, editor, *Special Purpose Computers*, pages 183–232. Academic Press, San Diego, 1988.

[3] B. R. Brooks, R. E. Bruccoleri, B. D. Olafson, D. J. States, S. Swaminathan, and M. Karplus. CHARMM: A program for macromolecular energy, minimization, and dynamics calculations. *Journal of Computational Chemistry*, 4:187–217, 1983.

[4] D. A. Case, T. Darden, T. E. Cheatham III, C. Simmerling, J. Wang, R. E. Duke, R. M. Luo, K. M. Merz, B. Wang, D. A. Pearlman, M. Crowley, S. Brozell, V. Tsui, H. Gohlke, J. M. Morgan, V. Hornak, G. Cui, P. Beroza, C. Schafmeister, J. W. Caldwell, W. S. Ross, and P. A. Kollman. *AMBER 8*. University of California, San Francisco, 2004.

[5] C. J. Fennell and J. Daniel Gezelter. Is the Ewald summation still necessary? Pairwise alternatives to the accepted standard for long-range electrostatics. *J. Chem. Phys.*, 124:234104, 2006.

[6] R. Fine, G. Dimmler, and C. Levinthal. FASTRUN: A special purpose, hardwired computer for molecular simulation. *PROTEINS: Structure, Function and Genetics*, 11:242–253, 1991.

[7] T. Fukushige, J. Makino, T. Ito, S. K. Okumura, T. Ebisuzaki, and

D. Sugimoto. A special purpose computer for particle dynamics simulations based on the Ewald method: WINE-1. In V. Milutinovix and B. D. Shriver, editors, *Proceedings of the 26th Hawaii International Conference on System Sciences*, pages 124–133, Los Alamitos, 1992. IEEE Computer Society Press.

[8] T. Fukushige, M. Taiji, J. Makino, T. Ebisuzaki, and D. Sugimoto. A highly-parallelized special-purpose computer for many-body simulations with arbitrary central force: MD-GRAPE. *Astrophysical J.*, 468:51–61, 1996.

[9] T. Ito, J. Makino, T. Ebisuzaki, S. K. Okumura, and D. Sugimoto. A special-purpose computer for N-body simulations: GRAPE-2A. *Publ. Astron. Soc. Japan*, 45:339, 1993.

[10] A. Kawai and J. Makino. Pseudoparticle multipole method: A simple method to implement a high-accuracy tree code. *Astrophysical J.*, 550:L143–L146, 2001.

[11] A. Kawai, J. Makino, and T. Ebisuzaki. Performance analysis of high-accuracy tree code based on the pseudoparticle multipole method. *The Astrophysical Journal Supplement Series*, 151:13–33, 2004.

[12] G. Kikugawa, R. Apostolov, N. Kamiya, M. Taiji, R. Himeno, H. Nakamura, and Y. Yonezawa. Application of MDGRAPE-3, a special purpose board for molecular dynamics simulations, to periodic biomolecular systems, 2007. Submitted.

[13] J. Makino. Treecode with a special-purpose processor. *Publ. Astron. Soc. Japan*, 43:621–638, 1991.

[14] J. Makino, K. Hiraki, and M. Inaba. GRAPE-DR: 2-Pflops massively-parallel computer with 512-core, 512-Gflops processor chips for scientific computing. In *Proceedings of Supercomputing 2007*, November 2007. http://grape-dr.adm.s.u-tokyo.ac.jp/index-en.html.

[15] J. Makino, E. Kokubo, and M. Taiji. HARP: A special-purpose computer for N-body simulations. *Publ. Astron. Soc. Japan*, 45:349, 1993.

[16] J. Makino and M. Taiji. *Scientific Simulations with Special-Purpose Computers*. John Wiley & Sons, Chichester, 1998.

[17] T. Narumi, Y. Ohno, N. Okimoto, T. Koishi, A. Suenaga, N. Futatsugi, R. Yanai, R. Himeno, S. Fujikawa, M. Ikei, and M. Taiji. A 185 TFLOPS simulation of amyloid-forming peptides from Yeast prion Sup35 with the special-purpose computer system MDGRAPE-3. In *Proceedings of Supercomputing 2006*, 2006.

[18] T. Narumi, R. Susukita, T. Ebisuzaki, G. McNiven, and B. Elmegreen. Molecular dynamics machine: Special-purpose computer for molecular dynamics simulations. *Molecular Simulation*, 21:401–415, 1999.

[19] T. Narumi, R. Susukita, T. Koishi, K. Yasuoka, H. Furusawa, A. Kawai, and T. Ebisuzaki. 1.34 Tflops molecular dynamics simulation for NaCl with a special-purpose computer: MDM. In *Proceedings of Supercomputing*, 2000.

[20] Y. Ohno, E. Nishibori, T. Narumi, T. Koishi, T. H. Tahirov, H. Ago, M. Miyano, R. Himeno, T. Ebisuzaki, M. Sakata, and M. Taiji. A 281 Tflops calculation for x-ray protein structure analysis with special-purpose computers MDGRAPE-3. In *Proceedings of Supercomputing 2007*, 2007.

[21] Y. Ohno, M. Taiji, A. Konagaya, and T. Ebisuzaki. MACE : MAtrix Calculation Engine. In *Proc. 6th World Multiconference on Systemics, Cybernetics and Informatics SCI*, pages 514–517, 2002.

[22] M. Parrinello and A. Rahman. Polymorphic transitions in single crystals: A new molecular dynamics method. *J. Appl. Phys.*, 52:7182–7190, 1981.

[23] K. Sankaralingam, R. Nagarajan, H. Liu, C. Kim, J. Huh, N. Ranganathan, D. Burger, S. W. Keckler, R. G. McDonald, and C. R. Moore. TRIPS: A polymorphous architecture for exploiting ILP, TLP, and DLP. *ACM Trans. Archit. Code Optim.*, 1(1):62–93, 2004.

[24] A. Suenaga, N. Okimoto, N. Futatsugi, Y. Hirano, T. Narumi, A. Kawai, R. Susukita, T. Koishi, H. Furusawa, K. Yasuoka, N. Takada, M. Taiji, T. Ebisuzaki, and A. Konagaya. A high-speed and accurate molecular dynamics simulation system with special-purpose computer: MDM - application to large-scale biomolecule -. in preparation, 2003.

[25] D. Sugimoto, Y. Chikada, J. Makino, T. Ito, T. Ebisuzaki, and M. Umemura. A special-purpose computer for gravitational many-body problems. *Nature*, 345:33, 1990.

[26] M. Taiji. MDGRAPE-3 chip: A 165 Gflops application specific LSI for molecular dynamics simulations. In *Proceedings of Hot Chips 16*. IEEE Computer Society, 2004.

[27] M. Taiji, J. Makino, T. Ebisuzaki, and D. Sugimoto. GRAPE-4: A teraflops massively-parallel special-purpose computer system for astrophysical *N*-body simulations. In *Proceedings of the 8th International Parallel Processing Symposium*, pages 280–287, Los Alamitos, 1994. IEEE Computer Society Press.

[28] M. Taiji, J. Makino, A. Shimizu, R. Takada, T. Ebisuzaki, and D. Sugimoto. MD-GRAPE: A parallel special-purpose computer system for classical molecular dynamics simulations. In R. Gruber and M. Tomassini, editors, *Proceedings of the 6th Conference on Physics Computing*, pages 609–612. European Physical Society, 1994.

[29] M. Taiji, T. Narumi, Y. Ohno, N. Futatsugi, A. Suenaga, N. Takada, and A. Konagaya. Protein explorer: A petaflops special-purpose computer system for molecular dynamics simulations. In *Proceedings of Supercomputing 2003*, 2003.

[30] M. B. Taylor, J. Kim, J. Miller, D. Wentzlaff, F. Ghodrat, B. Greenwald, H. Hoffman, P. Johnson, J.-W. Lee, W. Lee, A. Ma, A. Saraf, M. Seneski, N. Shnidman, V. Strumpen, M. Frank, S. Amarasinghe, and A. Agarwal. The Raw microprocessor: A computational fabric for software circuits and general-purpose programs. *IEEE Micro*, 22(2):25–35, 2002.

[31] The Blue Gene/L Team. An overview of the Blue Gene/L supercomputer. In *Proceedings of Supercomputing 2002*, 2002.

[32] Global Scientific Information and Computing Center, Tokyo Institute of Technology. http://www.gsic.titech.ac.jp/index.html.en.

[33] D. Wolf, P. Keblinski, S. R. Phillpot, and J. Eggebrecht. Exact method for the simulation of Coulombic systems by spherically truncated, pairwise r^{-1} summation. *J. Chem. Phys.*, 110:8254–8282, 1999.

[34] X. Wu and B. R. Brooks. Isotropic periodic sum: A method for the calculation of long-range interactions. *J. Chem. Phys.*, 122:044107, 2005.

Chapter 11

Simulating Biomolecules on the Petascale Supercomputers

Pratul K. Agarwal

Oak Ridge National Laboratory &
University of Tennessee, Knoxville
agarwalpk@ornl.gov

Sadaf R. Alam

Oak Ridge National Laboratory
alamsr@ornl.gov

Al Geist

Oak Ridge National Laboratory
gst@ornl.gov

11.1   Introduction

Computing continues to make a significant impact on biology. A variety of computational techniques have allowed rapid developments in design of experiments as well as collection, storage and analysis of experimental data. These developments have and are leading to novel insights into a variety of biological processes. The strength of computing in biology, however, comes from the ability to investigate those aspects of biological processes that are either difficult or are beyond the reach of experimental techniques. Particularly in the last three decades, availability of increasing computing power has had a significant impact on the fundamental understanding of the biomolecules at the molecular level. Molecular biochemists and biophysicists, through theoretical multi-scale modeling and computational simulations, have been able to obtain atomistic level understanding of biomolecular structure, dynamics, folding and function. The protein-folding problem, in particular, has attracted considerable interest from a variety of researchers and simulation scientists. However, it still remains an unsolved problem of modern computational biology. The lack of sufficient computing power has been commonly cited as the main factor holding back progress in the area of computational molecular biochemistry/biophysics.

Simulations of biomolecules are based on molecular dynamics (MD), energy minimization (EM) and related techniques; in combination with atomistic level description of the biomolecular system [27, 31]. MD simulations can be defined as computer-simulation methodology where the time evolution of a set of interacting particles is modeled by integrating the equation of motion. The underlying MD technique is based on the law of classical mechanics, most notably Newton's law, $F = ma$; and a mathematical description of the molecular energy surface. This mathematical function, known as the potential function, describes the interactions between various particles in the system as a function of geometric variables. MD simulations consist of three calculations: determining energy of a system and forces on atoms' centers, moving the atoms according to forces, and adjusting temperature and pressure. The force on each atom is represented as the combination of the contribution from forces due to atoms that are chemically bonded to it and nonbond forces due to all other atoms. For EM and related techniques, the stable conformations of the biomolecules are searched by minimizing the potential function to reach local (or global) minima. The lower energy states are associated with native and functioning conformation of biomolecules, and in the case of interaction of protein-ligands (or other biomolecules) the lower energy structure represents favorable interactions, therefore, structures of interest in docking-type studies. A wide variety of software packages are available for biomolecular simulations. CHARMM [15], AMBER [16], LAMMPS [37], and NAMD [26] are some of the popular codes that are in use by a variety of simulation scientists.

In the 1970s, MD simulations were used by Karplus, McCammon and coworkers to perform simple simulations of a protein in vacuum for about 10 picoseconds (10^{-11} seconds), a very short duration by today's standards [32]. These pioneering simulations provided new insights into proteins being dynamic entities that constantly undergo internal fluctuations — a shift from the rigid nature of proteins as made familiar by the ball-and-stick models and paper drawings. Over the next two decades advances in computing led to simulations several nanoseconds ($>10^{-9}$ seconds) in duration and consisting of an entire protein in explicit water with thousands of atoms [39]. In 1998, Kollman and coworkers reported a 1 microsecond (10^{-6} second) simulation of folding of a small peptide (36-residue) in explicit water [19]. This simulation represented a major development; it took over two months of computing time on 256-processor Cray T3D followed by another two months on 256-processor Cray T3E-600, and provided insights into events during the protein-folding process.

In 2001, it was suggested that as much as 1 PFLOPS computing power for an entire year will be required to simulate the folding process of a medium-size protein [11]. For biologically relevant length and timescales, it has been suggested that the computing power available typically falls 4–6 orders of magnitude short of what is desired. For example, a common subject of investigation is enzyme catalysis; enzymes catalyze their target reactions on a variety of timescales [3]. The fastest enzymes catalyze the reaction on the nanosecond timescale (10^{-9} seconds), while slower enzymes can take several seconds or much longer ($>10^0$ seconds). The current achievable timescale combining the best supercomputing power and most scalable software allows only about 100 nanoseconds, for a biologically relevant protein in a realistic wall-clock time. Therefore, there is a significant gap between the desired and achievable simulations. With the arrival of petascale machines, the amount of computing power available will increase substantially.

In 2007, the petascale machines are around the corner. As indicated by the "Top500" list published in November 2006, the top machine is able to deliver 280 TFLOPS with two additional machines that are close to the 100 TFLOPS range [33]. It is expected that the National Center for Computational Sciences (NCCS) at Oak Ridge National Laboratory will have one of the first petascale machine in the near future [34, 35]. In December 2006, NCCS housed a 54 TFLOPS supercomputer XT3 built by Cray, which grew to 119 TFLOPS in March 2007. The NCCS roadmap suggests that this computer is expected to grow to 250 TFLOPS in December 2007. Finally in December 2008, it will deliver 1 PFLOPS of computing power [35]. Availability of this unprecedented computing power provides new opportunities and also brings to light new challenges for the applications routinely in use by the wide community of computational molecular biochemists/biophysicists. The software for these applications was conceived and developed several decades ago, even though it has been considerably updated over the years, but it has failed to keep up with the emerging supercomputer architecture. This is particularly evident with the arrival of petascale computers.

This chapter briefly presents opportunities and challenges in the biomolecular simulation area to efficiently utilize the computing power at the petascale. In Section 11.2, we mention several new opportunities that will be opened up by the availability of petascale computing power and how it will impact the discoveries in biomolecular investigations. In Section 11.3 we discuss issues pertaining to computer hardware and software technology, as well as from the simulations methodology that will need to be addressed. Finally in Section 11.4, we provide a summary and outlook for the biomolecular simulations on petascale machines. We note that the material presented here may reflect some of our own biases on this topic and that opinions of many other researchers in the field would have also sufficed.

11.2   Opportunities

Biomolecular simulations enable the study of complex, dynamic processes that occur in biological systems and biochemical reactions. MD and related methods are now routinely used to investigate the structure, dynamics, functions, and thermodynamics of biological molecules and their complexes. The types of biological activities that have been investigated using MD simulations include protein folding, enzyme catalysis, conformational changes associated with biomolecular function, and molecular recognition of proteins, DNA, and biological membrane complexes. Biological molecules exhibit a wide range of time and length scales over which specific processes occur, hence the computational complexity of an MD simulation depends greatly on the length and timescales considered. The amount of computational work in these simulations is dependent on the number of particles (atoms) in the systems as well as the number of time-steps that are simulated. The use of explicit or implicit solvent conditions significantly impacts the number of atoms in the system as well. Explicit solvents are more realistic but come at an increased computational cost due to the addition of more atoms, while implicit models are fast but do not provide important details about the nature of solvent around the biomolecule. With an explicit solvation model, typical system sizes of interest range from 20,000 atoms to more than 1 million atoms; if the solvation is implicit, sizes range from a few thousand atoms to about 100,000. The simulation time period can range from picoseconds (10^{-12} seconds) to a few microseconds or longer ($>10^{-6}$ seconds) on contemporary platforms.

11.2.1   Ability to investigate bigger biomolecular systems

Biomolecules vary considerably in size ranging from a few hundred to millions of atoms. Moreover biomolecules rarely function alone; therefore, realistic simulations require simulations of entire complexes including protein–protein complexes, protein–DNA complexes, protein–lipid membrane complexes and other assemblies including entire viral particles. For several decades the simulations were limited to systems with a few hundred thousand atoms. In recent years, success has been reported by Schulten and coworkers in simulating an entire assembly of satellite tobacco mosaic virus (Figure 11.1) [20]. This complex consisted of over 1-million atoms and was simulated using NAMD. These simulations provided insights into the electrostatic potential, dynamical correlations and structural features of the viral particle over a timescale of 50 nanoseconds. Further, the Sanbonmatsu research group at Los Alamos National Laboratory has also succeeded in simulating the entire ribosome complex consisting of 2.64 million atoms for 20 nanoseconds [38]. Table 11.1 provides the range of biomolecular complex sizes investigated using biomolecular simulations.

FIGURE 11.1: (See color insert following page 18.) **Biomolecular simulations of a system with >1 million atoms.** This figure shows a schematic representation of satellite tobacco mosaic virus particle. The viral particle was solvated in a water box of dimensions 220 Å × 220 Å × 220 Å, consisting of about 1.06 million atoms. The protein capsid (green) is enveloping the RNA and part of the capsid is cut out to make the RNA core of the particle visible. The backbone of RNA is highlighted in red; ions were added to make the system charge neutral. Figure courtesy of Theoretical and Computational Biophysics Group. Reprinted with permission from P.L. Freddolino, et al., *Structure (2006)*, 14, 437-449. ©Elsevier 2006.

Petascale computing power will provide the ability to simulate multimillion atom simulations for larger complexes in a realistic timeframe. Moreover, multiple simulation trajectories for these large systems will provide valuable insights into the mechanistic details through investigations in different conditions including the studies of proteins with mutations or DNA with different sequences. Regular simulations of multimillion atom complexes will have significant impact on the understanding of complex processes involving membrane-based systems (such as ion-channels and G-protein-coupled receptors), DNA-replication and the translation process with an entire ribosome. In the past, these systems have been considered beyond the reach of biomolecular simulations due the requirement of large computing power.

TABLE 11.1: **Range of biomolecular system sizes.** The listed simulation systems include explicit solvents and the simulation times are based on a time-step of 1 femtosecond (10^{-15} s). Simulation runs were performed on the Cray XT3/4 unless otherwise mentioned.

Number of atoms	Type of system	Example	Time to simulate 10 nanoseconds
20 K - 30 K	Small protein 100–200 amino-acid residues	Enzyme dihydrofolate reductase with substrate [26,490 atoms]	1 day (PMEMD on 128 cores)
50 K- 100 K	Medium-size protein 300–400 amino-acid residues or a small protein–DNA complex	Protein–DNA complex with M. HhaI methyltransferase and 12 base-pair DNA [61,641 atoms]	36 hours (PMEMD on 512 cores)
100 K - 300 K	Large/multi-chain protein complex with $\geq$1,000 amino-acid residues	Ribulose-1,5-bisphosphate carboxylase/oxygenase (RuBisCO) enzyme [290,220 atoms]	40 hours (LAMMPS on 4096 cores)
300 K -800 K	Multi-biomolecule complexes	Cellulase on cellulose surface; membrane protein complex	
1-2 million	Entire viral particles	Satellite tobacco mosaic virus [~1.06 million atoms]	9 days [20] (NAMD on 256 nodes SGI Altix)
2-5 million	Large complexes or cellular organelles	Entire ribosome [~2.64 million atoms]	

Structure ⟷ Dynamics ⟷ Function

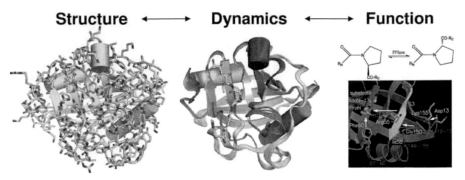

FIGURE 11.2: (See color insert following page 18.) **The dynamic personality of proteins.** An integrated view of protein structure, dynamics and function is emerging where proteins are considered dynamically active molecular machines. Biomolecular simulations spanning multiple timescales are providing new insights into the working of protein systems. Computational modeling of enzymes is leading to the discovery of network of protein vibrations promoting enzyme catalysis in several systems including cyclophilin A, which is shown here.

11.2.2 Ability to investigate longer timescales

Biochemical processes span multiple scales of time ranging from a few picoseconds to hours. Simulations covering a wide range of timescales are required for understanding the mechanistic details of these processes. For example, there is wide interest in understanding the link between protein structure, dynamics and function. Protein structure consists of arrangement of amino-acid residues in a 3-dimensional shape and intrinsic protein dynamics refers to the internal motions that occur within the protein at different timescales, ranging from femtosecond (10^{-15} second) to second and longer. The internal motions of the proteins have been implicated in the protein function such as enzyme catalysis. Enzymes are a class of proteins that catalyze chemical changes that occur in the cell. Experimental techniques continue to provide some details at selective timescales for protein dynamics and its link to protein function.

Multi-scale computational modeling and simulations have provided fascinating details about protein molecules. Novel insights into the detailed mechanism of several enzymes have been obtained through simulations. Based on the results of these computational studies, an integrated view of protein structure, dynamics and function is emerging, where proteins are considered as dynamically active assemblies and internal motions are closely linked to function such as enzyme catalysis (Figure 11.2). Multi-scale modeling of enzyme cyclophilin A over the entire reaction path has led to an interesting discovery that the internal protein dynamics of this enzyme is linked to its

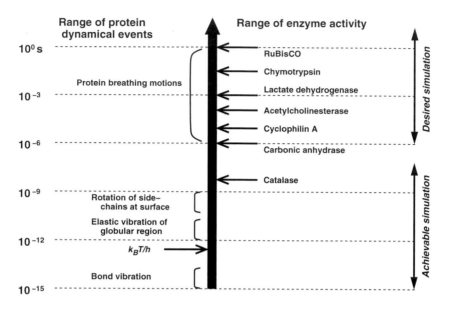

FIGURE 11.3: **Multi-scale modeling and simulations in biology.** Computational biochemists/biophysicists are now regularly using molecular simulations to investigate enzyme complexes. The structure, dynamics and function of enzyme complexes span multiple scales of time and length. Enzymes catalyze biochemical reactions as fast as billions of times per second on the right side of the range, while on the left side they can take seconds or longer for one catalytic cycle. The wide range of internal protein motions occur on 10^{-15} to $>10^{0}$ seconds, which are linked to a variety of protein functions (such as enzyme catalysis) on similar timescales. However, the current simulations fall short by several orders of magnitude. The typical state-of-the-art simulations can only reach 10^{-7} seconds at best for a real biological system, while the desired timescale is 10^{-6} to 10^{0} seconds or higher.

catalytic activity [6, 1, 2]. These modeling studies have identified protein vibrational modes that occur at the timescale of the reaction and play a role in promoting catalysis. Similar results have also been obtained for other enzymes including the enzyme dihydrofolate reductase [5]. Further, the role of hydration-shell and bulk solvent as well as temperature effects in enzyme mechanisms are now being understood [2].

Figure 11.3 depicts the wide range of timescales for activity of several enzymes. The fastest enzyme performs its function over a billion times per second, while slower enzymes can take seconds or longer to complete one cycle. It is interesting to note that a wide range of internal protein motions also occur on similar timescales as the enzyme function; therefore, raising the interesting question whether enzyme dynamics and function are interrelated or not. Currently, computational biochemists and biophysicists can only simulate a fraction of the biologically relevant timescales for most enzymes. Most common MD simulations on a single workstation or using a small PC-cluster explore nanosecond (10^{-9} s) timescales for a medium-size protein in an aqueous environment consisting of 5,000-20,000 atoms. Supercomputers can simulate around 100 nanoseconds for multimillion atom systems. These simulations continue to provide novel insights into enzymes. The combination of petascale hardware and scalable software is required to bridge the gap that exists between desired and achievable simulations.

Another aspect of proteins that has generated considerable interest from computational molecular biologists is the determination of a 3-dimensional protein structure from the primary sequence. Moreover, the protein-folding mechanism, the process by which proteins fold into their native or functional shape, has also been widely simulated for a number of proteins. The process of protein folding also involves a wide range of timescales, with faster events of local structure folding occurring at picosecond-nanosecond timescales while the overall process takes between milliseconds to seconds. As Figure 11.3 indicates, the commonly investigated timescale for MD simulations is the nanosecond, which falls 4–6 orders of magnitude short of the desired timescale of biological activity. It has been suggested that the computational requirements for multi-scale modeling of a medium-size protein can be as high as 1 PFLOPS for an entire year [11].

11.2.3   Hybrid quantum and classical (QM/MM) simulations

Understanding of the electronic and quantum effects within the biomolecules will be significantly impacted by the availability of petascale computing power. As compared to the MD simulations, quantum mechanical modeling requires significantly more computing. Methods such as hybrid quantum and classical simulation (QM/MM) methods will allow for combining the speed of classical methods with the accuracy of quantum methods. The classical methods including the MD simulations are fast but limited in their abilities. These methods do not allow simulation of breakage and formation of covalent bonds or multiple electronic states. Quantum methods will provide the ability to investigate systems with multiple electronic states as well as the breakage/ formation of covalent bonds that commonly occur in biochemical processes. In recent years, several groups have reported success with use of QM/MM methods to investigate the role of hydrogen tunneling in enzymes including liver alcohol dehydrogenase and dihydrofolate reductase [13, 4, 21]. It is expected that more biomolecular systems will be routinely investigated using these methods.

Full quantum calculations of entire proteins also will be made possible. The role of long-range interactions in biomolecules is an intensely debated topic;

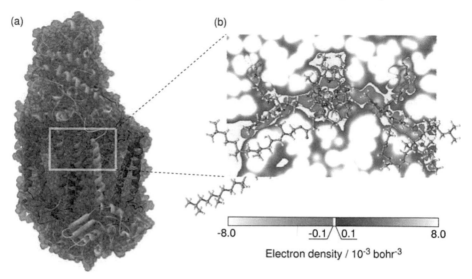

FIGURE 11.4: (See color insert following page 18.) **Full quantum calculation of a protein with 20,581 atoms.** Electron densities of the photosynthetic system were computed at the quantum level (RHF/6-31G*) with the FMO method: (a) an electron density of the whole system; and (b) a differential electron density around the special pair. Reprinted with permission from T. Ikegami, et al., *Proceedings of the 2005 ACM/IEEE Conference on Supercomputing* 2005.

they have been suggested as a factor playing a role in the function of enzymes, allosteric effects in biomolecules [41]. The calculation of full electronic density around the biomolecules with the use of quantum methods would provide valuable insights in the long-range interactions. Recent progress in theoretical methods and software developments, such as the development of the Fragment Molecular Orbitals (FMO) method [28], has already demonstrated that in combination with the available computing power, it is possible to investigate proteins with >10,000 atoms at the full quantum level. In the FMO method the system is divided into fragments and the electronic state for each fragment is separately calculated (special treatment is given for the bonds between the fragments). The electronic state calculation of a fragment is performed under the electrostatic environment posed by the other fragments. The fragment electron densities are obtained from these calculations, from which the environment is computed. The calculation is iterated until both fragment electron densities and the environment become mutually consistent. Using the FMO method the full electronic state for the photosynthetic reaction center of *Rhodopseudomonas viridis* has been computed (Figure 11.4) [25]. The system contained 20,581 atoms and 77,754 electrons. The calculation performed at the RHF/6-31G* level of theory took 72.5 hours with 600 CPUs. With petascale computing power it would be possible to perform quantum calculations on bigger systems, investigate quantum effects in multiple conformations of the proteins as well as investigate dynamics of the systems at the quantum level.

11.2.4 More accurate simulations

Biomolecules sample many conformations even in the natural state of functioning. The accurate estimate of various energetic components of a biochemical process requires better sampling of the conformations [29], which had been severely limited due to availability of limited computing power in the past. Petascale computing power will allow much more accurate estimates through better simulations. Methods such as umbrella sampling and empirical valence bond methods that allow computation of free energy changes during biochemical reactions require multiple simulations along the reaction pathway [21, 12]. Petascale computing along with new methodologies to perform simultaneous sampling of the various states along the reaction pathway, will provide quantitatively accurate information.

11.3 Challenges

Biomolecular simulation codes have been developed by a wide community of researchers, over a period that has spanned more than three decades. The earliest versions of the codes were designed for single processor machines and over the years they have been modified to utilize the computing power of supercomputers. Performance evaluations studies continue to bring to light several challenges that these codes will face on the petascale.

11.3.1 Scaling the biomolecular simulations code on >100K processors

Petascale machines bring a landmark change in the computing architecture paradigm. In the past, biomolecular simulations have benefited considerably from the increase in speed of the individual processors. In the last ten years alone, over a tenfold increase in processor speed has enabled an order of magnitude longer simulations, without any significant modifications to the code and programming paradigms. The petascale machines are expected to combine computing power of 100,000 or more individual processors without a substantial increase in the processor speed. The common MD software codes that have been developed over the last few decades were designed for a different programming paradigm. Unfortunately, these software codes have failed to keep up with the rapid changes in high-end parallel computer architectures. Therefore, it has become clear that software scalability and subsequently achievable performance will be an enormous challenge at the petascale [9].

The commonly used software in the biomolecular simulations community includes CHARMM [15], AMBER [16], LAMMPS [37], and NAMD [26]. CHARMM and AMBER have been developed over several decades and have a wide user base as they provide wide functionalities for simulations and analysis. The potential function and associated parameters (collectively referred as *force field*) for these packages has been developed and tested for more than 20 years. However, these codes scale only to a few hundred processors (128–256). LAMMPS and NAMD on the other hand are more recent efforts that have been designed keeping the emerging computer architecture in mind; therefore, these codes have been demonstrated to scale to several thousand nodes (see Figure 11.5). [Note, PMEMD that is a newer simulation code available in AMBER, can scale to about 1024 nodes.] The use of the particle mesh Ewald (PME) method for calculation of long-range interactions has allowed significant improvements in simulation performance [18, 17]. PME utilizes fast Fourier transform (FFT) and is commonly used when simulations are performed in explicit solvent conditions. LAMMPS and NAMD are expected to scale to a few tens of thousands of nodes; however, the ability to scale these codes to effectively utilize the computing power of >100K cores still remains an open question. In recent years, several proprietary MD frameworks targeting the development of scalable software have also emerged [22, 14].

Bigger sized biomolecular systems. Larger systems are expected to show better scaling for MD simulations due to an increase in the computational density (amount of computation on a processor core between communication steps) as the number of atoms is increased. Particularly, for systems with >1 million atoms it is expected that reasonable performance will be achieved on ten thousand nodes and above (see Figure 11.5). On the quantum side, with the use of methods such as the FMO method, it is expected that bigger biomolecules and protein complexes will efficiently utilize petascale computing power as well.

Longer timescales. As discussed above, new insights are expected from simulations that routinely achieve longer timescales and correspond to the timescale of biomolecular process. Researchers therefore require simulation that can provide trajectories reaching milliseconds or longer in realistic wallclock time. Longer timescales simulations require the number of steps per simulation per wall-clock second to be increased significantly. The interconnect latency is one aspect of the hardware design that is becoming apparent as the most limiting factor to achieving longer timescale simulations. It presents a' significant bottleneck to performance, as the ratio between computation and communication decreases with an increase in the number of parallel tasks. As an example, let us consider the ability to simulate millisecond timescales in a

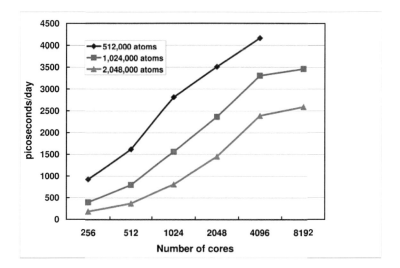

FIGURE 11.5: **Scaling behavior of LAMMPS.** Biomolecular systems with 512,000, 1,024,000 and 2,048,000 atoms were simulated on an increasing number of processor cores. Particle–particle–particle mesh (PPPM), closely related to the PME method, was used for the calculations. The results show that a performance metric of several picoseconds/day can be achieved.

day. The problem of biomolecular MD requires that the application synchronizes after every simulated time-step. During this synchronization step the individual computing tasks share the energy/forces and new coordinates are distributed. As an individual time-step corresponds to 1 femtosecond (10^{-15} seconds), a millisecond simulation per day would typically require about 10^7 time-steps/second or 0.1 microsecond/time-step! This places a strong emphasis on the interconnect latency. Currently, the fastest machines have latency typically around 2–5 microseconds. Therefore, latency will impact the scalability of codes, in turn significantly impacting the outcome of science in this area.

In order to understand the impact of interconnect latencies on a contemporary massively parallel system, Cray XT, we performed a range of scaling experiments to thousands of dual-core processors. Note that the interconnect latency for the Cray XT system is 6 microseconds, compensated by a high bandwidth interconnect [40]. For our evaluation, we selected NAMD as a representative MD framework that has been designed to scale on massively-parallel processors (MPPs) [30]. NAMD is a parallel, object-oriented MD program designed for simulation of large biomolecular systems [26]. NAMD employs the prioritized message-driven execution model of the Charm++/Converse parallel runtime system, in which collections of C++ objects remotely invoke methods on other objects with messages, allowing parallel scaling on both supercomputers and workstation clusters. We consider two representative petascale biological systems with approximately one (1M) and five million (5M) atoms, respectively. The simulation performance is measured in time per simulation step, which should theoretically decrease as processing units are added. Figure 11.6 shows the Cray XT4 performance for both the 1M- and 5M-atom simulations. Using the XT4 system in dual-core execution mode, the 1M-atom simulations scale to 4,096 cores and achieve ~13 ms/step, while the 5M-atom simulations scale to 12,000 XT4 cores, maintaining ~13 ms/step performance. Note that the IBM Blue Gene/L system, which has a lower interconnect latency than the Cray XT systems, achieves approximately 10 ms/step for a simulation of approximately 300 K atoms on 16K processing cores [30]. Unlike the Blue Gene/L runs, no system-specific optimization and instrumentation is applied for the XT4 simulation experiments. The analysis of NAMD trace files enabled us to identify that the scaling for the 1M-atom system is restricted by the size of the underlying FFT grid computations; hence, it does not scale beyond 2K processor cores. The larger system, 5M atoms, scales to 12K cores; however, the parallel efficiency is significantly reduced beyond 4K processor cores due to the aforementioned reason.

11.3.2 Adapting to the changes in the hardware

Memory. The memory requirements of MD codes are modest as they require only storage of the atomic coordinates and the topology-related data (charges, bond, angle constants and other force-field related information) and

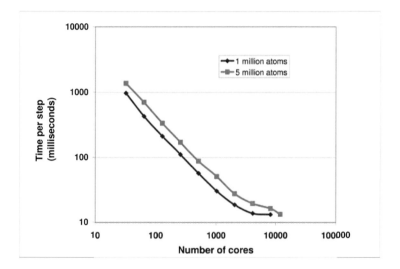

FIGURE 11.6: **Wall-clock time per simulation time-step as a function of the increasing number of cores.** These simulations were performed on the Cray XT4 system using the NAMD simulation framework. Two different systems with 1-million and 5-million atoms approximately were used for these studies.

pair-lists. The available memory of the next generation supercomputers, however, may also pose some challenges. Memory per core is not expected to increase considerably, as it is expected that the memory will typically go from 2 Gb to 8 Gb per node, while the number of cores per node increases from 2 to 4 or even as high as 80. Fortunately, as mentioned above, the demand placed on memory is not typically high. This is particularly true for codes which have implemented domain decomposition methods. However, in other codes, particularly the *legacy codes* which implement the replicated memory model, the available memory may limit the size of the system that can be simulated. This has been previously discussed for AMBER (*sander*) simulation on IBM Blue Gene/L [9]. However, most code developers feel that the memory available appears to be sufficient for most simulations currently used but may become an issue when better force fields (particularly polarizable force fields) and if fault-tolerance strategies are implemented (see below for further discussions).

Multicore processors and the need new programming models. The central calculation during a single MD time-step involves calculation of bonded and nonbonded energy terms and forces. This leads to a high computational density on processors as compared to little or no communication between processors during individual time-steps. As mentioned above, the interconnect latency and other factors impacting the data communication between nodes will play a major role in scalability of the calculations. The emergence of multicore technology introduces new trade-offs as increasingly more computations could be kept on a single node which reduces the amount of communication that needs to done between other nodes [10, 7]. The availability of 4 or 8 (and more) cores per node will help increase computational density in comparison to the network communications. multicore technology, however, also brings new challenges, particularly relating to the memory and programming models. In addition to the reduced amount of memory available per core, the memory bandwidth is also expected to continually decrease with the increase in the number of cores. Unless other solutions are made available, this may impact performance of MD codes. On the programming side, the level of parallelization on a single node and between nodes will pose complications for the parallel programming model. It appears that the most promising avenue in this regards is possibly the use of hybrid programming models, for instance, OpenMP shared-memory within a node and message-passing MPI among nodes.

In order to investigate the impact of shared resources among multiple cores within a processor, we attempt to quantify the slowdown in the dual-core execution mode with respect to the single-core execution times. We collected runtime scaling data on the XT3 platforms in single- and dual-core execution modes. Figure 11.7 show the percentage slowdown for two test systems, about

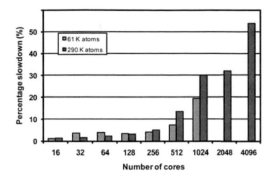

FIGURE 11.7: **Slowdown in MD performance on dual-core processors.** These studies were performed using a LAMMPS simulation framework with two different system sizes on the Cray XT3 system, in dual-core execution mode and compared to single-core execution mode.

61K and 290K atoms, respectively. Percentage slowdown is measured as:

$$slowdown(\%age) = \frac{100.0 \times (Time_{dual-core} - Time_{single-core})}{Time_{single-core}} \qquad (11.1)$$

The results show that the rate of the slowdown increases with the increase in the number of MPI tasks, which points to a possible impact of two cores sharing HyperTransport resources for message-passing MPI operations. Also, the slowdown percentage is higher for larger systems, i.e., systems with a large number of atoms. This could be an influence of higher memory and data transfer requirements on the shared-memory controller. Since the current software stack on the Cray XT systems does not support hybrid execution models [40], we have been unable to study the impact of hybrid programming on MD applications on large-scale systems. Currently, we are investigating OpenMP and MPI scaling, and memory and processor affinity schemes on stand-alone dual- and quad-core systems.

11.3.3 Fault-tolerance

As the MD codes start to utilize an increasing number of nodes and cores on petascale machines, with >100K processors, it is anticipated that hardware failure will not only hinder the performance of the simulation codes but could also affect the reliability of simulation results. The MD application frameworks have very little built-in fault-tolerance or check-pointing strategies. The well-known check-pointing strategy is saving the coordinates, velocities and atomic forces on the disk after a pre-determined number of steps. Even though restarting the simulation from this check-pointed data is trivial, the time spent in queues for restarting these simulations would severely

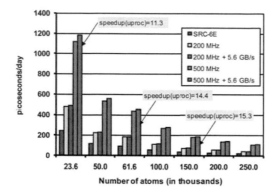

FIGURE 11.8: **Performance projections for MD simulation on FPGA devices.** The projections are performed by varying application (number of atoms) and device (clock frequency of FPGA devices and memory bandwidth to the host) parameters.

impact productivity. Moreover, in other cases when a node fails and the application stalls while waiting to receive data from the failed node. If such a situation occurs, it will also lead to a loss in productivity. There is a need for built-in fault-tolerance strategies and checks that allow self-health monitoring by the MD code and decision making in the case of detection of a node failure. One possible solution consists of check-pointing the simulation on the fly by saving critical data in a neighboring node's memory (assuming that enough memory is available without impacting simulation performance). In case of failure the application continues by reassigning the computation to a new node by sending the data from the neighboring node's memory. Such a strategy may require some interaction with the system-level tools that allow detection of node and other hardware failure.

11.3.4 Multi-paradigm hardware including reconfigurable computing

From the perspective of overcoming challenges due to interconnect latency, emerging multi-paradigm hardware, including systems with field programmable gate arrays (FPGAs) accelerators will allow localization of more atoms on individual nodes. However this poses programming challenges as discussed above. Moreover, the role of multi-paradigm hardware on emerging supercomputing machines also remains an open question. From the perspective of smaller calculations on workstations, the FPGAs could allow significant speedup as has been recently shown by porting AMBER's MD engine (*sander*). Figure 11.8 shows the performance projection results that are based on FPGA acceleration of the direct PME calculations in sander on SRC 6E systems using Fortran programming. We altered two FPGA-enabled system parameters for our experiments: the clock frequency and data bandwidth between the FPGA device and the host processor [8]. The clock frequency of our current FPGA implementation was 100 MHz and the sustained payload bandwidth is 2.8 GB/s (utilizing input and output 1.4 GB/s bandwidth). The clock speed and data transfer rates have different performance implications on small and large biological systems (speedup with respect to the microprocessor runtimes are highlighted in Figure 11.8). Overall, the clock speeds influence the performance gains of the PME calculations. For smaller biological systems, the change in the data transfer rates influences the runtime performance of the application. By contrast, the performance of the larger systems (100K and more atoms) nearly doubles by doubling the clock speed of the FPGA devices, while the data transfer rates do not impact the runtime performance of larger biological systems. Note that a 150K-atom system only achieves ∼12 picoseconds/day on a dual 2.8 GHz Xeon system today. An FPGA-enabled system using our current PME implementation can sustain over 75 picoseconds/day with 200 MHz and over 180 picoseconds/day with 500 MHz and a host bandwidth of 5.6 GB/s. FPGA devices are also being targeted by other research groups as means to speedup MD simulations [23].

11.3.5   New simulation methodologies enabled by petascale

Development of novel methods for insights into longer timescales. As discussed above, there is wider interest in simulating biomolecules and biochemical processes at longer timescales (microseconds and longer). It is becoming clearer that on the petascale machines addressing the time domain is a more challenging problem than addressing the length domain problem. Simulating longer length scales equates to larger numbers of atoms in the system. For a fixed number of processors, a rise in the number of atoms: increases the computational density on individual processors leading to better scaling; and also requires a larger FFT grid for the PME method, which is better handled on a larger number of processors in combination with domain decomposition methods. However, equivalent strategies to address the challenges arising for simulations that target longer timescales currently are not available. Therefore, interest also exists in developing alternate computational and simulation methods to reach longer timescale or methods that provide information and insights equivalent to longer timescales. A possible solution may be offered by simulations with multiple time-stepping, where the fast and slow degrees of the system are separated and certain tricks allow taking longer time-steps. In addition, methods that can be run as several closely coupled simulations performed simultaneously to sample conformations, may provide information from multiple timescales. Other strategies may include ways to independently explore separate *windows* in the longer timescales and then combining them through novel methods to obtain information from longer timescales. For example, umbrella sampling method for investigating reaction pathways and replica-exchange methods for conformational searches including protein folding have been proposed to address this problem on scalable systems. The availability of petascale computing will allow more trajectories to be computed quickly, therefore allowing access to longer timescales.

Development of new force fields. Another aspect of biomolecular simulations that requires mention is related to the accuracy of the force field parameters [24]. This issue does not directly relate to high-performance computing; however, it may impact the efficient utilization of computing resources. Even though over the last 30 years biomolecular simulations have provided a wealth of information regarding biomolecules, in recent years, it has become evident that the most common force fields have defects that are limiting new discoveries. In particular, the search for the global minimum of a protein, as a part of simulating the protein folding mechanism, has been considerably limited by force field defects. To address this challenge new polarizable force fields currently are being developed. These force fields are expected to require more computational resources to nonpolarizable counterparts. The availability of increasing computing power can possibly automate the procedure of improvement of force fields, as has been recently demonstrated [36].

11.4   Summary and Outlook

Petascale computing power brings new opportunities to biology. The quest for understanding the molecular basis of life has been aided considerably by computing through atomistic-level modeling of biomolecules. The tremendous increase in the computer power available to the scientist holds great promises. On one hand, petascale computing will provide new insights by allowing simulations to reach longer timescales as well as allow simulation of larger and more realistic systems. On the other hand, novel methods would allow simulation of challenging problems, such as protein folding and enzyme catalysis, by using a number of tightly coupled runs. A diverse community of simulation and computer scientists has contributed to a rich set of application frameworks available in this area. Developed and enhanced over the last three decades, these frameworks have some challenges to utilize the computing power of petascale machines. The petascale machines will combine computing power of >100 K individual processors, an architecture that was not envisioned during the course of development of current biomolecular simulation codes. New methodologies including the hybrid classical/quantum or fully quantum simulations may be able to benefit quickly from the tremendous increase in computing.

It is anticipated that in the coming years, the combination of algorithm and hierarchical software development together with petascale computing power will lead to vital breakthroughs in our understanding of biomolecular structure, dynamics, folding and function. Vital insights into the electronic environment, structural details, dynamic movements of the biomolecules and mechanistic level details of the biochemical processes will be made available. Petascale biomolecular simulations, therefore, will have direct and indirect impact on developments in health, energy and environmental research.

11.5   Acknowledgments

This research used resources of the Center for Computational Sciences at Oak Ridge National Laboratory (ORNL), which is supported by the Office of Science of the U.S. Department of Energy under contract no. DE-AC05-00OR22725. We also acknowledge the computing time allocation on the Cray XT3/XT4 of the National Center for Computational Sciences (NCCS) under the project BIO014.

References

[1] P. K. Agarwal. Computational studies of the mechanism of cis/trans isomerization in HIV-1 catalyzed by cyclophilin A. *Proteins: Structure, Function and Bionformatics*, 56:449–463, 2004.

[2] P. K. Agarwal. Role of protein dynamics in reaction rate enhancement by enzymes. *Journal of the American Chemical Society*, 127:15248–15246, 2005.

[3] P. K. Agarwal. Enzymes: An integrated view of structure, dynamics and function. *Microbial Cell Factories*, 5, 2006.

[4] P. K. Agarwal, S. R. Billeter, and S. Hammes-Schiffer. Nuclear quantum effects and enzyme dynamics in dihydrofolate reductase catalysis. *Journal of Physical Chemistry B*, 106:3283–3293, 2002.

[5] P. K. Agarwal, S. R. Billeter, P. T. R. Rajagopalan, S. J. Benkovic, and S. Hammes-Schiffer. Network of coupled promoting motions in enzyme catalysis. *Proceedings of the National Academy of Sciences U.S.A.*, 99:2794–2799, 2002.

[6] P. K. Agarwal, A. Geist, and A. A. Gorin. Protein dynamics and enzymatic catalysis: Investigating the peptidyl-prolyl cis/trans isomerization activity of cyclophilin A. *Biochemistry*, 43:10605–10618, 2004.

[7] S. R. Alam and P. K. Agarwal. On the path to enable multi-scale biomolecular simulations on PetaFLOPS supercomputer with multi-core processors. In *HiCOMB 2007, Sixth IEEE International Workshop on High Performance Computational Biology*, March 2007.

[8] S. R. Alam, P. K. Agarwal, D. Caliga, M. C. Smith, and J. S. Vetter. Using FPGA devices to accelerate biomolecular simulations. *IEEE Computer*, 40:66–73, 2007.

[9] S. R. Alam, P. K. Agarwal, J. S. Vetter, and A. Geist. Performance characterization of molecular dynamics techniques for biomolecular simulations. In *Proceeding of Principles and Practices of Parallel Programming (PPoPP)*, 2006.

[10] S. R. Alam, R. F. Barrett, J. A. Kuehn, P. C. Roth, and J. S. Vetter. Characterization of scientific workloads on systems with multi-core processors. In *IEEE International Symposium on Workload Characterization*, 2006.

[11] F. Allen, G. Almasi, W. Andreoni, D. Beece, B. J. Berne, A. Bright, J. Brunheroto, C. Cascaval, J. Castanos, P. Coteus, P. Crumley, A. Curioni, M. Denneau, W. Donath, M. Eleftheriou, B. Fitch, B. Fleischer, C. J. Georgiou, R. Germain, M. Giampapa, D. Gresh, M. Gupta, R. Haring, H. Ho, P. Hochschild, S. Hummel, T. Jonas, D. Lieber, G. Martyna, K. Maturu, J. Moreira, D. Newns, M. Newton, R. Philhower, T. Picunko, J. Pitera, M. Pitman, R. Rand, A. Royyuru, V. Salapura, A. Sanomiya, R. Shah, Y. Sham, S. Singh, M. Snir, F. Suits, R. Swetz, W. C. Swope, N. Vishnumurthy, T. J. C. Ward, H. Warren, R. Zhou, and I. B. G. Team. Blue Gene: A vision for protein science using petaflop supercomputer. *IBM Systems Journal*, 40:310–327, 2001.

[12] S.J. Benkovic and S. Hammes-Schiffer. A perspective on enzyme catalysis. *Science*, 301:1196–1202, 2003.

[13] S. R. Billeter, S. P. Webb, P. K. Agarwal, T. Iordanov, and S. Hammes-Schiffer. Hydride transfer in liver alcohol dehydrogenase: Quantum dynamics, kinetic isotope effects, and the role of enzyme motion. *Journal of the American Chemical Society*, 123:11262–11272, 2001.

[14] K.J. Bowers, E. Chow, H. Xu, R. O. Dror, M. P. Eastwood, B. A. Gregersen, J. L. Klepeis, I. Kolossvary, M. A. Moraes, F. D. Sacerdoti, J. K. Salmon, Y. Shan, and D. E. Shaw. Molecular dynamics—scalable algorithms for molecular dynamics simulations on commodity clusters. In *Proceedings of the 2006 ACM/IEEE Conference on Supercomputing*, 2006.

[15] B. R. Brooks, R. E. Bruccoleri, B. D. Olafson, D. J. States, S. Swaminathan, and M. Karplus. CHARMM: A program for macromolecular energy, minimization, and dynamics calculations. *Journal of Computational Chemistry*, 4:187–217, 1983.

[16] D. A. Case, T. E. Cheatham, T. Darden, H. Gohlkeand, R. Luo, K. M. Merz, A. Onufriev, C. Simmerling, B. Wang, and R. J. Woods. The amber biomolecular simulation programs. *Journal of Computational Chemistry*, 26:668–1688, 2005.

[17] M. Crowley, T. A. Darden, T. E. Cheatham, and D. W. Deerfield. Adventures in improving the scaling and accuracy of a parallel molecular dynamics program. *Journal of Supercomputing*, 11:255–278, 1997.

[18] T. Darden, D. York, and L. Pederson. Particle mesh Ewald: An $N \log(N)$ method for Ewald sums in large systems. *Journal of Chemical Physics*, 98:10089–10092, 1993.

[19] Y. Duan and P. A. Kollman. Pathways to a protein folding intermediate observed in a 1-microsecond simulation in aqueous solution. *Science*, 282:740–744, 1998.

[20] P. L. Freddolino, A. S. Arkhipov, S. B Larson, A. McPherson, and K. Schulten. Molecular dynamics simulations of the complete satellite tobacco mosaic virus. *Structure*, 14:437–449, 2006.

[21] M. Garcia-Viloca, J. Gao., M. Karplus, and D. G. Truhlar. How enzymes work: Analysis by modern rate theory and computer simulations. *Science*, 303:186–195, 2004.

[22] R. S. Germain, B. Fitch, A. Rayshubskiy, M. Eleftheriou, M. C. Pitman, F. Suits, M. Giampapa, and T. J. C. Ward. Blue Matter on Blue Gene/L: Massively parallel computation for biomolecular simulation. In *Proceedings of the 3rd IEEE/ACM/IFIP International Conference on Hardware/Software Codesign and System Synthesis*, pages 207–212, 2005.

[23] Y. Gu, T. VanCourt, and M.C. Herbordt. Improved interpolation and system integration for FPGA-based molecular dynamics simulations. In *Field Programmable Logic and Applications*, 2006.

[24] V. Hornak, R. Abel, A. Okur, B. Strockbine, A. Roitberg, and C. Simmerling. Comparison of multiple Amber force fields and development of improved protein backbone parameters. *Proteins: Structure, Function and Bioinformatics*, 3:712–725, 2006.

[25] T. Ikegami, T. Ishida, D. G. Fedorov, K. Kitaura, Y. Inadomi, H. Umeda, M. Yokokawa, and S. Sekiguchi. Full electron calculation beyond 20,000 atoms: Ground electronic state of photosynthetic proteins.

In *Proceedings of the 2005 ACM/IEEE Conference on Supercomputing*, 2005.

[26] L. Kale, R. Skeel, M. Bhandarkar, R. Brunner, A. Gursoy, N. Krawetz, J. Phillips, A. Shinozaki, K. Varadarajan, and K. Schulten. NAMD2: Greater scalability for parallel molecular dynamics. *Journal of Computational Physics*, 151:283–312, 1999.

[27] M. Karplus and J. A. McCammon. Molecular dynamics simulations of biomolecules. *Nature Structural Biology*, 99:646–652, 2002.

[28] K. Kitaura, E. Ikeo, T. Asada, T. Nakano, and M. Uebayasi. Fragment molecular orbital method: An approximate computational method for large molecules. *Chemical Physical Letters*, 313:701–706, 1999.

[29] P. Kollman. Free energy calculations: Applications to chemical and biochemical phenomena. *Chemical Reviews*, 93:2395–2417, 1993.

[30] S. Kumar, C. Huang, G. Almasi, and L.V. Kale. Achieving strong scaling with NAMD on Blue Gene/L. In *IEEE International Parallel & Distributed Processing Symposium (IPDPS)*, Rhodes Island, Greece, April 2006.

[31] A. R. Leach. *Molecular Modeling: Principles and Applications*. Prentice Hall, 2001.

[32] J. A. McCammon, B. R. Gelin, and M. Karplus. Dynamics of folded proteins. *Nature*, 267:585–590, 1977.

[33] H.W. Meuer, E. Strohmaier, J.J. Dongarra, and H.D. Simon. TOP500 Supercomputer Sites. `http://www.top500.org`.

[34] National Center for Computational Sciences (NCCS). `http://info.nccs.gov/`.

[35] Roadmap of the National Center for Computational Sciences (NCCS). `http://info.nccs.gov/resources/jaguar/jaguar_roadmap`.

[36] A. Okur, B. Strockbine, V. Hornak, and C. Simmerling. Using PC clusters to evaluate the transferability of molecular mechanics force fields for proteins. *Journal of Computational Chemistry*, 24:21–31, 2003.

[37] S. J. Plimpton. Fast parallel algorithms for short-range molecular dynamics. *Journal of Computational Physics*, 117(1-19), 1995.

[38] K. Y. Sanbonmatsu, S. Joseph, and C. S. Tung. Simulating movement of tRNA into the ribosome during decoding. *Proceeding of the National Academy of Sciences U.S.A.*, 102:15854–15859, 2005.

[39] K. Y. Sanbonmatsu and C. S. Tung. High performance computing in biology multimillion atom simulations of nanoscale systems. *Journal of Structural Biology*, 157:470–480, 2007.

[40] J.S. Vetter, S.R. Alam, T. Dunigan, M. Fahey, P. Roth, and P. Worley. Early evaluation of the Cray XT3. In *20th IEEE International Parallel & Distributed Processing Symposium (IPDPS)*, 2006.

[41] A. Warshel and A. Papazyan. Electrostatic effects in macromolecules: fundamental concepts and practical modeling. *Current Opinions in Structural Biology*, 8:211–7, 1998.

Chapter 12

Multithreaded Algorithms for Processing Massive Graphs

Kamesh Madduri

Georgia Institute of Technology

David A. Bader

Georgia Institute of Technology

Jonathan W. Berry

Sandia National Laboratories

Joseph R. Crobak

Rutgers University

Bruce A. Hendrickson

Sandia National Laboratories

12.1 Introduction

In this chapter, we consider the applicability of a non-traditional massively multithreaded architecture, the Cray MTA-2 [13], as a platform for graph algorithms. Graph-theoretic problems have emerged as a prominent computational workload in the petascale computing era, and are representative of fundamental kernels in biology, scientific computing, and applications in national security. However, they pose serious challenges on current parallel machines due to non-contiguous, concurrent accesses to global data structures with low degrees of locality [35]. We present multithreaded algorithms for two fundamental graph problems – single source shortest paths and connected components – that are designed for processing large-scale, unstructured graph instances.

Distributed memory, message-passing computers dominate the high performance computing landscape. These machines range from tightly coupled systems with proprietary interconnects like the Cray XT3 through commodity clusters with lower network performance and correspondingly lower price tags. But in all of these machines, each node has its own local memory and data is exchanged between nodes via a message-passing system (usually MPI). By using commodity processors and often commodity networks, these machines are attractively priced. They have proven to be very successful at performing a wide range of scientific computations.

Despite these many successes, distributed memory machines have widely recognized shortcomings that limit their applicability.

- The latency associated with accessing non-local data is relatively high compared to local accesses. Programmers try to mitigate this problem in several ways. Careful attention to data partitioning reduces the quantity of remote accesses. A bulk synchronous programming style ensures that multiple remote accesses are bundled, thereby amortizing latency. Some degree of pre-fetching or overlapping of computation and communication can mask the latency. But as we discuss below, there are important applications for which none of these techniques can be applied.

- Load balancing can be problematic for applications with adaptive or dynamic computational requirements. A familiar example is adaptive meshing computations in which the necessary cost of periodic repartitioning can significantly degrade parallel efficiency. But the problem is even more severe for applications with finer granularities or more highly dynamic computational requirements.

- Message-passing software is complicated to write and maintain. While this cost may be bearable for large scientific applications with long lifetimes, it is a significant impediment for applications that could benefit from high performance but which have a limited lifespan.

Fortunately, these shortcomings have not precluded great success on a wide variety of scientific applications. The inherent locality of most physical phenomena permits successful partitioning, and the structure of most scientific computations allows for alternating compute/communicate steps which allows latency to be amortized.

Yet, many important problems are very difficult to solve efficiently on distributed memory machines. Within the scientific computing community these include sparse direct methods, many agent-based models, graph algorithms and more. Outside of scientific computing, communities such as machine learning have not been able to take much advantage of current parallel machines, despite a need for high performance.

As we detail in Section 12.2, the MTA-2 addresses the latency challenge in a novel manner — tolerating latency by allowing a processor to remain busy performing useful work while waiting for data. This latency-tolerance mechanism enables the MTA-2 to support unstructured memory-access patterns and highly variable computations — features that describe many graph computations. The MTA-2 also provides very lightweight synchronization mechanisms to facilitate fine-grained parallelism.

12.1.1   The trouble with graphs

A graph consists of a set of entities called *vertices* and a set of pairwise linkages between vertices called *edges*. Graph abstractions and computations are foundational to many areas of scientific and other computational applications. Familiar scientific examples include sparse direct methods, mesh generation and systems biology [1, 45]. Graphs are also central to placement and layout in VLSI [34], data mining [27, 30], and network analysis [11, 33].

In some contexts, graphs can have a great deal of structure. For instance, the connectivity patterns associated with meshes (even so-called unstructured grids) have structure associated with the geometry that underlies the mesh. But in more data-centric settings, such as Internet analysis, business intelligence solutions, or systems biology, graphs can be highly unstructured. Such graphs cannot be easily partitioned into nearly disjoint pieces, and vertices display a highly variable number of neighbors. Algorithms on these kinds of graphs are particularly difficult to parallelize on distributed memory machines.

Despite the dearth of successful distributed memory parallelizations, many graph algorithms are known to exhibit a high degree of concurrency. Evidence for this is provided by the extensive literature on efficient PRAM algorithms for graph problems [28]. However, the parallelism in these algorithms tends to be very fine grained and dynamic. It maps more naturally to the massively multithreading paradigm than to the constraints of distributed memory machines. In the succeeding sections, we discuss our parallel implementations of several graph algorithms on the MTA-2, comparing against traditional parallel machines as appropriate. We draw some general conclusions from this work in Section 12.5.

12.1.2   Limits on the scalability of distributed-memory graph computations

Despite the challenges facing distributed memory graph algorithm computations, there is a research community pursing this direction. In particular, the Parallel Boost Graph Library (PBGL) [24] has been used to produce implementations of the Δ-stepping algorithm described in Section 12.3.2 that demonstrate strong scaling on instances of Erdös-Renyi random graphs with up to one billion edges [35]. This implementation owes its performance to the technique of using *ghost nodes*. Local copies of the neighbors of the set of vertices owned by a single processor are stored on that processor. In social networks with a small diameter and a nontrivial number of high-degree nodes, however, the size of the local information that needs to be stored on some processors may approach the number of nodes in the graph, exceeding the size of a local memory. Other work [46] dispenses with ghost nodes and uses a two-dimensional decomposition of the adjacency matrix in which the adjacency list of each vertex is distributed across $\sqrt{n}$ processors. This approach is more memory scalable, but loses much of the performance that allowed PBGL to achieve strong scaling on single source shortest paths.

Let us consider the feasibility of the basic ghost node strategy as the graph size grows to the terascale and beyond. Suppose that G is a graph with n vertices that was constructed using some method for generating randomized graph instances. The particular method is not important. Suppose further that upon construction of G, a subset S of k vertices of $V(G)$ has q neighbors in $G - S$. Now, suppose that the same graph construction process is continued in such a way to double the number of vertices, while holding the average degree constant. Since the same process has been used, we expect the set S to gain q additional neighbors not in S. In other words, we expect the number of required ghost nodes per processor to grow linearly with the number of vertices in the graph. As we do not expect the local memory per processor to grow as fast as the total memory in a supercomputer, we have doubts that the ghost node strategy is scalable to the instance sizes of concern in this book.

In the sections that follow, we give preliminary evidence that massively multithreaded supercomputers offer the potential to achieve both memory and runtime scalability.

12.2 The Cray MTA-2

The Cray MTA-2 [13] is a novel multithreaded architecture with no data cache and hardware support for synchronization. The computational model for the MTA-2 is *thread-centric*, not processor-centric. A thread is a logical entity comprised of a sequence of instructions that are issued in order. An MTA-2 processor consists of 128 hardware *streams* and one instruction pipeline. A stream is a physical resource (a set of 32 registers, a status word, and space in the instruction cache) that holds the state of one thread. Each stream can have up to eight outstanding memory operations. Threads from the same or different programs are mapped to the streams by the runtime system. A processor switches among its streams every cycle, executing instructions from non-blocked streams. As long as one stream has a ready instruction, the processor remains fully utilized. No thread is bound to any particular processor.

System memory size and the inherent degree of parallelism within the program are the only limits on the number of threads a program can use. The interconnection network is a modified Cayley graph capable of delivering one word per processor per cycle. The system has 4 GBytes of memory per processor. Logical memory addresses are hashed across physical memory to avoid stride-induced hotspots. Each memory word is 68 bits: 64 data bits and 4 tag bits. One tag bit (the full-empty bit) is used to implement synchronous load and store operations. A thread that issues a synchronous load or store remains blocked until the operation completes, but the processor that issued the operation continues to issue instructions from non-blocked streams.

The MTA-2 is closer to a theoretical PRAM machine than a shared memory symmetric multiprocessor system is. Since the MTA-2 uses concurrency to tolerate latency, algorithms must often be parallelized at very fine levels to expose sufficient parallelism. However, it is not necessary that all parallelism in the program be expressed such that the system can exploit it; the goal is simply to saturate the processors. The programs that make the most effective use of the MTA-2 are those which express the concurrency of the problem in a way that allows the compiler to best exploit it.

12.2.1 Expressing parallelism

The MTA-2 compiler automatically parallelizes *inductive* loops of three types: parallel loops, linear recurrences and reductions. A loop is inductive if it is controlled by a variable that is incremented by a loop-invariant stride during each iteration, and the loop-exit test compares this variable with a loop-invariant expression. An inductive loop has only one exit test and can only be entered from the top. If each iteration of an inductive loop can be executed completely independently of the others, then the loop is termed parallel. To attain the best performance, one needs to write code (and thus design algorithms) such that most of the loops are implicitly parallelized.

There are several compiler directives that can be used to parallelize various sections of a program. The three major types of parallelization schemes available are:

- Single-processor (*fray*) parallelism: The code is parallelized in such a way that just the 128 streams on the processor are utilized.

- Multiprocessor (*crew*) parallelism: This has higher overhead than single-processor parallelism. However, the number of streams available is much larger, bounded by the size of the whole machine rather than the size of a single processor. Iterations can be statically or dynamically scheduled.

- Future parallelism: The *future* construct (detailed below) is used in this form of parallelism. This does not require that all processor resources used during the loop be available at the beginning of the loop. The runtime growth manager increases the number of physical processors as needed. Iterations are always dynamically scheduled.

A *future* is a powerful construct to express user-specified explicit concurrency. It packages a sequence of code that can be executed by a newly created thread running concurrently with other threads in the program. Futures include efficient mechanisms for delaying the execution of code that depends on the computation within the future, until the future completes. The thread that spawns the future can pass information to the thread that executes the future via parameters. Futures are best used to implement task-level parallelism and the concurrency in recursive computations.

12.2.2   Support for fine-grained synchronization

Synchronization is a major limiting factor to scalability in the case of practical shared-memory implementations. The software mechanisms commonly available on conventional architectures for achieving synchronization are often inefficient. However, the MTA-2 provides hardware support for fine-grained synchronization through the full-empty bit associated with every memory word. The compiler provides a number of generic routines that operate atomically on scalar variables. We list a few useful constructs that appear in the algorithm pseudo-codes in subsequent sections:

- The `int_fetch_add` routine (`int_fetch_add(&v, i)`) atomically adds integer i to the value at address v, stores the sum at v, and returns the original value at v (setting the full-empty bit to full).

- `readfe(&v)` returns the value of variable v when v is full and sets v empty. This allows threads waiting for v to become empty to resume execution. If v is empty, the read blocks until v becomes full.

- `writeef(&v, i)` writes the value i to v when v is empty, and sets v back to full. The thread waits until v is set empty.

- `purge(&v)` sets the state of the full-empty bit of v to empty.

12.3   Case Study: Shortest Paths

In this section, we present an experimental study of multithreaded algorithms for solving the single source shortest path (SSSP) problem on large-scale graph instances. SSSP is a well-studied combinatorial problem with a variety of practical applications such as network routing and path planning in transportation networks. Most of the recent advances in shortest path algorithms have been for point-to-point computations in road networks [32, 5].

However, these algorithms cannot be applied to arbitrary graph instances due to their reliance on the Euclidean structure of the graph.

In addition to applications in combinatorial optimization, shortest path algorithms are finding increasing relevance in the domain of complex network analysis. Popular graph-theoretic analysis metrics such as betweenness centrality [21, 7, 29, 3] are based on shortest path algorithms. In contrast to transportation networks, real-world information networks are typically characterized by low-diameter, heavy-tailed degree distributions modeled by power laws [4, 20], and self-similarity. They are often very large, with the number of vertices and edges ranging from several hundreds of thousands to billions. Our primary focus in this section is on parallel algorithms and efficient implementations for solving SSSP on large-scale unstructured graph instances.

Parallel algorithms for solving the SSSP problem have been extensively reviewed by Meyer and Sanders [38, 40]. There are no known PRAM algorithms that run in sub-linear time and $O(m + n \log n)$ work. Parallel priority queues [18, 9] for implementing Dijkstra's algorithm have been developed, but these linear work algorithms have a worst-case time bound of $\Omega(n)$, as they only perform edge relaxations in parallel. Several matrix-multiplication-based algorithms [25, 22], proposed for the parallel All-Pairs Shortest Paths (APSP), involve running time and efficiency trade-offs. Parallel approximate SSSP algorithms [31, 12, 41] based on the randomized breadth-first search algorithm of Ullman and Yannakakis [44] run in sub-linear time. However, it is not known how to use the Ullman-Yannakakis randomized approach for exact SSSP computations in sub-linear time.

We identify two well-studied algorithms, Δ-stepping [40] and Thorup's algorithm [43], that exploit concurrency in traversal of unstructured, low-diameter graphs. Meyer and Sanders give the Δ-stepping [40] SSSP algorithm that divides Dijkstra's algorithm into a number of *phases*, each of which can be executed in parallel. For random graphs with uniformly distributed edge weights, this algorithm runs in sub-linear time with linear average case work. Several theoretical improvements [39, 37] are given for Δ-stepping (for instance, finding shortcut edges, adaptive bucket-splitting), but it is unlikely that they would be faster than the simple Δ-stepping algorithm in practice, as the improvements involve sophisticated data structures that are hard to implement efficiently. We present our parallel implementation of the Δ-stepping algorithm in Section 12.3.2.

Nearly all SSSP algorithms are based on the classical Dijkstra's [17] algorithm. However, Thorup [43] presents a sequential linear-time SSSP algorithm for undirected graphs with positive integer weights that differs significantly from Dijkstra's approach. To accomplish this, Thorup's algorithm encapsulates information about the input graph in a data structure called the *component hierarchy (CH)*. Based upon information in the *CH*, Thorup's algorithm identifies vertices that can be settled in arbitrary order. This strategy is well suited to a shared-memory environment since the component hierarchy can be constructed only once, then shared by multiple concurrent SSSP computations. We perform an experimental study of a parallel implementation of Thorup's original algorithm. In order to achieve good performance, our implementation uses simple data structures and deviates from some theoretically optimal algorithmic strategies. We discuss the details of our multithreaded implementation of Thorup's algorithm in Section 12.3.3.

12.3.1 Preliminaries

Let $G = (V, E)$ be a graph with n vertices and m edges. Let $s \in V$ denote the source vertex. Each edge $e \in E$ is assigned a nonnegative real weight by the length function $l : E \rightarrow \mathbb{R}$. Define the *weight of a path* as the sum of the weights of its edges. The single source shortest paths problem with nonnegative edge weights (NSSP) computes $\delta(v)$, the weight of the *shortest* (minimum-weighted) path from s to v. $\delta(v) = \infty$ if v is unreachable from s. We set $\delta(s) = 0$.

Most shortest path algorithms maintain a *tentative distance* value for each vertex, which is updated by *edge relaxations*. Let $d(v)$ denote the tentative distance of a vertex v. $d(v)$ is initially set to ∞ and is an upper bound on $\delta(v)$. *Relaxing* an edge $\langle v, w \rangle \in E$ sets $d(w)$ to the minimum of $d(w)$ and $d(v) + l(v, w)$. Based on the manner in which the tentative distance values are updated, most shortest path algorithms can be classified into two types: *label-setting* or *label-correcting*. Label-setting algorithms (for instance, Dijkstra's algorithm) perform relaxations only from *settled* ($d(v) = \delta(v)$) vertices, and compute the shortest path from s to all vertices in exactly m edge relaxations. Based on the values of $d(v)$ and $\delta(v)$, at each iteration of a shortest path algorithm, vertices can be classified into *unreached* ($d(v) = \infty$), *queued* ($d(v)$ is finite, but v is not settled) or *settled*. Label-correcting algorithms (e.g., Bellman-Ford) relax edges from unsettled vertices also, and may perform more than m relaxations. Also, all vertices remain in a *queued* state until the final step of the algorithm. Δ-stepping belongs to the label-correcting class, whereas Thorup's algorithm belongs to the label-setting type of shortest path algorithms.

12.3.2 Δ-stepping algorithm

The Δ-stepping algorithm (see Algorithm 12.1) is an "approximate bucket implementation of Dijkstra's algorithm" [40]. It maintains an array of buckets B such that $B[i]$ stores the set of vertices $\{v \in V : v$ is queued and $d(v) \in [i\Delta, (i+1)\Delta)\}$. Δ is a positive real number that denotes the "bucket width."

In each *phase* of the algorithm (the inner *while* loop in Algorithm 12.1, lines 9–14, when bucket $B[i]$ is not empty), all vertices are removed from the current bucket, added to the set S, and *light* edges ($l(e) \leq \Delta$, $e \in E$) adjacent to these vertices are relaxed (see Algorithm 12.2). This may result in new vertices being added to the current bucket, which are deleted in the next

---

Algorithm 12.1: Δ-stepping algorithm

Input: $G(V, E)$, source vertex s, length function $l : E \to \mathbb{R}$
Output: $\delta(v), v \in V$, the weight of the shortest path from s to v

1 **foreach** $v \in V$ **do**
2 | $heavy(v) \longleftarrow \{\langle v, w \rangle \in E : l(v, w) > \Delta\}$;
3 | $light(v) \longleftarrow \{\langle v, w \rangle \in E : l(v, w) \leq \Delta\}$;
4 | $d(v) \longleftarrow \infty$;
5 relax$(s, 0)$;
6 $i \longleftarrow 0$;
7 **while** B *is not empty* **do**
8 | $S \longleftarrow \phi$;
9 | **while** $B[i] \neq \phi$ **do**
10 | $Req \longleftarrow \{(w, d(v) + l(v, w)) : v \in B[i] \wedge \langle v, w \rangle \in light(v)\}$;
11 | $S \longleftarrow S \cup B[i]$;
12 | $B[i] \longleftarrow \phi$;
13 | **foreach** $(v, x) \in Req$ **do**
14 | relax(v, x);
15 | $Req \longleftarrow \{(w, d(v) + l(v, w)) : v \in S \wedge \langle v, w \rangle \in heavy(v)\}$;
16 | **foreach** $(v, x) \in Req$ **do**
17 | relax(v, x);
18 | $i \longleftarrow i + 1$;
19 **foreach** $v \in V$ **do**
20 | $\delta(v) \longleftarrow d(v)$;

Algorithm 12.2: The *relax* routine in the Δ-stepping algorithm

---

Input: v, weight request x
Output: Assignment of v to appropriate bucket

1 **if** $x < d(v)$ **then**
2 | $B\,[\lfloor d(v)/\Delta \rfloor] \leftarrow B\,[\lfloor d(v)/\Delta \rfloor] \setminus \{v\}$;
3 | $B\,[\lfloor x/\Delta \rfloor] \leftarrow B\,[\lfloor x/\Delta \rfloor] \cup \{v\}$;
4 | $d(v) \leftarrow x$;

---

phase. It is also possible that vertices previously deleted from the current bucket may be reinserted, if their tentative distance is improved. *Heavy* edges ($l(e) > \Delta$, $e \in E$) are not relaxed in a phase, as they result in tentative values outside the current bucket. Once the current bucket remains empty after relaxations, all heavy edges out of the vertices in S are relaxed at once (lines 15–17 in Algorithm 12.1). The algorithm continues until all the buckets are empty.

Observe that edge relaxations in each phase can be done in parallel, as long as individual tentative distance values are updated atomically. The number of phases bounds the parallel running time, and the number of *reinsertions* (insertions of vertices previously deleted) and *re-relaxations* (relaxation of their outgoing edges) costs an overhead over Dijkstra's algorithm. The performance of the algorithm also depends on the value of the bucket-width Δ. For $\Delta = \infty$, the algorithm is similar to the Bellman-Ford algorithm. It has a high degree of parallelism, but is inefficient compared to Dijkstra's algorithm. Δ-stepping tries to find a good compromise between the number of parallel phases and the number of reinsertions. For graph families with random edge weights and a maximum degree of d, we can show that $\Delta = \theta(1/d)$ is a good compromise between work efficiency and parallelism. The sequential algorithm performs $O(dn)$ expected work divided between $O(\frac{d_c}{\Delta} \cdot \frac{\log n}{\log \log n})$ phases *with high probability*. In practice, in the case of graph families for which d_c is $O(\log n)$ or $O(1)$, the parallel implementation of Δ-stepping yields sufficient parallelism in each phase.

Parallel Implementation Details

The bucket array B is the primary data structure used by the parallel Δ-stepping algorithm. We implement individual buckets as *dynamic arrays* that can be resized when needed and iterated over easily. To support constant time insertions and deletions, we maintain two auxiliary arrays of size n: a mapping of the vertex ID to its current bucket, and a mapping from the vertex ID to the position of the vertex in the current bucket. All new vertices are added to the end of the array, and deletions of vertices are done by setting the corresponding locations in the bucket and the mapping arrays to -1. Note that once bucket i is finally empty after a light edge relaxation phase, there will be no more insertions into the bucket in subsequent phases. Thus, the memory can be reused once we are done relaxing the light edges in the current bucket. Also observe that all the insertions are done in the relax routine, which is called once in each phase, and once for relaxing the heavy edges.

We implement a timed preprocessing step to *semi-sort* the edges based on the value of Δ. All the light edges adjacent to a vertex are identified in parallel and stored in contiguous virtual locations, and so we visit only light edges in a phase. The $O(n)$ work preprocessing step scales well in parallel on the MTA-2.

We also support fast parallel insertions into the request set R. R stores $\langle v, x \rangle$ pairs, where $v \in V$ and x is the requested tentative distance for v.

We add a vertex v to R only if it satisfies the condition $x < d(v)$. We do not store duplicates in R. We use a sparse set representation similar to one used by Briggs and Torczon [8] for storing vertices in R. This sparse data structure uses two arrays of size n: a *dense* array that contiguously stores the elements of the set, and a *sparse* array that indicates whether the vertex is a member of the set. Thus, it is easy to iterate over the request set, and membership queries and insertions are constant time operations. Unlike other Dijkstra-based algorithms, we do not relax edges in one step. Instead, we inspect adjacencies (light edges) in each phase, construct a request set of vertices, and then relax *vertices* in the relax step.

12.3.3 Thorup's algorithm

Thorup's algorithm uses the *component hierarchy (CH)* to identify vertices for which $d(v) = \delta(v)$. These vertices can then be visited in arbitrary order. *CH* is a tree structure that encapsulates information about the graph. Each *CH-node* represents a subgraph of G called a *component*, which is identified by a vertex v and a level i. *Component(v,i)* is the subgraph of G composed of vertex v, the set S of vertices reachable from v when traversing edges with weight $< 2^i$, and all edges adjacent to $\{v\} \cup S$ of weight less than 2^i. Note that if $w \in Component(v,i)$, then $Component(v,i) = Component(w,i)$. The root *CH-node* of the *CH* is a component containing the entire graph, and each leaf represents a singleton vertex. The children of *Component(v,i)* in the *CH* are components representing the connected components formed when removing all the edges with weight $> 2^{i-1}$ from *Component(v,i)*. See Figure 12.1 for an example *CH*.

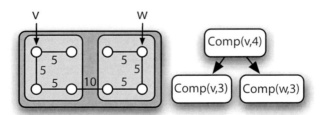

FIGURE 12.1: An example component hierarchy. *Component(v,4)*, the root of this hierarchy, represents the entire graph.

The major insight in Thorup's algorithm is presented in the following lemma.

Lemma 12.1 (from Thorup [43]) *Suppose the vertex set V divides into* *disjoint subsets $V_1, \ldots, V_k$ and that all edges between subsets have weight of at least ω. Let S be the set of settled vertices. Suppose for some i, such that $v \in V_i \backslash S$, that $d(v) = min\{d(x) | x \in V_i \backslash S\} \leq min\{d(x) | x \in V \backslash S\} + \omega$. Then $d(v) = \delta(v)$ (see Figure 12.2).*

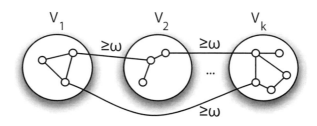

FIGURE 12.2: The vertex set V divided into k subsets.

Based upon this lemma, Thorup's algorithm identifies vertices that can be visited in arbitrary order. Let $\alpha = \log_2 \omega$. Component V buckets each of its children $V_1 \ldots V_k$ according to $min\{d(x) | x \in V_i \backslash S\} \gg \alpha$. Note that $(min\{d(x) | x \in V_i \backslash S\} \gg \alpha) \leq (min\{d(x) | x \in V \backslash S\} \gg \alpha)$ implies that $(min\{d(x) | x \in V_i \backslash S\}) \leq (min\{d(x) | x \in V \backslash S\} + \omega)$. Consider bucket $B[j]$ such that j is the smallest index of a non-empty bucket. If $V_i \in B[j]$ then $min\{d(x) | x \in V_i \backslash S\} \gg \alpha = min\{d(x) | x \in V \backslash S\} \gg \alpha$. This implies that $min\{d(x) | x \in V_i \backslash S\} \leq min\{d(x) | x \in V \backslash S\} + \omega$. Thus, each $v \in V_i \backslash S$ minimizing $D(v)$ can be visited by Lemma 12.1.

This idea can be applied recursively for each component in the *CH*. Each *component(v,i)* buckets each child V_j based upon $min\{d(x) | x \in V_j \backslash S\}$. Beginning at the root, Thorup's algorithm *visits* its children recursively, starting with those children in the bucket with the smallest index. When a leaf component l is reached, the vertex v represented by l is visited (all of its outgoing edges are relaxed). Once a bucket is empty, the components in the next highest bucket are visited and so on. We direct the reader to Thorup [43] for details about correctness and analysis.

Parallel Implementation Details

We define $minD(c)$ for component c as $min(d(x) | x \in c \backslash S)$. The value of $minD(c)$ can change when the $d(v)$ decreases for vertex $v \in c$, or it can change when a vertex $v \in c$ is *visited* (added to S). Changes in a components $minD$-value might also affect ancestor component's in the *CH*. Our implementation updates $minD$ values by propagating values from leaves towards the root. Our implementation must lock the value of $minD$ during an update since multiple vertices are visited in parallel. Locking on $minD$ does not create contention

```
int index=0;
#pragma mta assert nodep
for (int i=0; i<numChildren;
i++) {
  CHNode *c = children_store[i];
  if (bucketOf[c->id] == thisBucket) {
    toVisit[index++] = child->id;
  }
}
```

FIGURE 12.3: Parallel code to populate the toVisit set with children in the current bucket.

between threads because $minD$ values are not propagated very far up the CH in practice.

Conceptually, each component c at level i has an array of buckets. Each child c_k of c is in the bucket indexed $minD(c_k) \gg i$. Rather that explicitly storing an array of buckets, each component c stores $index(c)$, which is c's index into its parents' buckets. Child c_k of component c is in bucket j if $index(c_k) = j$. Thus, inserting a component into a bucket is accomplished by modifying $index(c)$. Inserting multiple components into buckets and finding the children in a given bucket can be done in parallel.

Traversing the Component Hierarchy in parallel

The component hierarchy is an irregular tree, in which some nodes have several thousand children and others only two. Additionally, it is impossible to know how much work must be done in a sub-tree because as few as one vertex might be visited during the traversal of a sub-tree. These two facts make it difficult to efficiently traverse the CH in parallel. To make traversal of the tree efficient, we have split the process of recursively visiting the children of a component into a two-step process. First, we build up a list of components to visit. Second, we recursively visit these nodes.

Throughout execution, Thorup's algorithm maintains a *current bucket* for each component (in accordance with Lemma 12.1). All of those children (virtually) in the *current bucket* compose the list of children to be visited, called the *toVisit* set. To build this list, we look at all of node n's children and add each child that is (virtually) in the current bucket to an array. The MTA supports automatic parallelization of such a loop with the *reduction* mechanism. On the MTA, code to accomplish this is shown in Figure 12.3.

Executing a parallel loop has two major expenses. First, the runtime system must set up for the loop. In the case of a *reduction*, the runtime system must fork threads and divide the work across processors. Second, the body of the loop is executed and the threads are abandoned. If the number of iterations is large enough, then the second expense far outweighs the first. Yet, in the case of the *CH*, each node can have between two and several hundred thousand children. In the former case, the time spent setting up for the loop far outweighs the time spent executing the loop body. Since the *toVisit* set must be built several times for each node in the *CH* (and there are $O(n)$ nodes in the *CH*), we designed a more efficient strategy for building the *toVisit* set.

Based upon the number of iterations, we either perform this loop on all processors, a single processor, or in serial. That is, if *numChildren > multi_par_threshold* then we perform the loop in parallel on all processors. Otherwise, if *numChildren > single_par_threshold* then we perform the loop in parallel on a single processor. Otherwise, the loop is performed in serial. We determined the thresholds experimentally by simulating the *toVisit* computation. In the next section, we present a comparison of the naïve approach and our approach.

12.3.4   Experimental Results

We report parallel performance results on a 40-processor Cray MTA-2 system with 160 GB uniform shared memory. Each processor has a clock speed of 220 MHz and support for 128 hardware threads. The Δ-stepping code is written in C with MTA-2 specific pragmas and directives for parallelization. The Thorup algorithm implementations are in C++ and leverage the fledgling MultiThreaded Graph Library (MTGL) [6] to perform operations such as finding connected components and extracting induced subgraphs. We compile the codes using the MTA-2 C/C++ compiler (Cray Programming Environment (PE) 2.0.3) with -O3 and -par flags.

Problem Instances

We evaluate the parallel performance on two graph families that represent unstructured data. The two families are among those defined in the 9th DIMACS Implementation Challenge [16]:

- *Random graphs*: These are generated by first constructing a cycle, and then adding $m - n$ edges to the graph at random. The generator may produce parallel edges as well as self-loops.

- *Scale-free graphs (R-MAT)*: We use the R-MAT graph model [10] to generate scale-free instances. This algorithm recursively fills in an adjacency matrix in such a way that the distribution of vertex degrees obeys an inverse power law.

For each of these graph classes, we fix the number of edges m to $4n$. We use undirected graphs for evaluating performance of Thorup's algorithm, and both directed and undirected graphs for Δ-stepping. In our experimental design, we vary two other factors: C, the maximum edge weight, and the weight distribution. The latter is either uniform in $[1, ..., C]$ (UWD) or *poly-logarithmic* (PWD). The *poly-logarithmic* distribution generates integer weights of the form 2^i, where i is chosen uniformly over the distribution $[1, \log C]$. Δ-stepping is designed for graphs with real-weighted edges, so we normalize integer edge weights to fall in the interval $[0, 1]$. In addition to these graph classes, we also conducted extensive experimental studies on high-diameter road networks and regular mesh graphs (see [36] and [15] for details). In the following figures and tables, we name data sets with the convention: <class>-<dist>-<n>-<C>.

We compare the sequential performance of our implementations to the serial performance of the "DIMACS reference solver," an implementation of Goldberg's multilevel bucket shortest path algorithm, which has an expected running time of $O(n)$ on random graphs with uniform weight distributions [23]. We do this comparison to establish that our implementations are portable and that they do not perform much extra work. It is reasonable to compare these implementations because they operate in the same environment, use the same compiler, and use a similar graph representation. Note that our implementations are not optimized for serial computation. For instance, in Δ-stepping, the time taken for semi-sorting and mechanisms to reduce memory contention on the MTA-2 both constitute overhead on a sequential processor. Regardless of this, both our Thorup and Δ-stepping computations are reasonably close to the reference SSSP solver – the solver is 1.5 to 2 times faster than Δ-stepping for large problem instances in each family, while the execution time of Thorup's algorithm is within 2–4X that of the solver.

Parallel Performance

On the MTA-2, we compare our implementation running times with the execution time of a multithreaded level-synchronized breadth-first search [3], optimized for low-diameter graphs. The multithreaded breadth-first search (BFS) scales as well as Δ-stepping for all the graph instances considered, and the execution time serves as a lower bound for the shortest path running time.

We define the speedup on p processors of the MTA-2 as the ratio of the execution time on 1 processor to the execution time on p processors. The Δ-stepping implementation performs impressively for low-diameter graphs with randomly distributed edge weights (see Figure 12.4). We achieve a speedup of approximately 31 on 40 processors for a directed random graph of nearly a billion edges, and the ratio of the BFS and Δ-stepping execution time is a constant factor (about 3–5) throughout. The implementation performs equally well for scale-free graphs, that are more difficult for partitioning-based parallel computing models to handle due to the irregular degree distribution. The execution time on 40 processors of the MTA-2 for the scale-free graph instance is within 9% (a difference of less than one second) of the running time for a random graph and the speedup is approximately 30 on 40 processors. To our knowledge, these are the first results to demonstrate near-linear speedup for such large-scale unstructured graph instances.

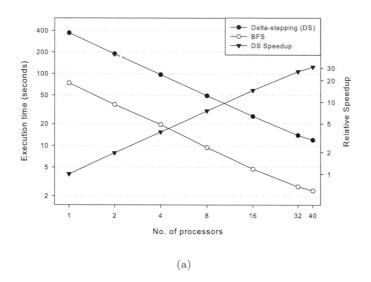

(a)

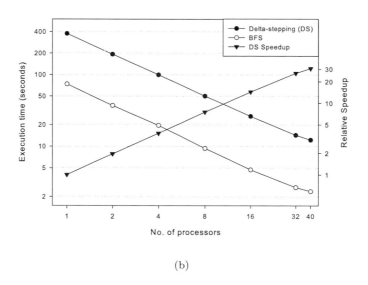

(b)

FIGURE 12.4: Δ-stepping execution time and relative speedup on the MTA-2 for (a) a Rand_directed-UWD-2^{28}-2^{28} graph instance and (b) a R-MAT_directed-UWD-2^{28}-2^{28} instance

We also ran Thorup's algorithm on graph instances from the random and R-MAT graph families, with uniform and poly-log weight distributions, and with small and large maximum edge weights. Both the Component Hierarchy construction and SSSP computations scale well on the instances studied (see Figure 12.5). We also observe that Δ-stepping performs better in all of the single source runs presented.

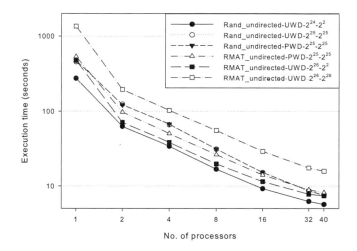

FIGURE 12.5: Scaling of Thorup's algorithm on the MTA-2 for various problem instances.

For Thorup's implementation, we explore the idea of allowing many SSSP computations to share a common component hierarchy and its performance compared to a sequence of parallel (but single source) runs of Δ-stepping. Figure 12.6 presents results of simultaneous Thorup SSSP computations on random graphs with a uniform weight distribution. When computing for a modest number of sources simultaneously, our Thorup implementation outpaces the baseline Δ-stepping computation.

In the previous section, we showed our strategy for building the *toVisit* set. This task is executed repeatedly for each component in the hierarchy. As a result, the small amount of time that is saved by selectively parallelizing this loop translates to an impressive performance gain. As seen in Table 12.1, the improvement is nearly two-fold for most graph instances. In the current programming environment, the programmer can only control if a loop executes on all processors, on a single processor, or in serial. We conjecture that better control of the number of processors employed for a loop would lead to a further speedup in our implementation.

TABLE 12.1:   Comparison of naïve strategy (Thorup A) to our strategy (Thorup B) for building *toVisit* set on 40 processors.

Family	Thorup A	Thorup B
R-MAT_undirected-UWD-2^{26}-2^{26}	28.43s	15.86s
R-MAT_undirected-PWD-2^{25}-2^{25}	14.92s	8.16s
R-MAT_undirected-UWD-2^{25}-2^{2}	9.87s	7.57s
Rand_undirected-UWD-2^{25}-2^{25}	13.29s	7.53s
Rand_undirected-PWD-2^{25}-2^{25}	13.31s	7.54s
Rand_undirected-UWD-2^{24}-2^{2}	4.33s	5.67s

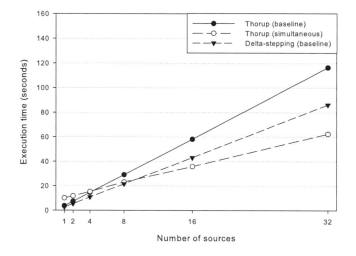

FIGURE 12.6: Simultaneous Thorup SSSP runs from multiple sources using a shared *CH*.

12.4   Case Study: Connected Components

The problem of finding the connected components of a graph is fundamental, and particularly illustrative of the issues facing large-scale parallel graph algorithm designers. In finding connected components of a graph, we wish to assign a label to each vertex such that if a path exists between a pair of vertices, then they share a common label. There are simple and efficient classical PRAM algorithms to find connected components that have been accepted by the community for decades. However, the advent of machines implementing the massive multithreading paradigm has caused algorithm designers to revisit these ideas.

12.4.1 Traditional PRAM algorithms

We will consider perhaps the most familiar of PRAM algorithms for connected components: the Shiloach-Vishkin algorithm [42]. This elegant algorithm works by manipulating rooted tree structures induced by the component attributes of a graph's vertices. These tree structures are grafted onto one another, and compressed using pointer jumping. The end result of this iterative process is that two vertices are in the same connected component if and only if they have the same parent in the final forest of rooted tree structures. We will not give the algorithm in detail here, but JaJa [28] has a thorough description.

The rooted tree manipulations of the Shiloach-Vishkin algorithm serve the purpose of selecting a single representative for each connected component. In recent years, it has been discovered that many real graphs, such as the Internet and graphs of sexual relationships among large communities of people, tend to contain a large connected component that encompasses much of the graph [19]. As an implementation of Shiloach-Vishkin forms this large component, the selected representative becomes a memory *hotspot*, a location in memory that is accessed with a high degree of concurrency.

The levels of multithreading in traditional parallel machines have not highlighted this hotspot. Even the smaller MTA-2 machines did not experience a slowdown due to this factor. However, the largest MTA-2 ever built – a 40 processor machine – did indeed demonstrate the seriousness of hotspots like this and the importance of designing algorithms that avoid them.

12.4.2 Kahan's multi-level algorithm

A familiar way to deal with large-scale graph problems is to use *multi-level algorithms* that collapse the graph in some sense, operate on the collapsed graph, then expand the result. This approach has been used extensively, for example, in graph partitioning [26]. Kahan recently demonstrated that it is an effective way to deal with hotspots in multithreaded graph algorithms. We describe his algorithm in general terms here; see [6] for a formal description.

The first step in a multi-level algorithm is to compute a coarse-grain representation of the input graph. For example, we might find a matching in the original graph, then collapse the matching edges. Kahan's algorithm leverages massively multithreaded architectures to find a coarse-grain representation via asynchronous, concurrent graph searches.

Conceptually, searches are started from many randomly selected root nodes, which are analogous to the representative nodes of the Shiloach-Vishkin PRAM algorithm. The neighbor list of each visited node is processed in parallel, if its size warrants. As these searches expand, their frontiers sometimes meet. Meetings between searches rooted at vertices v and w cause the pair (v, w) to be added to a hash table of colliding searches, and terminate the respective threads of these searches.

Once the searches have completed, the pairs added to the hash table induce the coarse-grain graph to be processed. Each edge in this graph connects two prospective component representatives. A PRAM-connected components algorithm such as Shiloach-Vishkin (SV) is then run on this coarse graph, and the component representatives found during this computation are true component representatives for the original graph. The final phase of Kahan's algorithm consists of concurrent searches started from each of these representatives to label all vertices in their respective components.

Implementation details are important in order to achieve performance with Kahan's algorithm, and other multi-level algorithms. For example, when processing graphs with inverse power-law degree distributions, even the MTA-2 machines can become overwhelmed with threads if too many concurrent searches are started at once. It is necessary to manage this problem, and one successful strategy is to process the high degree nodes *in serial*, starting a parallel search from each one in turn. When this sub-phase is complete, the remaining vertices may be processed with large numbers of searches from low-degree vertices. Such heuristics abound in multithreaded programming, and a consistent methodology for applying them would be a contribution.

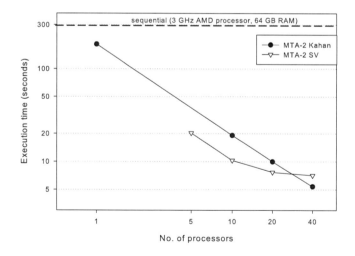

FIGURE 12.7: Scalability of the basic Shiloach-Vishkin algorithm for connected components with Kahan's multi-level algorithm

12.4.3   Performance comparisons

Figure 12.7 compares the strong scaling of the basic Shiloach-Vishkin algorithm, as implemented for the MTA-2 in [2], with a C implementation of Kahan's algorithm run on the same platform. Kahan's algorithm on a single processor of the MTA-2 is nearly twice as fast as a 3-GHz AMD workstation with 64-GB memory. The plots cross in favor of Kahan's algorithm after roughly 30 MTA-2 processors. There is reason to believe that the halt in scalability of the basic Shiloach-Vishkin algorithm will become even more abrupt in the future as larger machines permit the processing of larger inputs. The hotspot in such cases will be all the more severe, and future architectures such as the Cray XMT will have less network bandwidth than the MTA-2 did, and will lack the adaptive network routing of this aging platform. Unfortunately, this means that machines themselves will become less tolerant of hotspots than the MTA-2 was, and hotspot-free algorithms will become even more important.

12.5   Conclusion

Despite the many undeniable successes of mainstream parallel architectures, they have significant limitations. These shortcomings preclude successful parallelization of some important applications. In this chapter, we have showcased the ability of massively multithreaded machines to address graph algorithms, a prototype of a wider class of highly unstructured applications. Despite its slow clock, the MTA-2 is able to deliver impressive performance on problems that have eluded parallelization on traditional platforms.

When devising algorithms for distributed memory applications the developer focuses attention upon the maximization of locality to minimize interprocessor communication. Most distributed memory algorithms fit into a bulk synchronous processing framework in which local computation alternates with global data exchanges. In contrast, algorithm developers on the MTA-2 pay no attention to locality or to data exchanges. Instead, the key objective is to identify and express the innate, perhaps low-level, operations that can be performed at the same time. This requires a very different way of thinking about parallelism and parallel programming, but one that we have found to be highly productive and liberating.

It is our belief that the future will see a growth in the kinds of unstructured and dynamic computations that are well suited to massive multithreading. Traditional scientific computations are growing in complexity through the inclusion of adaptivity, multiscale and multiphysics phenomena. These trends will likely reduce the efficiencies achievable on message-passing machines. Further, we anticipate a continued growth in data-driven science whether from new experiments like telescopes or accelerators, from the output of large simulations or from inherently data-driven sciences like biology. Data analysis can be much less structured than traditional scientific simulations and so high-performance analysis may require the flexibility of a massively multithreaded architecture. We also hope to see a broadening of the traditional high-performance computing community to include new domains like machine learning or information processing. Massively multithreaded machines may address problems in these fields that existing computers cannot. Cray's XMT [14], formerly called Eldorado and the follow-on to the MTA-2, may enable a range of additional communities to embrace high-performance computing to the benefit of all.

Even within existing applications, the continuing growth in the relative cost of memory accesses will further erode the already-dismal utilization of computational resources. Processing speed is increasingly irrelevant as the vast majority of time is spent accessing data. Fortuitously, the seemingly inexorable march of Moore's Law has created an unprecedented opportunity for architectural innovation. For the first time in history, designers have a surplus of transistors with which to work. Although the current focus is merely on multiple cores, we anticipate a creative renaissance of architectural exploration in this new world of plenty. As they strive to address the inherent difficulties of memory-limited applications, we hope that our work on the MTA-2 can provide an informative case study.

12.6 Acknowledgments

This work was supported in part by NSF grants CAREER CCF-0611589, NSF DBI-0420513, ITR EF/BIO 03-31654, IBM Faculty Fellowship and Microsoft Research grants, NASA grant NP-2005-07-375-HQ, and DARPA contract NBCH30390004. Sandia is a multipurpose laboratory operated by Sandia Corporation, a Lockheed-Martin Company, for the U.S. Department of Energy under contract DE-AC04-94AL85000.

References

[1] D. A. Bader. High-performance algorithm engineering for large-scale graph problems and computational biology. In S. E. Nikoletseas, editor, *Proc. 4th Int'l Workshop on Efficient and Experimental Algorithms*

(WEA 2005), volume 3503 of *Lecture Notes in Computer Science*, pages 16–21, Santorini Island, Greece, May 2005. Springer-Verlag.

[2] D. A. Bader, G. Cong, and J. Feo. On the architectural requirements for efficient execution of graph algorithms. In *Proc. 34th Int'l Conf. on Parallel Processing (ICPP)*, Oslo, Norway, June 2005. IEEE Computer Society.

[3] D.A. Bader and K. Madduri. Parallel algorithms for evaluating centrality indices in real-world networks. In *Proc. 35th Int'l Conf. on Parallel Processing (ICPP)*, Columbus, OH, August 2006. IEEE Computer Society.

[4] A.-L. Barabási and R. Albert. Emergence of scaling in random networks. *Science*, 286(5439):509–512, 1999.

[5] H. Bast, S. Funke, P. Sanders, and D. Schultes. Fast routing in road networks with transit nodes. *Science*, 316(5824):566, April 2007.

[6] J. W. Berry, B. A. Hendrickson, S. Kahan, and P. Konecny. Software and algorithms for graph queries on multithreaded architectures. In *Proc. Workshop on Multithreaded Architectures and Applications*, Long Beach, CA, March 2007.

[7] U. Brandes. A faster algorithm for betweenness centrality. *J. Mathematical Sociology*, 25(2):163–177, 2001.

[8] P. Briggs and L. Torczon. An efficient representation for sparse sets. *ACM Lett. Program. Lang. Syst.*, 2(1-4):59–69, 1993.

[9] G. S. Brodal, J. L. Träff, and C. D. Zaroliagis. A parallel priority queue with constant time operations. *Journal of Parallel and Distributed Computing*, 49(1):4–21, 1998.

[10] D. Chakrabarti, Y. Zhan, and C. Faloutsos. R-MAT: A recursive model for graph mining. In *Proc. 4th SIAM Intl. Conf. on Data Mining (SDM)*, Orlando, FL, April 2004. SIAM.

[11] T. Coffman, S. Greenblatt, and S. Marcus. Graph-based technologies for intelligence analysis. *Communications of the ACM*, 47(3):45–47, 2004.

[12] E. Cohen. Using selective path-doubling for parallel shortest-path computation. *J. Algs.*, 22(1):30–56, 1997.

[13] Cray, Inc. Cray MTA-2 system: HPC technology initiatives. `http://www.cray.com/products/programs/mta_2/`, 2006.

[14] Cray, Inc. Cray XMT platform. `http://www.cray.com/products/xmt/`, 2007.

[15] J. R. Crobak, J. W. Berry, K. Madduri, and D. A. Bader. Advanced shortest path algorithms on a massively-multithreaded architecture. In

Proc. Workshop on Multithreaded Architectures and Applications, Long Beach, CA, March 2007.

[16] C. Demetrescu, A. Goldberg, and D. Johnson. 9th DIMACS implementation challenge – Shortest paths. `http://www.dis.uniroma1.it/~challenge9/`, 2006.

[17] E. W. Dijkstra. A note on two problems in connexion with graphs. *Numerische Mathematik*, 1:269–271, 1959.

[18] J. R. Driscoll, H. N. Gabow, R. Shrairman, and R. E. Tarjan. Relaxed heaps: An alternative to fibonacci heaps with applications to parallel computation. *Communications of the ACM*, 31(11):1343–1354, 1988.

[19] S. Dill, et al. Self-similarity in the web. *ACM Trans. Inter. Tech.*, 2(3):205–223, 2002.

[20] M. Faloutsos, P. Faloutsos, and C. Faloutsos. On power-law relationships of the Internet topology. In *Proc. ACM SIGCOMM*, pages 251–262, Cambridge, MA, August 1999. ACM.

[21] L. C. Freeman. A set of measures of centrality based on betweenness. *Sociometry*, 40(1):35–41, 1977.

[22] A. M. Frieze and L. Rudolph. A parallel algorithm for all-pairs shortest paths in a random graph. In *Proc. 22nd Allerton Conference on Communication, Control and Computing*, pages 663–670, 1985.

[23] A. V. Goldberg. A simple shortest path algorithm with linear average time. In *9th Ann. European Symp. on Algorithms (ESA 2001)*, volume 2161 of *Lecture Notes in Computer Science*, pages 230–241, Aachen, Germany, 2001. Springer.

[24] D. Gregor and A. Lumsdaine. The parallel BGL: A generic library for distributed graph computations. In *Proc. 5th workshop on Parallel/High-Performance Object-Oriented Scientific Computing (POOSC)*, Glasgow, U.K., July 2005.

[25] Y. Han, V. Pan, and J. Reif. Efficient parallel algorithms for computing the all pair shortest paths in directed graphs. *Algorithmica*, 17(4):399–415, 1997.

[26] B. Hendrickson and R. Leland. A multilevel algorithm for partitioning graphs. In *Proc. Supercomputing '95*, San Diego, CA, December 1995.

[27] A. Inokuchi, T. Washio, and H. Motoda. An apriori-based algorithm for mining frequent substructures from graph data. In *Proc. 4th European Conf. on Principles of Data Mining and Knowledge Discovery (PKDD)*, pages 13–23, Lyon, France, September 2000.

[28] J. JáJá. *An Introduction to Parallel Algorithms*. Addison-Wesley Publishing Company, New York, 1992.

[29] H. Jeong, S.P. Mason, A.-L. Barabási, and Z.N. Oltvai. Lethality and centrality in protein networks. *Nature*, 411:41–42, 2001.

[30] G. Karypis, E. Han, and V. Kumar. Chameleon: Hierarchical clustering using dynamic modeling. *IEEE Computer*, 32(8):68–75, 1999.

[31] P. N. Klein and S. Subramanian. A randomized parallel algorithm for single-source shortest paths. *J. Algs.*, 25(2):205–220, 1997.

[32] S. Knopp, P. Sanders, D. Schultes, F. Schulz, and D. Wagner. Computing many-to-many shortest paths using highway hierarchies. In *Proc. the 9th Workshop on Algorithm Engineering and Experiments (ALENEX07)*, New Orleans, LA, January 2007.

[33] V. E. Krebs. Mapping networks of terrorist cells. *Connections*, 24(3):43–52, 2002.

[34] T. Lengauer. *Combinatorial Algorithms for Integrated Circuit Layout.* John Wiley & Sons, Inc., New York, 1990.

[35] A. Lumsdaine, D. Gregor, B. Hendrickson, and J. Berry. Challenges in parallel graph processing. *Parallel Processing Letters*, 17(1):5–20, 2007.

[36] K. Madduri, D. A. Bader, J. W. Berry, and J. R. Crobak. An experimental study of a parallel shortest path algorithm for solving large-scale graph instances. In *Proc. the 9th Workshop on Algorithm Engineering and Experiments (ALENEX07)*, New Orleans, LA, January 2007.

[37] U. Meyer. Buckets strike back: Improved parallel shortest-paths. In *Proc. 16th Int'l Parallel and Distributed Processing Symp. (IPDPS)*, pages 1–8, Fort Lauderdale, FL, April 2002. IEEE Computer Society.

[38] U. Meyer. *Design and Analysis of Sequential and Parallel Single-Source Shortest-Paths Algorithms.* PhD thesis, Universität Saarlandes, Saarbrücken, Germany, October 2002.

[39] U. Meyer and P. Sanders. Parallel shortest path for arbitrary graphs. In *Proc. 6th International Euro-Par Conference (Euro-Par 2000)*, volume 1900 of *Lecture Notes in Computer Science*, pages 461–470, Munich, Germany, 2000. Springer-Verlag.

[40] U. Meyer and P. Sanders. Δ-stepping: A parallelizable shortest path algorithm. *J. Algs.*, 49(1):114–152, 2003.

[41] H. Shi and T. H. Spencer. Time-work tradeoffs of the single-source shortest paths problem. *J. Algorithms*, 30(1):19–32, 1999.

[42] Y. Shiloach and U. Vishkin. An $O(\log n)$ parallel connectivity algorithm. *J. Algs.*, 3(1):57–67, 1982.

[43] M. Thorup. Undirected single-source shortest paths with positive integer weights in linear time. *Journal of the ACM*, 46(3):362–394, 1999.

[44] J. Ullman and M. Yannakakis. High-probability parallel transitive closure algorithms. In *Proc. 2nd Ann. Symp. Parallel Algorithms and Architectures (SPAA-90)*, pages 200–209, Crete, Greece, July 1990. ACM.

[45] A. Vazquez, A. Flammini, A. Maritan, and A. Vespignani. Global protein function prediction in protein-protein interaction networks. *Nature Biotechnology*, 21(6):697–700, 2003.

[46] A. Yoo, E. Chow, K. Henderson, W. McLendon, B. Hendrickson, and Ü. V. Çatalyürek. A scalable distributed parallel breadth-first search algorithm on Bluegene/L. In *Proc. Supercomputing (SC 2005)*, Seattle, WA, November 2005.

Chapter 13

Disaster Survival Guide in Petascale Computing: An Algorithmic Approach

Jack J. Dongarra

University of Tennessee
Oak Ridge National Laboratory
University of Manchester
dongarra@cs.utk.edu

Zizhong Chen

Jacksonville State University
zchen@jsu.edu

George Bosilca

Innovative Computing Laboratory
University of Tennessee, Department of Electrical Engineering and Computer Science
bosilca@cs.utk.edu

Julien Langou

University of Colorado at Denver and Health Sciences Center,
Mathematical Sciences Department
julicn.langou@cudenver.edu

As the unquenchable desire of today's scientists to run ever-larger simulations and analyze ever-larger data sets drives the size of high-performance computers from hundreds, to thousands, and even tens of thousands of processors, the mean-time-to-failure (MTTF) of these computers is becoming significantly shorter than the execution time of many current high performance computing applications.

Even making generous assumptions on the reliability of a single processor or link, it is clear that as the processor count in high-end clusters grows into the tens of thousands, the mean-time-to-failure of these clusters will drop from a few years to a few days, or less. The current DOE ASCI computer (IBM Blue Gene L) is designed with 131,000 processors. The mean-time-to-failure of some nodes or links for this system is reported to be only six days on average [8].

In recent years, the trend of high-performance computing [17] has been shifting from the expensive massively parallel computer systems to clusters of commodity off-the-shelf systems [17]. While commodity off-the-shelf cluster systems have an excellent price–performance ratio, the low reliability of the off-the-shelf components in these systems leads a growing concern with the fault tolerance issue. The recently emerging computational grid environments [14] with dynamic resources have further exacerbated the problem.

However, driven by the desire of scientists for ever-higher levels of detail and accuracy in their simulations, many computational science programs are now being designed to run for days or even months. Therefore, the next generation computational science programs need to be able to tolerate failures.

Today's long-running scientific applications typically deal with faults by writing checkpoints into stable storage periodically. If a process failure occurs, then all surviving application processes are aborted and the whole application is restarted from the last checkpoint. The major source of overhead in all stable-storage-based checkpoint systems is the time it takes to write checkpoints to stable storage [21]. The checkpoint of an application on a, say, 10,000d-processor computer implies that all critical data for the application on all 10,000 processors have to be written into stable storage periodically, which may introduce an unacceptable amount of overhead into the checkpointing system. The restart of such an application implies that all processes have to be recreated and all data for each process have to be reread from stable storage into memory or regenerated by computation, which often brings a large amount of overhead into restart. It may also be very expensive or unrealistic for many large systems such as grids to provide the large amount of stable storage necessary to hold all process states of an application of thousands of processes. Therefore, due to the high frequency of failures for next generation computing systems, the classical checkpoint/restart fault tolerance approach may become a very inefficient way to deal with failures. Alternative fault-tolerance approaches need to be investigated.

In this chapter, we study an alternative approach to build fault-tolerant high performance computing applications so that they can survive a small number of simultaneous processor failures without restarting the whole application. Based on diskless checkpointing [21] and FT-MPI, a fault-tolerant version of MPI we developed [9, 10], our fault-tolerance approach removes stable storage from fault tolerance and takes an application-level approach, which gives the application developer an opportunity to achieve as low of a fault-tolerance overhead as possible according to the specific characteristics of an application. Unlike in the traditional checkpoint/restart fault-tolerance paradigm, in our fault-tolerance framework, if a small number of application processes failed, the survival application processes will not be aborted. Instead, the application will keep all survival processes, and adapt itself to failures.

The rest of the chapter is organized as follows. Section 13.1 gives a brief introduction to FT-MPI from the user point of view. Section 13.2 introduces floating-point arithmetic encodings into diskless checkpointing and discusses several checkpoint-encoding strategies, both old and new, with detail. In Section 13.3, we give a detailed presentation on how to write a fault-survivable application with FT-MPI by using a conjugate-gradient equation solver as an example. In Section 13.4, we evaluate both the performance overhead of our fault tolerance approach and the numerical impact of our floating-point arithmetic encoding. Section 13.5 discusses the limitations of our approach and possible improvements. Section 13.6 concludes the chapter and discusses future work.

13.1 FT-MPI: A Fault Tolerant MPI Implementation

Current parallel-programming paradigms for high-performance computing systems are typically based on message-passing, especially on the message-passing interface (MPI) specification [16]. However, the current MPI specification does not deal with the case where one or more process failures occur during runtime. MPI gives the user the choice between two possibilities of how to handle failures. The first one, which is also the default mode of MPI, is to immediately abort all the processes of the application. The second possibility is just slightly more flexible, handing control back to the user application without guaranteeing, however, that any further communication can occur.

13.1.1 FT-MPI overview

FT-MPI [10] is a fault-tolerant version of MPI that is able to provide basic system services to support fault-survivable applications. FT-MPI implements the complete MPI-1.2 specification, some parts of the MPI-2 document and extends some of the semantics of MPI for allowing the application the possibility to survive process failures. FT-MPI can survive the failure of n-1 processes in an n-process job, and, if required, can re-spawn the failed processes. However, the application is still responsible for recovering the data structures and the data of the failed processes.

Although FT-MPI provides basic system services to support fault-survivable applications, prevailing benchmarks show that the performance of FT-MPI is comparable [11] to the current state-of-the-art MPI implementations.

13.1.2 FT-MPI: A fault tolerant MPI implementation

FT-MPI provides semantics that answer the following questions:

1. What is the status of an MPI object after recovery?

2. What is the status of ongoing communication and messages during and after recovery?

When running an FT-MPI application, there are two parameters used to specify which modes the application is running.

The first parameter, the "communicator mode," indicates what the status is of an MPI object after recovery. FT-MPI provides four different communicator modes, which can be specified when starting the application:

- ABORT: like any other MPI implementation, FT-MPI can abort on an error.

- BLANK: failed processes are not replaced, all surviving processes have the same rank as before the crash and MPI_COMM_WORLD has the same size as before.

- SHRINK: failed processes are not replaced, however the new communicator after the crash has no "holes" in its list of processes. Thus, processes might have a new rank after recovery and the size of MPI_COMM_WORLD will change.

- REBUILD: failed processes are re-spawned, surviving processes have the same rank as before. The REBUILD mode is the default, and the most used mode of FT-MPI.

The second parameter, the "communication mode", indicates how messages, which are on the "fly" while an error occurs, are treated. FT-MPI provides two different communication modes, which can be specified while starting the application:

- CONT/CONTINUE: all operations which returned the error code MPI_SUCCESS will finish properly, even if a process failure occurs during the operation (unless the communication partner has failed).

- NOOP/RESET: all ongoing messages are dropped. The assumption behind this mode is, that on error the application returns to its last consistent state, and all currently ongoing operations are not of any further interest.

13.1.3 FT-MPI usage

Handling fault-tolerance typically consists of three steps: 1) failure detection, 2) notification, and 3) recovery. The only assumption the FT-MPI specification makes about the first two points is that the runtime environment discovers failures and all remaining processes in the parallel job are notified about these events. The recovery procedure is considered to consist of two steps: recovering the MPI library and the runtime environment, and recovering the application. The latter one is considered to be the responsibility of the application. In the FT-MPI specification, the communicator-mode discovers the status of MPI objects after recovery; and the message-mode ascertains the status of ongoing messages during and after recovery. FT-MPI offers for each of those modes several possibilities. This allows application developers to take the specific characteristics of their application into account and use the best-suited method to handle fault-tolerance.

13.2 Application-Level Diskless Checkpointing

In order to build fault-survivable applications with FT-MPI, application developers have to design their own recovery schemes to recover their applications after failure. Checkpointing, message-logging, algorithm-based check point-free schemes such as the lossy approach [2, 7] or combinations of these approaches may be used to reconstruct the required consistent state to continue the computation. However, due to its generality and performance, the diskless checkpointing technique [21] is a very promising approach to build fault-survivable applications with FT-MPI.

Diskless checkpointing is a technique to save the state of a long-running computation on a distributed system without relying on stable storage. With diskless checkpointing, each processor involved in the computation stores a copy of its state locally, either in memory or on a local disk. Additionally, encodings of these checkpoints are stored in local memory or on local disks of some processors which may or may not be involved in the computation. When a failure occurs, each live processor may roll its state back to its last local checkpoint, and the failed processor's state may be calculated from the local checkpoints of the surviving processors and the checkpoint encodings. By eliminating stable storage from checkpointing and replacing it with memory and processor redundancy, diskless checkpointing removes the main source of overhead in checkpointing on distributed systems [21].

To make diskless checkpointing as efficient as possible, it can be implemented at the application level rather than at the system level [19]. There are several advantages to implementing checkpointing at the application level. Firstly, application-level checkpointing can be placed at synchronization points in the program, which achieves checkpoint consistency automatically. Secondly, with application-level checkpointing, the size of the checkpoint can be minimized because the application developers can restrict the checkpoint to the required data. This is opposed to a transparent checkpointing system which has to save the whole process state. Thirdly, transparent system-level checkpointing typically writes binary memory dumps, which rules out a heterogeneous recovery. On the other hand, application-level checkpointing can be implemented such that the recovery operation can be performed in a heterogeneous environment as well.

In typical long-running scientific applications, when diskless checkpointing is taken from the application level, what needs to be checkpointed is often some numerical data [15]. These numerical data can either be treated as bit-streams or as floating-point numbers. If the data are treated as bit-streams, then bit-stream operations such as parity can be used to encode the checkpoint. Otherwise, floating-point arithmetic such as addition can be used to encode the data.

However, compared with treating checkpoint data as numerical numbers, treating them as bit-streams usually has the following disadvantages:

1. To survive general multiple-process failures, treating checkpoint data as bit-streams often involves the introduction of Galois field arithmetic in the calculation of checkpoint encoding and recovery decoding [18]. If the checkpoint data are treated as numerical numbers, then only floating-point arithmetic is needed to calculate the checkpoint encoding and recovery decoding. Floating-point arithmetic is usually simpler to implement and more efficient than Galois field arithmetic.

2. Treating checkpoint data as bit-streams rules out a heterogeneous recovery. The checkpoint data may have different bit-stream representation on different platforms and even have different bit-stream lengths on different architectures. The introduction of a unified representation of the checkpoint data on different platforms within an application for checkpoint purposes scarifies too much performance and is unrealistic in practice.

3. In some cases, treating checkpoint data as bit-streams does not work. For example, in [15], in order to reduce memory overhead in fault-tolerant dense matrix computation, no local checkpoints are maintained on computation processors, only the checksum of the local checkpoints is maintained on the checkpoint processors. Whenever a failure occurs, the local checkpoints on surviving computation processors are reconstructed by reversing the computation. Lost data on failed processors are then re-constructed through the checksum and the local checkpoints obtained are from the reverse computation. However, due to round-off errors, the local checkpoints obtained from reverse computation are not the same bit-streams as the original local checkpoints. Therefore,

in order to be able to reconstruct the lost data on failed processors, the checkpoint data have to be treated as numerical numbers and the floating-point arithmetic has to be used to encode the checkpoint data.

The main disadvantage of treating the checkpoint data as floating-point numbers is the introduction of round-off errors into the checkpoint and recovery operations. Round-off errors are a limitation of any floating-point number calculations. Even without checkpoint and recovery, scientific computing applications are still affected by round-off errors. In practice, the increased possibility of overflows, underflows, and cancellations due to round-off errors in numerically stable checkpoint and recovery algorithms is often negligible.

In this chapter, we treat the checkpoint data as floating-point numbers rather than bit-streams. However, the corresponding bit-stream version schemes could also be used as long as the application developer thinks they are more appropriate. In the following subsections, we discuss how the local checkpoint can be encoded so that applications can survive failures.

13.2.1 Neighbor-based checkpointing

In neighbor-based checkpointing, a neighbor processor is first defined for each computation processor. Then, in addition to keeping a local checkpoint in its memory, each computation processor stores a copy of its local checkpoint in the memory of its neighbor processor. Whenever a computation processor fails, the lost local checkpoint data can be recovered from its neighbor processor.

The performance overhead of the neighbor-based checkpointing is usually very low. The checkpoints are localized to only two processors: a computation processor and its neighbor. The recovery only involves the failed processors and their neighbors. There are no global communications or encoding/decoding calculations needed in the checkpoint and recovery.

Because no floating-point operations are involved in the checkpoint and recovery, no round-off errors are introduced in neighbor-based checkpointing.

Depending on how we define the neighbor processor of a computation processor, there are three neighbor-based checkpointing schemes

13.2.1.1 Mirroring

The mirroring scheme of neighbor-based checkpointing is originally proposed in [21]. In this scheme, if there are n computation processors, another n checkpoint processors are dedicated as neighbors of the computation processors. The ith computation processor simply stores a copy of its local checkpoint data in the ith checkpoint processor (see Figure 13.1).

Up to n processor failures may be tolerated, although the failure of both a computation processor and its neighbor processor cannot be tolerated. If we assume that the failure of each processor is independent and identically distributed, then the probability that the mirroring scheme survives k processor failures is

$$\frac{C_n^k 2^k}{C_{2n}^k}$$

When k is much smaller than n, the probability to survive k failures can be very close to 1.

The disadvantage of the mirroring scheme is that n additional processors are dedicated as checkpoint processors, therefore,they cannot be used to do computation.

13.2.1.2   Ring neighbor

In [23], a ring neighbor scheme was discussed by Silva et al. In this scheme, there are no additional processors used. All computation processors are organized in a virtual ring. Each processor sends a copy of its local checkpoint to the neighbor processor that follows on the virtual ring. Therefore, each processor has two checkpoints maintained in memory: one is the local checkpoint of itself, another is the local checkpoint of its neighbor (see Figure 13.1).

The ring neighbor scheme is able to tolerate at least one and up to $\lfloor \frac{n}{2} \rfloor$ processor failures in an n processor job depending on the distribution of the failed processors.

Compared with the mirroring scheme, the advantage of the ring neighbor scheme is that there is no processor redundancy in the scheme. However, two copies of checkpoints have to be maintained in the memory of each computation processor. The degree of fault tolerance of the ring neighbor scheme is also lower than the mirroring scheme.

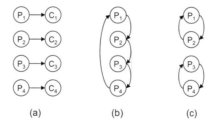

(a)   (b)   (c)

FIGURE 13.1: Neighbor-based schemes.

13.2.1.3   Pair neighbor

Another possibility is to organize all computation processors as pairs (assuming that there are an even number of computation processors). The two processors in a pair are neighbors of each other. Each processor sends a copy of its local checkpoint to its neighbor processor (see Figure 13.1).

Like the ring neighbor scheme, there is no processor redundancy used in the paired neighbor scheme and two copies of checkpoints have to be maintained in the memory of each computation processor.

However, compared with the ring neighbor scheme, the degree of fault tolerance for the pair neighbor scheme is improved. Like the mirroring scheme, if we assume that the failure of each process is independent and identically distributed, then the probability that the pair neighbor scheme survives k failures in an n processor job is

$$\frac{C_{n/2}^k 2^k}{C_n^k}.$$

13.2.2 Checksum-based checkpointing

Checksum-based checkpointing is a modified version of the parity-based checkpointing proposed in [20]. In checksum-based checkpointing, instead of using parity, the floating-point number addition is used to encode the local checkpoint data. By encoding the local checkpoint data of the computation processors and sending the encoding to some dedicated checkpoint processors, checksum-based checkpointing introduces a much lower memory overhead into the checkpoint system than neighbor-based checkpointing. However, due to the calculating and sending of the encoding, the performance overhead of checksum-based checkpointing is usually higher than neighbor-based checkpoint schemes. There are two versions of the checksum-based checkpointing schemes.

13.2.2.1 Basic checksum scheme

The basic checksum scheme works as follow. If the program is executing on N processors, then there is an $N+1$st processor called the checksum processor. At all points in time a consistent checkpoint is held in the N processors in memory. Moreover, a checksum of the N local checkpoints is held in the checksum processor (see Figure 13.2). Assume P_i is the local checkpoint data in the memory of the ith computation processor. C is the checksum of the local checkpoint in the checkpoint processor. If we look at the checkpoint data as an array of real numbers, then the checkpoint encoding actually establishes an identity

$$P_1 + \ldots + P_n = C \tag{13.1}$$

between the checkpoint data P_i on computation processors and the checksum data C on the checksum processor. If any processor fails then the identity in Equation (13.1) becomes an equation with one unknown. Therefore, the data in the failed processor can be reconstructed through solving this equation.

Due to the floating-point arithmetic used in the checkpoint and recovery, there will be round-off errors in the checkpoint and recovery. However, the checkpoint involves only additions and the recovery involves additions and only one subtraction. In practice, the increased possibility of overflows, underflows, and cancellations due to round-off errors in the checkpoint and recovery algorithm is negligible.

The basic checksum scheme can survive only one failure. However, it can be used to construct a one-dimensional checksum scheme to survive certain multiple failures.

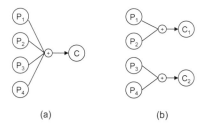

(a)   (b)

FIGURE 13.2: Checksum based schemes.

13.2.2.2   One Dimensional Checksum Scheme

The one-dimensional checksum scheme works as follow. Assume the program is running on mn processors. Partition the mn processors into m groups with n processors in each group. Dedicate one checksum processor for each group. At each group, the checkpoints are done using the basic checksum scheme (see Fig. 13.2).

The advantage of this scheme is that the checkpoint are localized to a subgroup of processors, so the checkpoint encoding in each subgroup can be done parallelly. Therefore, compared with the basic checksum scheme, the performance of the one-dimensional checksum scheme is usually better. If we assume that the failure of each process is independent and identically distributed, then the probability that the one-dimensional checksum scheme survives k $(k < m)$ failures is

$$\frac{C_m^k (n+1)^k}{C_{m(n+1)}^k}$$

13.2.3   Weighted-checksum-based checkpointing

The weighted-checksum scheme is a natural extension to the checksum scheme to survive multiple failures of arbitrary patterns with minimum processor redundancy. It can also be viewed as a version of the Reed-Solomon erasure-coding scheme [18] in the real number field. The basic idea of this scheme works as follow: Each processor takes a local in-memory checkpoint, and m equalities are established by saving weighted checksums of the local checkpoint into m checksum processors. When f failures happen, where $f \leq m$, the m equalities becomes m equations with f unknowns. By appropriately choosing the weights of the weighted checksums, the lost data on the f failed processors can be recovered by solving these m equations.

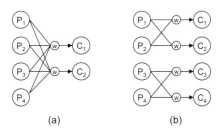

(a)   (b)

FIGURE 13.3: Weighted checksum schemes.

13.2.3.1 The basic weighted-checksum scheme

Suppose there are n processors used for computation. Assume the checkpoint data on the ith computation processor is P_i. In order to be able to reconstruct the lost data on failed processors, another m processors are dedicated to hold m encodings (weighted checksums) of the checkpoint data (see Figure 13.3). The weighted checksum C_j on the jth checksum processor can be calculated from

$$\begin{cases} a_{11}P_1 + \ldots + a_{1n}P_n & = C_1 \\ & \vdots \\ a_{m1}P_1 + \ldots + a_{mn}P_n & = C_m \end{cases} \tag{13.2}$$

where a_{ij}, $i = 1, 2, ..., m$, $j = 1, 2, ..., n$, is the weight we need to choose. Let $A = (a_{ij})_{mn}$. We call A the checkpoint matrix for the weighted checksum scheme.

Suppose that k computation processors and $m - h$ checkpoint processors have failed, then there are $n - k$ computation processors and h checkpoint processors survive. If we look at the data on failed processors as unknowns, then Equation (13.2) becomes m equations with $m - (h - k)$ unknowns.

If $k > h$, then there are less equations than unknowns. There is no unique solution for Equation (13.2). The lost data on the failed processors cannot be recovered.

However, if $k < h$, then there are more equations than unknowns. By appropriately choosing A, a unique solution for Equation (13.2) can be guaranteed. Therefore, the lost data on the failed processors can be recovered by solving Equation (13.2).

Without loss of generality, we assume: (1) the computational processor j_1, $j_2, ..., j_k$ failed and the computational processor $j_{k+1}, j_{k+2}, ..., j_n$ survived; (2) the checkpoint processor $i_1, i_2, ..., i_h$ survived and the checkpoint processor $i_{h+1}, i_{h+2}, ..., i_m$ failed. Then, in Equation (13.2), $P_{j_1}, ..., P_{j_k}$ and $C_{i_{h+1}}, ..., C_{i_m}$ become unknowns after the failure occurs. If we restructure (13.2), we can get

$$\begin{cases} a_{i_1 j_1} P_{j_1} + ... + a_{i_1 j_k} P_{j_k} &= C_{i_1} - \sum_{t=k+1}^{n} a_{i_1 j_t} P_{j_t} \\ \quad\vdots \\ a_{i_h j_1} P_{j_1} + ... + a_{i_h j_k} P_{j_k} &= C_{i_h} - \sum_{t=k+1}^{n} a_{i_h j_t} P_{j_t} \end{cases} \quad (13.3)$$

and

$$\begin{cases} C_{i_{h+1}} &= a_{i_{h+1}1} P_1 + ... + a_{i_{h+1}n} P_n \\ \quad\vdots \\ C_{i_m} &= a_{i_m 1} P_1 + ... + a_{i_m n} P_n \end{cases} \quad (13.4)$$

Let A_r denote the coefficient matrix of the linear system (Equation (13.3)). If A_r has full column rank, then $P_{j_1}, ..., P_{j_k}$ can be recovered by solving Equation (13.3), and $C_{i_{h+1}}, ..., C_{i_m}$ can be recovered by substituting $P_{j_1}, ..., P_{j_k}$ into Equation (13.4).

Whether we can recover the lost data on the failed processes or not directly depends on whether A_r has full column rank or not. However, A_r in Equation (13.3) can be any sub-matrix (including minor) of A depending on the distribution of the fail processors. If any square sub-matrix (including minor) of A is non-singular and there are no more than m process failed, then A_r can be guaranteed to have full column rank. Therefore, to be able to recover from any more than m failures, the checkpoint matrix A has to satisfy *any square sub-matrix (including minor) of A is non-singular*.

How can we find such matrices? It is well known that some structured matrices such as the Vandermonde matrix and Cauchy matrix satisfy any square sub-matrix (including minor) of the matrix is non-singular.

However, in computer floating-point arithmetic where no computation is exact due to round-off errors, it is well known [2] that, in solving a linear system of equations, a condition number of 10^k for the coefficient matrix leads to a loss of accuracy of about k decimal digits in the solution. Therefore, in order to get a reasonably accurate recovery, the checkpoint matrix A actually has to satisfy *any square sub-matrix (including minor) of A is well-conditioned*.

It is well-known [6] that Gaussian random matrices are well-conditioned. To estimate how well conditioned Gaussian random matrices are, we have proved the following theorem:

THEOREM 13.1

Let $G_{m \times n}$ be an $m \times n$ real random matrix whose elements are indepen-dent and identically distributed standard normal random variables, and let $\kappa_2(G_{m \times n})$ be the 2-norm condition number of $G_{m \times n}$. Then, for any $m \geq 2$, $n \geq 2$ and $x \geq |n - m| + 1$, $\kappa_2(G_{m \times n})$ satisfies

$$P\left(\frac{\kappa_2(G_{m \times n})}{n/(|n - m| + 1)} > x\right) < \frac{1}{\sqrt{2\pi}}\left(\frac{C}{x}\right)^{|n-m|+1}$$

and

$$E(\ln \kappa_2(G_{m \times n})) < \ln \frac{n}{|n - m| + 1} + 2.258$$

where $0.245 \leq c \leq 2.000$ and $5.013 \leq C \leq 6.414$ are universal positive con-stants independent of m, n and x.

Due to the length of the proof for Theorem 13.1, we omit it here and refer interested readers to [3] for the complete proof.

Note that any sub-matrix of a Gaussian random matrix is still a Gaussian random matrix. Therefore, a Gaussian random matrix would satisfy any sub-matrix of the matrix is well-conditioned with high probability.

Theorem 13.1 can be used to estimate the accuracy of recovery in the weighted-checksum scheme. For example, if an application uses 100,000 pro-cessors to perform computation and 20 processors to perform checkpointing, then the checkpoint matrix is a 20 by 100,000 Gaussian random matrix. If 10 processors fail concurrently, then the coefficient matrix A_r in the recovery algorithm is a 20 by 10 Gaussian random matrix. From Theorem 13.1, we can get

$$E(\log_{10} \kappa_2(A_r)) < 1.25$$

and

$$P(\kappa_2(A_r) > 100) < 3.1 \times 10^{-11}$$

Therefore, on average, we will lose about one decimal digit in the recovered data and the probability to lose 2 digits is less than 3.1×10^{-11}.

13.2.3.2 One-dimensional weighted-checksum scheme

The one-dimensional weighted-checksum scheme works as follows. Assume the program is running on mn processors. Partition the mn processors into m groups with n processors in each group. Dedicate another k checksum processors for each group. At each group, the checkpoints are done using the basic weighted checksum scheme (see Figure 13.3). This scheme can survive k processor failures at each group. The advantage of this scheme is that the checkpoints are localized to a subgroup of processors, so the checkpoint encoding in each subgroup can be done parallelly. Therefore, compared with the basic weighted-checksum scheme, the performance of the one-dimensional weighted-checksum scheme is usually better.

13.3   A Fault-Survivable Iterative Equation Solver

In this section, we give a detailed presentation on how to incorporate fault tolerance into applications by using a preconditioned conjugate gradient equation solver as an example.

13.3.1   Preconditioned conjugate-gradient algorithm

The preconditioned conjugate gradient (PCG) method is the most commonly used algorithm to solve the linear system $Ax = b$ when the coefficient matrix A is sparse and symmetric-positive definite. The method proceeds by generating vector sequences of iterates (i.e., successive approximations to the solution), residuals corresponding to the iterates, and search directions used in updating the iterates and residuals. Although the length of these sequences can become large, only a small number of vectors needs to be kept in memory. In every iteration of the method, two inner products are performed in order to compute update scalars that are defined to make the sequences satisfy certain orthogonality conditions. The pseudo-code for the PCG is given in Figure 13.4. For more details of the algorithm, we refer the reader to [1].

Compute $r^{(0)} = b - Ax^{(0)}$ for some initial guess $x^{(0)}$
for $i = 1, 2, \ldots$
    solve $Mz^{(i-1)} = r^{(i-1)}$
    $\rho_{i-1} = r^{(i-1)^T} z^{(i-1)}$
    if $i = 1$
        $p^{(1)} = z^{(0)}$
    else
        $\beta_{i-1} = \rho_{i-1}/\rho_{i-2}$
        $p^{(i)} = z^{(i-1)} + \beta_{i-1}p^{(i-1)}$
    endif
    $q^{(i)} = Ap^{(i)}$
    $\alpha_i = \rho_{i-1}/p^{(i)^T} q^{(i)}$
    $x^{(i)} = x^{(i-1)} + \alpha_i p^{(i)}$
    $r^{(i)} = r^{(i-1)} - \alpha_i q^{(i)}$
    check convergence; continue if necessary
end

FIGURE 13.4: Preconditioned conjugate gradient algorithm.

13.3.2   Incorporating fault tolerance into PCG

We first implemented the parallel non-fault-tolerant PCG. The preconditioner M we use is the diagonal part of the coefficient matrix A. The matrix A is stored as sparse row compressed format in memory. The PCG code is implemented such that any symmetric, positive definite matrix using the Harwell Boeing format or the Matrix Market format can be used as a test problem. One can also choose to generate the test matrices in memory according to testing requirements.

We then incorporate the basic weighted-checksum scheme into the PCG code. Assume the PCG code uses n MPI processes to do computation. We dedicate another m MPI processes to hold the weighted checksums of the local checkpoint of the n computation processes. The checkpoint matrix we use is a pseudo-random matrix. Note that the sparse matrix does not change during computation, therefore, we only need to checkpoint three vectors (i.e., the iterate, the residual and the search direction) and two scalars (i.e., the iteration index and $\rho^{(i-1)}$ in Figure 13.4).

The communicator mode we use is the REBUILD mode. The communication mode we use is the NOOP/RESET mode. Therefore, when processes failed, FT-MPI will drop all ongoing messages and re-spawn all failed processes without changing the rank of the surviving processes.

An FT-MPI application can detect and handle failure events using two different methods: either the return code of every MPI function is checked, or the application makes use of MPI error handlers. The second mode gives users the possibility to incorporate fault tolerance into applications that call existing parallel numerical libraries which do not check the return codes of their MPI calls. In PCG code, we detect and handle failure events by checking the return code of every MPI function.

The recovery algorithm in PCG makes use of the *longjmp* function of the C-standard. In case the return code of an MPI function indicates that an error has occurred, all surviving processes set their state variable to RECOVER and *jump* to the recovery section in the code. The recovery algorithm consists of the following steps:

1. Re-spawn the failed processes and recover the FT-MPI runtime environment by calling a specific, predefined MPI function.

2. Determine how many processes have died and who has died.

3. Recover the lost data from the weighted checksums using the algorithm described in Section 13.2.

4. Resume the computation.

Another issue is how a process can determine whether it is a survival process or it is a re-spawned process. FT-MPI offers the user two possibilities to solve this problem:

- In the first method, when a process is a replacement for a failed process, the return value of its MPI_Init call will be set to a specific new FT-MPI constant (MPI_INIT_RESTARTED_PROCS).

- The second possibility is that the application introduces a static variable. By comparing the value of this variable to the value on the other processes, the application can detect whether every process has been newly started (in which case all processes will have the pre-initialized value), or whether a subset of processes has a different value, since each process modifies the value of this variable after the initial check. This second approach is somewhat more complex, however, it is fully portable and can also be used with any other non-fault-tolerant MPI library.

In PCG, each process checks whether it is a re-spawned process or a surviving process by checking the return code of its MPI_Init call.

The relevant section with respect to fault tolerance is shown in the source code below.

```
/* Determine who is re-spawned */
rc = MPI_Init( &argc, &argv );
if (rc==MPI_INIT_RESTARTED_NODE) {
  /* re-spawned procs initialize */
  ...
} else {
  /* Original procs initialize*/
  ...
}

/*Failed procs jump to here to recover*/
setjmp( env );

/* Execute recovery if necessary */
if ( state == RECOVER ) {

  /*Recover MPI environment*/
  newcomm = FT_MPI_CHECK_RECOVER;
  MPI_Comm_dup(oldcomm, &newcomm);

  /*Recover application data*/
  recover_data (A, b, r, p, x, ...);

  /*Reset state-variable*/
  state = NORMAL;
}

/*Major computation loop*/
```

```
do {

    /*Checkpoint every K iterations*/
    if ( num_iter % K == 0 )
        checkpoint_data(r, p, x, ...);

    /*Check the return of communication
      calls to detect failure. If failure
      occurs, jump to recovery point*/
    rc = MPI_Send ( ...)
    if ( rc == MPI_ERR_OTHER ) {
        state = RECOVER;
        longjmp ( env, state );
    }

} while ( not converge );
```

13.4   Experimental Evaluation

In this section, we evaluate both the performance overhead of our fault-tolerance approach and the numerical impact of our floating-point arithmetic encoding using the PCG code implemented in the last section.

We performed four sets of experiments to answer the following four questions:

1. What is the performance of FT-MPI compared with other state-of-the-art MPI implementations?

2. What is the performance overhead of performing checkpointing?

3. What is the performance overhead of performing recovery?

4. What is the numerical impact of round-off errors in recovery?

For each set of experiments, we test PCG with four different problems. The size of the problems and the number of computation processors used (not including checkpoint processors) for each problem are listed in Table 13.1.

All experiments were performed on a cluster of 64 dual-processor 2.4 GHz AMD Opteron nodes. Each node of the cluster has 2 GB of memory and runs the Linux operating system. The nodes are connected with a gigabit Ethernet. The timer we used in all measurements is MPI_Wtime.

TABLE 13.1:   Experiment configurations for each
problem

	Size of the Problem	Num. of Comp. Procs
Prob #1	164,610	15
Prob #2	329,220	30
Prob #3	658,440	60
Prob #4	1,316,880	120

13.4.1   Performance of PCG with different MPI implementations

The first set of experiments was designed to compare the performance of different MPI implementations and evaluate the overhead of surviving a single failure with FT-MPI. We ran PCG with MPICH-1.2.6 [13], MPICH2-0.96, FT-MPI, FT-MPI with one checkpoint processor and no failure, and FT-MPI with one checkpoint processor and one failure for 2000 iterations. For PCG with FT-MPI with checkpoint, we checkpoint every 100 iterations. For PCG with FT-MPI with recovery, we simulate a processor failure by exiting one process at the 1000th iteration. The execution times of all tests are reported in Table 13.2.

TABLE 13.2:   PCG execution time (in seconds) with
different MPI implementations

Time	Prob#1	Prob#2	Prob#3	Prob#4
MPICH-1.2.6	916.2	1985.3	4006.8	10199.8
MPICH2-0.96	510.9	1119.7	2331.4	7155.6
FT-MPI	480.3	1052.2	2241.8	6606.9
FT-MPI ckpt	482.7	1055.1	2247.5	6614.5
FT-MPI rcvr	485.8	1061.3	2256.0	6634.0

Figure 13.5 compares the execution time of PCG with MPICH-1.2.6, MPICH2-0.96, FT-MPI, FT-MPI with one checkpoint processor and no failure, and FT-MPI with one checkpoint processor and one failure for different sizes of problems. Figure 13.5 indicates that the performance of FT-MPI is slightly better than MPICH2-0.96. Both FT-MPI and MPICH2-0.96 are much faster than MPICH-1.2.6. Even with checkpointing and/or recovery, the performance of PCG with FT-MPI is still at least comparable to MPICH2-0.96.

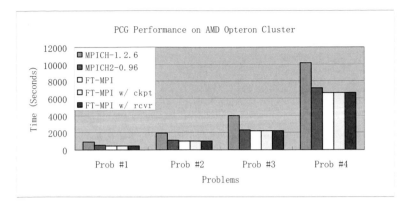

FIGURE 13.5: PCG performance with different MPI implementations.

13.4.2 Performance overhead of taking checkpoint

The purpose of the second set of experiments is to measure the performance penalty of taking checkpoints to survive general multiple simultaneous processor failures. There are no processor failures involved in this set of experiments. At each run, we divided the processors into two classes. The first class of processors is dedicated to perform PCG computation work. The second class of processors are dedicated to performing checkpoint. In Table 13.3 and 13.4, the first column of the table indicates the number of checkpoint processors used in each test. If the number of checkpoint processors used in a run is zero, then there is no checkpoint in this run. For all experiments, we ran PCG for 2000 iterations and performed checkpointing of every 100 iterations.

TABLE 13.3: PCG execution time (in seconds) with checkpoint

Time	Prob #1	Prob #2	Prob #3	Prob #4
0 ckpt	480.3	1052.2	2241.8	6606.9
1 ckpt	482.7	1055.1	2247.5	6614.5
2 ckpt	484.4	1057.9	2250.3	6616.9
3 ckpt	486.5	1059.9	2252.4	6619.7
4 ckpt	488.1	1062.2	2254.7	6622.3
5 ckpt	489.9	1064.3	2256.5	6625.1

Table 13.3 reports the execution time of each test. In order to reduce the disturbance of the noise of the program execution time to the checkpoint time, we measure the time used for checkpointing separately for all experiments.

TABLE 13.4:   PCG checkpointing time (in seconds)

Time	Prob #1	Prob #2	Prob #3	Prob #4
1 ckpt	2.6	3.8	5.5	7.8
2 ckpt	4.4	5.8	8.5	10.6
3 ckpt	6.0	7.9	10.2	12.8
4 ckpt	7.9	9.9	12.6	15.0
5 ckpt	9.8	11.9	14.1	16.8

Table 13.4 reports the individual checkpoint time for each experiment. Figure 13.6 compares the checkpoint overhead (%) of surviving different numbers of simultaneous processor failures for different sizes of problems.

Table 13.4 indicates, as the number of checkpoint processors increases, the time for checkpointing in each test problem also increases. The increase in time for each additional checkpoint processor is approximately the same for each test problem. However, the increase of the time for each additional checkpoint processor is smaller than the time for using only one checkpoint processor. This is because from no checkpoint to checkpointing with one checkpoint processor, PCG has to first set up the checkpoint environment and then do one encoding. However, from checkpointing with k (where $k > 0$) processors to checkpointing with $k + 1$ processors, the only additional work is performing one more encoding.

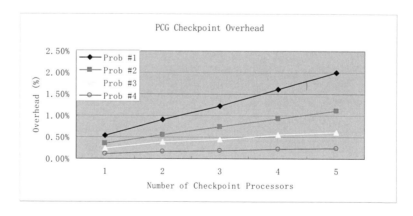

FIGURE 13.6: PCG checkpoint overhead.

Note that we are performing checkpointing every 100 iterations and run PCG for 2000 iterations, therefore, from Table 13.3, we can calculate the checkpoint interval for each test. Our checkpoint interval ranges from 25 seconds (Prob #1) to 330 seconds (Prob #4). In practice, there is an optimal checkpoint interval which depends on the failure rate, the time cost of each checkpoint and the time cost of each recovery. A great deal of literature about the optimal checkpoint interval [12, 22, 25] is available. We will not address this issue further here.

From Figure 13.6, we can see, even if we checkpoint every 25 seconds (Prob #1), the performance overhead of checkpointing to survive five simultaneous processor failures is still within 2% of the original program execution time, which actually falls into the noise margin of the program execution time. If we checkpoint every 5.5 minutes (Prob #4) and assume a processor fails one after another (one checkpoint processor case), then the overhead is only 0.1%.

13.4.3 Performance overhead of performing recovery

The third set of experiments is designed to measure the performance overhead to perform recovery. All experiment configurations are the same as in the previous section except that we simulate a failure of k (k equals the number of checkpoint processors in the run) processors by exiting k processes at the 1000th iteration in each run.

Table 13.5 reports the execution time of PCG with recovery. In order to reduce the disturbance of the noise of the program execution time to the recovery time, we measure the time used for recovery separately for all experiments. Table 13.6 reports the recovery time in each experiment. Figure 13.7 compares the recovery overhead (%) from different numbers of simultaneous processor failures for different sizes of problems.

TABLE 13.5: PCG execution time (in seconds) with recovery

Time	Prob #1	Prob #2	Prob #3	Prob #4
0 proc	480.3	1052.2	2241.8	6606.9
1 proc	485.8	1061.3	2256.0	6634.0
2 proc	488.1	1063.6	2259.7	6633.5
3 proc	490.0	1066.1	2262.1	6636.3
4 proc	492.6	1068.8	2265.4	6638.2
5 proc	494.9	1070.7	2267.5	6639.7

From Table 13.6, we can see that the recovery time increases approximately linearly as the number of failed processors increases. However, the recovery time for a failure of one processor is much longer than the increase of the recovery time from a failure of k (where $k > 0$) processors to a failure of $k + 1$ processors. This is because, from no failure to a failure with one failed

TABLE 13.6:   PCG recovery time (in seconds)

Time	Prob #1	Prob #2	Prob #3	Prob #4
1 proc	3.2	5.0	8.7	18.2
2 proc	3.7	5.5	9.2	18.8
3 proc	4.0	6.0	9.8	20.0
4 proc	4.5	6.5	10.4	20.9
5 proc	4.8	7.0	11.1	21.5

processor, the additional work the PCG has to perform includes first setting up the recovery environment and then recovering data. However, from a failure with k (where $k > 0$) processors to a failure with $k + 1$ processors, the only additional work is to recover data for an additional processor.

From Figure 13.7, we can see that the overheads for recovery in all tests are within 1% of the program execution time, which is again within the noise margin of the program execution time.

13.4.4   Numerical impact of round-off errors in recovery

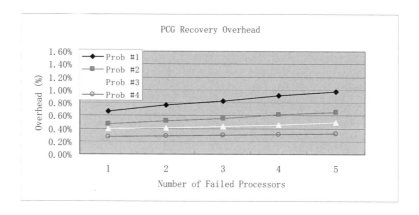

FIGURE 13.7: PCG recovery overhead.

As discussed in Section 13.2, our diskless-checkpointing schemes are based on floating-point arithmetic encodings, therefore introducing round-off errors into the checkpointing system. The experiments in this subsection are designed to measure the numerical impact of the round-off errors in our checkpointing system. All experiment configurations are the same as in the previous section except that we report the norm of the residual at the end of each computation.

Note that if no failures occur, the computation proceeds with the same computational data as without checkpoint. Therefore, the computational results are affected only when there is a recovery in the computation. Table 13.7 reports the norm of the residual at the end of each computation when there are 0, 1, 2, 3, 4, and 5 simultaneous process failures.

TABLE 13.7: Numerical impact of round-off errors in PCG recovery

Residual	Prob #1	Prob #2	Prob #3	Prob #4
0 proc	3.050e-6	2.696e-6	3.071e-6	3.944e-6
1 proc	2.711e-6	4.500e-6	3.362e-6	4.472e-6
2 proc	2.973e-6	3.088e-6	2.731e-6	2.767e-6
3 proc	3.036e-6	3.213e-6	2.864e-6	3.585e-6
4 proc	3.438e-6	4.970e-6	2.732e-6	4.002e-6
5 proc	3.035e-6	4.082e-6	2.704e-6	4.238e-6

From Table 13.7, we can see that the norm of the residuals is different for different numbers of simultaneous process failures. This is because, after recovery, due to the impact of round-off errors in the recovery algorithm, the PCG computations are performed based on different recovered data. However, Table 13.7 also indicates that the residuals with recovery do not have much difference from the residuals without recovery.

13.5 Discussion

The size of the checkpoint affects the performance of any checkpointing scheme. The larger the checkpoint size is, the higher the diskless-checkpoint overhead would be. In the PCG example, we only need to checkpoint three vectors and two scalars periodically, therefore the performance overhead is very low.

Diskless checkpointing is good for applications that modify a small amount of memory between checkpoints. There are many such applications in the high-performance computing field. For example, in typical iterative methods for sparse matrix computation, the sparse matrix is often not modified during the program execution, only some vectors and scalars are modified between checkpoints. For this type of application, the overhead for surviving a small number of processor failures is very low.

Even for applications which modify a relatively large amount of memory between two checkpoints, decent performance results to survive single processor failure were still reported in [15].

The basic weighted-checksum scheme implemented in the PCG example has a higher performance overhead than other schemes discussed in Section 13.2. When an application is executed on a large number of processors, to survive general multiple simultaneous processor failures, the one-dimensional weighted-checksum scheme will achieve a much lower performance overhead than the basic weighted-checksum scheme. If processors fail one after another (i.e., no multiple simultaneous processor failures), the neighbor-based schemes can achieve even lower performance overhead. It was shown in [5] that neighbor-based checkpointing was an order of magnitude faster than parity-based checkpointing, but takes twice as much storage overhead.

Diskless checkpointing could not survive a failure of all processors. Also, to survive a failure occurring during checkpoint or recovery, the storage overhead would double. If an application needs to tolerate these types of failures, a two-level recovery scheme [24] which uses both diskless checkpointing and stable-storage-based checkpointing is a good choice.

Another drawback of our fault-tolerance approach is that it requires the programmer to be involved in the fault-tolerance. However, if the fault tolerance schemes are implemented into numerical software such as LFC [4], then transparent fault tolerance can also be achieved for programmers using these software tools.

13.6   Conclusion and Future Work

We have presented how to build fault-survivable high-performance computing applications with FT-MPI using diskless checkpointing. We have introduced floating-point arithmetic encodings into diskless checkpointing and discussed several checkpoint-encoding strategies with detail. We have also implemented a fault-survivable example application (PCG) which can survive general multiple simultaneous processor failures. Experimental results show that FT-MPI is at least comparable to other state-of-the-art MPI implementations with respect to performance and can support fault-survivable MPI applications at the same time. Experimental results further demonstrate that our fault-tolerance approach can survive a small number of simultaneous processor failures with low performance overhead and little numerical impact.

For the future, we will evaluate our fault-tolerance approach on systems with larger numbers of processors. We would also like to evaluate our fault-tolerance approach with more applications and more diskless-checkpointing schemes.

References

[1] R. Barrett, M. Berry, T. F. Chan, J. Demmel, J. Donato, J. Dongarra, V. Eijkhout, R. Pozo, C. Romine, and H. Van der Vorst. *Templates for the Solution of Linear Systems: Building Blocks for Iterative Methods, 2nd Edition.* SIAM, Philadelphia, PA, 1994.

[2] G. Bosilca, Z. Chen, J. Dongarra, and J. Langou. Recovery patterns for iterative methods in a parallel unstable environment. Technical Report UT-CS-04-538, University of Tennessee, Knoxville, TN, 2004. to appear in SIAM SISC.

[3] Z. Chen and J. Dongarra. Condition numbers of Gaussian random matrices. *SIAM Matrix Analysis and Applications*, 27(3):603–620, 2005.

[4] Z. Chen, J. Dongarra, P. Luszczek, and K. Roche. Self-adapting software for numerical linear algebra and LAPACK for clusters. *Parallel Computing*, 29(11-12):1723–1743, November-December 2003.

[5] T. Chiueh and P. Deng. Evaluation of checkpoint mechanisms for massively parallel machines. In *FTCS*, pages 370–379, 1996.

[6] A. Edelman. Eigenvalues and condition numbers of random matrices. *SIAM J. Matrix Anal. Appl.*, 9(4):543–560, 1988.

[7] C. Engelmann and G. A. Geist. Super-scalable algorithms for computing on 100,000 processors. In *Proceedings of the 5th International Conference on Computational Science (ICCS), Part I*, volume 3514 of *Lecture Notes in Computer Science*, pages 313–320, Atlanta, GA, May 2005.

[8] N. R. Adiga, et al. An overview of the BlueGene/L supercomputer. In *Proceedings of the Supercomputing Conference (SC'2002), Baltimore MD,*, pages 1–22, 2002.

[9] G. E. Fagg and J. Dongarra. FT-MPI: Fault tolerant MPI, supporting dynamic applications in a dynamic world. In *PVM/MPI 2000*, pages 346–353, 2000.

[10] G. E. Fagg, E. Gabriel, G. Bosilca, T. Angskun, Z. Chen, J. Pjesivac-Grbovic, K. London, and J. J. Dongarra. Extending the MPI specification for process fault tolerance on high performance computing systems. In *Proceedings of the International Supercomputer Conference*, Heidelberg, Germany, 2004.

[11] G. E. Fagg, E. Gabriel, Z. Chen, T. Angskun, G. Bosilca, J. Pjesivac-Grbovic, and J. J. Dongarra. Process fault-tolerance: Semantics, design

and applications for high performance computing. *International Journal of High Performance Computing Applications*, 19(4):465–477, 2005.

[12] E. Gelenbe. On the optimum checkpoint interval. *J. ACM*, 26(2):259–270, 1979.

[13] W. Gropp, E. Lusk, N. Doss, and A. Skjellum. A high-performance, portable implementation of the MPI message passing interface standard. *Parallel Computing*, 22(6):789–828, September 1996.

[14] I. Foster and C. Kesselman. *The Grid: Blueprint for a New Computing Infrastructure.* Morgan Kauffman, San Francisco, 1999.

[15] Y. Kim. *Fault Tolerant Matrix Operations for Parallel and Distributed Systems.* Ph.D. dissertation, University of Tennessee, Knoxville, June 1996.

[16] Message Passing Interface Forum. MPI: A Message Passing Interface Standard. Technical Report UT-CS-94-230, University of Tennessee, Knoxville, TN, 1994.

[17] H.W. Meuer, E. Strohmaier, J.J. Dongarra, and H.D. Simon. TOP500 Supercomputer Sites. `http://www.top500.org`.

[18] J. S. Plank. A tutorial on Reed-Solomon coding for fault-tolerance in RAID-like systems. *Software – Practice & Experience*, 27(9):995–1012, September 1997.

[19] J. S. Plank, Y. Kim, and J. Dongarra. Fault-tolerant matrix operations for networks of workstations using diskless checkpointing. *J. Parallel Distrib. Comput.*, 43(2):125–138, 1997.

[20] J. S. Plank and K. Li. Faster checkpointing with $n+1$ parity. In *FTCS*, pages 288–297, 1994.

[21] J. S. Plank, K. Li, and M. A. Puening. Diskless checkpointing. *IEEE Trans. Parallel Distrib. Syst.*, 9(10):972–986, 1998.

[22] J. S. Plank and M. G. Thomason. Processor allocation and checkpoint interval selection in cluster computing systems. *J. Parallel Distrib. Comput.*, 61(11):1570–1590, November 2001.

[23] L. M. Silva and J. G. Silva. An experimental study about diskless checkpointing. In *EUROMICRO'98*, pages 395–402, 1998.

[24] N. H. Vaidya. A case for two-level recovery schemes. *IEEE Trans. Computers*, 47(6):656–666, 1998.

[25] J. W. Young. A first order approximation to the optimal checkpoint interval. *Commun. ACM*, 17(9):530–531, 1974.

Chapter 14

The Road to TSUBAME and Beyond

Satoshi Matsuoka

Tokyo Institute of Technology

14.1 Introduction — the Road to TSUBAME

It has been 12 years since Tom Sterling and Don Becker proposed the "Beowulf"-style commodity PC cluster in 1994 [12]. The performance of the first cluster ever built (called "Wigraf") was a mere few tens of megaflops (MFLOPS), but enormous progress has been made, with the fastest systems reaching 10 teraflops (TFLOPS) or beyond. However, compared to specialized and dedicated supercomputers such as the Earth Simulator [3], it was unknown then whether scaling of 10s to 100s of thousands of processors with commodity PC clusters would be possible in order to attain 100s of TFLOPS and even petaflops (PFLOPS), not just in terms of peak performance but actual stability in operation as well as wide applicability.

TSUBAME (Tokyo-Tech Supercomputer and Ubiquitously Accessible Mass-storage Environment) is a new supercomputer cluster installed at Tokyo Institute of Technology in Tokyo, Japan in April 2006, boasting over 85 TFLOPS of peak compute power with acceleration, 21.7 terabytes of memory, 1.6 petabytes of online disk storage, and "Fat Node" as well as fast parallel interconnect — architectural principles based on traditional supercomputers. TSUBAME became the fastest and largest supercomputer in Asia in terms of peak performance, memory and storage capacity, etc., when it became operational in April 2006. At the same time, being PC-architecture-based, TSUBAME can also be regarded as a large PC cluster server, being able to provide much broader services than traditional supercomputers. As a result, TSUBAME allows coexistence of so-called capability computing and capacity computing, in that it satisfies both requirements at the same time. This is in contrast to previous large-scale supercomputers and/or servers where satisfaction of either of the properties was sacrificed to the other. We term this architectural and operational property of TSUBAME as "everybody's supercomputer," as opposed to traditional supercomputers with very limited numbers of users, thus making their financial justifications increasingly difficult.

Particularly in Japan, supercomputers have not followed the trend of explosive PC and Internet expansion and growth into an industrial ecosystem of over 200 billion dollars annually, but rather, have confined themselves to a narrow market marred by legacy software and hardware. Such a trend has been continuing in most of the supercomputer centers in Japan, including those in the universities, whose missions are to serve a wide range of users.. This is causing a continuous decrease in the number of users, as the users observe decreasing benefit in overcoming the high hurdle of leaving their familiar PC-based notebook and workstation environments to conduct their science and engineering. This effect is not only limited to low-end users, but rather has the longer-ranging effect that the incubation of next generation high-end users will be on the decrease. In particular, users observe the following technical hindrances on "attempting" to use a supercomputer for the first time:

- Divergent operating system (OS) and middleware platforms from familiar ones such as Linux, Windows, etc.

- Lack of continuity in familiar services, such as file sharing, single sign-on identity management

- Missing independent software vendor (ISV) as well as open source tools and applications they are familiar with in their own environment

- Shortcomings in the programming environment especially modern integrated visual source management, debugging such as Visual Studio

Faced with such a situation, it was decided at Tokyo Institute of Technology that we would depart from such a legacy, and design a supercomputer for everyone on campus, or "everybody's supercomputer", serving as a core of Internet-style IT consolidation infrastructure to serve a variety of needs within the organization, linking to and hosting the IT services from research, education, computing, archival, as well as administrative departments. This allows all IT users to be easily cognizant of TSUBAME (Figure 14.1), in that it is an infrastructure they will be in touch with almost every day.

The design challenge, then, was how to design a supercomputer that would scale to 10,000 CPUs and 100 TFLOPS with room for growth to a petaflop machine, while embodying the flexibility to be a multipurpose server. The

FIGURE 14.1: The TSUBAME 100 TeraScale supercomputer as installed at Tokyo Institute of Technology, Tokyo, Japan in the Spring of 2006.

purpose of this article is to provide the architectural as well as operational overview of TSUBAME as "everybody's supercomputer," its upgrade growth path to a petaflop machine, as well as looking forward to the successor machine, TSUBAME 2.0, in the 2010-11 time frame.

14.2 Architectural Requirements of TSUBAME

In the initial design stages of TSUBAME, the following requirements were identified as constituting "everybody's supercomputer":

- Become the top-performing machine at the time of its installation as of March-April 2006, in achieving (1) over 40 TFLOPS of total (peak) compute power and a large memory, (2) over 1 petabyte of online secondary storage capacity, (3) Interconnection by a high bandwidth network that provide as terabits of aggregate performance, achieved through high-end but commodity technologies.

- Provide ubiquitous access as an entity of a (single sign-on) grid in that all the above mentioned resources be accessible from both internal and external virtual organizations, including not only academic research institutions but also private corporations, in a seamless and easy fashion.

- Provide various (web-based and other) services not only for expert, high-capability users, but also for so-called "light" users to incubate new users and stem growth of overall supercomputing needs

- Design an architecture with high-cost performance, low-power consumption, and high dependability/stability/reliability, facilitating recognition of Tokyo Tech's GSIC (Global Scientific Information and Computing Center) as one of the leading supercomputer centers in the world.

In particular, the following technical requirements were identified for compute nodes and the machine interconnect:

1. The CPU in the compute node would be a high-performance 64-bit processor with x86 software compatibility. For modern supercomputing, high 64-bit floating points as well as integer performance are crucial; at the same time, generality of the architecture for hosting a variety of operating systems, middleware, as well as applications is also essential. Modern x86 processors provide such benefits as well as other favorable properties such as low cost and low power. To be more specific:

   - high floating point performance with multicore CPUs to solve the cost, power consumption, and stability issues, with a lower parts count

   - general purpose x86 processor to utilize leading-edge process technology to attain high performance and low power consumption

   - native x86 CPU for efficient execution of various OS, middleware, and applications

2. Also, based on our 20-year supercomputing experience at the GSIC center, the individual compute node architecture should be shared memory and facilitate many CPUs and abundant memory, i.e., "Fat Node" architecture. Many traditional supercomputers typically embody shared memory nodes of 8–32 CPUs, and memory capacity of several 10s of gigabytes per node, or the so-called "Fat Node" architecture. These are due to the following reasons:

   - Ease of large-scale, parallel programming in a shared memory environment and a programming model such as OpenMP

   - In-core execution of data-intensive and search applications such as databases and large hash tables for genomics research

- Lowering of node counts for improving on various system metrics such as low power, better reliability (due to a lower parts count), simplification of the overall interconnect topologies for high bandwidth, etc.

3. The interconnect itself must satisfy fairly standard desirable properties such as scalability, high bandwidth, low latency, high-cost performance, high reliability, and sufficient reachability to span multiple floors. In addition, the following requirements were identified:

  - Future upgradability in performance and scaling: As compute nodes and storage could be upgraded in the near future, the upgrade of the network itself with relative ease and low expense were required.

  - Compliance to standards and general connectivity: Connectivity to both our multi-gigabit campus network (Super TITANET) as well as to the outside world (10Gbps and beyond to the academic backbone SuperSINET and JGN — Japan Gigabit Network) were deemed essential for TSUBAME to serve as a core for future national research grid and other network-based infrastructures.

  - Multi-protocol support including TCP/IP: In order to attain performance and reliability at lower acquisition and maintenance cost, the TSUBAME network was designed to accommodate a variety of high-performance TCP/IP and UDP/IP based protocols, co-existing with more propriety high-performance protocols on the same wire, especially the storage protocols which have been SAN-oriented for traditional clusters. Such consolidation of compute, storage, and other traffic was being practiced in some supercomputers already (such as the Cray XT3); the added value in TSUBAME would be the generality, flexibility and the reliability offered by IP-protocol families. Indeed, it has been our extensive experience that disk I/O traffic is commonly "bursty," but overall minuscule compared to MPI traffic, so it makes sense for I/O to utilize the high-performance network for message passing instead of facilitating its own network.

  - Fat switches allowing for indirect networks with low cable count: Employing Fat nodes allows for low overall node count but will mandate multilane network configuration to attain the necessary high bandwidth. As such, using smaller switches (e.g., 24–96 ports) would require a switch array consisting of 10s to 100s of switches and the resulting inter-switch cabling, resulting in an undesirable increase in floor space, power, and most importantly, high cable count and high switch administrative overhead. We required a network fabric which would allow for 100s of ports at low cost, which would reduce the overall switch and cable count, and allow symmetrical, indirect networks such as fat trees and CLOS.

4. In order to support a much wider range of users, support for powerful and flexible storage was considered the most important technical attribute of TSUBAME. Indeed, in traditional Japanese supercomputer centers, storage, in particular its capacity sadly was often neglected and insufficient. Based on various operational and archival storage systems at the Earth Simulator Center as well as at the San Diego Supercomputer Center, one petabyte was deemed as the minimal storage capacity requirement. Also, the required bandwidth was estimated to be approximately 40–50Gbytes/s. In terms of reliability, since the overall system lifetime would be planned as 4 years, high reliability up to that point would be necessary, but would be acceptable if it could degrade significantly beyond that period. In addition, the storage system had to be very cost effective, with very low cost being added over the cost of the bare drive itself. Finally, low power consumption and a small footprint were deemed important operationally. Overall, the goal was stated as a combination of all the properties.

   The problem was that, many of the standard, SAN-based commodity or enterprise-class storage systems did not necessarily realize these properties collectively — either they were low capacity, expensive, slow, too big and/or power consuming, unreliable, or a combination. Instead, we opted to employ a smaller number of large-capacity/high-bandwidth/high-performance networkd-attached storage (NAS) systems, and attach directly to the high-performance interconnect as mentioned above. In particular, for 10Gbps-class networks such as 10GbE or Infiniband, the internal (RAID-ed) disk array must exhibit over 1GB/s of streaming I/O bandwidth, which would surmount to nearly 50 channels of SATA disks. The problem was, most NAS systems only could accommodate 14 disks per each NAS node, far below the requirement to sustain the network bandwidth. In the next section we demonstrate how our infrastructure solves this problem.

5. Acceleration via low power/high density accelerator: The performance of current generations of x86 processors is now on a par with the fastest RISC processors in application-level performance metrics, for instance, as measured by the SpecFP2000 benchmark. Still, for the 100 TFLOPS of the current day (circa Spring 2006 when TSUABAME was to be installed) and for future PFLOPS by the next generation TSUBAME 2.0, the power projections with standard processors will require 2–3 megawatts (MW) of electrical power including cooling, far exceeding the capacity provided by the building. In fact, a post-installation stress test of TSUBAME revealed that the maximum power consumption of a single node would exceed 1300W, and to achieve a near 100 TFLOPS performance would have required over 1,000 nodes. Adding the power for storage, networks and cooling, the total requirement would have easily exceeded 2MW.

In order to "off-load" some of the kernel workloads in a power-efficient manner, we have decided to invest in SIMD-vector style commodity acceleration. Acceleration is now regarded as one of the key technologies for future supercomputing for achieving dramatic improvements in the FLOPS/power ratio, and have seen successful implementations in limited application in systems such as GRAPE [4]. However, in order for the accelerator to be much more applicable to wide-ranging applications, we have imposed the following requirements:

(a) It will accelerate mainstream numerical libraries such as basic linear algebra (BLA) and fast Fourier transform (FFT) in double precision IEE754 format, transparently from the users by simple relinking of the library

(b) It will effectively accelerate popular ISV and public-domain applications such as Matlab and Amber.

(c) For advanced users, it will allow them to program with programming language and/or software development kit (SDK) support of SIMD-vector acceleration.

The usages 1. and 2. above will benefit users immediately, while 3. was deemed as a longer-term goal. In fact, because the accelerated computations are fundamentally difficult to scale for tightly coupled applications, due to their greater computation to communication ratios, it would be better to use conventional CPUs for those purposes; on the other hand, for simple, non-scaling applications, the use of conventional CPUs would be a waste in terms of power, cost, etc. Although we are undertaking research to allow effective, scalable, and combined use of conventional and accelerated CPUs in our research [2], in production for the current moment we identify the largest benefit of acceleration as having the simple kernels to be off-loaded to be "out of their way" so as to free the general-purpose CPUs to be applied to more complex, harder scaling problems.

14.3 The Hatching of TSUBAME

Given the requirements thus presented, TSUBAME was designed, procured, and installed at the end of March 2006. The contract was awarded to NEC, who with Sun Microsystems jointly built and installed the entire machine, and also provide on-site engineering to operate the machine. Other commercial partners, such as AMD (Opteron CPUs), Voltaire (Infiniband), ClearSpeed (accelerator), CFS (LUSTRE parallel filesystem [1]), Novell (SUSE Linux),

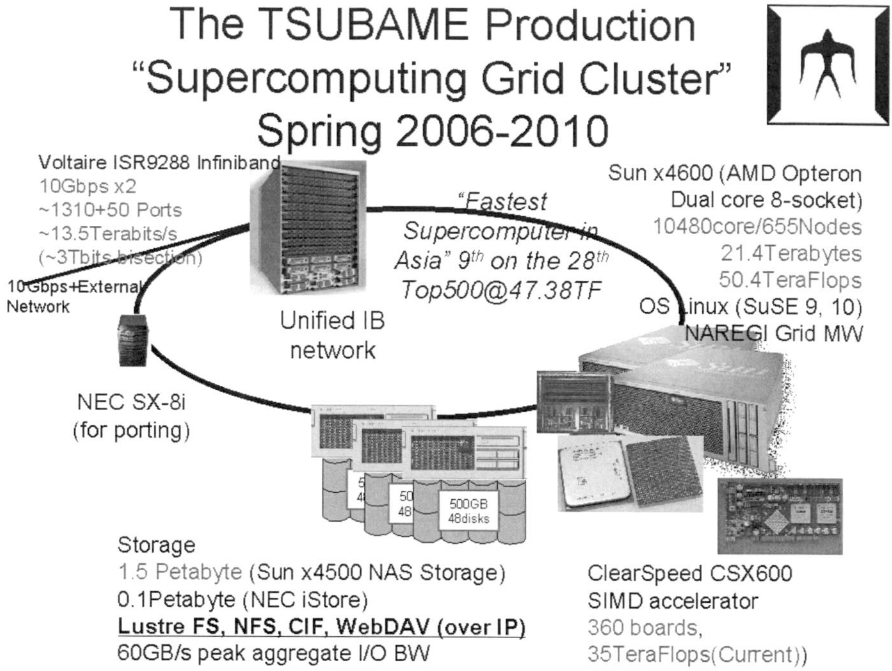

The TSUBAME Production "Supercomputing Grid Cluster" Spring 2006-2010

Voltaire ISR9288 Infiniband
10Gbps x2
~1310+50 Ports
~13.5Terabits/s
(~3Tbits bisection)

10Gbps+External Network

"Fastest Supercomputer in Asia" 9th on the 28th Top500@47.38TF

Unified IB network

NEC SX-8i (for porting)

Sun x4600 (AMD Opteron Dual core 8-socket)
10480core/655Nodes
21.4Terabytes
50.4TeraFlops
OS Linux (SuSE 9, 10)
NAREGI Grid MW

500GB
48disks

Storage
1.5 Petabyte (Sun x4500 NAS Storage)
0.1Petabyte (NEC iStore)
Lustre FS, NFS, CIF, WebDAV (over IP)
60GB/s peak aggregate I/O BW

ClearSpeed CSX600
SIMD accelerator
360 boards,
35TeraFlops(Current))

FIGURE 14.2: Architectural overview of TSUBAME.

provided their own products and expertise as building blocks. It was installed in just three weeks, and when its operation began on April 3, 2006, it became the largest academic machine in the world to be hosted by a university.

The overall architecture of TSUBAME is shown in Figure 14.2, and is described below:

- Total 5,120 socket/10,480 CPU dual core opterons running at 2.4GHz (some nodes are 2.6GHz).

- The 8 CPU sockets/16 CPU cores interconnected with Coherent HyperTransport provide fat-node SMP shared memory characteristics per node allowing packaging in a 4U rack-mount server package. The resulting 655 Sun X4600 nodes constitute approximately 50.4 TFLOPS worth of computing power.

- For memory, most nodes are 32GBytes/node, with some nodes being 64GBytes and 128GBytes, totaling 21.7 terabytes, This is more than twice the memory capacity of the Earth Simulator at 10 terabytes, allowing executions of large memory supercomputing applications to be readily ported to TSUBAME.

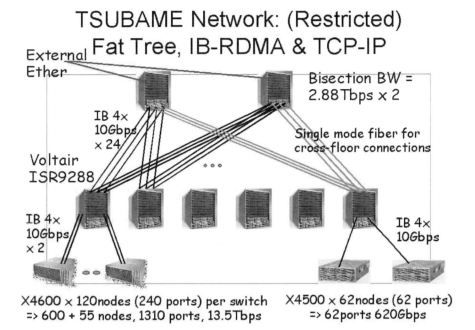

TSUBAME Network: (Restricted) Fat Tree, IB-RDMA & TCP-IP

X4600 × 120nodes (240 ports) per switch   X4500 × 62nodes (62 ports)
=> 600 + 55 nodes, 1310 ports, 13.5Tbps   => 62ports 620Gbps

FIGURE 14.3: Tsubame's network — an Infiniband-based interconnect that consolidates both parallel computing and storage networks

- As for the interconnect (Figure 14.3), we host dual HCA lane Infiniband per each node, providing 20Gbps bandwidth per node. (Storage nodes are single lane.) We employed the largest switch available at the time, which was the Voltaire ISR 9288 with 288 ports Infiniband 4x ports. The resulting switching fabric uses just 8 switches, and is configured in a restricted fat-tree topology, where we have 1,352 end points, each edge switch hosting 120 x 2 node connections, and 24 x 2 links to the upper-level core switch. The resulting edge bandwidth is approximately 13.1 terabits/sec, whereas the bisection bandwidth is approximately 2.88 terabits/sec, resulting in approximately a 1:5 ratio. Users enjoy full bisection up to 1920+ processors, and the bandwidth beyond is restricted but fine for most cases in production so far.

- TSUBAME's storage consists of two subsystems, one being a fleet of Sun Microsystems x4500, which effectively is a large 4U NAS storage node with 48 500GB SATA HDDs, totaling 24 terabytes of raw storage capacity, controlled by dual-socket, dual-core AMD Opterons. The 4 HyperTransport channels available for I/O are bridged to 48 SATA links, providing substantial internal bandwidth. TSUBAME initially started with 42 nodes/1 petabyte, and has been expanded to 62 units, or approximately 1.5 petabytes of raw storage. Each node is interconnected to the unified Infiniband fabric in the same fashion as the compute nodes, providing measured Livermore IOR benchmark speed of up to approximately 1GB/s, or about 60GB/s, combined. The other storage subsystem is the NEC's iStore S1800AT storage system that provides hardware RAID 6 capabilities and controller redundancy for high reliability, and is primarily used for the user home directory.

TSUBAME's flexibility in storage lies in the massive storage bandwidth over Infiniband, as well as the flexibility of x4500's ability to host multiple storage services and file-system types, thanks to its large NAS and fast Infiniband I/O structure. The storage protocols and services supported on TSUBAME are as follows:

- The LUSTRE parallel filesystem [1], which provides very fast parallel I/O primarily for work and scratch directories.
- NFS, for home directories (both on NEC and Sun x4500 systems, the latter over the local ZFS file system on Solaris)
- WebDAV, for general remote access from various end terminals from both inside and outside campus
- CIF, for supporting Microsoft Windows client access
- Tape emulation, for disk-to-disk backup
- We also facilitate running storage-centric services, such as databases, directly on the Sun x4500

Each x4500 is configured specifically to provide (possibly combinations of) the above services. The resulting storage system exhibits very high capacity, very high performance, is highly reliable with considerable flexibility and variety in the services it provides on its own.

Such a storage system contributes significantly to providing motivating experiences such that users prefer using TSUBAME over using their dedicated workstations and clusters. We have found that many applications could be accelerated by an order of magnitude compared to the user code running on their own machines due to dramatic storage bandwidth improvements — for example, some Gaussian 03 runs exhibited a 10–30 times speedup due to increased storage bandwidth along with shared memory parallelization.

- As for the accelerator on each node, we facilitate the ClearSpeed Advanced Accelerator Board by ClearSpeed Inc. Each accelerator board plugs into the PCI-X slot of TSUBAME, and sports two CSX600 chips, each of which are SIMD-vector processors with 96 processing elements, in addition to the onboard dedicated 1 gigabyte of external memory. The chip and the board exhibit an extremely high FLOPS/power ratio, each chip running at a maximum 250MHz, or the entire board embodying theoretical peak performance of 96 gigaflops (GFLOPS), while only consuming approximately 25 watts of power, or approximately 9KW of power, less than 7 TSUBAME nodes. Of the 655 TSUBAME nodes, 360 nodes have one ClearSpeed board each, while the remaining ones are to be added in the future.

The usage mode of ClearSpeed has been largely that of scenario 1. and 2. as explained in the previous section, predominantly off-loading kernel library workloads as well as some existing applications with embedded ClearSpeed support. We have, however, been conducting research so that high-scaling applications could also utilize the accelerators in a combined fashion with the general-purpose processors (Opterons on TSUBAME). The technical difficulty is that TSUBAME is vastly heterogeneous, not only within the node (intra-node heterogeneity) with a Opterons–ClearSpeed combination, but also between the nodes (inter-node heterogeneity). By devising methodologies and algorithms for coping with such heterogeneity, we have been able to successfully modify and extend HPL (high-performance Linpack) [10], so as to increase the full-machine Linpack performance from 37.18 TFLOPS in June 2006, when we only used Opterons, to 47.38 TFLOPS in November 2006 with their combined use, or about a 25% increase in performance as we observe in Figure 14.4. With expected kernel BLAS performance on the ClearSpeed with the new version of the numerical libraries, we expect additional an performance increase to 50+ TFLOPS in Spring 2007 without changes in the hardware. We are also working to accelerate more practical applications, such as Amber and Gaussian.

Overall, TSUBAME's installation space is approximately $350m^2$ including the service area, out of the available $600m^2$ available area in our computer room. There are approximately 80 compute/storage/network racks, as well as 32 CRC units for cooling. The total weight of TSUBAME exceeds 60 tons, requiring minor building reinforcements as the current building was designed for systems of a much smaller scale. TSUBAME occupies three rooms, where room-to-room Infiniband connections are achieved via optical fiber connection, whereas CX4 copper cables are used within a room.

The total power consumption of TSUBAME, including cooling, is about 1.2 megawatts peak. By all means we have and still continue to conduct research in power efficient computing and achieving cooling efficiency. For example, one challenging factor in the design was high density and cooling efficiently. Although TSUBAME is fairly power efficient, sometimes by several times compared to supercomputers and clusters of similar scale, nevertheless the overall heat density around the rack area was computed to be approximately 700watts/ft^2, which is several times beyond the normal cooling capability of large commercial data centers, that are usually spec'd at 100watts/ft^2. In order to effectively cool the machine while achieving premium power efficiency,

(Accelerated) Linpack Performance on TSUBAME

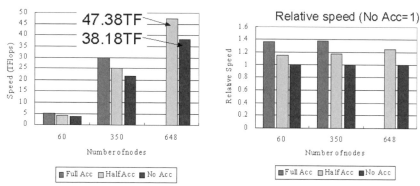

- **47.38TF with 648 nodes and 360 Accelerators**
  - Half Acceleration @ 648 nodes is the TSUBAME configuration
  - +24 % improvement over No Acceleration (38.18TF)
  - +25.5GFlops per accelerator
  - Matrix size N=1148160, 5.9hours
- **#1 in Asia Pacific, #9 overall --- improvements expected**

FIGURE 14.4: TSUBAME Linpack performance in November 2006 — no. 1 in Asia-Pacific.

we custom-designed the cooling configuration instead of retrofitting existing cooling systems, and with various innovative endeavors, some as exemplified in [9] and some our own, we were able to perform extremely efficient cooling (Figures 14.5 and 14.6), with additional capacity to spare for future machines. We are also conducting research into better use of various power-saving technologies, such as DVS (dynamic voltage scaling) on Opterons, detecting idle CPUs and putting them in a low state to save power, for example, without sacrificing performance.

These are plans to update TSUBAME's current OS Linux SUSE 9 to future versions of SUSE when they are judged to be stable. At the lowest-level batch scheduling we customize the Sun GridEngine (SGE) to add various applications support, accommodate our new innovative scheduling and accounting policies, etc. We have also started accommodating portal and other services, and integrate with the campus-wide PKI infrastructure that had been launched at the same time as TSUBAME. In the summer of 2007, we will be testdeploying the *beta*2 version of NAREGI grid middleware which is being developed at the National Institute of Informatics as the core middleware of the Japanese national research grid. We are also investigating the

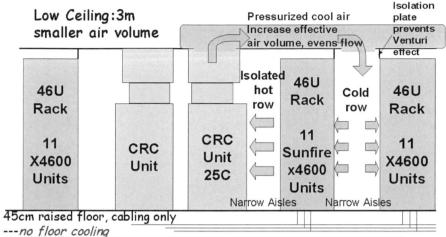

FIGURE 14.5: TSUBAME cooling and airflow.

feasibility of hosting a virtual private cluster environment [7] to accommodate customized user configurations as well as other operating systems such as Solaris and Windows CCS (compute cluster system) as requested by the user.

14.4 The Flight of TSUBAME — Performance, and Its Operations So That Everybody Supercomputes

TSUBAME became the fastest supercomputer in the Asia Pacific area and seventh in the world in June 2006 according to the Top500 [5] list at 38.18 TFLOPS, besting the Earth Simulator [3] that had held its crown since 2002; the performance is currently 47.38 TFLOPS with expected upcoming improvements as mentioned earlier. In production operation, users regularly launch MPI jobs beyond 500 CPUs up to approximately 1000 CPUs without any manual allocation provisioning from the operation. Also, there have been considerable large-scale parameter-sweep runs of low-parallel codes, with users submitting up to 20,000 jobs simultaneously. As such, many users and their applications have experienced a qualitative change in computing capability

FIGURE 14.6: TSUBAME cooling and airflow photographs — the woman's hair is blown vigorously with very strong pressurized cooling air from the ceiling plenum above.

by using TSUBAME, well exceeding our expectations.

At the same time, TSUBAME is "everybody's supercomputer" as seen in Figure 14.7, in that it gives a low usage account to everyone who is a member of Tokyo Tech, as well as its research partners. In fact, we are utilizing TSUBAME as the "core" of overall IT consolidation of the entire institution, as is typically done in commercial enterprises, but at a much lower scale. This serves three purposes, one is the vast cost savings achieved by the consolidation, not just in terms of hardware acquisition costs, but also maintenance and other operational costs — the standard motivation for enterprise IT consolidation, as seen in Figure 14.9. The second is to make the students and staff "feel at home" in everyday usage of supercomputers, so that there will be increased adaption and usage to advance their research. Finally, such consolidation allows tighter coupling of Web and other remote services, both mutually and with back-end supercomputing services, and moreover, enjoyment of the massive processing and data capability of TSUBAME. We believe such a model will greatly help to incubate the new generation of users who are accustomed to Internet-age computing, where services are linked and browser-based interactive UI is the norm, instead of cryptic batch scheduling and textual output.

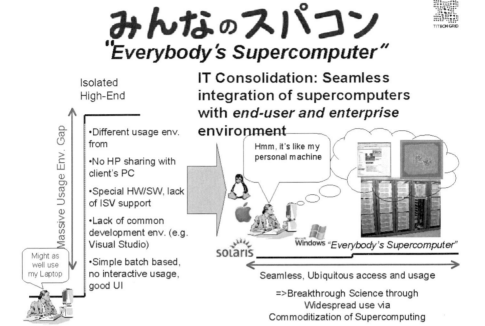

FIGURE 14.7: TSUBAME as "everybody's supercomputer".

To be more concrete, as of this writing the above-mentioned campus authentication/authorization system is partially hosted by TSUBAME, as well as other services such as e-mail and Microsoft Windows terminal server for university administration usage. Virtual machine hosting is employed for smaller-scale services, such as the campus information system and educational OCW (Open CourseWare). The next generation library system is planned to be hosted on TSUBAME as well. Hosting of such services only requires a small portion — approximately 1–2% — of TSUBAME's entire capability. Given the small sacrifice and the resulting great benefit we are experiencing, we believe that synergy of supercomputers with IT infrastructure could become predominant in the future, instead of a massive machine serving only 10–20 users. Such is possible because the building blocks, both hardware and middleware, are now part of the larger IT ecosystem.

One issue is how to make the traditional high-end usage and such everyday, low-demanding usage coexist. In particular, what is the scheduling and brokering policy of resources, which will be ultimately limited despite the massive size of TSUBAME? Our approach is to conduct such resource allocation so that it will have social metaphors with which both novice and expert users will be familiar. We also provide a disciplined "market economy"-based approach, in that as much open information as possible will be provided, along

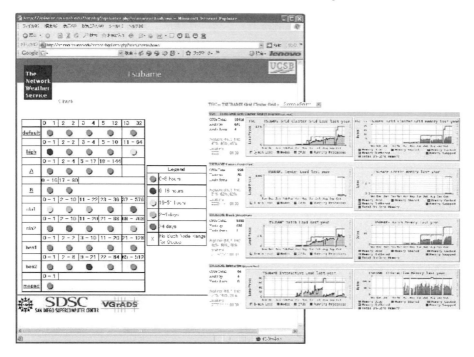

FIGURE 14.8: User monitoring tools on TSUBAME — University of California, Berkeley, Ganglia [11] and University of California, Santa Barbara, batch queue predictor [8] are available as well as other tools to give all users fair access to as much machine information as possible, so that users can make their own responsible judgments of resource requests and usages.

with a guarantee of a level playing field, so that individual users or groups can make their own intelligent decisions, sometimes winning and sometimes losing out, and will ultimately be satisfied as we provide a very fair and level playing field for everyone. Such a structure is (deliberately) quite synonymous to the historically proven market economy, and ultimately the entire "social benefit" will be optimized in the global view, despite small wins and losses at a microscopic level.

More concretely, resource allocation on TSUBAME is currently categorized into three classes of services at different QoS SLA (service level agreements):

1. Small Usage (Free) — This is the default account provided on initial signup. Up to 16 CPUs could be used at any one time, with limited storage. The overall amount of CPUs and storage allocated to the service are small, and there is no guarantee on job execution except on an first-come, first-served (FCFS) basis.

2. Best Effort — Users are charged allocation fees per "unit" of usage, each

みんなのスパコン

TSUBAME General Purpose DataCenter Hosting
As a core of IT Consolidation
All University Members == Users

- **Campus-wide AAA Sytem (April 2006)**
  - 50TB (for email), 9 Galaxy1 nodes
- **Campus-wide Storage Service (NEST)**
  - 10s GBs per everyone on campus
    PC mountable, but accessible directly from TSUBAME
  - Research Repository
- **CAI, On-line Courses
  (OCW = Open CourseWare)**
- **Administrative Hosting (VEST)**

I can backup ALL my data☺

FIGURE 14.9: TSUBAME as the core of university-wide IT consolidation, offering not only supercomputing services but hosting many others.

unit corresponding to a maximum flat fee usage of a certain number of CPUs per month (currently 64 CPUs). Users could "purchase" multiple units per month to increase the number of CPUs. There is no service guarantee — nodes could be shared with other users, and a job may be killed at any time (although this is rarely done). The allocation fee the user is charged is flat rate per unit, and very inexpensive.

3. Service Level Guarantee — Users are provided a much higher quality of service, such as exclusive usage of allocated nodes, favored scheduling and backfill for large jobs. The users in turn are charged higher allocation fees, and for per time metered usage.

In order to allow users to individually make strategic decisions on their own, we provide as much information as possible, even at which has been considered "private" to centers in the past. We provide not only past and current dynamic system status with various monitoring tools such as Ganglia [11], but also provide benchmark numbers of various applications, every instance of past failures and service records, as well as future forecasts such as the batch queue predictor in collaboration with a group at UCSB [8]. Finally, we declare that scheduling policies conform to the above without any backdoor policies or special hidden preferences that may compromise their decisions — this is a very important principle, to win trust so that users are motivated to make efforts to optimize their usage. Although this might be obvious if one would consider stock markets where unfair "insider" treatments are unlawful, one would be surprised to learn that such impromptu delegation of privileges is quite a common practice for many centers. This might work if the user base is small, but will not work for serving a large population with competing goals and needs.

14.5   Conclusion and into the Future—TSUBAME 2.0

TSUBAME demonstrated that a 100 TFLOPS-scale supercomputer that accommodates both high-end users and a widespread user base in terms of organizational IT consolidation can be realized to the scale, with appropriate selection and combinations of commodity building blocks, advanced research into various aspects of the machine, as well as novel operational policies. The philosophical underpinning of TSUBAME is "everybody's supercomputer," in that it not only accommodates the ever-increasing needs of high-end supercomputer users, but also establishes a widespread user base, facilitating a "positive cycle" of a growing user base, in user incubation, training, and adoption, for these uses to become experts in the future.

TSUBAME's lifetime was initially designed to be 4 years, until Spring of 2010. This could be extended up to a year with interim upgrades, such as an upgrade to future quad-core processors or beyond (Figure 14.10). However, eventually the lifetime will expire, and we are already beginning plans for designing "TSUBAME 2.0." But what are the requirements, the constraints, and the resulting architectural principles necessary in next-generation supercomputers? One thing clear is that the success of "everybody's supercomputer" should be continued; however, simply waiting for processor improvements end relying on CPU vendors would not be sufficient to meet the growing demands, as a result of the success of "everybody's supercomputer," in the growth of the supercomputing community itself, not just individual needs.

Although we are still conducting considerable research into the next generation of TSUBAME 2.0, which will become a petaScale machine, one tough requirement is not to increase the power or the footprint requirement of the current machine. Our current estimate would only allow us to scale to at most a petaflop, perhaps less. TSUBAME's current MFLOPS/watt ratio is approximately 100MFLOPS/watt; This is about half of IBM Blue Gene/L, and is vastly superior to similar machines on the Top500. Still, we must

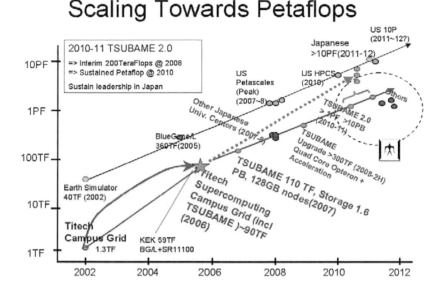

FIGURE 14.10: Future planned timeline of TSUBAME peformance, until and beyond TSUBAME 2.0 in 2010–2011.

improve this by a factor of 10 for a petaflop machine, to 1 GFLOPS/watt, including memory, network, and storage, marred by various power losses in the system including power supply and transmission. A multi-petascale machine will entail an even better MFLOPS/watt ratio. This cannot be achieved by simple scaling of VLSI process rules, as it will only achieve ×6 improvements in 4 years, according to Moore's law.

One hint is in our advanced investment into acceleration technologies, which will provide a vastly improved MFLOPS/watt ratio. By all means as we have mentioned, acceleration has its merits in this regard, but currently has its limitations as well, and as such will only facilitate smaller numbers of users. So, creating a general-purpose supercomputer whose dominant computing power is derived from current-day acceleration technology would not be appropriate as a next generation TSUBAME 2.0 architecture; rather, we must generalize the use of acceleration via advances in algorithm and software technologies, as well as design a machine with the right mix of various heterogeneous resources, including general-purpose processors, and various types of accelerators (Figure 14.11). Another factor is storage, where multi-petabyte storage with high bandwidth must be accommodated. Challenges are in devising more efficient cooling, better power control (as we have done in research, such as [6]), etc., etc.... Various challenges abound, and it will require advances in a multi-disciplinary fashion to meet this challenge. This is not a mere pursuit of

Future Multi-Petascale Designs

- Assuming Upper bound on Machine Cost
- A homogeneous machine entails compromises in all applications
- Heterogeneous sets of of large resources would allow multiple design points to coexist
- Especially effective for future "coupled" applications

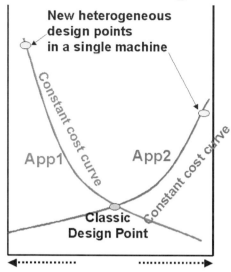

New heterogeneous design points in a single machine

App1   App2

Classic Design Point

◄·············   ·············►
More FLOPS   More Storage/BW

FIGURE 14.11: Rationale for heterogeneous acceleration usage in TSUBAME 2.0.

FLOPS, but rather, "pursuit of FLOPS usable by everyone" — a challenge worth taking for those of us who are computer scientists. And the challenge will continue beyond TSUBAME 2.0 for many years to come.

References

[1] Cluster Filesystems Inc. Lustre: A scalable, high-performance file system, November 2002. http://www.lustre.org/docs/whitepaper.pdf.

[2] T. Endo and S. Matsuoka. Methodology for coping with heterogeneity of modern accelerators on a massive supercomputing scale. Technical Report C-246, Dept. of Mathematical and Computing Science, Tokyo Institute of Technology, 2007. http://www.is.titech.ac.jp/research/research-report/C/C-246.pdf.

[3] Shinichi Habata, Kazuhiko Umezawa, Mitsuo Yokokawa, and Shigemune Kitawaki. Hardware system of the Earth Simulator. In *Parallel Computing*, volume 30-12, pages 1287–1313. Elsevier Science Publishers, 2004.

[4] J. Makino and *et al.* A 1.349 Tflops simulation of black holes in a galaxy center on GRAPE-6. In *Proc. IEEE Supercomputing 2000.* IEEE Press, Nov. 2000.

[5] H.W. Meuer, E. Strohmaier, J.J. Dongarra, and H.D. Simon. TOP500 Supercomputer Sites. http://www.top500.org.

[6] H. Nakashima, H. Nakamura, M. Sato, T. Boku, S. Matsuoka, D. Takahashi, and Y. Hotta. MegaProto: 1 TFlops/10kw rack is feasible even with only commodity technology. In *Proceedings of ACM/IEEE Supercomputing 2005.* IEEE Press, November 2005.

[7] H. Nishimura, N. Maruyama, and S. Matsuoka. Virtual clusters on the fly—fast, scalable, and flexible installation. In *Proceedings of the IEEE Computing Cluster and the Grid*, Rio de Janeiro, May 2007.

[8] D. Nurmi, A. Mandal, J. Brevik, C. Keolbel, R. Wolski, and K. Kennedy. Evaluation of a workflow scheduler using integrated performance modelling and batch queue wait time prediction. In *Proceedings of ACM/IEEE Supercomputing 2006.* IEEE Press, November 2006.

[9] C. D. Patel and *et al.* Thermal considerations in cooling large scale high compute density data centers. In *Itherm2002—8th Intersociety Conf. on Thermal and Thermomechanical Phenomena in Electronic Systems*, San Diego, 2002.

[10] A. Petitet, R. C. Whaley, J. Dongarra, and A. Cleary. HPL - a portable implementation of the high-performance Linpack benchmark for distributed-memory computers, 2004. http://www.netlib.org/benchmark/hpl/.

[11] Federico D. Sacerdoti, Mason J. Katz, Matthew L. Massie, and David E. Culler. Wide area cluster monitoring with Ganglia. In *Proc. IEEE Cluster 2003 Conf.*, Hong Kong, 2003. IEEE Press.

[12] T. L. Sterling, J. Salmon, and D. J. Becker. *How to build a Beowulf: A Guide to the Implementation and Application of PC Clusters.* MIT Press, Inc., Cambridge, MA, 1999.

Chapter 15

Petaflops Basics - Performance from SMP Building Blocks

Christian Bischof

Institute for Scientific Computing and Center for Computing and Communication
RWTH Aachen University
bischof@rz.rwth-aachen.de

Dieter an Mey

Center for Computing and Communication
RWTH Aachen University
anmey@rz.rwth-aachen.de

Christian Terboven

Center for Computing and Communication
RWTH Aachen University
terboven@rz.rwth-aachen.de

Samuel Sarholz

Center for Computing and Communication
RWTH Aachen University
sarholz@rz.rwth-aachen.de

15.1 Introduction

The number of processor cores in commodity architectures and in particular in the nodes of high-end computers is increasing. Today only few architectures in the current TOP500 list, like the IBM Blue Gene/L and some of the Cray

XT3 systems, employ nodes which are restricted to the MPI-only parallelization paradigm. But IBM is building its petaflop machine in the context of
the DARPA High Productivity Computing Systems (HPCS) program based
on Power7 processors [28] and Cray is already providing dual-core Opteron
processors for its leading edge systems.

Other current high-end machines like the TSUBAME at the Tokyo Institute of Technology (TITech) and the ASCI Purple machine at Lawrence Livermore National Lab employ standard rack-mounted mid-range systems with
16 and 8 processor cores, respectively, and the opportunity for an increasing degree of shared-memory programming within the nodes. The 8-socket
dual-core Opteron nodes of the TITech machine even exploit a high degree of
ccNUMA-ness (see Section 15.2), which in the past has predominantly been
a challenge for scalable multithreading on larger symmetric multiprocessor
(SMP) machines such as those offered by SGI.

It can be expected that in the future the number of cores per processor chip
will increase further and, in addition, on each of these cores multiple threads
may be running simultaneously.

By the end of 2005, Sun Microsystems presented the UltraSPARC T1 processor — code named Niagara — containing 8 cores, which are able to run 4
threads each. More advanced studies on multicore architectures include the
IBM Cyclops-64 with 80 cores and 160 threads on a single chip [8, 16].

Thus, future high-end machines will support substantial multithreading in
combination with message passing. And with these upcoming node architectures, OpenMP, which is the predominant parallelization paradigm for SMP
systems in scientific programming, will gain steam, as it is much more convenient than explicit thread programming.

What it comes down to, is that all the techniques of OpenMP programming
which have been explored in the past on larger SMP machines, will be available
for exploitation on nodes of future high-end systems with a growing number
of cores and even threads per core.

Of course the MPI-only programming style works on SMP nodes as well
and there seem to be good reasons to stick with MPI-only on SMP clusters:

- Programming complexity. Sticking with one parallelization paradigm
  simplifies development and maintenance of parallel codes.

- Efficiency. The naïve approach of hybrid parallelization restricts MPI
  communication to the serial program regions leaving all slave threads
  idle [27]. But communication may be overlapped with (OpenMP-parallel)
  computation with asynchronous MPI transfers or by providing additional threads for communication with nested OpenMP parallelization,
  if the MPI implementation is thread-safe.

However upon closer inspection, there are good reasons for adding the complexity of OpenMP parallelization to the MPI approach:

- Scalability. As soon as the scaling of an MPI program levels off, there may be an opportunity to speedup the MPI processes by adding another level of parallelism using OpenMP.

- Network limitations. The more MPI processes run on the single nodes, the higher the requirement for network bandwidth gets, and multiple simultaneous communications may increase the latency. A hybrid approach may help to alleviate bandwidth and latency problems, if it reduces the number or the size of MPI messages.

- Memory limitations. Main memory chips account for a considerable part of the total hardware costs and power consumption. Shared-memory programming can help to reduce the overall memory requirement.

- Application limitations. The structure of the application may limit the number of MPI processes. Additional levels of parallelism may be easier to be exploited by shared-memory parallelization. Fine-grained parallelism cannot be exploited with MPI.

- Programming productivity. It is much easier to apply SMP instead of distributed-memory parallelization to some algorithms.

- Dynamic thread balancing. If a hybrid application suffers from load imbalances on the MPI level, the number of threads can be increased in order to speedup busy MPI processes or decreased to slow down idle MPI processes, provided these processes reside on the same SMP node [31, 30].

Thus, as soon as the granularity of a partition of the computation fits within an SMP node, OpenMP is a vital alternative to MPI.

This chapter is organized as follows:

In Section 15.2 we shortly discuss some architectural aspects of the machines we are referring to in this chapter.

Then we address OpenMP parallelization on the loop-level in Section 15.3. Even with loop-level parallelization, an amazing speedup can be obtained if the OpenMP overhead can be kept to a minimum by making use of OpenMP's orphaning capability.

It turns out that C++ programmers need to be aware of certain pitfalls and machines with nonuniform memory access can easily spoil the fun. We cover some of the C++ aspects in Section 15.4.

On nodes with many cores, OpenMP can nicely be exploited on multiple levels. Future OpenMP specifications will have to provide some more support, but we report on first successes in Section 15.5. Again ccNUMA architectures may need special attention.

15.2   Architectures for OpenMP Programming

OpenMP as a programming model has no notion of the hardware a program is running on. This is adequate on machines having a uniform memory access (UMA), but may cause performance losses on those with a cache-coherent nonuniform memory access (ccNUMA) if data is not allocated in the memory partitions which are close to the threads which most frequently use them.

For our performance studies we used the machines described in Table 15.1. The 144-way Sun Fire 25K (SFE25K), and even more so the Opteron-based 8-way Sun Fire V40z (SFV40z) and 16-way Sun Fire X4600 (SFX4600) have such a ccNUMA architecture which makes the OpenMP performance of these machines very dependent on a good thread/data affinity.

On the SFE25K the two-stage cache coherence protocol and the limited bandwidth of the backplane lead to a reduction of the global memory bandwidth and to an increased latency when data is not local to the accessing process. The machine has 18 processor boards with 4 dual-core UltraSPARC IV processors and local memory, thus each locality domain consists of 8 processor cores.

On the dual-core Opteron-based SFV40z and SFX4600 machines, data and cache coherence information is transferred using the HyperTransport links. Whereas access to the memory which is local to each processor chip is very fast, multiple simultaneous remote accesses can easily lead to grave congestion of the HyperTransport links.

TABLE 15.1:   List of the computer systems for the performance studies

Machine model (abbrev.)	Processors	Remark
Sun Fire E25K (SFE25K)	72 UltraSPARC IV 1.05 GHz dual core	ccNUMA
Sun Fire 15K (SF15K)	72 UltraSPARC III 0.9 GHz	ccNUMA
Sun Fire E6900 (SFE6900)	24 UltraSPARC IV 1.2 GHz dual core	UMA
Sun Fire E2900 (SFE2900)	12 UltraSPARC IV 1.2 GHz dual core	UMA
Sun Fire V40z (SFV40z)	4 Opteron 875 2.2 GHz dual core	ccNUMA
Sun Fire X4600 (SFX4600)	8 Opteron 885 2.6 GHz dual core	ccNUMA
NEC SX-8 (NECSX8)	8 NEC SX-8 2 GHz vector unit	UMA

All UltraSPARC- and Opteron-processor-based machines ran Solaris and the Sun Studio compilers. The Solaris operating system includes the Memory Placement Optimization facility (MPO) [26] which helps to improve the scalability of OpenMP applications on ccNUMA architectures. The Solaris system calls to pin threads to processor cores and to migrate pages to where they are used (next-touch mechanism) allow for precise control of the affinity between data and threads.

On Linux we frequently apply the `taskset` command to bind the threads of an OpenMP program to a subset of the available processors. Unfortunately, Linux does not support explicit page migration yet.

15.3 Loop-Level Parallelization with OpenMP

The obvious targets for parallelization are loops. Efficient loop-level parallelization requires techniques similar to vectorization. In the case of loop nests, the optimal loop for parallelization has to be selected, loops may have to be interchanged, cache effects and scalability have to be taken into account. The upcoming OpenMP version will eventually include the parallelization of perfectly nested loops, which will help to improve the scalability and flexibility of many loop constructs [5].

Another issue that is laborious with OpenMP is the scoping of variables, i.e., the declaration whether a variable is shared between threads or replicated as private for each thread. To ease this burden on the programmers, a feature called auto-scoping has been implemented into Sun Studio compilers. A compiler that supports auto-scoping determines the appropriate scopes of variables referenced in a parallel region, based on its analysis of the program and on a set of rules. Special knowledge that the programmer has about certain variables can be specified explicitly by using additional data-sharing attribute clauses, thus supporting the compiler in its analysis. Put to use on the 3D Navier-Stokes solver PANTA [36], auto-scoping automatically generated correct OpenMP clauses for scoping 1376 variables in some 200 OpenMP directives, only 13 variables had to be taken care of manually [23].

Attempts to automatically augment a given Fortran program with OpenMP directives in a preprocessing step culminated in the ParaWise toolkit by Parallel Software Products (PSP) [17, 22]. ParaWise is designed with the understanding that user interaction is desirable and often essential for the production of effective parallel code. Furthermore, runtime profiling information provided by the Sun Performance Analyzer can be fed into ParaWise to guide further user efforts for improving scalability [21, 20].

To further enhance OpenMP performance, one can extend the parallel region to contain not just one parallel loop, but multiple or even all loops.

This technique relies on the so-called orphaning aspect of OpenMP, which allows worksharing constructs — like parallel-loop constructs — to reside in routines which are called in parallel regions. Extending parallel regions requires special care by the programmer, however, as variables in the syntactical context of the parallel construct are shared between all threads by default, local variables in routines called within parallel regions are private, unless they are explicitly declared to be static. Static variables within parallel regions can easily introduce data races, if their usage is not protected by critical regions.

This approach has been implemented in the context of parallelizing Thermoflow [11], a finite element solver used for simulating the heat distribution in a rocket combustion chamber: A sole parallel region encompassed the entire compute-intensive program part, where 69 inner loops were parallelized. On the 72-way Sun Fire 15K a speedup of 40 was achieved with 64 threads [2].

15.4   C++ and OpenMP

OpenMP has initially been specified for Fortran (1994), but the C/C++ specification followed a year later. Anyhow, the coverage of C++-specific language features still has some shortcomings, which will hopefully be overcome with the next version of OpenMP. As a consequence OpenMP is not yet widespread in C++ production codes.

In this section we present selected C++-specific problems which we experienced with several C++ codes. For most topics, we relate to the Navier-Stokes solver DROPS [35] as an example. DROPS is developed at the IGPM at RWTH Aachen University to investigate two-phase flow phenomena [10, 29].

15.4.1   Iterator loops

The Standard Template Library's (STL) iterator loops cannot be parallelized with a for-worksharing construct in OpenMP, because the loop is not in the required canonical form.

We considered four ways [33] to parallelize these loops: (1) placing the for-loop into a parallel region and the loop-body in a single work-sharing construct with the "nowait" clause; (2) using the task-queuing feature of the Intel C++ compilers; (3) storing the iterator pointers in an array that can then be processed in parallel; and (4) assigning the value of the incremented iterator in a critical region to a private iterator which is then used for computation. Depending on the relation of the number of loop iterations to the amount of work inside the loop body, the scalability of the four methods varies. In general, we found method (3) to be the most efficient one. The upcoming OpenMP specification 3.0 [5] will most likely contain a new feature named tasking, which will have noticeably less administrative overhead than the mentioned task-queuing construct.

15.4.2   ccNUMA issues

STL data types like `std::valarray` are commonly used in scientific programs. As `std::valarray` guarantees that all elements are initialized to zero, on a ccNUMA architecture which supports the first touch principle, the data will be placed in the memory close to the CPU on which the initial zeroing thread is running.

A typical solution to this problem of an initialization step predetermining memory placement is a parallel initialization with the same memory access pattern as in the computation. The problem with the `std::valarray` container is that the initialization is already done when an object is constructed. Therefore the programmer cannot take care of proper placement in a parallel initialization loop.

We implemented three approaches [32] to address this issue:

1. Use of a modification of `std::valarray` so that the initialization is done in parallel with a given memory-access pattern. This approach is not portable, as typically every compiler provides its own STL implementation.

2. Use of `std::vector` instead of `std::valarray` with a custom allocator. We implemented an allocator that uses `malloc()` and `free()` for memory allocation and initializes the memory in a loop parallelized with OpenMP.

3. Use of page migration functionality provided currently only by the Solaris operating system. The `madvise()` function gives advice to the virtual memory system for a given memory range, e.g., advising to physically migrate the memory pages to the memory of that CPU which will access these pages next.

To control data placement of general C++ classes a mix-in [4] can be used. Thereby it is possible to overwrite the `new` and `delete` operators for a given class without modifying or having access to the class source code. By using additional template parameters, the same flexibility as with a custom allocator can be reached.

15.4.3   Parallelizing OO-codes

In C++ programs making extensive use of object-oriented programming, a large amount of computing time may be spent inside of member functions

```
1 PCG(const Mat& A, Vec& x,const Vec& b,...) {
2    Vec p(n), z(n), q(n), r(n); Mat A(n, n);
3    [...]
4    for (int i = 1; i <= max_iter; ++i) {
5       q = A * p;
6       double alpha = rho / (p * q);
7       x += alpha * p;
8       [...]
```

FIGURE 15.1: PCG code from DROPS: high-level C++.

of variables of class type. An example is shown in Figure 15.1 where a part of the preconditioned conjugate gradient (PCG)-type linear equation solver of DROPS is shown. The data types Vec and Mat represent vector or matrix implementations, respectively, and hide the implementation by providing an abstract interface. To parallelize such codes there are two choices:

1. Internal parallelization: a complete parallel region is embedded in the member function. Thus, the parallelization is completely hidden by the interface but there is no chance to enlarge the parallel region or to reduce the number of barriers.

2. External parallelization: the parallel region starts and ends outside of member functions (in the PCG code example it could span the whole loop body) and inside member functions orphaned work-sharing constructs are used. This can reduce the overhead of thread creation and termination. The disadvantage is that the interface is changed implicitly, because the parallelized member functions may only be called by a serial program part or out of a parallel region, but not out of another work-sharing construct and there is no way for a member function to find out if it is called in a work-sharing construct.

For performance reasons and from a library writer's point of view (imagine C++ templates) the external parallelization is preferable. The ability to check for calling situations out of a work-sharing construct would broaden the applicability of that programming style.

15.4.4 Thread safety

The OpenMP specification, in section 1.5, states "All library, intrinsic and built-in routines provided by the base language must be thread-safe in a compliant implementation." Neither the C nor the C++ standard even contain the word "thread," so the languages themselves do not provide any guarantees. Hence the level of thread safety is implementation-defined. There is an ongoing discussion so that hopefully this issue will be clarified for OpenMP 3.0.

Our experience is the following:

- C++ STL: Simultaneous access to distinct containers are safe, and simultaneous read access to shared containers are safe. If multiple threads access a single container, and at least one thread may potentially write, then the user is responsible for ensuring mutual exclusion between the threads during their access to that container.

- C library routines: Most routines can be assumed to be thread-safe. Most vendors nowadays provide multiple versions, including a thread-safe one, of their C library. Still, there are some functions that are not reentrant by their interface (e.g., `strtok()`) and have to be used with caution.

15.5 Nested Parallelization with OpenMP

Nested parallelization has been included as an optional feature already in the first OpenMP specification, leaving the implementer the flexibility to execute parallel regions that are nested within an active outer parallel region with only one thread.

Here we describe experiences gained when employing nested parallelization using OpenMP in three different production codes:

1 FIRE is a C++ code for content-based image retrieval using OpenMP on two levels [34]. A nested OpenMP approach turned out to be easily applicable and highly efficient.

2 NestedCP is written in C++ and computes critical points in multi-block computational fluid dynamics (CFD) data sets by using a highly adaptive algorithm which profits from the flexibility of OpenMP to adjust the thread count and to specify suitable loop schedules on three parallel levels [9].

3 The multi-block Navier-Stokes Solver TFS written in Fortran90 is used to simulate the human nasal flow. OpenMP is employed on the block and on the loop level. This application puts a high burden on the memory system and thus is quite sensitive to ccNUMA effects [21].

15.5.1 Nested Parallelization in the current OpenMP specification

Programming nested parallelization is as easy as just nesting multiple parallel regions using the standard OpenMP parallel construct plus activating nested parallelism [25].

When at the outer level the initial thread encounters a parallel region, a team of threads is created and the initial thread becomes the master thread. Now when any or all of these threads of the outer team encounter another parallel region, and nested support is turned on, these threads create further (inner) teams of threads of which they are the masters. So the one thread which is the master of the outer parallel region may also become the master of an inner parallel region. But also the slave threads of the outer parallel region may become the masters of inner parallel regions.

The OpenMP runtime library of the Sun Studio compilers, for example, maintains a pool of threads that can be used as slave threads in parallel regions and provides environment variables to control the number of slave threads in the pool and the maximum depth of nested active parallel regions that require more than one thread [1].

15.5.2 Content-based image retrieval with FIRE

The Flexible Image Retrieval Engine (FIRE) [6] has been developed at the Human Language Technology and Pattern Recognition Group of the RWTH Aachen University for content-based image retrieval [34]. It is designed as a research system and is easily extensible and highly modular.

15.5.2.1 Image retrieval

Given a query image and the goal to find images from a database that are similar to the given query image, a score for each image from the database is calculated. The database images with the highest scores are returned. Three different layers can be identified that offer potential for parallelization:

- Queries tend to be mutually independent. Thus, several queries can be processed in parallel.

- The scores for the database images can be calculated in parallel as the database images are independent from each other.

- Parallelization is possible on the feature level, because the distances for the individual features can be calculated in parallel.

Only the first two layers have been parallelized so far, as the third may require larger changes in the code for some distance functions and we do not expect it to be profitable as the parallelization in the first two layers already leads to sufficient scalability.

TABLE 15.2:   Scalability of FIRE: comparing solely the outer and solely the inner parallel region to nested parallelism.

#Threads	SFX4600			SFE25K		
	outer	inner	nested	outer	inner	nested
4	3.9	3.9	3.9	-	3.8	
8	7.5	7.9	7.9	-	7.6	
16	15	15.7	15.8	14.8	14.1	15.4
32	-	-	-	29.6	28.9	30.6
72	-	-	-	56.52	-	67.6
144	-	-	-	-	-	133.3

15.5.2.2   Parallelization

Shared-memory parallelization is more suitable than message-passing for the image retrieval task, as the image database which can be several gigabyte in size can then be accessed by all threads and does not need to be replicated. The object-oriented programming paradigm as employed in the FIRE C++ code simplified the parallelization by preventing unintended data dependencies and facilitating the data-dependency analysis.

15.5.2.3   Scalability

Because the computations on both levels are almost independent, because the load is nicely balanced and because the required memory bandwidth is rather small and data is mainly read the code scales perfectly well even on ccNUMA architectures.

Experimental results with the FIRE code on a 16-core Opteron Sun Fire SFX4600 and a 144-core UltraSPARC-IV Sun Fire E25K exhibit excellent speedup, which is the ratio of the runtime of the program using one thread versus the runtime with n threads (see Table 15.2).

15.5.3   Computation of 3D critical points in multi-block CFD data sets

In order to interactively analyze results of large-scale flow simulations in a virtual environment, different features are extracted and visualized from the raw output data. One feature that helps describe the topology is the set of critical points, where the velocity is zero [9].

15.5.3.1   Critical point algorithm

The algorithm for critical point extraction is organized in three nested loops, which are all candidates for parallelization, as there is no data dependency among their iterations. The outermost loop iterates over the time steps of unsteady data sets. The middle loop deals with the blocks of multi-block data sets and the inner loop checks all grid cells within the blocks. As soon as the number of threads is larger than the iteration count of the single loops, nested parallelization is obviously appropriate to improve speedup.

We use a heuristic on each cell to determine whether it may contain a critical point. If so, the cell is recursively bisected and the heuristic is applied again on each subcell. Noncandidate (sub-) cells are discarded. After a certain recursion depth, the Newton-Raphson iteration is used to determine the exact position. The time needed to check different cells may vary considerably as a result. If critical points are lined up within a single cell, the computational cost related to this cell may even increase exponentially. Furthermore, the size of the blocks as well as the number of cells per time step may vary considerably. Therefore, this code really profits from the flexibility of the OpenMP loop-scheduling construct.

15.5.3.2 Loop scheduling in OpenMP

The `schedule` clause can be used to elegantly specify how iterations of a parallel loop are divided into contiguous subsets, called chunks, and how these chunks are assigned among threads of a team. In addition to specifying the chunk size, the programmer has a choice of three different schedule kinds: static, dynamic and guided. Whereas in a static schedule, the chunks are statically assigned to threads in a round-robin fashion, the chunks are dynamically assigned to threads in chunks, as the threads request them. The `guided` schedule offers a nice compromise by reducing the chunk size over time according to Equation (15.1):

$$chunk\_size = \#\,unassigned\_iterations/(c \cdot \#\,threads) \qquad (15.1)$$

Here the chunk-size parameter determines the minimum chunk-size. Furthermore the Sun compiler allows for adjustment of the weight parameter c in Equation (15.1) with an environment variable.

15.5.3.3 Results

We measured the runtime of the parallel feature extraction on the Sun Fire E25K, varying the loop schedules and the number of threads assigned to each of the parallelization levels and the loop schedules. Let n be the number of threads involved. The amount of threads for the time level is denoted as ti and for block level as bj. The remaining $n - i \cdot j$ threads, denoted as ck, are assigned to the cell level.

The achievable speedup heavily depends on the selected data set. Data sets which do not cause severe load imbalance display almost perfect scalability, such as the output data set of a supersonic shock simulation. A speedup of 115 can be reached by just using all 144 threads on the outer level and a `static` loop schedule. The speedup can be increased to 119.9 with nested parallelization (t24 b1 c6) and `static` scheduling.

The situation is quite different on another data set simulating the inflow and compression phase of a combustion engine. When the valves start to close and suddenly move in the opposite direction of the inflowing air, plenty of critical points close to the valve can be detected in the output data of the corresponding simulation. Applying static schedules on this heavily imbalanced data set limits the speedup to 11 at best.

By choosing an appropriate schedule on all parallelization levels, the speedup can be considerably increased. The `dynamic` schedule with a chunk size of one turned out to work best on both outer parallelization levels. However it was not possible to find a suitable chunk size for the `dynamic` schedule on the cell level for all data sets: it either caused too much overhead or the chunks were too large to satisfactorily improve the load balance. Here the guided schedule kind turned out to be optimal such that the critical point computation for the engine data set scaled up to 33.6 using 128 threads (t4 b4 c8). It turned out, however, that the weight parameter c in Equation (15.1), which defaults to 2 on Sun's OpenMP implementation, is not an optimal choice. We reached the best results using the dynamic schedule with a chunk size of 1 on both outer loops and the guided schedule with a minimum chunk size of 5 and weight parameter $c = 20$ on the innermost level, as shown in Figure 15.2. The speedup improved for all thread combinations reaching a maximum of 70.2 with t6 b4 c6 threads.

Even the speedup for the supersonic shock data set profits from these scheduling parameters: it increases to 126.4 with t144 c1 b1 threads (outer-level parallelization only) and to 137.6 with t12 b1 c12 case, which corresponds to an efficiency of 96%.

15.5.4   The TFS flow solver

The Navier-Stokes solver TFS developed by the Institute of Aerodynamics of the RWTH Aachen University is currently used in a multidisciplinary project to simulate the airflow through the human nose [15, 14]. TFS uses a multi-block structured grid with general curvilinear coordinates.

The ParaWise/CAPO automatic parallelization environment [17, 19] has been used to assist in the OpenMP parallelization of the TFS multi-block code and runtime information provided by the Sun Performance Analyzer fed into it guided further efforts for improving the scalability.

15.5.4.1   Intra-block parallelization

The initial OpenMP version produced by ParaWise/CAPO without any user interaction was improved in a second step with the help of the ParaWise GUI, which assists the user to investigate, add to and alter information to enhance the parallelization. In the TFS code a large work array is dynamically allocated in the beginning of the program and then frequently passed

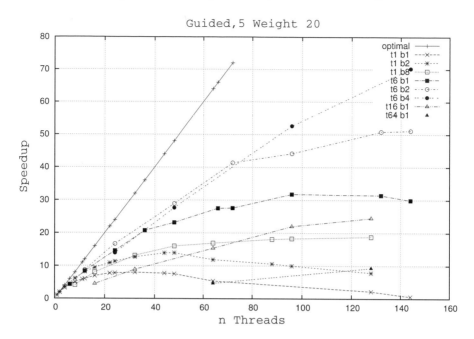

FIGURE 15.2: Computing the critical points for a combustion engine.

to subroutines through parameter lists and used throughout the program by indirect addressing. It is obviously not possible by any static program analysis to assure that there is no data race in accessing this work array, so the user must provide further information. The user can exploit his or her knowledge of the program to address the parallelization inhibitors determined by ParaWise and displayed in its browsers.

Additional manual code modifications lead to increased parallel regions and a reduction of the number of synchronizations.

15.5.4.2 Inter-block parallelization

Then a version where parallelism is exploited at the block level was generated with the previous parallel version as a starting point. The major step was to split the large work array into sections describing the mesh data which have to be shared among the threads, and sections used for temporary workspace which need to be privatized due to the reuse for every block.

15.5.4.3 The nested-parallel version

The nested-parallel version was created by merging the two previous inter-block and intra-block parallel versions. Unfortunately, the persistence of `threadprivate` data between parallel regions is not guaranteed when nested parallelization is employed, which caused quite some code changes. Instead, the affected variables had to be passed into the necessary routines as additional arguments and then defined as `private` for the outer (inter-block) parallel region, but shared by the subsequent inner (intra-block) parallel region. In the final nested-parallel version, 7 major parallel outer block loops out of 11 include inner-parallel loops. This version contains some 16,000 lines of Fortran code with 534 OpenMP directives in 79 parallel regions. Further manual tuning improved the scalability of the nested version by improving the load balancing and by taking care of ccNUMA effects.

15.5.4.4 Improving the load balance of the nested-parallel program

Sorting blocks by size. Because of the complex geometry of the human nose, the blocks of the computational grid vary considerably in size: The largest block has about 15 times more grid-points than the smallest block and accounts for about 10% out of the 2,200,000 grid-points. This fact limits the attainable speedup on the block level to no more than 10. The first approach of selecting a dynamic schedule for all of the block-level loops in order to handle the resulting load imbalance works reasonably well. But if a relatively large block is scheduled to one thread at the end of the loop, the other threads might be idle. Sorting the blocks in decreasing order, such that the smallest block is scheduled last, leads to an improvement in runtime of 5–13 %.

Grouping blocks. As the block sizes remain constant during the whole runtime of the program, the blocks can be explicitly grouped and accordingly distributed to a given number of threads on the outer-parallel level in order to reduce the overhead of the dynamic schedule and to avoid idle threads. Surprisingly this did not lead to a measurable performance improvement on the SFE25K. Further investigations using hardware counters to measure the number of L2-cache misses revealed that threads working on smaller blocks profit more from the large size (8 MB) of the external L2 caches of the UltraSPARC IV-based machines than threads working on larger blocks and therefore ran at a much higher speed.

Grouping and distributing the blocks was profitable on the SFV40z as the varying block size did not impact the MFLOPS rate, because of the smaller L2 cache (1 MB) of the Opteron processor. The performance improved by 6.7% when using 8 threads.

The nested approach leads to an increased speedup for all of the larger SMP machines.

There is no single strategy which performs best in all cases. Figure 15.3 depicts the best effort speedup of the nested parallel versions on the machines listed in Table 15.1. On the SFE25K, our target platform, the speedup of the nested version is close to 20 with 64 threads, whereas the speedup of both single-level approaches is less than 10 in all cases.

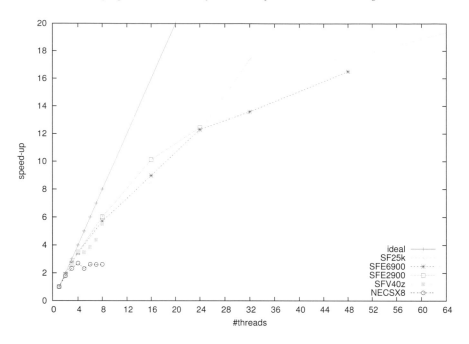

FIGURE 15.3: Speedup of the nested-parallel version of TFS using Sun's regular multitasking library.

15.5.4.5 Improving the memory locality of the nested-parallel program

Hardware counter measurements indicated that L2-cache misses led to a high percentage of remote misses and that the global memory bandwidth consumption of the code on the SFE25K was close to the maximum value, which we observed when stressing the memory system with the STREAM benchmark [24] in earlier experiments with disadvantageous memory placement. We concluded that an improvement in the memory locality would have a positive impact on the performance of TFS on the SFE25K.

In order to improve memory locality, threads were bound to processors and also pages were migrated to where they are used (next-touch mechanism) with the Solaris `madvise()` system call after a warm-up phase of the program.

Surprisingly, this was only profitable when applied to a single level of parallelism.

Unfortunately, applying these techniques to the nested-parallel version was not profitable at all: As described above, the current implementation of nested OpenMP parallelization in the Sun Studio compilers employs a pool of threads. Because these threads are dynamically assigned whenever an inner team is forked, they lose their data affinity frequently.

Compiler engineers from Sun Microsystem provided an experimental version of the threading library `libmtsk` which improves the thread affinity by maintaining the mapping between threads of the pool and the members of the inner teams. The combination of thread affinity, processor binding and explicit data migration finally led to a speedup of 25 for 64 threads, an improvement in scalability of about 25 % on the SFE25K. This finally is a satisfying result taking into account

- that each locality group (processor board) has 8 cores and thus using 8 threads is a sweet spot for the intra-block version delivering a speedup of 5–6

- that the largest block dominates the inter-block version with more than 8 threads thus limiting the speedup to about 6

- that there are some serial parts and some parts only suited for one level of parallelization

15.6   Conclusions and Outlook

Message passing with MPI will be the dominating parallelization paradigm of high-end computing in the foreseeable future. With nodes of future high-end machines having a growing number of cores, techniques which have been used to program large SMP machines in the past will be applicable and may be very appropriate to increase the scalability of hybrid codes combining message passing and multithreading (see [7]).

Already today, OpenMP is the de-facto standard for shared memory parallelization in the context of scientific programming and has proven to be much more easily applicable than explicit thread programming in most cases. Still, there are many opportunities to make OpenMP even more useful in the near future. In [5] the issues which are currently under consideration for the upcoming version 3.0 of the OpenMP specification have been disclosed. A new task concept will greatly facilitate parallelization of control structures other than simple loops, the support for nested parallelism will be improved, and features to make OpenMP ccNUMA-aware are extensively discussed, to mention only a few.

Future multicore node architectures will make the playground for OpenMP programs even more diverse. The memory hierarchy will grow with more caches on the processor chips. Whereas the four cores on the future Opteron processor will share one single L3 cache [12], pairs of two threads share one L2 cache in the Intel Clovertown quad-core processor [13]. As a consequence, placing two cooperating threads close together may be a performance benefit on the latter, if they share data [18]. As multicore processor chips will be more and more characterized by sharing resources, the speedup of such nodes cannot be expected to scale linearly with the number of cores.

And then there may be more threads per core: All 8 cores of the Ultra-SPARC T1 processor access a common L2 cache, whereas 4 threads running on each core share one L1 cache. Interestingly, memory load operations of four threads running on a single core can be nicely overlapped, an important capability for memory bandwidth hungry codes [3]. Techniques which previously have been developed on proprietary hardware platforms like the Denelcore HEP and the Tera machines to bridge the gap between processor and memory speed are now available on a single chip which is fully compatible to the popular UltraSPARC architecture. When the chip is fully loaded with 32 threads, the memory latency only increases by a factor of 1.4, as measured by a pointer chasing benchmark. Such an architecture cuts down the price per thread and presents a flat memory which can be nicely programmed with OpenMP.

References

[1] Sun Studio 11. OpenMP API user's guide, chapter 2, nested parallelism. `http://docs.sun.com/source/819-3694/2_nested.html`.

[2] D. an Mey and T. Haarmann. Pushing loop-level parallelization to the limit. In *EWOMP*, Rome, September 2002.

[3] D. an Mey, C. Terboven, and S. Sarholz. OpenMP on multicore architectures. In *IWOMP*, 2007.

[4] G. Bracha and W. Cook. Mixin-based inheritance. In *Proc. Object-Oriented Programming: Systems, Languages and Applications (OOPSLA)*. ACM Press, 1990.

[5] M. Bull. The status of OpenMP 3.0. In *SC06, OpenMP BoF*, November 2006.

[6] T. Deselaers, D. Keysers, and H. Ney. Features for image retrieval – a quantitative comparison. In *DAGM 2004, Pattern Recognition, 26th DAGM Symposium, Number 3175 in Lecture Notes in Computer Science*, pages 228–236, Tübingen, Germany, 2004.

[7] C. Freundl, T. Gradl, U. Rüde, and B. Bergen. Towards petascale multilevel finite element solvers. In D.A. Bader, editor, *Petascale Computing: Algorithms and Applications*. Chapman & Hall / CRC Press, 2007.

[8] G.R. Gao. Landing OpenMP on multi-core chips - challenges and opportunities. In *Keynote Presentation at IWOMP 2006*, Reims, France, 2006.

[9] A. Gerndt, S. Sarholz, M. Wolter, D. an Mey, T. Kuhlen, and C. Bischof. Nested OpenMP for efficient computation of 3D critical points in multiblock CFD datasets. In *SC*, Tampa, FL, November 2006.

[10] S. Gross, J. Peters, V. Reichelt, and A. Reusken. The DROPS package for numerical simulations of incompressible flows using parallel adaptive multigrid techniques. `ftp://ftp.igpm.rwth-aachen.de/pub/reports/pdf/IGPM211_N.pdf`.

[11] T.M. Haarmann and W.W. Koschel. Numerical simulation of heat loads in a cryogenic H_2/O_2 rocket combustion chamber. *ZAMM: J. Applied Mathematics and Mechanics*, 2(1):360–361, 2003.

[12] Heise News. Fall processor forum. `http://www.heise.de/newsticker/meldung/79279`, 11 October 2006.

[13] Heise News. Quad-Core-Xeon, mehr rechenkerne für server. `http://www.heise.de/newsticker/meldung/80976`, 11 November 2006.

[14] I. Hörschler, C. Brücker, W. Schröder, and M. Meinke. Investigation of the impact of the geometry on the nose flow. *European Journal of Mechanics B/Fluids*, 25(4):471–490, 2006.

[15] I. Hörschler, M. Meinke, and W. Schröder. Numerical simulation of the flow field in a model of the nasal cavity. *Computers and Fluids*, 32:3945, 2003.

[16] Z. Hu, J. del Cuvillo, W. Zhu, and G.R. Gao. IBM Cyclops-64: Challenges and experiences. In *Europar*, Dresden, Germany, 2006.

[17] PSP Inc. ParaWise automatic parallelization environment. `http://www.parallelsp.com`.

[18] L. Jerabkova, C. Terboven, S. Sarholz, T. Kuhlen, and C. Bischof. Exploiting multicore architectures for physically based simulation of deformable objects in virtual environments. In *4. Workshop "Virtuelle und Erweiterte Realität" der GI-Fachgruppe VR/AR*, Weimar, 2007.

[19] H. Jin, M. Frumkin, and J. Yan. Automatic generation of OpenMP directives and its application to computational fluid dynamics codes. In *International Symposium on High Performance Computing*, volume 440, Tokyo, 2000.

[20] S. Johnson and C. Ierotheou. Parallelization of the TFS multi-block code from RWTH Aachen using the ParaWise/CAPO tools. In *PSP Inc, TR-2005-09-02*, 2005.

[21] S. Johnson, C. Ierotheou, A. Spiegel, D. an Mey, and I. Hörschler. Nested parallelization of the flow solver TFS using the ParaWise parallelization environment. In *IWOMP*, Reims, France, 2006.

[22] S.P. Johnson, M. Cross, and M. Everett. Exploitation of symbolic information. Interprocedural dependence analysis. *Parallel Computing*, 22:197–226, 1996.

[23] Y. Lin, C. Terboven, D. an Mey, and N. Copty. Automatic scoping of variables in parallel regions of an OpenMP program. In *WOMPAT 04*, Houston, May 2004.

[24] J.D. McCalpin. STREAM: Sustainable Memory Bandwidth in High Performance Computers. `http://www.cs.virginia.edu/stream/`.

[25] OpenMP Architecture Review Board. OpenMP application program interface, v2.5. `http://www.openmp.org`, 2005.

[26] Sun Technical White Paper. Solaris memory placement optimization and Sun Fire servers. `http://www.sun.com/servers/wp/docs/mpo_v7_CUSTOMER.pdf`.

[27] R. Rabenseifner. Hybrid parallel programming on parallel platforms. In *EWOMP*, Aachen, Germany, 2003.

[28] Press release. DARPA selects IBM for supercomputing grand challenge. `http://www-03.ibm.com/press/us/en/pressrelease/20671.wss`.

[29] A. Reusken and V. Reichelt. Multigrid methods for the numerical simulation of reactive multiphase fluid flow models (DROPS). `http://www.sfb540.rwth-aachen.de/Projects/tpb4.php`.

[30] A. Spiegel and D. an Mey. Hybrid parallelization with dynamic thread balancing on a ccNUMA system. In *EWOMP*, Stockholm, Sweden, 2004.

[31] A. Spiegel, D. an Mey, and C. Bischof. Hybrid parallelization of CFD applications with dynamic thread balancing. In J. Dongarra, K. Madsen, and J. Wasniewski, editors, *Proc. PARA04 Workshop, Lyngby, Denmark, June 2004, LNCS 3732*, pages 433–441. Springer Verlag, 2006.

[32] C. Terboven. *Shared-Memory Parallelisierung von C++ Programmen*. Diploma thesis, RWTH Aachen University, Aachen, Germany, 2006.

[33] C. Terboven and D. an Mey. OpenMP and C++. In *IWOMP 2006*, Reims, France, June 2006.

[34] C. Terboven, T. Deselaers, C. Bischof, and H. Ney. Shared-memory parallelization for content-based image retrieval. In *Workshop on Computation Intensive Methods for Computer Vision*, 2006.

[35] C. Terboven, A. Spiegel, D. an Mey, S. Gross, and V. Reichelt. Parallelization of the C++ Navier-Stokes solver DROPS with OpenMP. In

ParCo, Vol 33 in the NIC book series, Malaga, Spain, September 2005. Research Centre Jülich, Germany.

[36] T. Volmar, B. Brouillet, H.E. Gallus, and H. Benetschik. Time accurate 3D Navier-Stokes analysis of a 1 1/2 stage axial flow turbine. Technical Report 1998-3247, AIAA, 1998.

Chapter 16

Performance and its Complexity on Petascale Systems

Erich Strohmaier

Lawrence Berkeley National Laboratory

16.1 Introduction

The unofficial race for the first petaflops computer is on. MDGRAPE-3, a system with hardware specialized for molecular dynamics simulations was announced in June 2006 as the first system with a theoretical peak performance at this level [12]. An informal survey at the 2006 Dagstuhl workshop on Petascale Algorithms and Application showed that the first petaflop Linpack performance is widely expected by 2008. At the same time the DARPA High Productivity Computing Systems (HPCS) program leads the charge to develop more productive petascale systems, which are easier to program and should achieve higher effective performance levels. In this chapter we are looking at the general question of how to evaluate performance and productivity of current and future systems.

For this we look look at current trends and extrapolate from them in Section 16.2 to predict some baseline parameters of potential petascale systems. We give a brief overview on the current state of high-performance computing (HPC) performance analysis in Section 16.3 and describe the synthetic performance probe APEX-Map as an example of a modern, parameterized performance probe in Section 16.4. In Section 16.5 we present a consistent methodology for calculating average performance values and associated performance complexity numbers, which capture how transparent achieved performance levels are.

16.2   Architectural Trends and Concurrency Levels for Petascale Systems

In the early 2000s it has become obvious that performance improvements due to Moore's Law cannot be sustained with previously used strategies of increasing the chip frequencies or the complexity of single processors [1]. To avoid the potential stall of processor speed, microprocessor companies have started to integrate multiple processing-cores on single chips. There are currently several different strategies for this integration: In multicore systems, a few cores of the previous processor generation are used and slightly adapted to operate in parallel (IBM Power, AMD Opteron, and Intel Woodcrest — currently up to two cores each). In many-core systems, a comparable large number of cores with new slimmed-down architectures designed to maximize performance and power benefits are placed on a single chip (Intel Polaris - 80 cores, Clearspeed CSX6400 - 96 cores, nVidia G80 - 128 cores, or CISCO Metro - 188 cores). This approach will lead to thousands of cores on single chips within a few years [1]. Concurrency levels will jump by an additional factor of about 50 if high-end HPC systems start utilizing such many-core chips. In hybrid architectures, a conventional complex core is combined with an array of smaller, more efficient cores for parallel operations (IBM Cell or the planned AMD Opteron with integrated ATI graphics).

The HPC community itself is currently using systems with 10^4–10^5 chip-sockets. The data of the TOP500 project shown in Figure 16.1 clearly reflect the recent stall in chip frequencies and the related accelerated growth rates of concurrency levels after 2004 [13]. If we extrapolate these recent trends we arrive at concurrency levels of 10^6–10^7 cores for the largest systems in the 2010–2015 time frame. These extrapolations are based on current multicore technologies. There is a good chance that we will face a switch from multi- many-core chip technologies, as many-core chips promise higher peak-performance and better power and space efficiencies. This could easily lead to an additional increase in concurrency levels of the order of 10–100.

In the first half of the next decade, we can therefore expect petascale systems to have on the order of 10^7–10^8 cores or even more! These systems will have strongly hierarchical architectures with large differences in communication properties between cores close to each other on the same chip and cores on sockets far from each other in the system. Past attempts to use hierarchical parallel-programming models have failed miserably. As a result, this leaves the HPC community unprepared for the programming challenges it will face in only a few years. To prepare for the challenges of programming such a type of system key questions such as the following need to be addressed: Which programming paradigms allow efficient coding for different algorithms for many-core chips? How should HPC systems with strong hierarchical architectures be programmed? How can feedback to chip and systems developers

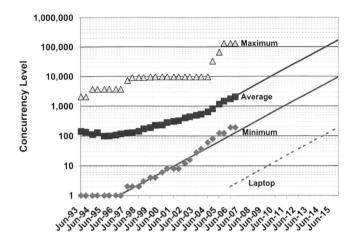

FIGURE 16.1: Development and projection of the concurrency levels as tracked by the TOP500 project [13]. Shown are the maximum, average, and minimum concurrency levels for each list together with a projection for multicore-based PCs and laptops.

about the utility of design alternatives for algorithms be provided? How can different classes of applications be mapped for efficient execution to various system architectures?

To answer such questions, a set of relevant applications should be used to analyze the interaction of hardware features, programming paradigms, and algorithms. However, currently there is no such set of application benchmarks available, which would be widely accepted as common application references by the HPC community.

16.3 Current Situation in Performance Characterization and Benchmarking

During the last few decades the variety and complexity of architectures used in HPC computing have increased steadily. At the same time we have seen an increasing user community with new and constantly changing applications. The space of performance requirements of HPC applications is growing more varied and complex. A single architecture can no longer satisfy the need for cost-effective high-performance execution of applications for the whole scientific HPC community. Despite this variety of system architectures and application requirements, benchmarking and performance evaluation are still dominated by spot measurements using isolated benchmarks based on specific scientific kernels or applications. Several benchmarking initiatives have attempted to remedy this situation using the traditional approach of constructing benchmark suites. These suites contain a variety of kernels or applications, which ideally represent a broad variety of requirements. Relating these multiple spot measurements to other applications is however still difficult, since no general application performance characterization methodology exists. Using real applications as benchmarks is also very time consuming, and collections of results are limited and hard to maintain. Due to the difficulty of defining performance characteristics of applications, many synthetic benchmarks are instead designed to measure specific hardware features. While this allows understanding of hardware features in great detail, it does not help to understand overall application performance.

Due to renewed interest in new architectures and new programming paradigms, there is also a growing need for flexible and hardware-independent approaches to benchmarking and performance evaluation across an increasing space of architectures. These benchmarks also need to be general enough so they can be adapted to many different parallel-programming paradigms.

16.3.1 Benchmarking initiatives

The memory wall between the peak performance of microprocessors and their memory performance has become the prominent performance bottleneck for many scientific application codes. Despite this development, many benchmarking efforts in scientific computing have in the past focused on measuring the floating-point computing capabilities of a system and have often ignored or downplayed the memory subsystem and processor interconnect. One prominent example is the Linpack benchmark, which is used to rank systems in the TOP500 project [13]. This type of benchmark can provide guidance for the performance of some compute-intensive applications, but fails to provide reasonable guidance for the performance of any memory bound real applications. On most platforms, Linpack can achieve over 70% of peak performance, while on the same systems real applications typically achieve substantially lower performance rates. Nevertheless, Linpack is still the most widely used and cited benchmark for HPC systems and its use is a first test for the stability and accuracy of any new system.

Despite this situation, there is still no standard or widely accepted way to measure progress in our ability to access globally distributed data. STREAM [11] is often used to measure memory bandwidth but its use is limited to at the most a single-shared memory node. In addition, it emphasizes exclusively regular stride one access to main memory.

During recent years several new approaches for benchmarking HPC systems have been explored. The HPC Challenge benchmark [7] is a major community-driven benchmarking effort backed by the DARPA HPCS program. It pursues the goal to help define the performance boundaries of future petascale computing systems, and is built on the traditional idea of spanning the range of possible application performances with a set of different benchmarks. HPC Challenge is a suite of kernels with memory-access patterns more challenging than those of the High Performance Linpack (HPL) benchmark used in the TOP500 list. Thus, the suite is designed to provide benchmarks that bound the performance of many real applications as a function of memory-access characteristics, e.g., spatial and temporal locality, and providing a framework for including additional tests. In particular, the suite is composed of several well-known computational kernels (STREAM, HPL, matrix multiply - DGEMM, parallel-matrix transpose - PTRANS, FFT, RandomAccess, and bandwidth/latency tests - b_{eff}) that attempt to span high and low spatial and temporal locality space. Other than STREAM, the RandomAccess benchmark comes closest to being a data-access benchmark by measuring the rate of integer random updates possible in global memory. Unfortunately, the structure of the RandomAccess benchmark cannot easily be related to scientific applications and thus does not help much for applications performance prediction. The major problem with the HPC Challenge benchmark faces is how to become widely accepted and used as a reference benchmark.

The DARPA HPCS program itself pursues an extensive traditional layered approach to benchmarking by developing benchmarks on all levels of complexity from simple kernels to full applications. Due to the goal of the HPCS program of developing a new highly productive HPC system in the petaflops performance range by the end of the decade, it will also need methodologies for modeling performance of nonexisting workloads on nonexisting machines.

At the Berkeley Institute for Performance Studies (BIPS) [2], several small synthetic tunable benchmark probes have been developed. These probes focus on exploring the efficiencies of specific architectural features and are designed with a specific limited scope in mind. Sqmat [5] is a sequential probe to explore the influence of spatial locality and of different aspects of computational kernels on performance. It is limited to sequential execution only and has no concept for temporal locality but has tackled the problem on how to characterize the detail of computation for performance characterization purposes.

In many situations application performance is the best measure of performance, however using full applications for benchmarking is very time consuming and large collections of comparable results are also very hard to get and to maintain [14, 15]. Applications benchmarks are also not suitable for simulators and thus hard to use for the evaluation of future systems.

16.3.2   Application performance characterization

Different characterizations of applications with concepts such as spatial and temporal data access locality have recently been proposed. APEX-Map [16, 17, 19, 21, 20] is a parameterized, synthetic benchmark which measures global data-access performance. It is designed based on parameterized concepts for temporal and spatial locality and generates a global data-access stream according to specified levels of these measures of locality. This allows exploring the whole range of performance for all levels of spatial and temporal locality. Parameterized benchmarks like this have the advantage of being able to perform parameter sweeps and to generate complete performance surfaces. APEX-Map stresses a machine's memory subsystem and processor interconnect according to the parameterized degrees of spatial and temporal locality. By selecting specific values for these measures for temporal and spatial locality, it can serve as a performance proxy for a specific scientific kernel [19]. APEX-Map will be described in detail in the next section of this chapter.

16.3.3   Complexity and productivity measures of performance

There is a rich collection of research available on the subject of software complexity. Complexity measures discussed range from code size expressed in lines of code (LOC), which is used in various places such as the DARPA HPCS program [4]; Halstead Software Science metrics [6] based on counts of operators and operands used; McCabe cyclomatic complexity measure [9] based on the number of linearly independent paths through a program module, and variants thereof such as design complexity [10]. These software complexity metrics have also been applied and extended to the context of parallel computing [22].

An active area of research within the DARPA HPCS program is productivity metrics, which focus on capturing the complexity of the task of coding itself [4, 8, 3]. The approach presented in Section 16.5 complements research in software complexity and productivity by considering performance complexity (PC), which represents a measure quite different from the former two as it characterizes code execution behavior in a second dimension orthogonal to performance itself. PC is based on performance model accuracy, which has the advantage of depending only on performance measurements and is not based on and does not require code inspection or supervision of coding itself.

16.4   APEX-Map

Within the Application Performance Characterization (APEX) project we developed a characterization for global data-access streams together with a

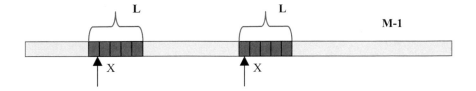

FIGURE 16.2: The data access model of APEX-Map.

related synthetic benchmark called memory-access probe (APEX-Map). This characterization and APEX-Map allowed us to capture the performance behavior of several scientific kernels across a wide range of problem sizes and computational platforms [19]. We further extended these performance characterization and benchmarking concepts for parallel execution [21, 20]. In this section we describe briefly the principles behind APEX-Map, our experience implementing it, and the results we obtained using it. We will also give a brief overview of ongoing work and possible future extensions.

16.4.1 Design principles of APEX-Map

The synthetic memory access probe APEX-Map is designed based on parameterized concepts for temporal and spatial locality. It uses a blocked data access to a global array of size M to simulate the effects of spatial locality. The block length L is used as a measure for spatial locality and L can take any value between 1 (single-word access) and M. A nonuniform random selection of starting addresses for these blocks is used to simulate the effects of temporal locality (Figure 16.2). A power function distribution is selected as nonuniform random distribution and nonuniform random numbers X are generated based on uniform random numbers r with the generating function $X = r^{(1/\alpha)}$. The characteristic parameter α of the generating function is used as a measure for temporal locality and can take values between 0 and 1. A value of $\alpha = 1$ generates uniform random numbers while small values of α close to zero generate random numbers centered towards the starting address of the global-data array. The effect of the temporal locality parameter α on cache hit and miss rates is illustrated in Figure 16.3 for a ratio of cache sizes to utilized memory of 1:256.

APEX-Map uses the same three main parameters for both the sequential version and parallel version. These are the global memory size M, the measure of temporal locality α, and of spatial locality L. These three parameters are related to our methodology to characterize applications. A detailed discussion of the locality concepts used in APEX-Map can be found in [21, 20].

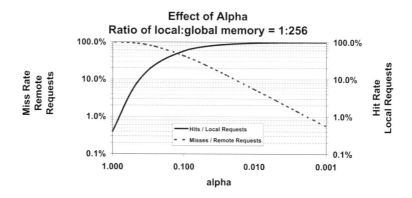

FIGURE 16.3: Effect of alpha on hit/miss rates and on the ratio of local to remote data requests.

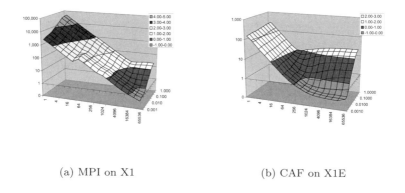

(a) MPI on X1   (b) CAF on X1E

FIGURE 16.4: APEX-Map performance in cycles/data-access on 256 processors. The first graph shows MPI on the X1 system, the second graph shows CAF on the X1E system. Note the difference in scale.

16.4.2 Comparison of parallel-programming paradigms with APEX-MAP

The parallel version of APEX-MAP was initially developed in MPI and we have since ported it to SHMEM, UPC, CAF. We have measured performances of all these versions on a variety of systems, some of which were reported earlier [21, 20]. Figure 16.4 shows as an example the performance surfaces for execution in MPI and CAF on the Cray X1 system on 256 processors. Please notice the differences in the vertical scale.

We have also investigated a series of simple performance models for APEX-Map [18]. A simple model with a two-level system hierarchy and a separate linear latency and bandwidth-access-time model for each level fits experimental results best on most parallel systems and for most programming languages. Residual errors of this model allow for investigating performance peculiarities and anomalies in detail.

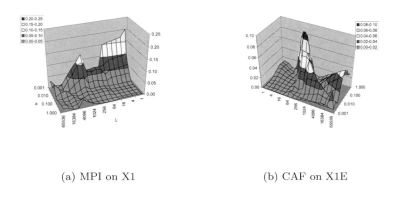

(a) MPI on X1 (b) CAF on X1E

FIGURE 16.5: Residual errors after model subtraction for parallel execution on 256 processors. The first graph shows MPI on the X1 system, the second graph shows CAF on the X1E system.

Figure 16.5 shows the residual errors after model subtraction for MPI and CAF on the Cray X1 for 256 processors. On the X1 system with MPI the dominant performance anomaly occurs for high temporal localities (small α) as the system behavior for local memory access is quite different from message exchange and not resolved in a model with a two-level hierarchy for this system. For CAF on the X1E, the dominant anomaly occurs for access length exceeding the vector register length of the system of 64 doubles. The dominant performance effects are thus very different for CAF and MPI.

16.5   How to Characterize Performance Complexity

Performance evaluation of code execution focuses on determining performance and efficiency levels for specific application scenarios. However, there is no measure characterizing how complex it is to achieve performance and how transparent performance results are. In this section we present an execution time metric called performance complexity (PC) which we developed to capture these important aspects [18]. PC is based on performance results from a set of benchmark experiments and related performance models reflecting the behavior of a programmer. Residual modeling errors are used to derive PC as a measure for how transparent a program performance is and how complex the performance appears to the programmer. PC is independent from performance (P) itself, which allows for plotting system behavior in a performance-complexity map (P-C map). We present a detailed description for calculating compatible P and PC values and use results from a parametric benchmark to illustrate the utility of PC for analyzing systems and programming paradigms.

Performance of a system can only be measured relative to a workload description. This is usually achieved by selecting a set of benchmarks representative for a particular workload of interest. Beyond this, there cannot be a meaningful, absolute performance measure valid for all possible workload. In contrast to performance itself, there are no methodologies for the characterization of performance transparency or the complexity of programming for performance. The complexity of the task of programming codes that perform well on a given system also depends on the set of codes in question. Therefore, a concept for performance complexity (PC) can also be defined only relative to a workload and not in absolute terms!

However, the situation is even more complex when we consider different programmers coding the same algorithm in different languages and with different programming styles. They might easily generate codes, whose performances are not identical. Even worse, they might experience quite different levels of control and understanding of the achieved performance levels.

Programming for performance is easy if we understand the influence of system and code features on performance behavior and if any unexplained performance artifacts and variations are relatively small. The level of performance transparency, however, does not indicate any level of performance itself. It only specifies that we understand performance. This understanding of performance behavior of a code on a system implies that we can develop an accurate performance model for it. Therefore our approach is to use the accuracy of one or a suitable set of performance models to quantify the transparency of performance on a system and of performance complexity (PC) in general.

In our approach the transparency of performance is expressed in the generated performance models and, as a consequence, derived PC values will depend on the selected performance model. This dependency also reflects the fact that a programmer willing to consider more restrictions on a larger variety of code features, hopefully, can understand and control performance behavior better than a programmer who codes without considerations for system architectures.

In the ideal case the selected performance model should incorporate the effects of all code features easily controllable in a given programming language and not include any architectural features inaccessible to the programmer. In essence, the performance models for calculating a PC value should reflect the (possible) behavior of a programmer, in which case PC reflect the performance transparency and complexity of performance control this programmer will experience.

This concept for a PC measure is orthogonal to performance itself — high-/low performance does not imply high/low complexity. Thus, PC is an ideal complement to performance itself and we can use these two values to plot performance and complexity of single benchmarks, benchmarks sets, and systems in a two-dimensional P-C map.

16.5.1 Definition of performance complexity

Goodness of fit measures for modeling measurements are based on the sum of squared errors (SSE). For a system, S_i, the performance, P_{ij}, of a set of n codes, C_j, is measured. Different measurements, j, might also be obtained with the same code executed with different problem parameters such as problem sizes and resource parameters such as concurrency levels. A performance model, M_{kl}, is used to predict performance, M_{ikl}, for the same set of experiments. The coefficient of determination, R^2, is one widely used measure for quality of fit between a model and a set of observations.

R^2 is a dimensionless and scale-free number between 0 and 1 and a model with perfect agreement to measurements would achieve a score of 1. By eliminating the scale of the original performance variation in such a way, we would lose the basis to compare different codes. Programs with the larger R^2 value might have an absolutely larger performance variation. Using R^2 for our purposes would be like using speedup for comparing different computer systems. Such comparisons have their validity in certain situations, but cannot be used for general situations such as ranking of systems.

To arrive at a PC measure with the same metric as performance, we have to use standard deviation (SD), which is the square root of average SSE and cannot use the more widely used total sum. The average SSE_{kl} for each system i and model k across all codes j is then given by $\overline{SSE_{ik}} = \frac{1}{n} \sum_j (P_{ij} - M_{ijk})^2$. While these SSE values are still absolute numbers (they carry dimension), they are easily transformed into relative numbers by dividing them by the similarly defined sum of squares (SS) of the measured data

P: $\overline{SS} = \sum_k (P_k - \bar{P})^2$ and $R^2 = \overline{SSE}/\overline{SS}$.

These SSE values are based on absolute errors and, to be meaningful, the original range of values of P and M should be as small as possible. Otherwise the difference between absolute and relative errors would become significant. This can be achieved by a careful selection of the original metric for P and M. We consider an absolute flat system for which all operations take the same time as least complex to program. The appropriate metric for such a system would therefore be of the dimension of "operations/cycle" (or its inverse).

TABLE 16.1:   Calculation steps for the values of average performance P and performance complexity PC (absolute and relative) along with a dimensional analysis. The index i for different systems is suppressed.

Operations	Dimension[*]
Initial data: P_j, M_j	[ops/sec]
Transform P_j, M_j to ideal flat metric	[ops/cycle]
Log-transformation: $P'_j = \log(P_i)$, $M'_j = \log(M_i)$	[log(ops/cycle)]
Basic Calculations	
$\bar{P} = \frac{1}{n}\sum_j P'_j$	[log(ops/cycle)]
$\overline{SS'} = \frac{1}{n}\sum_j (P'_j - \bar{P}')^2$	[(log(ops/cycle))^2]
$\overline{SSE'} = \frac{1}{n}\sum_j (P'_j - M'_j)^2$	[(log(ops/cycle))^2]
$1 - R^2 = \overline{SSE'}/\overline{SS'}$	[]
Back Transformations	
$\bar{P} = exp(\bar{P}') = \sqrt[n]{\prod_j P_j}$	[ops/cycle]
Absolute $PC_a = exp(\sqrt{\overline{SSE'}}) - 1$	[ops/cycle]
Relative $PC_r = exp(\sqrt{\overline{SSE'}/\overline{SS'}} - 1)$	[]
Back-transformation of $\bar{P}$ and PC_a to original scale and metrics	[ops/sec]

[*]Inverse dimensions could be chosen as well.

To further reduce the range of performance values, a log-transformation should be applied to the original data. This effectively bases the calculated SSE values on relative errors instead of absolute errors. It also has the added benefit, that the initial choice between a time metric [time/ops] or a performance metric [ops/time] is irrelevant as they log-transform into the same absolute values. To obtain numbers in regular dimensions, we transform the final values of SSE back with an exponential function. The resulting numbers turn out to be known as geometric standard deviation, representing multiplicative instead of additive, relative values, which are larger or equal to one. For convenience we transform these values into the usual range of larger than zero by subtracting one. For calculating average performance values on a comparable scale, a similar sequence of log, mean, and exponential transformations has to be followed. In this case, the resulting performance value turns out to be the geometric mean of the individual measurements. The full sequence of the calculation step is summarized in Table 16.1.

The outlined methodology can be applied equally to absolute performance metrics or to relative efficiency metrics. PC_a represents an absolute metric while PC_r would be the equivalent, relative efficiency metric. PC_r is in the range [0,1] and reflects the percentage of the original variation not resolved in the performance model. Hence, while using PC_r, care has to be taken when comparing different systems, as a system with larger relative complexity PC_r might have lower absolute complexity PC_a. This is the same problem as comparing the performance of different systems using efficiency metrics such as ops/cycle or speedup.

16.5.2   Performance model selection

We now illustrate the outlined calculations using APEX-Map. Executions for different parameters are used as different performance experiments and replace for this study the experiments with different codes. The derived performance and complexity numbers are therefore based on only a single restricted benchmark, which limits their expressiveness, and their absolute values should be considered carefully.

In our methodology the selection of the constituent features for performance models is as important as the selection of individual benchmarks for performance measurement. It is widely accepted that there cannot be a single measure for performance, which does not relate to specific codes. For analogous reasons, we argue that there cannot be a measure of performance transparency and performance complexity independent without relation to specific codes. For embarrassing parallel codes, most systems will exhibit only small performance complexity, while this is obviously not the case for tightly coupled, irregular scientific problems. In addition, the performance complexities programmers experiences on a system also depends on the programming languages and coding styles they use. These differences can be reflected in different performance models and it is therefore only natural — and cannot be avoided — that performance complexity values depend on the chosen performance models.

Ideally, the features of the performance models should reflect characteristics of our programming languages, which the user can easily control and use to influence performance behavior. An example would be vector length as it is easily expressed in most languages as a loop count or array dimension and thus is user controllable. Unfortunately, many programming languages do not complement system architectures well, as they do not have appropriate means of controlling hardware features. This situation is exacerbated as many hardware features are actually designed not to be user controllable, which makes performance optimization often a painful exercise in trial and error until a desired behavior can be achieved. Cache usage would be a typical example here as programming languages have little means to control directly which data should reside in the cache or not. However, in developing performance models we often have to revert to using such noncontrollable hardware features to achieve satisfactory accuracy.

For interpreting complexity values derived with our methodology, we have to carefully consider what the components of the used performance model are, as they greatly influence any interpretation.

APEX-Map is designed to measure the dependency of global-data access performance on spatial and temporal locality. This is achieved by a sequence of blocked data accesses of unit stride with a pseudo-random-starting address. From this we can expect that any good performance model for it should contain the features of access length, and for access probabilities to different levels of memory hierarchy. The former is a loop length and easily controlled in programs, the latter depends on cache hit rates, which are usually not directly controllable by the programmer. The metric for APEX-Map performance is [data-access/second] or any derivative thereof.

In a previous study we analyzed the influence of using different performance models on the calculated complexity values by using a sequence of four models [18]. In this section we present results for the most complex model and parallel execution only. This performance model assumes a linear timing model for each level in a memory hierarchy with two levels:

$$T = P\left(\frac{c}{M}\right) * \left(l_1 + g_1 * (L-1)\right)/L + \left(1 - P\left(\frac{c}{M}\right)\right) * \left(l_2 + g_2 * (L-1)\right)/L$$

with the probability to find data in the closer memory level c as:

$$P\left(\frac{c}{M}\right) = \left(\frac{c}{M}\right)^{\alpha}$$

M, L and α are APEX-Map parameters and c is a system parameter reflecting the most important memory or system hierarchy level, which has to be set to different appropriate levels for different systems and execution conditions. The probability to find data in c is given by the temporal locality parameter α as $P(c/M) = (c/M)^a$. l_1, g_1, l_2, g_2 are latency and gap parameters for the two levels of memory hierarchy modeled. These parameters are back-fitted by minimizing SSE values and thus the prediction error of the model. Backfitting these effective parameter values, also allows us to compare systems and programming paradigms in different ways. To check the validity of the generated model and the overall approach, we inspect residual error plots and fitted latency and gap values to rule out any fictitious models or parameter values.

For complex parallel system hierarchies, it is also not always clear what the second most important hierarchy level is and what the value of c should be. It is therefore advisable to at least use a backfitting approach to confirm an initial choice, or to probe the performance signature of a system, to determine which level is most influential on performance.

16.5.3 PC analysis of some parallel systems

Analyzing parallel APEX-Map results for 256 processors and a memory consumption of 512MB/process, we face the problem of a large range of raw performance values, which span 5 orders of magnitude. This would not be feasible without a sound statistical procedure, such as avoiding the log and exponential transformations. This analysis is of special interest as we had obtained performance results for implementations with different parallel-programming paradigms such as MPI, SHMEM, and the two PGAS languages UPC, and CoArray Fortran (CAF) on the Cray X1 and X1E systems [20]. In UPC, two different implementations are compared; one for accessing a global-shared array element by element and one for a block-transfer access to remote data.

It represents a programmer, who optimizes for long loops, large messages, and high-data locality. The relative lowest complexities PC are now calculated for blocked access in UPC and SHMEM on the X1 followed by the SP Power4, SX6, SP Power3, and Power5, Itanium-Quadrics cluster, CAF on the X1E, Opteron Infiniband cluster, UPC on the X1 and X1E, and, with somewhat higher PC values, the X1 with MPI, and the Blue Gene/L system.

The most important level in the memory hierarchy of a system is expressed by the local memory size of c, which for most systems needs to be set to the symmetric multiprocessor (SMP) memory rather than process memory to achieve the best model accuracy. While this is not surprising, there are a fair number of systems for which best model fit is achieved for even larger values of c. This is an indication that network contention might be an important factor for the randomized, non-deterministic communication pattern of APEX-Map, which is not included at all in any of our models. We tested several more elaborate models with a third level of hierarchy and with constant overhead terms for parallel code overhead, but none of these models produced significant improvements and back-fitted parameter values often did not make sense.

Figure 16.6 shows back-fitted values for the latency l_1, l_2, and gap parameters g_1, and g_2 of the performance model. Latencies for PGAS languages are noticeable lower than for other languages even on the same architecture. In contrast to this, remote gap values seem to be mostly determined by architecture. The P-C maps for all four models are shown in Figure 16.7. Due to the rather large performance difference between PGAS- and MPI-based systems, we use a logarithmic scale for the horizontal efficiency axis P. We

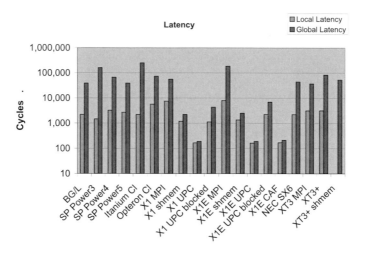

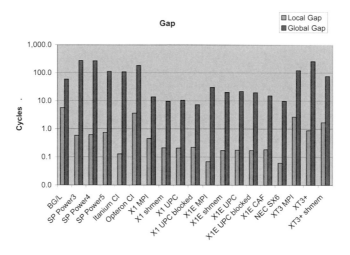

FIGURE 16.6: Back-fitted latency l_1, l_2, and gap parameters g_1, and g_2 for both memory hierarchies for parallel executions.

P-C Map for a Two Level Memory, Linear Timing Model

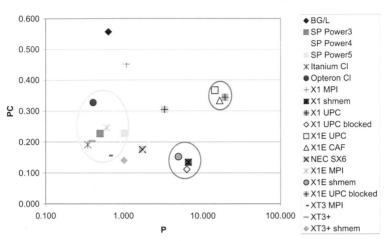

FIGURE 16.7: (See color insert following page 18.) P-C map for parallel systems based on efficiencies [accesses/cycle]. Horizontal axis is performance P in [accesses/100 cycles] and the vertical axis is complexity PC in [accesses/cycles]. Systems fall with few exceptions into 4 categories: PGAS languages (UPC, CAF), one-sided block-access (SHMEM, UPC block mode), MPI-vector (X1-MPI and SX6), and superscalar-based MPI systems.

notice grouping of systems in three major clusters across all models:

PGAS languages: UPC and CAF on the X1 show higher architectural efficiencies (horizontal separation). They end up with higher PC values than one-sided block-access paradigms, as they expose users to more idiosyncrasies of the architecture in the effort to achieve higher performance.

One-sided block-access paradigms: SHMEM- and UPC-blocked-access mode on the X1 also show horizontal separation, but most all end up with the lowest complexity values PC. The exceptions here are UPC on the X1E showing higher complexity and UPC on the T3E+ showing lower performance.

MPI systems-vector-based: NEC SX6 and the Cray X1 show different changes in complexity between models than superscalar-based systems.

MPI systems-superscalar-based: They show the lowest performance and medium complexity, with the exception of Blue Gene/L for model 3, which was discussed in [18].

This matches the results of a software complexity comparison [22], which found programs written in message-passing paradigms (MPI, PVM) to have consistently higher software complexity than SHMEM, High Performance Fortran (HPF), or shared-memory programs. Outliers from these clusters typically show some performance anomalies, which can be revealed and explained by inspections of the residual errors [18].

Figure 16.7 also shows very clearly that performance and performance complexity of a specific algorithm on a single system can depend heavily on the programming languages used and that P and PC values are indeed independent. For the Cray X1 system family, lowest PC values are achieved with one-sided message-passing languages, which show medium performance P values. The highest performing PGAS languages exhibit medium to larger performance complexity.

16.6   Conclusions

Petascale systems in the middle of the next decade will have on the order of 10^7–10^8 cores. These systems will have strongly hierarchical architectures with quite different communication capabilities between cores close to each other on the same chip and between different cores on different chip sockets far from each other in the system. Parallel programming in general and performance optimization in particular will present new challenges to the HPC community. Unfortunately, there is still no set of scalable, flexible, but simple benchmarks available, which could test such systems realistically.

In this chapter we presented a synthetic parameterized benchmark probe, called APEX-Map. Such parameterized probes are easy to execute or simulate and allow us to map the performance response of a system for a variety of execution conditions through simple parameter sweeps. Analyzing system behavior with performance models proves very beneficial by amplifying any unusual system responses.

We also described a concept for a quantitative measure of performance complexity (PC). The transparency of performance behavior is linked to the complexity of programming and optimizing for performance and can be characterized by a measure for the accuracy of appropriately chosen performance models. We presented a definition and detailed description on how to calculate average performance (P) and complexity numbers based on performance numbers from a set of benchmark and accompanying performance predictions from a performance model.

PC is a measure characterizing code execution behavior on a system. In the first order, it is independent from performance P itself and serves as a second dimension to evaluate systems and programming paradigms. Having P and PC as two orthogonal code execution metrics, allows for plotting different systems in a performance-complexity map. This P-C map permits a high level analysis of performance and complexity determining commonalities between systems and programming languages used in the experiments.

This combination of advanced parameterized performance probes, performance models, and new high level performance metrices promises to be very helpful in making future petascale systems succesful tools for scientific computing.

References

[1] K. Asanovic, R. Bodik, B.C. Catanzaro, J.J. Gebis, P. Husbands, K. Keutzer, D.A. Patterson, W.L. Plishker, J. Shalf, S.W. Williams, and K.A. Yelick. The landscape of parallel computing research: a view from Berkeley. Technical Report UCB/EECS-2006-183, EECS Department, University of California, Berkeley, 2006. http://www.eecs.berkeley.edu/Pubs/TechRpts/2006/EECS-2006-183.html.

[2] The Berkeley Institute for Performance Studies. http://crd.lbl.gov/html/bips.html.

[3] J. Carver, S. Asgari, V. Basili, L. Hochstein, J. K. Hollingsworth, F. Shull, and M. Zelkowitz. Studying code development for high performance computing: The hpcs program. In *Proceedings of the First International Workshop on Software Engineering for High Performance Computing System Applications at ICSE*, pages 32–36, Edinburgh, Scotland, May 2004.

[4] A. Funk, V. Basili, L. Hochstein, and J. Kepner. Application of a development time productivity metric to parallel software development. In *SE-HPCS '05: Proceedings of the Second International Workshop on Software Engineering for High Performance Computing System Applications*, pages 8–12, New York, 2005. ACM Press.

[5] G. Griem, L. Oliker, J. Shalf, and K. Yelick. Identifying performance bottlenecks on modern microarchitectures using an adaptable probe. In *Proc. 3rd International Workshop on Performance Modeling, Evaluation, and Optimization of Parallel and Distributed Systems (PMEO-PDS)*, Santa Fe, NM, April 26-30, 2004.

[6] M. H. Halstead. *Elements of Software Science*. Elsevier North-Holland, New York, 1977.

[7] HPC Challenge Benchmark. `http://icl.cs.utk.edu/hpcc/index.html`.

[8] J. Kepner. High performance computing productivity model synthesis. *International Journal of High Performance Computing Applications: Special Issue on HPC Productivity*, 18(4):505–516, November 2004.

[9] T. J. McCabe. A complexity measure. *IEEE Transactions on Software Engineering*, 2(4):308–320, December 1976.

[10] T. J. McCabe and C. W. Butler. Design complexity measurement and testing. *Communications of the ACM*, 32(12):1415–1425, December 1989.

[11] J.D. McCalpin. STREAM: Sustainable Memory Bandwidth in High Performance Computers. `http://www.cs.virginia.edu/stream/`.

[12] MDGRAPE-3 Peta Computing Institute. `http://www.peta.co.jp/index-en.html`, April 2007.

[13] H.W. Meuer, E. Strohmaier, J.J. Dongarra, and H.D. Simon. TOP500 Supercomputer Sites. `http://www.top500.org`.

[14] Modern vector architecture. `http://ftg.lbl.gov/twiki/bin/view/FTG/ModernVectorarch`.

[15] L. Oliker, J. Carter, M. Wehner, A. Canning, S. Ethier, A. Mirin, D. Parks, P. Worley, S. Kitawaki, and Y. Tsuda. Leading computational methods on scalar and vector HEC platforms. In *SC '05: Proceedings of the 2005 ACM/IEEE Conference on Supercomputing*, page 62, Washington, DC, 2005. IEEE Computer Society.

[16] H. Shan and E. Strohmaier. Performance characteristics of the Cray X1 and their implications for application performance tuning. In *ICS '04: Proceedings of the 18th Annual International Conference on Supercomputing*, pages 175–183, New York, NY, USA, 2004. ACM Press.

[17] H. Shan and E. Strohmaier. MPI, SHMEM, and UPC performance on the Cray X1: A case study using APEX-Map. In *Proc. of Cray Users Group Meeting (CUG 2005)*, May 2005.

[18] E. Strohmaier. Performance complexity: An execution time metric to characterize the transparency and complexity of performance. *Cyberinfrastructure Technology Watch Quarterly*, 2(4B), November 2006.

[19] E. Strohmaier and H. Shan. Architecture independent performance characterization and benchmarking for scientific applications. In *International Symposium on Modeling, Analysis, and Simulation of Computer and Telecommunication Systems*, Volendam, The Netherlands, October 2004.

[20] E. Strohmaier and H. Shan. APEX-Map: A global data access benchmark to analyze HPC systems and parallel programming paradigms. In *SC '05: Proceedings of the 2005 ACM/IEEE conference on Supercomputing*, page 49, Washington, DC, 2005. IEEE Computer Society.

[21] E. Strohmaier and H. Shan. APEX-Map: A synthetic scalable benchmark probe to explore data access performance on highly parallel systems. In *EuroPar2005*, Lisbon, Portugal, August 2005.

[22] S. VanderWiel, D. Nathanson, and D. Lilja. Performance and program complexity in contemporary network-based parallel computing systems. Technical Report HPPC-96-02, University of Minnesota, 1996.

Chapter 17

Highly Scalable Performance Analysis Tools

Michael Gerndt

Technische Universität München

Faculty of Informatics, Boltzmannstr. 3, 85748 Garching, Germany

gerndt@in.tum.de

Karl Fürlinger

University of Tennessee

Innovative Computing Laboratory, 1122 Volunteer Blvd, Knoxville, Tennessee
37996-3450

karl@cs.utk.edu

17.1 Introduction

Performance analysis of applications is and certainly will be of even higher importance on future petascale systems. The reason is that these architectures will exploit multiple levels of parallelism and deep memory hierarchies to reach several petaflops (PFLOPS) of peak performance without having automatic programming tools available.

Current designs for petascale systems employ specialized coprocessors, such as the Roadrunner system and the system to be built in Japan's Next Generation program. Roadrunner, with a peak performance of 1.6 PFLOPS, will be a hybrid system of 16.000 AMD Opteron cores and 16.000 IBM Cell Broadband Engine processors. It will be installed at Los Alamos National Laboratory. Japan's Next Generation supercomputer targets 10 PFLOPS in 2012 and will also exploit a combination of several types of processors ranging from vector to special-purpose coprocessors.

Cray will install at Oak Ridge National Laboratory a Cray XT4 which will deliver over 1 PFLOPS of peak performance at the end of 2008. Compared to the previous systems, it will be based on multicore Opteron processors without coprocessors.

Designing parallel programs exploiting all levels of parallelism efficiently, taking into account memory access and communication will be extremely difficult. Therefore program tuning guided by the observed performance behavior of test runs will be very important. In contrast to program debugging, performance tuning has to be based on experiments on the full number of processors. Measuring and analyzing the performance of runs on multiple thousands of processors imposes strong requirements on the scalability of performance analysis tools.

This chapter first revisits the general techniques used in performance analysis tools in Section 17.2. It then presents four tools with different approaches to tool scalability. These tools are Paradyn 17.3, SCALASCA 17.4, Vampir NG 17.5, and Periscope 17.6. The techniques used in these tools are summarized in the concluding section.

17.2   Performance Analysis Concepts Revisited

Performance analysis tools are based on an abstraction of the execution, an *event model*. Events happen at a specific point in time in a process or thread. Events belong to event classes, such as enter and exit events of a user-level function, start and stop events of a send operation in MPI programs, iteration assignment in work-sharing constructs in OpenMP, and cache misses in sequential execution.

In *profiling tools*, information on events of the same class is aggregated during runtime. For example, cache misses are counted, the time spent between start and exit of a send operation is accumulated as communication time of the call site, and the time for taking a scheduling decision in work-sharing loops is accumulated as parallelization overhead. Representatives of profiling tools are the Unix tools prof/gprof and system-specific profilers, such as Xprof on IBM systems and Vtune on Intel platforms. Profiling tools are also available for MPI (mpiP [22]) and OpenMP (ompP [6]).

In *tracing tools*, for each event specific information is recorded in a trace file. The event record written to the file at least contains a timestamp and an identification of the executing process or thread. Frequently, additional information is recorded, such as the message receiver and the message length or the scheduling time for an iteration assignment. Tracing tools have been developed in the context of parallel programming where the dynamic behavior of multiple processes and threads is important. This class includes Tau [21],

KOJAK [23], as well as Vampir [2]. Tracing tools typically employ elaborate visualization techniques to illustrate the application's dynamic behavior.

Especially for tracing tools, but also for profiling tools, scalability is a very important issue on large systems. The following sections introduce MRNet and the Distributed Performance Consultant of Paradyn, the parallel analysis of traces in SCALASCA and Vampir Next Generation, as well as the distributed performance bottleneck search of Periscope.

17.3 Paradyn

Paradyn [15] is a tool for automated performance analysis developed at the Universities of Wisconsin and Maryland, with a history dating back to the early 1990s. A characteristic feature of Paradyn is the usage of dynamic instrumentation to add, customize, and remove instrumentation code patches in applications while they are running. To perform the runtime code patching, Paradyn relies on the Dyninst API. The analysis of observered performance data is done online to avoid the space and time overhead of trace-based post-mortem analysis tools.

A number of new ideas have recently been developed in the context of Paradyn. *Deep Start* [18] is an approach for augmenting Paradyn's automated search strategy with call stack samples. Deep starters are defined to be functions that arise frequently in call stack samples (either because they are called frequently or because they are long-running functions) and are thus likely to be performance bottlenecks. The *Deep Start* strategy gives precedence to searching performance bottlenecks in deep starters and their callees, which results in a quicker search and in a detection of more bottlenecks.

A second recent improvement of Paradyn is *MRNet* [16, 17], which is a software-based scalable communication infrastructure for connecting the tool's daemons to the front end. The original Paradyn approach uses a flat model, where each daemon process communicates directly with the front end. With an increasing number of daemons (i.e., performance analysis on larger machines), the frontend becomes a bottleneck and can be flooded with daemon data. MRNet therefore introduces a multicast-reduction network of internal processes arranged as a tree, to distribute the tool's activities and to keep the load on the front end manageable. MRNet consists of a library (which becomes part of the tool daemons) and a set of internal processes that implement data aggregation and synchronization operations as filters on one or more streams of data.

Even with efficient means for data collection provided by MRNet, the analysis and the decision process is still performed centrally by the Performance Consultant in Paradyn. To remedy the bottleneck this centralized approach presents for large-scale machines, work towards a Distributed Performance Consultant (DPC) [19] was recently conducted. The DPC supports a distributed bottleneck search strategy, where host-specific decisions on instrumentation and refinement of hypotheses are made by local search agents running on the local nodes, while global application behavior is examined by the front end. A subgraph-folding algorithm was additionally developed to visualize the search history graphs generated by local search agents. The folding algorithm places application processes into classes of qualitatively similar behavior, effectively "folding" similar search history graphs (for different processes) into one, which can be visualized readily by the user.

17.4 SCALASCA

SCALASCA [7] is a tool for scalable trace-based performance analysis of parallel applications. While currently being limited to MPI-1 applications, it is the follow-up project to Expert which performs an automated post-mortem performance analysis of C/C++ and Fortran applications using MPI, OpenMP, SHMEM, or a combination thereof. For both SCALASCA and Expert, the application is instrumented semiautomatically at compile-time using a combination of Opari (for OpenMP), the MPI-profiling interface, and TAU (for user-defined functions). Executing the instrumented application generates a set of trace files (one per process) in the Epilog format which are then analyzed after program termination (off-line analysis).

The sequential analysis (of merged trace files) performed by Expert has been found to be a major obstacle in applying this approach to large-scale machines. Hence, in SCALASCA the analysis itself is performed in parallel using a technique termed communication replay. The analysis engine is a parallel MPI program consisting of the same number of processes that have been used in the execution of the target application. Each analysis process analyzes its local trace file (no trace file merge happens) and performance problems (inefficiency patterns) such as "late sender" are detected by exchanging performance data among the involved processes using the same kind of communication operation. That is, to detect inefficiencies in point-to-point operations point-to-point operations are used by the analysis engine as well. Similarly, collective operations are used to combine and exchange performance data for the detection of collective communication inefficiencies, hence the term communication replay.

Both Kojak and SCALASCA output their analysis results as an XML file that can be viewed using a graphical viewer called CUBE with three tree-like hierarchies. The first hierarchy shows the type of inefficiency ("barrier waiting" is a descendant of "synchronization overhead," for example), the second hierarchy lists the program's resources (files, functions, and regions), and the third presents the machine organization (nodes, processes, threads). Color coding indicates how severe each discovered inefficiency is for the respective program resource and thread.

17.5   Vampir Next Generation

Vampir NG (VNG) [2] is the new and parallel implementation of the successful trace analysis tool Vampir (now Intel Trace Analyzer). Trace files are recorded while the application executes and then visually and manually analyzed using a time-line display and various other statistical views on the trace data.

VNG decouples the analysis part from the visual display component of Vampir by employing a distributed software architecture. The analysis is conducted by a parallel analysis server that can be located on a part of the actual parallel production system, while the visualization component is executed on a standard workstation computer. The two components communicate using a sockets-based networking protocol that can optionally be encrypted. The analysis server is itself parallelized using a combination of MPI and POSIX threads (Pthreads). MPI is chosen for its proven high performance and scalability properties while a multithreaded implementation is required to support multiple concurrent visualization clients and to support the interactive cancelation of outstanding analysis requests.

The analysis server consists of a master process and multiple worker processes. The master organizes the communication with the visualization clients and each worker holds a part of the overall tracefile to be analyzed. The two major benefits of this approach are an increased amount of overall main memory (summed over all worker processes) to hold the tracefile and the increased computing power for analysis tasks. The visualization client offers the same kind of functionality as the original sequential Vampir tool, however the separation of analysis and visualization avoids the transportation of large quantities of performance data. Instead, the data can be analyzed close to the place where it is generated.

17.6   Periscope

Periscope [8, 13, 5, 9] is a distributed online performance analysis tool for hybrid MPI/OpenMP applications currently under development at Technische

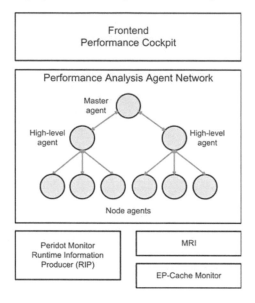

FIGURE 17.1: Periscope currently consists of a front end, a hierarchy of analysis agents, and two separate monitoring systems.

Universität München. It uses a set of autonomous agents that search for performance bottlenecks in a subset of the application's processes and threads. The agents request measurements of the monitoring system, retrieve the data, and use the data to identify performance bottlenecks. The types of bottlenecks searched are formally defined in the APART Specification Language (ASL) [3, 4].

17.6.1 Architecture

Periscope consists of a front end called the Performance Cockpit, a hierarchy of analysis agents, and two separate monitoring systems (Figure 17.1).

The user interface allows the user to start up the analysis process and to inspect the results. The agent hierarchy performs the actual analysis. The node agents autonomously search for performance problems which have been specified with ASL. Typically, a node agent is started on each symmetric multiprocessor (SMP) node of the target machine [10]. This agent is responsible for the processes and threads on that node. Detected performance problems are reported to the master agent that communicates with the performance cockpit.

The node agents access a performance-monitoring system for obtaining the performance data required for the analysis. Periscope currently supports two different monitors, the *Peridot monitor* or Runtime Information Producer

(RIP) [5] developed in the Peridot project focusing on OpenMP and MPI performance data, and the *EP-cache monitor* [14] developed in the EP-cache project focusing on memory hierarchy information.

The node agents perform a sequence of experiments. Each experiment lasts for a program phase, which is defined by the programmer, or for a predefined amount of execution time. Before a new experiment starts, an agent determines a new set of hypothetical performance problems based on the predefined ASL properties and the already found problems. It then requests the necessary performance data for proving the hypotheses and starts the experiment. After the experiment, the hypotheses are evaluated based on the performance data obtained from the monitor.

17.6.2 Specification of performance properties with ASL

Periscope's analysis is based on the formal specification of performance properties in ASL. The specification determines the condition, the confidence value, and the severity of performance properties.

```
PROPERTY ImbalanceInParallelLoop(OmpPerf pd, Experiment exp) {
 LET
      imbal = pd.exitBarT[0]+...+
              pd.exitBarT[pd.threadC-1];

 IN
   condition  : (pd.reg->type==LOOP ||
                  pd.reg->type==PARALLEL_LOOP) &&
                 (imbal > 0);
   confidence : 1.0;
   severity   : imbal / RB(exp);
}
```

This property captures the situation of imbalance in the OpenMP work-sharing constructs parallel **FOR** or **DO**. The waiting time at the implicit synchronization point appears in **exitBarT** and refers only to the waiting time for the particular construct. The property parameters identify the data structure containing the OpenMP performance data and the current experiment. The severity is the time for imbalance relative to the overall execution time. The confidence value of properties might be less than 1.0 if the condition simply gives a hint instead of a proof.

17.6.3 The Periscope node agents

The node agents form the lowest level of the Periscope agent hierarchy. Their task is the analysis of performance data on a single node of a multi-node system.

The main target systems for Periscope are clusters of symmetric multiprocessor (SMP) nodes and an agent is responsible for one SMP node. However,

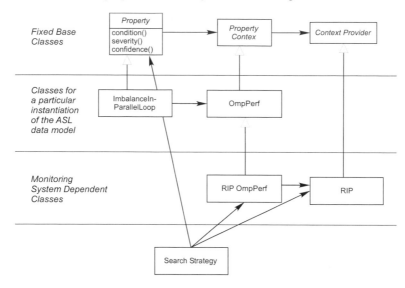

FIGURE 17.2: The layered design of the Periscope node agents.

Periscope can also be used on other machine configurations. For example, on large-scale shared memory nodes, several node agents can be used on the same node.

Node agents consist of multiple components. The "Agent Core" component encapsulates the main performance properties search functionality while performance data is acquired by using one or more monitors. A separate monitor is instantiated for each application process (which can itself be composed of several threads) which executes on the node. A node agent also contains a component to handle the communication with other agents by using a custom communication protocol which is described in Section 17.6.6.

The performance properties knowledge base of Periscope is implemented as a set of dynamically loadable modules. That is, for each of the ASL performance properties a dynamically loadable module is created that implements the property as a C++ object. At startup, the node agent dynamically loads the set of properties that reside in a configurable directory. The separation of performance analysis knowledge from the implementation of the main tool is beneficial for several reasons. Most importantly, the functionality of the tool can easily be extended, without requiring any changes to the main tool. Hence, advanced users or performance analysis experts can easily experiment with new performance properties and develop a custom set of properties specifically tailored to their requirements.

The loose coupling of the performance properties knowledge base and the core component of the agent is facilitated by the layered design shown in a UML-like class diagram in Figure 17.2. The top layer ("Fixed Base Classes") contains abstract classes that represent base entities which are independent of a concrete instantiation of an ASL model. The class *Property* represents the concept of a performance property. A property offers methods to check whether the specified condition holds and to compute the severity and confidence values. Since the property only specifies *how* to compute these values and to check the condition, a concrete *Property Context* object is needed to instantiate a concrete property. The component that generates this context (i.e., the monitoring system) is represented by the abstract *Context Provider* base class.

The second layer contains classes that represent an instantiation of ASL for a particular usage scenario. For example, the formal specification of properties is transformed one-to-one in a set of classes deriving from the base layer. E.g., the formal specification for the `ImbalanceInParallelLoop` property given in Section 17.6.2 is translated into a C++ class that derives from the *Property* base class. Similarly, a C++ class for the data model `OmpPerf` is generated that derives from the *Property Context* base class. This translation of the ASL specification into C++ classes is currently performed manually, but since the process is completely mechanized, it would be feasible to develop an ASL compiler that performs this translation.

The third layer shown in Figure 17.2 contains classes that represent the connection to the actual monitoring system used. For example *RIP OmpPerf* derives from the class *OmpPerf* and implements the data model by relying on the Peridot monitor.

The presented layered design offers flexibility and potential for software reuse in two respects. Firstly, components can partially be reused in a potential application of ASL in a different area. If, for example, an ASL data model and accompanying performance properties are developed for a new programming paradigm, the base-class layer and the *Search Strategy* object can stay the same. Naturally, a specific monitoring solution will be required to support the collection of performance data for the new programming paradigm. Similarly, the classes in the monitoring-system-specific layer have to be adapted.

Secondly, the layered design allows new properties for an existing data model to be integrated into the Periscope system without any change to the main tool. As mentioned, Periscope node agents dynamically load the set of properties on startup. This is possible because, as shown in Figure 17.2, the *Search Strategy* object only references the abstract base class *Property* but references to specific properties are never used.

17.6.4 Search for performance properties

When a node agent receives the request to search for performance properties via the `ACC_CHECK` command (see Section 17.6.6), it invokes a search strategy object to conduct the actual search. The search strategy object encapsulates the sequence of steps needed to instantiate the performance properties. Currently, a `TestAll` search strategy is used which simply tries to instantiate all performance properties for all property contexts (i.e., regions).

The `TestAll` search strategy first employs the monitor to instantiate the ASL data model. That is, for each program region, a corresponding `OmpPerf` or `MpiPerf` data structure is created and filled with performance data. The ASL data model represents the *current* view of the program, i.e., the counts and times contained in the data structures represent the duration from program start until the invocation of the properties search.

When the ASL data model has been created, the `TestAll` search strategy tries to instantiate each loaded property for each property context (i.e., each `OmpPerf` or `MpiPerf` structure). The `condition()` method is invoked to check if the specified property holds for each property context and if, yes, the `severity()` and `confidence()` methods are subsequently used to determine the severity and confidence values, respectively.

For each detected property, a report is sent to the parent agent that specifies the name of the property, the severity and confidence values, and the property context (i.e., the type and location of the region in the application's source code). This report of detected properties is passed up the agent hierarchy and the front end displays the result to the user.

17.6.5   The Periscope high-level agents

The task of the Periscope high-level agents is the efficient and scalable collection of performance analysis results from the node agents as well as the dissemination of commands from the front end to the node agents. As shown in Figure 17.1, the high-level agents are arranged in a tree-like hierarchy. The root of the tree is represented by a single *master agent*, which represents the connection to the Periscope front end.

Periscope high-level agents have always exactly one parent (which is either another high-level agent or the front end) and one or more child agents (which are either high-level agents or node agents). The smallest instantiation of Periscope therefore consists of the front end, a master agent, and a single-node agent.

The agents and the front end use a custom sockets-based protocol as a means of communication, by exchanging so-called ACC (agent command and control) messages. The ACC protocol (described in detail in Section 17.6.6) developed for Periscope offers platform-independent messaging for maximum flexibility with respect to the placement of the agents and the front end. The high-level agents and front end can be executed on any node on a high-performance computing system or even anywhere on the Internet, as long as the required sockets-based communication links can be established. However, the Periscope front end is typically started on a node of the target computing system which provides interactive access, while the high-level agents reside on nodes that have been reserved for performance analysis.

17.6.6 Agent communication infrastructure

As mentioned, the agents and the Periscope front end employ a custom sockets-based communication protocol by exchanging ACC (agent command and control) messages. The ACC messages are sent synchronously (the sender waits for an acknowledgment from the receiver) and cause the invocation of a message-specific handler function at the receiver side. ACC messages (and also their acknowledgments) can carry message-specific data that is encoded using CDR (common data representation), which allows a platform and operating-system-independent messaging between the agents and the front end.

While high-performance computing systems typically offer a homogenous compute partition (or several homogenous partitions that are used independently), it is not uncommon that the platform used for interactive access is different from the platform used for the compute partition. The common data representation used by Periscope therefore offers the benefit of allowing the front end to run on any system that can establish communication links to the agents, even on the user's personal computer. Platform independency is also important for a possible future application of Periscope in grid computing. Grids often combine several different computer architectures and support for heterogeneity is therefore an important requirement for any performance tool used in a grid environment.

The ACC messages used by the agents and by the Periscope front end are:

ACC_INIT: This message is used to initialize the communication link between peers. The data exchanged with this message is an identification string such as "**Periscope Frontend**" or "**Periscope nodeagent**" that specifies the particular role of the component in the Periscope performance analysis system.

ACC_QUIT: This message is used to tear down the communication link between peers.

ACC_HEARTBEAT: This message is sent periodically from an agent to its parent to indicate its status. An agent that receives an ACC_HEARTBEAT message knows that the child agent sending the message and the entire sub-tree rooted at that child are in a healthy state.

ACC_START: This message is used during the coordinated startup procedure of Periscope and the target application. The target application is started first, but then halted by the Periscope monitoring library until the agent network has been instantiated. Once the agents are up and running, the ACC_START message is broadcast from the front end to the node agents to resume the execution of the target application.

ACC_CHECK: This message is sent from the front end to the node agents to initiate the search process for performance properties. High-level agents that receive this message simply forward it to their child agents, until it arrives at the node agents. At the node agents the message causes the instantiation of the ASL data model and the search for performance properties as described in Section 17.6.3.

ACC_PROPERTY_FOUND: This message is propagated from the node agents up the agent hierarchy towards the front end. A node agent sends this message when a performance property was detected. The data carried by the message include the type of property discovered (i.e., its name), the node on which it was detected, the severity and confidence values, as well as the context of the property. The property context is typically the name of the region for which the property was detected; for OpenMP locks the context gives the address of the lock instead. For example, the message

```
ACC_PROPERTY_FOUND WaitAtBarrier "opt33" 0.34 1.0 "ssor.f 186"
```

indicates that the WaitAtBarrier property was detected for an explicit (programmer-added) OpenMP barrier at line 186 of file "ssor.f" on node "opt33." The detected severity and confidence values are 0.34 and 1.0, respectively.

When a high-level agent receives an ACC_PROPERTY_FOUND message, it tries to aggregate and combine all received properties before passing the results on to its parent agent. Properties with the same property context and with similar severity and confidence values are combined. For example, the set of properties

```
(1) ACC_PROPERTY_FOUND WaitAtBarrier "opt33" 0.34 1.0 "ssor.f 186"
(2) ACC_PROPERTY_FOUND WaitAtBarrier "opt33" 0.33 1.0 "ssor.f 207"
(3) ACC_PROPERTY_FOUND WaitAtBarrier "opt34" 0.30 1.0 "ssor.f 186"
(4) ACC_PROPERTY_FOUND WaitAtBarrier "opt34" 0.01 1.0 "ssor.f 207"
(5) ACC_PROPERTY_FOUND LockContention "opt34" 0.01 1.0
        "0xBF232343"
```

is passed on as

```
ACC_PROPERTY_FOUND WaitAtBarrier "opt33,opt34" 0.32 1.0
"ssor.f 186"
ACC_PROPERTY_FOUND WaitAtBarrier "opt33" 0.33 1.0 "ssor.f 207"
ACC_PROPERTY_FOUND WaitAtBarrier "opt34" 0.01 1.0 "ssor.f 207"
ACC_PROPERTY_FOUND LockContention "opt34" 0.01 1.0 "0xBF232343"
```

That is, properties (1) and (3) are combined into a compound property with averaged severity and confidence values, while (1) and (4) are not combined, because they have different property contexts. Properties (2) and (4) are not combined, because their difference in severity exceeds the predefined threshold of 10%.

The ACE (adaptive communication environment) framework [1, 20, 11] is used in Periscope to implement the sockets-based ACC communication protocol outlined above. ACE is a cross-platform, object-oriented framework designed to simplify the usage of operating systems mechanisms such as inter-process communication, threading, and memory management.

The Periscope agents and the front end use ACE's *reactor* framework, which allows an application to handle several tasks concurrently without relying on multiple threads or processes. Hence, overheads associated with thread or process creation can be avoided and the complexity of the source code can be reduced (there is no need to coordinate the access to shared resources, for example).

17.6.7 Evaluation

This section tests Periscope on eight applications (BT, CG, EP, FT, IS, LU, MG, and SP) from the NAS parallel benchmark suite [12]. The NAS benchmark applications exist in many versions utilizing various programming interfaces (OpenMP, MPI, High Performance Fortran [HPF], and Java, among others). The rest of this section tests the MPI and OpenMP versions of the benchmarks.

17.6.7.1 MPI

The applications were executed with benchmark class "C" on 64 processes of our InfiniBand cluster on 16 nodes. The following text lists the performance properties for some application benchmarks detected after program termination.

BT			
Property Name	Location	Processes	Severity
WaitAtMpiBarrier	MPI_BARRIER	all	0.001300
PointToPointCommOvhd	MPI_RECV	all	0.001191
CollectiveCommOvhd	MPI_BCAST	all	0.000166
CollectiveCommOvhd	MPI_REDUCE	all	1.160e-06
PointToPointCommOvhd	MPI_SEND	all	7.411e-08

CG			
Property Name	Location	Processes	Severity
PointToPointCommOvhd	MPI_SEND	26	0.350885
PointToPointCommOvhd	MPI_SEND	30	0.285780
PointToPointCommOvhd	MPI_SEND	8	0.036432
WaitAtMpiBarrier	MPI_BARRIER	all	0.000126
PointToPointCommOvhd	MPI_RECV	all	0.000157
CollectiveCommOvhd	MPI_REDUCE	all	3.276e-06

EP			
Property Name	Location	Processes	Severity
PointToPointCommOvhd	MPI_RECV	24	0.388950
WaitAtMpiBarrier	MPI_BARRIER	32	0.388440
PointToPointCommOvhd	MPI_RECV	40	0.000380
WaitAtMpiBarrier	MPI_BARRIER	32	0.000362
PointToPointCommOvhd	MPI_SEND	all	9.071e-07

FT			
Property Name	Location	Processes	Severity
CollectiveCommOvhd	MPI_ALLTOALL	all	0.846625
PointToPointCommOvhd	MPI_RECV	all	0.004725
WaitAtMpiBarrier	MPI_BARRIER	all	0.004038
CollectiveCommOvhd	MPI_REDUCE	all	0.002983
CollectiveCommOvhd	MPI_BCAST	all	1.764e-05
PointToPointCommOvhd	MPI_SEND	all	6.805e-08

Evidently, most applications contain one dominant MPI call that accounts for a significant fraction of the total execution time, while other performance properties could only be identified with very low severity values. Also, in general, the applications show very similar behavior among processes. The only exceptions are the CG and EP benchmarks which show markedly dissimilar behavior among groups of processors. The exact cause of these dissimilarities could only be identified by performing a close inspection of the application's source code and underlying algorithmic structure.

17.6.7.2 OpenMP

In addition to the MPI version, the OpenMP implementation of the NAS benchmarks was tested with Periscope. The applications were executed on a 32-CPU SGI Altix system based on Itanium-2 processors with 1.6 GHz and 6MB L3 cache using a batch system. The number of OpenMP threads was set to eight and the Periscope node agent was executed on a separate CPU (i.e., nine CPUs were requested for the batch runs).

Table 17.3 shows the three most severe properties identified by Periscope for the NAS benchmarks. All properties have severity values below 9% and most are in the range of 3–4%.

17.7 Tool Comparison and Future Research

Developing efficient parallel programs for petascale architectures will require scalable performance analysis tools. This chapter presents four performance analysis tools that apply different techniques to improve scalability.

Benchmark	Property	Region	Severity
BT	ImbalanceInParallelLoop	rhs.f 177--290	0.0446
BT	ImbalanceInParallelLoop	y_solve.f 40--394	0.0353
BT	ImbalanceInParallelLoop	rhs.f 299--351	0.0347
CG	ImbalanceInParallelLoop	cg.f 556--564	0.0345
CG	ImbalanceInParallelRegion	cg.f 772--795	0.0052
CG	ImbalanceInParallelRegion	cg.f 883--957	0.0038
EP	ImbalanceInParallelRegion	ep.f 170--230	0.0078
EP	ImbalanceInParallelLoop	ep.f 129--133	0.0001
FT	ImbalanceInParallelLoop	ft.f 606--625	0.0676
FT	ImbalanceInParallelLoop	ft.f 653--672	0.0304
FT	ImbalanceInParallelLoop	ft.f 227--235	0.0269
IS	WaitAtBarrier	is.c 526	0.0272
IS	ImbalanceInParallelRegion	is.c 761--785	0.0087
IS	ImbalanceInParallelLoop	is.c 397--403	0.0020
LU	WaitAtBarrier	ssor.f 211	0.0040
LU	WaitAtBarrier	ssor.f 182	0.0032
LU	ImbalanceInParallelLoop	rhs.f 189--309	0.0011
MG	ImbalanceInParallelLoop	mg.f 608--631	0.0831
MG	ImbalanceInParallelLoop	mg.f 779--815	0.0291
MG	ImbalanceInParallelLoop	mg.f 536--559	0.0248
SP	ImbalanceInParallelLoop	x_solve.f 27--296	0.0285
SP	ImbalanceInParallelLoop	y_solve.f 27--292	0.0265
SP	ImbalanceInParallelLoop	z_solve.f 31--326	0.0239

FIGURE 17.3: The three most severe performance properties with source-code location and severity value, identified by Periscope (only two properties were found for EP).

These techniques are:

Distributed online aggregation. Instead of transferring individual data from all the processes to a front end, the information is aggregated in a distributed fashion. This approach is implemented by MRNet in Paradyn. A similar approach is used in Periscope on a much higher level. The properties detected by the agents are aggregated. If multiple processes have the same property only one piece of information is propagated up the tree of high-level agents.

Automatic search. The tool's analysis is done automatically. The search for performance bottlenecks is based on formalized knowledge about possible inefficiencies. This approach is taken by SCALASCA, Paradyn, and Periscope. Automation is a requirement for online analysis but it also improves off-line tools dramatically, e.g., by reducing the amount of information presented to the programmer.

Autonomous agents. Autonomous agents are used to search for performance bottlenecks in a subset of the application's processes. This distribution eliminates the bottleneck in current tools of a centralized analysis. Both Paradyn and Periscope apply this technique.

Parallel analysis back end. SCALASCA and Vampir NG off-load part of
the analysis process to a parallel back end. While SCALASCA performs
an automatic analysis of trace data on the same processors as the application within the same job, Vampir NG computes graphical views of
performance data on a parallel back end.

Scalable visualization. Trace analysis not only suffers from huge trace files
but also from the limitation of visualizing performance data on current
screens. Vampir applies clustering of time lines, zooming and scrolling
to circumvent this problem.

Classification of processes. Typically processes of a parallel application
can be classified into classes with similar behavior. Paradyn and Periscope
use this technique. Paradyn provides subgraph folding, while Periscope
combines performance properties of different processes in the agent tree.

Incremental search. Profiling and tracing both have their advantages. Profiling reduces the amount of information while tracing provides all the
details with the drawback of huge trace files. Paradyn and Periscope
search for properties incrementally assuming a cyclic behavior of the
application. During one execution of a phase coarse measurements are
performed, while in the next execution more precise measurements are
taken. Currently, both tools apply this technique only to profiling data.

The above list of techniques supporting scalability of performance tools
highlights that quite some work is performed in that area.

For the future, we believe that the use of autonomous agents and of process
classification will receive more attention. Autonomous agents can incrementally search for performance bottlenecks where the performance data are generated. They reduce information instead of communicating and storing huge
amounts of data. The classification of processes will be important as well
since even the automatic detection of performance bottlenecks in processes
will not be sufficient. Imagine that thousands of bottlenecks are presented to
the user even if in each process of a petascale machine only a few bottlenecks
were found.

Both techniques, autonomous agents and classification, can benefit from
additional knowledge about the parallel program's structure. Frequently, parallel programs are based on structures such as master worker, work pool, divide and conquer, or single-program, multiple-data (SPMD). This information
could be used to guide the automatic search, e.g., to identify repetitive regions
that can be incrementally analyzed, as well as for process classification, e.g.,
by specifying that there are two classes, the master and the workers.

References

[1] The ADAPTIVE Communication Environment (ACE), web page. `http://www.cs.wustl.edu/~schmidt/ACE.html`.

[2] H. Brunst and B. Mohr. Performance analysis of large-scale OpenMP and hybrid MPI/OpenMP applications with VampirNG. In *Proceedings of the First International Workshop on OpenMP (IWOMP 2005)*, Eugene, OR, May 2005.

[3] T. Fahringer, M. Gerndt, G. Riley, and J. Träff. Knowledge specification for automatic performance analysis. Apart Technical Report, Forschungszentrum Jülich, Jülich, Germany, 2001. `http://www.fz-juelich.de/apart`.

[4] T. Fahringer, M. Gerndt, G. Riley, and J. L. Träff. Specification of performance problems in MPI-programs with ASL. In *International Conference on Parallel Processing (ICPP'00)*, pages 51–58, 2000.

[5] K. Fürlinger. *Scalable Automated Online Performance Analysis of Applications Using Performance Properties*. Ph.d. thesis, Technische Universität München, 2006.

[6] K. Fürlinger and M. Gerndt. Performance analysis of shared-memory parallel applications using performance properties. In *Proceedings of the 2005 International Conference on High Performance Computing and Communications (HPCC-05)*, pages 595–604, Sorrento, Italy, September 2005.

[7] M. Geimer, F. Wolf, B. J. N. Wylie, and B. Mohr. Scalable parallel trace-based performance analysis. In *Proceedings of the 13th European PVM/MPI Users' Group Meeting on Recent Advances in Parallel Virtual Machine and Message Passing Interface (EuroPVM/MPI 2006)*, pages 303–312, Bonn, Germany, 2006.

[8] M. Gerndt, K. Fürlinger, and E. Kereku. Advanced techniques for performance analysis. In G.R. Joubert, W.E. Nagel, F.J. Peters, O. Plata, P. Tirado, and E. Zapata, editors, *Parallel Computing: Current & Future Issues of High-End Computing (Proceedings of the International Conference ParCo 2005)*, volume 33 of *NIC Series*, pages 15–26a, 2006.

[9] M. Gerndt and E. Kereku. Automatic memory access analysis with Periscope. In *Proceedings of the International Conference on Computational Science (ICCS 2007)*, Beijing, China, May 2007. Springer.

[10] M. Gerndt and S. Strohhäcker. Distribution of Periscope analysis agents on ALTIX 4700. In *Proceedings of the International Conference on Parallel Computing (ParCo 07)*, 2007.

[11] S. D. Huston, J. C. E. Johnson, and U. Syyid. *The ACE Programmer's Guide*. Pearson Education, 2003.

[12] H. Jin, M. Frumkin, and J. Yan. The OpenMP implementation of NAS parallel benchmarks and its performance. Technical Report, NAS-99-011, 1999.

[13] E. Kereku. *Automatic Performance Analysis for Memory Hierarchies and Threaded Applications on SMP Systems*. Ph.d. thesis, Technische Universität München, 2006.

[14] E. Kereku and M. Gerndt. The EP-cache automatic monitoring system. In *International Conference on Parallel and Distributed Systems (PDCS 2005)*, 2005.

[15] P. B. Miller, M. D. Callaghan, J. M. Cargille, J. K. Hollingsworth, R. B. Irvin, K. L. Karavanic, K. Kunchithapadam, and T. Newhall. The Paradyn parallel performance measurement tool. *IEEE Computer*, 28(11):37–46, 1995.

[16] P. C. Roth, D. C. Arnold, and B. P. Miller. MRNet: A software-based multicast/reduction network for scalable tools. In *Proceedings of the 2003 Conference on Supercomputing (SC 2003)*, AZ, November 2003.

[17] P. C. Roth, D. C. Arnold, and B. P. Miller. Benchmarking the MRNet distributed tool infrastructure: Lessons learned. In *Proceedings of the 18th International Parallel and Distributed Processing Symposium (IPDPS '04), High Performance Grid Computing Workshop*, page 272, 2004.

[18] P. C. Roth and B. P. Miller. Deep start: A hybrid strategy for automated performance problem searches. In *Proceedings of the 8th International Euro-Par Conference on Parallel Processing (Euro-Par '02)*, pages 86–96, Paderborn, Germany, August 2002. Springer-Verlag.

[19] P. C. Roth and B. P. Miller. The distributed performance consultant and the sub-graph folding algorithm: On-line automated performance diagnosis on thousands of processes. In *Proceedings of the ACM SIGPLAN Symposium on Principles and Practice of Parallel Programming (PPoPP'06)*, March 2006.

[20] D. C. Schmidt, S. D. Huston, and F. Buschmann. *C++ Network Programming Vol. 1: Mastering Complexity with ACE and Patterns*. Pearson Education, 2002.

[21] S. S. Shende and A. D. Malony. The TAU parallel performance system. *International Journal of High Performance Computing Applications*, 20(2):287–311, 2006. ACTS Collection Special Issue.

[22] J. S. Vetter and M. O. McCracken. Statistical scalability analysis of communication operations in distributed applications. *ACM SIGPLAN Notices*, 36(7):123–132, 2001.

[23] Felix Wolf and Bernd Mohr. Automatic performance analysis of hybrid MPI/OpenMP applications. In *Proceedings of the 11th Euromicro Conference on Parallel, Distributed and Network-Based Processing (PDP 2003)*, pages 13–22. IEEE Computer Society, February 2003.

Chapter 18

Towards Petascale Multilevel Finite-Element Solvers

Christoph Freundl, Tobias Gradl, Ulrich Rüde

Friedrich–Alexander–Universität, Erlangen, Germany

Benjamin Bergen

Los Alamos National Laboratory, USA

18.1   Introduction

18.1.1   Overview

High-end supercomputers are inevitably massively parallel systems and they often feature complex architectures for each compute node. Additionally, exploiting the performance potential even of a single CPU is becoming increasingly difficult, e.g., due to limited memory performance. Such architectural restrictions pose great challenges to the development of efficient computational methods and algorithms.

Many supercomputing applications are based on the finite-element (FE) method for the solution of partial differential equations. Finite-element methods require the solution of large sparse linear systems, but only few algorithms qualify as the basis of scalable FE solvers. For a scalable solution only algorithms can be used that obtain linear or almost linear computational complexity in the number of unknowns. Therefore, we will focus on multigrid methods that can achieve asymptotically optimal complexity. However, due to the multigrid structure, the efficient parallelization of the algorithms is not trivial (cf. [16] and the references listed there).

Our contribution addresses these problems by presenting two scalable multilevel finite element packages that were implemented within projects of the Bavarian KONWIHR supercomputing research consortium [9, 5].

18.1.2 Exemplary petascale architectures

Petascale architectures, as they will be available in a few years' time, will be parallel computers consisting of many thousands of processors similar to today's top performance computers. These are usually connected by a high-speed network with a bandwidth of at least 1 GBit per second. Often small numbers of processors are grouped together in *nodes* where the interconnection of processors on the same node is faster than the connection between the nodes. Additionally, a growing trend towards *multicore* processors combines several single CPU cores together on a chip. The memory configuration of these architectures can range from distributed memory for each processor, shared memory access for processors residing on the same node, to virtually shared memory access for all processors in the computer.

As a first exemplary architecture, we present the meanwhile outdated computer *Höchstleistungsrechner in Bayern I* (HLRB I), since it has been the motivation for the two software projects that will be described in the next section. The Hitachi SR8000-F1 model, which had been in operation at the Leibniz Computing Center Munich until summer 2006, consisted of 168 SMP nodes with 8 processors on each node. The processor design was a modified PowerPC architecture with an increased number of floating-point registers and a very powerful pre-fetch mechanism. The compilers on the machine were capable of making use of the COMPAS (cooperative microprocessors in single address space) feature which enabled the auto-parallelization of loops on the processors of a node that gave each node characteristics similar to a vector processor. A single processor had a clock speed of 375 MHz with a theoretical peak performance of 1.5 GFLOPS, resulting in a LINPACK performance of 1,645 GFLOPS for the whole machine. At the time of installation, the HLRB I started at rank 5 in the Top 500 list of June 2000.

In June 2006, the SR8000-F1 was shut down and replaced by HLRB II*, ranked number 18 in the Top 500 list of November 2006. This machine is an SGI Altix 4700 and currently consists of 9728 Intel Itanium2 CPUs and 39 terabytes of main memory. The NUMAlink network helps it to reach 56.5 TFLOPS LINPACK performance, 90% of its theoretical peak performance, while providing shared memory access to groups of up to 512 CPUs at a time.

Exploiting the performance potential of these supercomputers and future petascale systems also requires a focused effort in the development of algorithms and programs. In the following, we will highlight the goals of the HHG and ParExPDE projects, which have been motivated by the HLRB machines and the associated KONWIHR research effort. Both projects aim at developing highly scalable finite element solvers. ParExPDE is centered around the expression template programming paradigm that is being used to hide technical complexities — such as MPI parallelism — from the user of the software, while providing a intuitive high level user interface for developing

---

*http://www.lrz-muenchen.de/services/compute/ .

partial differential equation (PDE) solvers. The expression template mechanism is being used to generate efficient code on each processor by keeping the overhead from operator overloading low.

The HHG package is a prototype of a very memory- and CPU-efficient implementation of a FE solver. To achieve ultimate performance on a single node as well as on massively parallel systems, it employs mixed-language programming, combining C++ and Fortran 77. It provides a highly efficient multigrid solver capable of solving finite element problems of currently up to 10^{11} unknowns on 9170 processors.

18.2   Design Paradigms

18.2.1   Hierarchical hybrid grids

Regular refinement and grid decomposition

HHG has been introduced in [2] (also see [1]). Summarizing, the HHG approach can be described as follows: beginning with a purely unstructured input grid, regular refinement is applied by adding new vertices along each edge of the input grid and then connecting the appropriate vertices with new edges and faces. To illustrate this, Figure 18.1 shows an example of regular refinement of a two-dimensional input grid. The two-dimensional example is included because it is easier to visualize, but the current HHG implementation is designed to accommodate three-dimensional grids.

The refinement results in a new, still logically unstructured grid that is a superset of the input grid. This process is repeated successively on each new grid, forming a nested hierarchy of finite-element meshes. Clearly, the regular refinement process generates a system of *structured* sub-meshes. In a conventional FE approach, however, this structure would not be exploited, but HHG does exploit this structure both for defining the multigrid algorithm, and as the basis of a very efficient implementation in terms of overall memory consumption, parallelization, and performance of the solver in each processing node.

Each grid in the hierarchy is then decomposed into the *primitive types*: (3D-) elements, (2D-) faces, (1D-) edges, and vertices. This decomposition allows each class of primitive types to be treated separately during the discretization and solver phases of the simulation, so that the structure of the added points can be exploited: for example, instead of explicitly assembling a global stiffness matrix for the finite-element discretization element by element, we can define it implicitly using stencils. This works when the material parameters are constant within an element, since then the stencil for each element primitive is *constant for all unknowns that are interior to it* for a given

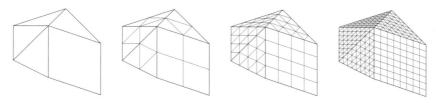

FIGURE 18.1: A regular refinement example for a two-dimensional input grid. Beginning with the input grid on the left, each successive level of refinement creates a new grid that has a larger number of interior points with structured couplings.

level of refinement. The same approach may be used for the face primitives and, to some extent, for the edges. Essentially, this results in a variant of block-structured meshes.

In order to exploit the structure of the refined primitives in terms of processor performance, data structures with contiguous memory must be allocated for *each* primitive. Then the primitives of each type with a regular structure of unknowns can be treated as independent structured grids. This allows the use of stencil-based discretization and solver techniques that are more efficient than those available for use with standard unstructured grid data structures.

HHG in practice

In applying HHG, we make the assumption that the input grid is fairly coarse and that it primarily resolves only the large-scale features of the domain. The most favorable problem type for this approach begins with an input grid which has patch-wise constant material parameters but requires several levels of regular refinement to properly resolve the fine-scale features of the solution. This type of problem leads to a grid hierarchy with large collections of structured unknowns that have a constant-coefficient discretization within each patch.

As an example, consider the exterior mirror of a car and the problem of an acoustic simulation that requires elliptic solvers. We might be interested in the noise level produced by the air current around the mirror, typically for a distance much larger than the dimensions of the mirror. Naturally, between the mirror and the "listener" there is a lot of free space without any geometric features that would need to be resolved by the grid. Nevertheless, the grid resolution must be very fine to resolve the sound waves. Such properties make this application a good candidate for HHG.

The HHG approach does also have limitations. It cannot be used easily for domains where even the geometry description requires extremely fine meshes everywhere, such as in a porous medium. The approach would have to be extended (in a straightforward way) to deal with material parameters that are not piece-wise constant. As much as possible, the philosophy of HHG requires uniform sub-grids, and therefore the flexibility for adaptive grid refinement is resticted.

HHG and multigrid

Implementing a multigrid algorithm on the HHG data structures is in principle quite straightforward, because the grid decomposition in the HHG framework immediately induces the necessary processes quite analogous to a structured multigrid implementation for the parts of the grid that are within the structured regions.

The question for the solver on the coarsest grid is nontrivial, since we consider the unstructured input grid as the coarsest one, and this may of course still involve quite a large system. The current HHG software implements several alternatives, including the use of simple iterative schemes, such as successive over-relaxation, that might be suitable if the input grid is really quite coarse. Alternatives are to use a (parallel) direct solver, or — possibly the most interesting alternative — to delegate the solution on the input grid to an algebraic multigrid method. In all cases the basic assumption is that the input grid has many fewer unknowns than the refined grids and, therefore, the computational cost for solving the problem is small.

Parallelization strategy

For a parallel implementation, the HHG framework is again an almost ideal starting point. We employ a distributed-memory parallel communication model using the *message-passing interface (MPI)*. Again, the idea is that the mesh distribution is essentially done on the level of the input grid, that is, with a grid size that can be handled efficiently by standard software, such as Metis or ParMetis [11].

In order to parallelize the computation, the grid is therefore distributed among the available MPI processes. Figure 18.2(a) shows a simple 2D example with just two triangular elements that are assigned to the two processes P_0 and P_1. The unknowns on the edge between the elements are coupled to both elements and are thus needed by both processes, which introduces communication (Figure 18.2(b)). This is equivalent to using ghost nodes, as is typical in parallel mesh algorithms [8]. The HHG edge data structure itself can be assigned to any one of the two processors. In HHG, every *primitive type* (element, face, edge or vertex) has specialized communication methods implemented for handling message buffers and issuing calls to the MPI library.

In order to avoid excessive latency during inter-node communication, each type of primitive is updated as a group. If, for example, two processes share not only one but several faces, the ghost values on all the common faces are packed into a single MPI message. So, rather than sending many small messages over the interconnect, only a few large ones have to be sent. This reduces communication latency and allows a more efficient implementation.

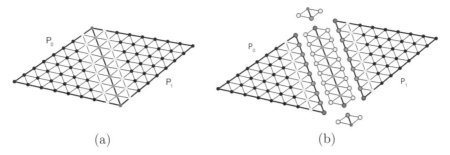

(a)   (b)

FIGURE 18.2: Grid distribution among processes. The encircled nodes are *ghost values.*

18.2.2 ParExPDE

Similar to the HHG solver, the ParExPDE (parallel expression templates for partial differential equations) framework aims at providing a highly scalable FE solver on massively parallel systems. Different from HHG, it is based on using expression templates in C++ and FE to provide a user-friendly, yet efficient framework for developing PDE-based parallel applications. Similar to HHG, ParExPDE relies on the regular refinement of a relatively coarse input grid only in it the geometric primitives are hexahedrons instead of tetrahedrons.

Expression templates

The term "expression templates" (ET) refers to a programming technique which uses the C++ template mechanism and allows one to generate efficient code for the evaluation of arithmetic expressions. It was invented in 1994 independently by Todd Veldhuizen and David Vandevoorde (see. [17]) and has been made popular by [7].

The ParExPDE library, inspired by [13] and [14], uses the ET programming technique because of its efficiency in performing vector arithmetics while providing a user-friendly interface at the same time, and is therefore pursuing a similar goal as the approach presented earlier in this book (see Chapter 19 [12]). The impacts of ET can be illustrated with a simple vector operation involving more than one operator:

```
Vector a, b, c, d;
d = a + b + c;
```

If a C++ user wants to make use of the object-oriented features, the obvious approach is

- to overload the "+" operator such that it adds two vectors and returns the resulting vector, and

- to implement the assignment operator in a straightforward way, i.e., it just takes a vector as a parameter and copies all elements to the result vector

This is, in fact, clean design, however, it has fatal consequences on the efficiency of the resulting code. In each call to the plus operator a temporary vector has to be created only to store the values that will be reused immediately afterwards when the next arithmetic operator or, finally, the assignment operator is called. So not only the creation and destruction of temporaries but also unnecessary copying of the temporary values will cost performance.

A C programmer, in contrast, who does not want to handle general arithmetic expressions would implement the above operation in a single loop:

```
loop over all components i of the vectors
    d[i] = a[i] + b[i] + c[i]
```

The goal of using ET is to produce this code from the abstract formulation in the first C++ listing. In order to achieve this, ET does not treat each operator independently but the expression on the right-hand side of the assignment operator as a whole. The syntax tree of the expression a + b + c is mapped to a nested template structure, and the implementation of the assignment operator loops over all components of the expression. When all necessary evaluation methods are declared "inline" the result is — at least theoretically — code that exhibits the same performance as handwritten C or Fortran code. This becomes possible since expression expansion and code optimization is performed at compile time. Results have shown that the expression template performance can in fact be even better (cf. [18]).

Parallel expression templates on hexahedral grids

In order to use ET for a petascale multilevel solver, it has to be embedded in a parallel context where each processor works just on its subdomain as defined during the partitioning phase. This does not pose a problem because the object containing the partitioning information should generally provide an iterator returning all parts of the whole domain belonging to a specific processor.

Furthermore, the parallel execution should ideally happen completely transparently to the user. Therefore, we distinguish two types of expressions: those which require a communication of ghost values (i.e., basically all expressions containing the application of a differential operator) and those which do not. The assignment operator is specialized for these two types and performs the necessary communication accordingly.

It is more difficult to treat the non-homogeneity of the hexahedral grid: while the regularly refined interior of a hexahedron is perfectly homogeneous the faces and especially the edges and vertices of the grid can have an arbitrary number of neighbor hexahedrons (Figure 18.3). These parts of the domain have to be treated completely different from the hexahedrons' interiors, e.g.,

FIGURE 18.3: Left picture: regular neighborhood (8 neighbors) in the interior of a quadrilateral. Middle picture: irregular neighborhood for corners of quadrilaterals. The thick dashed lines mark the boundaries of a quadrilateral. Right picture: ParExPDE grid describing a cylinder.

the first have arbitrary-sized stencils while the latter will always have 27-point stencils (the term "stencil" refers in this case to the assembled global stiffness matrix, expressing the dependencies of an unknown just as when using finite differences).

The introductory expression templates description in the previous section has assumed a completely regular data structure. For the application of expression templates in a hexahedral grid we have to provide means to specialize the expression referring to the whole grid to an expression which refers, e.g., only to a certain edge in the grid. The assignment operator in ParExPDE becomes in pseudo-code:

```
loop over all local hexahedrons in the grid
    specialize the expression for the hexahedron
    for all components (i,j,k) of the hexahedron
        evaluate the specialized expression at (i,j,k)
loop over all local faces in the grid
    specialize the expression for the face
    for all components (i,j) of the face
        evaluate the specialized expression at (i,j)
loop over all local edges in the grid
    specialize the expression for the edge
    for all components i of the edge
        evaluate the specialized expression at i
loop over all local vertices in the grid
    specialize the expression for the vertex
    evaluate the specialized expression at the vertex
```

18.3 Evaluation and Comparison

Multigrid Convergence Rates

Local Fourier analysis [19] shows that solving, for example, a 3D Poisson equation with a multigrid using $W(1, 1)$ cycles and a red-black Gauß-Seidel smoother should exhibit a convergence rate of about 0.19 (see [16], table 3.5). For the FE discretization, theoretical convergence rates are not readlily available in the literature, so using the cited values as a yardstick is not the worst thing to do. However, for the sake of parallel efficiency, some compromises may be necessary, and therefore the algorithm may not achieve the best convergence rates theoretically possible. For example, HHG ignores a few data dependencies in the Gauß-Seidel smoother, which, in turn, saves a significant amount of communication effort. To demonstrate that the multigrid method's excellent convergence behavior hasn't been sacrificed in the race for parallel efficiency and megaflops, we include experimentally determined convergence rates.

A solver exposes its *asymptotic convergence rate* typically after several V cycles. The convergence rates of HHG and ParExPDE, determined by a vector iteration, similar to the one described in [16, section 2.5.2], are shown in Figure 18.4. The asymptotic convergence rate is excellent in ParExPDE, which uses hexahedral elements and, therefore, 27-point stencils. HHG uses tetrahedral elements which lead to 15-point stencils. The convergence rates are still very good.

Serial Results

Before presenting parallel results, we first verify that our codes achieve a satisfactory performance already on a single processor, otherwise any discussion about speedups or scalability would be meaningless.

Our test case is the application of a constant-coefficient Gauß-Seidel smoothing algorithm on a single geometry element which is a tetrahedron in the context of HHG and a hexahedron in the context of ParExPDE. This test case represents the best possible situation of a purely structured grid with a constant stencil, and serves in determining the absolute maximum performance that can be obtained by the frameworks' data structures. Both programs traverse the mesh in a red-black ordering to eliminate data dependencies that could obstruct the computer's vectorization capabilities. We measure the execution performance for various mesh sizes.

The internal structures of HHG allowed us to mix C++ and Fortran 77 without sacrificing the usability and flexibility of the program to its performance. While the *framework* of HHG exploits the object-oriented features of C++, the *computational kernels* are implemented in Fortran 77, which has proved to deliver higher performance than C++ on a number of systems. However,

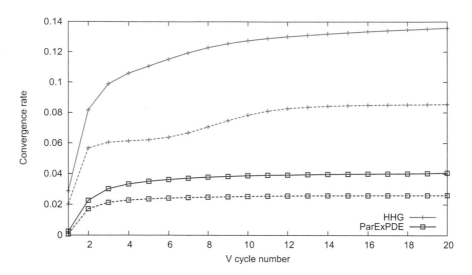

FIGURE 18.4: Convergence rates for HHG and ParExPDE when solving the Poisson problem with Dirichlet boundaries on the unit cube discretized with 129^3 points. Rates are shown for 20 consecutive cycles of the $V(2,2)$ (solid lines, —) and $V(3,3)$ (dashed lines, - - -) type.

ParExPDE relies fully on C++ and the expression template mechanism, even for the numerically intensive routines.

The serial performance measurements were carried out on a single CPU of the HLRB II. HHG's performance increases from about 800 MFLOPS for a tetrahedron with 15 unknowns on every edge to about 3500 MFLOPS for a tetrahedron with 511 unknowns on every edge. For ParExPDE we obtain about 1200 MFLOPS for a 32^3 hexahedron, and the performance increases to about 1600 MFLOPS for a 256^3 hexahedron. The lower performance compared to HHG is due to the fact that the compiler optimizations which are applied to the expression templates generate less efficient code than the one that is generated out of HHG's specialized Fortran 77 routines. Unfortunately, this is the price that must currently be paid for the flexibility and expressive power of a more modern language.

This behavior of modern high-end CPUs is quite similar to what one would expect from classical vector computers: clearly, larger structured regions are beneficial for performance. This may be unexpected, but has been observed uniformly (though to a varying degree) for virtually all current architectures. It has its fundamental reason in the on-chip (or instruction-level) parallelism that is essential for all state-of-the-art CPUs [10]. To further optimize the performance, it would be possible to use cache-blocking techniques, such as in [15, 4].

Parallel Scalability

The goals of our parallel tests are to show that extremely large problems can be solved with good efficiency using both the HHG and the ParExPDE frameworks. In reality, it is likely that we have solved the largest finite-element system to date with performance rarely achieved by most scientific codes.

In this section, we present the results for solving a Poisson problem with Dirichlet boundary conditions. For our tests we use domains consisting of several connected cubes. The choice of this setup is convenient for performing our scalability tests, because we can easily adjust the domain size to the number of CPUs used. In HHG, we additionally partition the cubes into tetrahedrons. However, the grids used for our tests are purely unstructured and no structural feature of the input grids is exploited in obtaining these results.

Figure 18.5 shows the weak-scaling results for our programs. The plotted *weak efficiency* is defined as

$$E_w(n,p) = \frac{T(p_0 \cdot n, p_0)}{T(p \cdot n, p)}$$

where $T(p_0 \cdot n, p_0)$ is the time elapsed during one V cycle on p_0 CPUs for a problem of size $p_0 \cdot n$, and $T(p \cdot n, p)$ is the time elapsed during a V cycle on p CPUs for a problem of size $p \cdot n$. To measure E_w, we adjust the domain to have as many cubes as CPUs used in each test run. p_0 is set to 4. The results show that HHG is faster than ParExPDE, but ParExPDE can sustain good parallel efficiency up to larger numbers of CPUs.

Figure 18.6 shows the strong-scaling results for our programs. *Strong efficiency* is defined as

$$E_s(n,p) = \frac{p_0 \cdot T(n, p_0)}{p \cdot T(n, p)}$$

In contrast to E_w, the problem size n is held constant in E_s. To measure E_s in ParExPDE, a domain of 510 connected hexahedrons, each discretized with 65^3 points, is chosen. In the HHG measurements the domain consists of 1008 hexahedrons, each split into 6 tetrahedrons and discretized with 128^3 points.

18.4   Conclusions

The weak-scaling results of Figure 18.5 demonstrate that both the ParExPDE and HHG solvers scale quite well up to approximately 500 Altix processors, but that the scalability deteriorates slightly beyond 1000 processors. The analogous behavior is seen also in the strong-scaling measurements in Figure 18.6.

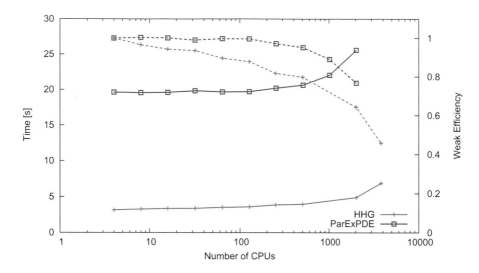

FIGURE 18.5: Weak scaling for HHG and ParExPDE. Solid lines (—) show the time for one *V*-cycle, dashed lines (- - -) show the (weak) parallel efficiency, defined as the quotient of the time for one CPU and the time for p CPUs with a p times larger problem.

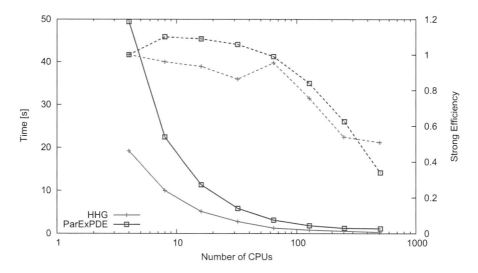

FIGURE 18.6: Strong scaling for HHG and ParExPDE. Solid lines (—) show the time for one *V*-cycle, dashed lines (- - -) show the (strong) parallel efficiency, defined as the quotient of the time for one CPU and p times the time for p CPUs with the same problem size.

Generally, the efficiency and scalability measurements in this chapter should be seen in the context of our solver. Both packages, HHG and ParExPDE, use a highly efficient multigrid solver. This algorithm class has a very low operation count of $\mathcal{O}(n)$ in terms of the number of unknowns and a complex hierarchical structure. Therefore, the ratio of computational work to communication is low and this makes it more difficult to implement it efficiently on large parallel systems. This is made even more challenging by our carefully optimized programs that achieve unusually high performance on each sequential processor.

We wish to point out explicitly that this is absolutely no argument against multigrid or our implementation technology. It is true that using less carefully tuned codes would likely make the speedups look more impressive, but it would also increase the wall-clock run-times. It is a triviality that any speedup result can be faked to look better by doing redundant work on each processor or (equivalently) slowing down the execution on each processor, but this is not of any help in practice.

Similarly, using algorithms other than multigrid ones would likely lead to effectively slower parallel solvers. But since they would use more operations, they would make it easier to amortize the communication cost and consequently they might have better values in relative metrics, such as speedup or scale-up. Nevertheless, they are unlikely to show any benefit in terms of time-to-solution, the metric that is ultimately the only relevant one.

Therefore it also is important to consider the absolute performance results achieved with our codes. In the weak-scaling experiments, the HHG solver is handling 33×10^6 unknowns per processor, and consequently it performs one multigrid V-cycle for 128×10^9 unknowns on 3,825 processors in 6.90 seconds, each cycle resulting in a convergence factor of 0.14 or better (see Figure 18.4). Even larger runs on almost 10,000 processors are available in a technical report [6].

For ParExPDE, the corresponding numbers are 33×10^9 unknowns on 2040 processors in 25.57 seconds per multigrid V-cycle. The somewhat slower performance of ParExPDE is partly compensated for by its superior convergence behavior.

In our opinion, the performance in absolute values shows the advantages of our approach quite impressively. Nevertheless, it is also important to consider the relative measures and the observed drop in efficiency when going to 1024 processors and beyond. At this time it is unclear what causes the deterioration. It could be some architectural bottleneck of the Altix system or a nonoptimal implementation of the communication on our part. Since both ParExPDE and HHG show similar problems, it is unlikely that this is a problem of the implementation alone, but it is clear that while the deterioration in efficiency is still acceptable in the processor ranges up to 4000, it may become highly problematic when we think of future true petascale systems with possibly hundreds of thousands of processor cores. Therefore, we are currently exploring this problem in detail and hope to present a clearer analysis and solutions in the near future. Another alternative is a hybrid parallelization using MPI and OpenMP; also see a previous chapter in this book by Bischof *et al.* (Chapter 15 [3]) using shared memory parallelism.

References

[1] B. Bergen, T. Gradl, F. Hülsemann, and U. Rüde. A massively parallel multigrid method for finite elements. *Computing in Science and Engineering*, 8(6):56–62, November 2006.

[2] B. Bergen and F. Hülsemann. Hierarchical hybrid grids: Data structures and core algorithms for multigrid. *Numerical Linear Algebra with Applications*, 11:279–291, 2004.

[3] C. Bischof, D. an Mey, C. Terboven, and S. Sarholz. Petaflops basics - Performance from SMP building blocks. In D.A. Bader, editor, *Petascale Computing: Algorithms and Applications*. Chapman & Hall/CRC Press, 2007.

[4] C.C. Douglas, J. Hu, M. Kowarschik, U. Rüde, and C. Weiß. Cache optimization for structured and unstructured grid multigrid. *Electronic Transactions on Numerical Analysis (ETNA)*, 10:21–40, February 2000.

[5] C. Freundl, B. Bergen, F. Hülsemann, and U. Rüde. ParExPDE: Expression templates and advanced PDE software design on the Hitachi SR8000. In A. Bode and F. Durst, editors, *High Performance Computing in Science and Engineering, Garching 2004*, pages 167–179. Springer Verlag, 2005.

[6] T. Gradl, C. Freundl, and U. Rüde. Scalability on all levels for ultra-large scale finite element calculations. Technical Report 07-5, Lehrstuhl für Informatik 10 (Systemsimulation), Friedrich-Alexander-Universität Erlangen-Nürnberg, 2007.

[7] S.W. Haney. Beating the abstraction penalty in C++ using expression templates. *Computers in Physics*, 10(6):552–557, 1996.

[8] F. Hülsemann, M. Kowarschik, M. Mohr, and U. Rüde. Parallel geometric multigrid. In A.M. Bruaset and A. Tveito, editors, *Numerical Solution of Partial Differential Equations on Parallel Computers*, volume 51 of *Lecture Notes for Computational Science and Engineering*, chapter 5, pages 165–208. Springer, 2005.

[9] F. Hülsemann, S. Meinlschmidt, B. Bergen, G. Greiner, and U. Rüde. Gridlib — A parallel, object-oriented framework for hierarchical-hybrid

grid structures in technical simulation and scientific visualization. In A. Bode and F. Durst, editors, *High Performance Computing in Science and Engineering, Garching 2004*, pages 117–128. Springer Verlag, 2005.

[10] S. Kamil, B. Kramer, L. Oliker, J. Shalf, H. Shan, E. Strohmaier, and K. Yelick. Science driven supercomputing architectures: Analyzing architectural bottlenecks with applications and benchmark probes. Technical Report LBNL-58914, Lawrence Berkeley National Library, 2005.

[11] G. Karypis and V. Kumar. Multilevel k-way partitioning scheme for irregular graphs. *Journal of Parallel and Distributed Computing*, 48(1):96–129, 1998.

[12] A. Logg, K.-A. Mardal, M.S. Alnæs, H.P. Langtangen, and O. Skavhaug. A hybrid approach to efficient finite element code development. In D. A. Bader, editor, *Petascale Computing: Algorithms and Applications*. Chapman & Hall/CRC Press, 2007.

[13] C. Pflaum. Expression templates for partial differential equations. *Computing and Visualization in Science*, 4(1):1–8, November 2001.

[14] C. Pflaum and R.D. Falgout. Automatic parallelization with expression templates. Technical Report UCRL-JC-146179, Lawrence Livermore National Laboratory, 2001.

[15] M. Stürmer, J. Treibig, and U. Rüde. Optimizing a 3D multigrid algorithm for the IA-64 architecture. In M. Becker and H. Szczerbicka, editors, *Simulationstechnique - 19th Symposium in Hannover, September 2006*, volume 16 of *Frontiers in Simulation*, pages 271–276. ASIM, SCS Publishing House, September 2006.

[16] U. Trottenberg, C. Oosterlee, and A. Schüller. *Multigrid*. Academic Press, 2001.

[17] T. L. Veldhuizen. Expression templates. *C++ Report*, 7(5):26–31, June 1995. Reprinted in C++ Gems, ed. Stanley Lippman.

[18] T. L. Veldhuizen and M. E. Jernigan. Will C++ be faster than Fortran? In *Proceedings of the 1st International Scientific Computing in Object-Oriented Parallel Environments (ISCOPE'97)*, Berlin, 1997. Springer-Verlag.

[19] R. Wienands and W. Joppich. *Practical Fourier Analysis for Multigrid Methods*. CRC Press, 2004.

Chapter 19

A Hybrid Approach to Efficient Finite-Element Code Development

Anders Logg

Simula Research Laboratory and Department of Informatics, University of Oslo

Kent-Andre Mardal

Simula Research Laboratory and Department of Informatics, University of Oslo

Martin Sandve Alnæs

Simula Research Laboratory and Department of Informatics, University of Oslo

Hans Petter Langtangen

Simula Research Laboratory and Department of Informatics, University of Oslo

Ola Skavhaug

Simula Research Laboratory and Department of Informatics, University of Oslo

19.1 Introduction

Scientific applications that demand petascale computing power often involve very complex mathematical models. For continuum models based on partial differential equations (PDEs), the model complexity also tends to significantly amplify the associated software complexity. This is particularly the case when the PDEs are solved by finite-element methods involving unstructured grids on parallel computers. The present chapter aims to present a new way of developing human- and machine-efficient finite-element software for challenging scientific applications.

Many approaches have been developed to deal with software complexity in PDE codes. Among the most successful ones are general libraries based on object-oriented or generic (template) programming [50, 5, 18, 11, 6, 30, 44, 10]. However, petascale computing makes extreme demands on the computational efficiency, and meeting these demands with general libraries is very challenging. Code developers seeking ultimate performance on the most recent supercomputers have therefore traditionally preferred to write their own domain-specific codes, where particular features of the physical and numerical problem at hand can be taken advantage of in order to tune the efficiency of arithmetic operations, memory access, and parallel speedup.

The contradictory strategies of writing a new code more or less from scratch versus building it on top of general libraries is a fundamental issue to resolve when targeting new supercomputer architectures. The ideal solution is a hybrid procedure where general libraries are reused whenever appropriate, but where large portions of the code can be highly tuned to the application at hand. Such a hybrid approach is the topic of the present chapter. Our suggested software development method makes it human-efficient to write a new PDE application, yet with computational efficiency that may outperform handwritten, domain-specific codes.

Our key ideas are threefold: (i) the application code is flexibly composed in a scripting language; (ii) computationally intensive, problem-specific parts of the code are automatically generated; (iii) highly tuned general libraries are used whenever appropriate. Point (i) makes it convenient to write new scientific applications. A particularly important feature of point (ii) is that the compiler which generates the code is restricted to a particular set of PDE problems and can therefore analyze the user's problem and generate highly optimized, specialized code in Fortran, C, or C++. A human would struggle to write similar complicated code free of bugs and with comparable efficiency. Finally, point (iii) contributes to further reliability and efficiency by reusing mature libraries developed and tuned by lots of skilled people over many years. Linear solver packages, such as PETSc [4], Trilinos [52], and Hypre [28], constitute examples on this type of code.

Using scripting languages, and Python [46] in particular, for serial and parallel scientific applications is an emerging and promising approach [29, 35, 15, 14]. To gain the required efficiency it is paramount to migrate computationally intensive loops to compiled code. The idea advocated in this paper is to automatically generate the source for certain parts of the compiled code. This is not a new idea, e.g., ELLPACK [48] offers the user a high-level PDE language from which efficient Fortran 77 code is automatically generated. Fastflo [21] and FreeFEM [22] are two other examples that adopt similar ideas and use their own PDE languages as the front end for users. However, employing a full-fledged scripting language as the user interface, provides great flexibility in the composition of the basic steps of the overall solution algorithm. Even more important is the fact that Python, with its rich collection of libraries, makes it easy and convenient to deal with all the administrative and non-numerical tasks that fill up large portions of scientific codes. In the following sections we explain how to automatically generate a PDE-specific layer of optimized C++ code, how this layer is combined with ready-made libraries for unstructured grids, linear solvers and preconditioners, and how we use Python "on top" to compose PDE application codes.

Section 19.2 presents some examples on how an end user of our software framework can write high-level PDE codes. A critical issue for the efficiency of the high-level codes is the generation of matrices and vectors in the discrete PDE problem. This topic is treated in Section 19.3, where we advocate problem-dependent, automatic code generation for performing intensive numerics. Section 19.4 describes the format of the generated code and its relation to external libraries. How the suggested software methodology fits well in a parallel computing context is outlined in the final section, along with concluding remarks.

19.2   High-Level Application Codes

Our goal is to have a user-friendly software framework where we can write algorithms for solving PDEs, while maintaining the speed of compiled languages. In a sense, we try to create an environment in which one can express solution algorithms for PDE problems just as easily as one can write algorithms for linear algebra·problems in Matlab.

Algorithm 19.1: Newton's method.

Given $\mathbf{F}$, $\mathbf{u}$, α, and tolerance
$\epsilon = \|\mathbf{F}(\mathbf{u})\|$

while $\epsilon >$ tolerance
$\quad \mathbf{e} = \mathbf{J}^{-1}\mathbf{F}(\mathbf{u})$
$\quad \mathbf{u} = \mathbf{u} - \alpha\mathbf{e}$
$\quad \epsilon = \|\mathbf{F}(\mathbf{u})\|$

```
def Newton(F, u, Jinv, tolerance=1.0e-6, alpha=1.0):
    iterations = 0
    Fu = F(u)
    epsilon = sqrt(inner(Fu,Fu))
    while epsilon > tolerance:
        e = Jinv(u)*Fu
        u - = alpha*e
        Fu = F(u)
        epsilon = sqrt(inner(Fu,Fu))
        iterations + = 1
    return u, epsilon, iterations
```

FIGURE 19.1: Python implementation of Newton's method (Algorithm 19.1).

As an introductory example, let us show how well Python is suited to implement a basic numerical algorithm in a readable, compact and yet efficient way. To this end, we consider Newton's method for solving a system of nonlinear equations $\mathbf{F}(\mathbf{u}) = 0$ (Algorithm 19.1) and the corresponding Python code (Figure 19.1). This implementation of Newton's method is more general than corresponding Matlab code because it will work for any objects having * and () operators and common array data structures. For example, when the nonlinear system is small, `Jinv(u)` can be a plain matrix representing the inverse of the Jacobian ($\mathbf{J}^{-1}$), while for large, sparse systems arising from discretizing PDEs, `e = Jinv(u)*F(u)` should imply that a linear system with the Jacobian $\mathbf{J}$ as coefficient matrix and $\mathbf{F}$ as the right-hand side is solved.*

If a user tries to use the above function with objects `F` and `Jinv` that have not implemented these operators, an informative exception is thrown. Hence, "templatization" as known from C++, comes naturally and for free in Python. The reason why we adopt Python instead of C++ for writing the application code is simply that the Python code is cleaner and easier to understand, and because we think Python is more convenient than C++ for all the nonnumerical tasks in the application code.

The example above is an illustration of a general Matlab-like code in Python that utilizes different objects in different occasions to maintain computational efficiency. In the rest of this section we will discuss how solvers for PDE problems can be implemented in our PyCC framework [45]. PyCC, which stands

---

*This solution approach is easily implemented in Python by letting `F` and `Jinv` be classes, with call operators (`__call__`, corresponding to `operator()` in C++) such that `Jinv(u)` assembles $\mathbf{J}$, while `F(u)` assembles $\mathbf{F}$. The multiplication operator (`__prod__` in Python, `operator*` in C++) in the class for `Jinv` then solves the linear system $\mathbf{Jx} = \mathbf{F}$ by some appropriate method attached to the `Jinv` object. The low-level implementation of the linear system solution process should take place in compiled Fortran, C or C++ code.

for Python Computing Components, aims at developing a user-friendly environment for solving PDEs in Python. Until now, our efforts have mostly been concentrated on implementing schemes for computing the electrical activity of the heart as described in [39], but we are now working with solvers for viscous flow and elasticity problems. In the following, we will present code snippets for a few viscous flow solvers. Incompressible flow is governed by the following Navier-Stokes equations:

$$\frac{\partial \mathbf{v}}{\partial t} + \mathbf{v} \cdot \nabla \mathbf{v} = -\nabla p + \mathrm{Re}^{-1} \Delta \mathbf{v} + \mathbf{b},$$

$$\nabla \cdot \mathbf{v} = 0$$

where $\mathbf{v}$ is the fluid velocity field, p is the pressure field, Re is the Reynolds number, and $\mathbf{b}$ denotes body forces. The first scheme is an explicit projection scheme

$$\mathbf{v}^* = \mathbf{v}^n - k\mathbf{v}^n \cdot \nabla \mathbf{v}^n + k\mathrm{Re}^{-1}\Delta \mathbf{v}^n - k\nabla p^n + k\mathbf{b}^n,$$

$$-\Delta \phi = -\frac{1}{k}\nabla \cdot \mathbf{v}^*,$$

$$p^{n+1} = p^n + \phi,$$

$$\mathbf{v}^{n+1} = \mathbf{v}^n - k\nabla \phi$$

The superscript n denotes the time step and k is the (local) time step length. Hereafter, we will present the schemes in their fully discretized "linear algebra" form. The above scheme can be written as

$$\mathbf{Mu}^* = \mathbf{Mu}^n - k\mathbf{C}(\mathbf{u}^n)\mathbf{u}^n - k\mathrm{Re}^{-1}\mathbf{Au}^n - k\mathbf{B}^T\mathbf{p}^n + k\mathbf{Mf}^n \quad (19.1)$$

$$\mathbf{D}\phi = -\frac{1}{k}\mathbf{Bu}^* \quad (19.2)$$

$$\mathbf{p}^{n+1} = \mathbf{p}^n + \phi, \quad (19.3)$$

$$\mathbf{u}^{n+1} = \mathbf{u}^n - k\mathbf{B}^T\phi \quad (19.4)$$

When a finite element method [8, 17, 34, 54] is used for the spatial discretization the matrices are given by

$$\mathbf{M}_{ij} = \int_\Omega \mathbf{N}_i \cdot \mathbf{N}_j \, dx \quad (19.5)$$

$$\mathbf{A}_{ij} = \int_\Omega \nabla \mathbf{N}_i : \nabla \mathbf{N}_j \, dx \quad (19.6)$$

$$\mathbf{C}(\mathbf{u}^{n+1})_{ij} = \int_\Omega (\nabla \cdot \sum_k \mathbf{v}_k^{n+1}\mathbf{N}_k)\mathbf{N}_i \cdot \mathbf{N}_j \, dx \quad (19.7)$$

$$\mathbf{B}_{ij} = \int_\Omega \nabla \cdot \mathbf{N}_i \, L_j \, dx \quad (19.8)$$

$$\mathbf{D}_{ij} = \int_\Omega \nabla L_i \cdot \nabla L_j \, dx \quad (19.9)$$

$$\mathbf{E}_{ij} = \int_\Omega L_i L_j \, dx \quad (19.10)$$

Here $\{\mathbf{N}_i\}$ and $\{L_i\}$ are the basis functions of the velocity and pressure, respectively.

```
gal = Gallery(grid)

M  = gal.assemble_mass_matrix(v_element) A  =
gal.assemble_stiffness_matrix(v_element) C  =
gal.assemble_convection_matrix(u_n, v_element)

B  = gal.assemble_div_matrix(v_element, p_element) BT =
B.transposed()

D  = gal.assemble_stiffness_matrix(p_element) E  =
gal.assemble_mass_matrix(p_element)
```

FIGURE 19.2: The creation of some predefined matrices in PyCC.

In PyCC, the Python code snippet for assembling the various matrices takes the form shown in Figure 19.2. The `Gallery` object supports the creation of many predefined matrices, such as the mass, stiffness, convection and divergence matrices. Additionally, it supports creation of matrices based on user-defined variational forms, which will be described in more detail in the next section.

Having constructed the matrices, the complete implementation of this explicit projection scheme is shown in Figure 19.3. This implementation works for any vector and matrix with addition and multiplication operators, and for any preconditioner with a preconditioner-vector product. In the present example the matrices are in PyCC's own format, while the preconditioners are created as a thin layer on top of the algebraic multigrid package ML in the open source parallel linear algebra package Trilinos [52].

Although we used ML as a preconditioner in the above example, there are several high-quality linear algebra packages, such as PETSc [4] and Hypre [28], that for some problems might be more suited. All these projects are fairly mature, have Python bindings, and are scalable on parallel computers. Since it is not clear which of these linear algebra libraries is the best in a given situation, we construct our software framework such that the high-level PDE codes are independent of the underlying linear algebra package. Interfacing a new linear algebra library is simple; we only have to provide basic operations for matrices, vectors, and preconditioners, and assembler functions (block insertion and addition) for matrices and vectors.

Modifying the above simulator to, for instance, the second-order implicit

```
precM = MLPrec(M);  precD = MLPrec(D)  # calc. preconditioners

# time loop T = 1; k = 0.1; Re1 = 1/100.0 while t < T:
    t = t + k

    # update from previous solution
    p_n, u_n = p, u

    # assemble right hand side
    f = gal.assemble_source_vector(body_force(t=t))
    C = gal.assemble_convection_matrix(u_n, v_element)
    rhs = M*u_n - k*(C*u_n + BT*p_n + Re1*A*u_n - f)

    # compute tentative velocity
    M, rhs = bc_velocity(M, rhs)  # impose boundary conditions
    u_star, iter = conjgrad(M, u_star, rhs, precM, tol=1e-9)

    # compute phi
    rhs = (-1.0/k)*B*u_star
    phi, iter = conjgrad(D, phi, rhs, precD, tol=1e-9)

    # update p and u
    p = p_n + phi
    u = u_n - k*BT*phi
```

FIGURE 19.3: PyCC code for a first-order explicit projection scheme, see Equation (19.1)–(19.4).

projection method found in [9] is easy. A variant of this scheme[†] reads

$$\mathbf{M}\mathbf{u}^* + \frac{k}{2}(\mathrm{Re}^{-1}\mathbf{A}\mathbf{u}^* + \mathbf{C}(\mathbf{u}^n))\mathbf{u}^* = \mathbf{M}\mathbf{u}^n - \frac{k}{2}\mathrm{Re}^{-1}\mathbf{A}\mathbf{u}^n \qquad (19.11)$$

$$-\frac{k}{2}\mathbf{C}(\mathbf{u}^n)\mathbf{u}^n + k\mathbf{M}\mathbf{f}^{n+1/2} \quad (19.12)$$

$$\mathbf{D}\phi = \mathbf{B}\mathbf{u}^* \qquad (19.13)$$

$$\mathbf{u}^{n+1} = \mathbf{u}^* + \mathbf{B}^T\phi \qquad (19.14)$$

$$\mathbf{p}^{n+1} = (\mathbf{M} + k\mathrm{Re}^{-1}\mathbf{A})\phi \qquad (19.15)$$

and is displayed in Figure 19.4 together with the Python code. Note that we

---

[†]The convective term is approximated as $\mathbf{C}(\mathbf{u}^{n-1/2}))\mathbf{u}^{n-1/2} \approx \frac{1}{2}(\mathbf{C}(\mathbf{u}^n)\mathbf{u}^* + \mathbf{C}(\mathbf{u}^n)\mathbf{u}^n)$, which is not a second-order approximation. Alternatively, we could include a nonlinear iteration and handle the $\mathbf{C}(\mathbf{u}^{n-1/2}))\mathbf{u}^{n-1/2}$ properly as we do in the final scheme in this section. In [9] it was computed by a second-order extrapolation.

```
    . . .
    rhs = M*u_n - k/2*A*u_n - k/2*C*u_n + k*M*f

    A1 = M + k/2*(Re1*A + C)
    A1, f = bc_velocity(A1, f)  # impose boundary conditions
    precA1 = MLPrec(A1)
    u_star, iter = bicgstab(A1, u_star, rhs, precA1, tol=1e-9)

    rhs = (1.0/k)*B*u_star
    phi, iter = conjgrad(D, phi, rhs, precD, tol=1e-9)
    u = u_star - k*BT*phi
    p = (M + k*Re1*A)*phi
```

FIGURE 19.4: Changes in the implementation of a second-order implicit projection scheme, see Equations (19.11)–(19.15).

display only the lines in the computer code that differ from the program in Figure 19.3.

The two projection schemes shown so far work for any kind of finite elements, but they often give poor accuracy close to the boundary. Saddle-point schemes circumvent this problem at the expense of using mixed finite elements, which lead to indefinite linear systems. Our next scheme is such a saddle-point scheme where the convection is handled explicitly:

$$\mathbf{M}\mathbf{u}^{n+1} + k\mathrm{Re}^{-1}\mathbf{A}\mathbf{u}^{n+1} + k\mathbf{B}^T\mathbf{p}^{n+1} = \mathbf{M}\mathbf{u}^n - k\mathbf{M}\mathbf{f}^{n+1} - k\mathbf{C}(\mathbf{u}^n)\mathbf{u}^n \quad (19.16)$$

$$k\mathbf{B}\mathbf{u}^{n+1} = \mathbf{0} \quad (19.17)$$

For this scheme there exist order optimal preconditioners of the form [12, 41, 53]:

$$\mathbf{K} \approx (\mathbf{M} + k\mathrm{Re}^{-1}\mathbf{A})^{-1} \quad (19.18)$$

$$\mathbf{L} \approx 1/(k\mathrm{Re}^{-1})\mathbf{E}^{-1} + \mathbf{D}^{-1} \quad (19.19)$$

$$\mathbf{N} = \mathrm{diag}(\mathbf{K}, \mathbf{L}) \quad (19.20)$$

where $\mathbf{K}$ and $\mathbf{L}$ can be constructed by standard elliptic preconditioners such as multigrid or domain decomposition. The scheme is now conveniently written in block form

$$\begin{bmatrix} \mathbf{M} + k\mathrm{Re}^{-1}\mathbf{A} & k\mathbf{B}^T \\ k\mathbf{B} & \mathbf{0} \end{bmatrix} \begin{bmatrix} \mathbf{u}^{n+1} \\ \mathbf{p}^{n+1} \end{bmatrix} = \begin{bmatrix} \mathbf{M}\mathbf{u}^n - k\mathbf{M}\mathbf{f}^{n+1} - k\mathbf{C}(\mathbf{u}^n)\mathbf{u}^n \\ \mathbf{0} \end{bmatrix}$$

$$(19.21)$$

with a preconditioning matrix

$$\begin{bmatrix} (\mathbf{M} + k\mathrm{Re}^{-1}\mathbf{A})^{-1} & \mathbf{0} \\ \mathbf{0} & 1/(k\mathrm{Re}^{-1})\mathbf{E}^{-1} + \mathbf{D}^{-1} \end{bmatrix} \quad (19.22)$$

```
# create block vector and block rhs x = BlockVector([u, p]);  rhs =
BlockVector([0, 0])

# create block matrix and block preconditioner Q = BlockMatrix([[M +
k*Re1*A, k*BT], [k*B, Null]]) Q, rhs = bc_velocity(Q, rhs, t=0)

precQ = BlockDiagMatrix([MLPrec(M + k*Re1*A),
                    1/(k*Re1)*MLPrec(E) + MLPrec(D)])
... # time loop: while t < T:
    ...
    rhs[0] = M*u_n - k*(M*f + C*u_n)
    Q, rhs = bc_velocity(Q, rhs, t=t)

    x, iter = minres(Q, x, rhs, precQ, tol=1e-9)
    ...
```

FIGURE 19.5: Implementation of a saddle-point scheme, see Equations (19.21)–(19.22), with convection handled explicitly.

The resulting preconditioned system is symmetric and indefinite and can be efficiently solved with the minimal residual method. A code snippet for this scheme is shown in Figure 19.5.

The above scheme handles convection explicitly, which is feasible only in very transient problems. A more robust scheme may apply a Newton iteration on the fully nonlinear system:

$$\mathbf{M}\mathbf{u}^{n+1} + k\mathrm{Re}^{-1}\mathbf{A}\mathbf{u}^{n+1} + k\mathbf{C}(\mathbf{u}^{n+1})\mathbf{u}^{n+1} + k\mathbf{B}^T\mathbf{p}^{n+1} = \mathbf{M}\mathbf{u}^n - k\mathbf{M}\mathbf{f}^{n+1} \quad (19.23)$$

$$k\mathbf{B}\mathbf{u}^{n+1} = 0 \quad (19.24)$$

The part of the Jacobian matrix and the right-hand side associated with the nonlinear convection term are given by[‡]:

$$\mathbf{J}_{ij}^c = \frac{\partial \mathbf{F}_i}{\partial u_i} \quad (19.25)$$

$$\mathbf{F}_i^c = \int_\Omega (\mathbf{v} \cdot \nabla)\mathbf{v} \cdot \mathbf{N}_i \, dx \quad (19.26)$$

The corresponding code is shown in Figure 19.6. This scheme is very stable in terms of the discretization parameters, but the resulting linear systems are hard to solve. The matrix is neither positive nor symmetric. The algebraic

---

[‡]The rest of the Jacobian matrix and the right-hand side consist of matrices and vectors described before, i.e., mass, stiffness, divergence matrices, etc.

```
# time loop T = 1; k = 0.1; t = 0 while t < T:
    t = t + k

    f = gal.assemble_source_vector(body_force(t=t))
    eps = sqrt(inner(F,F))
    while eps > tolerance and iter <= maxiter:

        # assemble right hand side corresponding to convection term
        FC = gal.assemble_vector(rhs_form, u)
        # assemble Jacobian part corresponding to convection term
        JC = gal.assemble_matrix(Jacobian_form, u)

        F = BlockVector([M*u + k*(FC + Re1*Au - BT*p + f), 0])

        J = BlockMatrix([[M + k*(JC + Re1*A), k*BT], [k*B, 0]])
        J, F = bc_velocity(J, F)    # impose boundary conditions

        K = MLPrec(J)
        precJ = BlockDiagMatrix(K, L)

        e, iter = bicgstab(J, e, F, precJ, tol=1e-9)

        x -= alpha*e
        eps = sqrt(inner(F,F))
```

FIGURE 19.6: Implementation of a fully implicit scheme, see Equations (19.23)–(19.24).

multigrid preconditioners in ML and BoomerAMG work well as preconditioners for $\mathbf{K}$ as long as the convection is not too dominating, in which case
renumbering strategies such as [7, 24] should be used. However, the second
component $\mathbf{L}$ is hard to construct efficiently. It is well known that it should
be a cheap approximation of

$$(\mathbf{B}\mathbf{J}^{-1}\mathbf{B}^T)^{-1} \tag{19.27}$$

Since such a preconditioner is currently not available, we use the preconditioner given in (19.19).

It is difficult to know in advance which of the above schemes will be most
efficient in a given physical problem, and it should therefore be easy to switch
between schemes. This is particularly the case in complex flow problems of
the size that may demand petascale computing power. A possibly attractive
approach is to combine the speed of the projection scheme with the accuracy
of the mixed approach by using projection to compute either a start vector
for the mixed approach, or to use it as a preconditioner, as outlined in [36].
Such trial-and-error-based construction of schemes is feasible only in a flexible
software framework, as the one shown. The computational efficiency of the
framework relies heavily on the speed of creating finite-element matrices and
vectors, and these quantities are tightly connected to the specific PDE problem
being solved. The next section deals with how to construct such matrices and
vectors efficiently.

19.3 Code Generation

In the previous section we demonstrated how to write high-level PDE codes
by mapping the time loop with linear algebra problems to Python code. A
fundamental step in these code examples is to generate the matrices and vectors involved. The current section presents ideas and generic tools for automatic, efficient generation of matrices and vectors arising from finite element
discretizations in space.

General systems for the numerical solution of PDEs are often met with
skepticism, since it is believed that the flexibility of such tools cannot be
combined with the efficiency of competing specialized computer programs that
only need to know how to solve a single equation. However, we will argue that
it is possible to design general systems that accept a wide range of inputs,
while competing with or even outperforming specialized and hand-optimized
codes.

19.3.1   Meta-programming

The key to combining generality with efficiency in scientific software is code generation, or *meta-programming*. Meta-programming refers to a collection of programming techniques for writing computer programs that generate other programs or parts of programs. Examples of meta-programming include using the C preprocessor `cpp` to expand macros into code, C++ template meta-programming, and using high-level language compilers. A more direct approach that we will discuss below is explicit generation of code in some general-purpose language like C or C++ from a set of simple instructions.

Code generation can be accomplished by inventing a new high-level domain-specific language and implementing a compiler for that language that translates a high-level description in the domain-specific language to code in a general-purpose language, or directly to machine code. Alternatively, one may embed the domain-specific language in an expressive general-purpose language. This has the advantage that one may focus on the domain-specific subsystem and build on top of a mature, well-known language. Another advantage is that end users can develop application codes in a familiar language. Choosing Python as this language exposes the end user to a clean, MATLAB-like syntax, which has proven to be attractive among computational scientists.

By meta-programming, it is possible to combine generality with efficiency without loss of either. Instead of developing a potentially inefficient general-purpose program that accepts a wide range of inputs, a suitable subset of the input is processed at compile-time[§] (at little extra cost) to generate a specialized program that takes a more limited set of inputs, as illustrated in Figure 19.7.

In the context of an automating system for the solution of PDEs, the subset of input processed at compile-time (input 1) may include the variational problem and finite element spaces, while the subset of input processed at run-time (input 2) may include the mesh and physical input parameters. One can then, for a given variational formulation of a PDE problem, generate a program that takes a mesh as input and computes the corresponding matrix $\mathbf{A}$ or vector $\mathbf{b}$ as output. The overall system may then be viewed as a system that takes a variational problem and a mesh as input and computes the matrix or vector. Thus, if the compiler generating the specialized code from the meta program accepts a wide range of inputs (variational problems), the overall system may be both general and efficient. The efficiency comes from the compiler's knowledge about the problem domain and the possibility of performing case-specific analysis of the variational problem at hand to introduce smart optimizations

---

[§]Note that the distinction between *compile-time* and *runtime* is blurred if the compiler that turns the meta-program into code is implemented as a *just-in-time* compiler, that may be called at runtime to generate code on the fly.

in the generated C, C++, or Fortran code.

Below, we present two particular instances of such a domain-specific compiler: FFC and SyFi. For a given finite-element variational problem, these compilers generate C++ code for a particular set of functions that get called during the assembly of the matrix **A** or vector **b**. In particular, code is generated for the computation of the element matrix (or element vector) on each element of a finite element mesh, from which the global matrix or vector may be assembled. Thus, FFC or SyFi take the role of the *generating machine* of Figure 19.7; the input (variational problem) is a subset of the total input (variational problem and mesh) and the output is computer code that may be called to produce the global matrix or vector from the remaining input (the mesh).

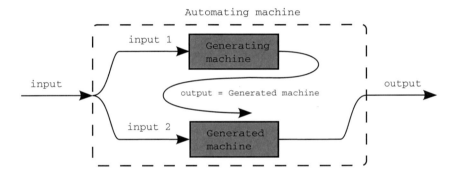

FIGURE 19.7: A machine (meta-program) accepting given input data and producing output by generating a specialized machine (computer program) from a subset (input 1) of input data. The generated machine accepts the remaining subset (input 2) of input data to produce output. The overall system may be viewed as a single system that accepts the full set of input data.

To limit the complexity of the automating system, it is important to identify a minimal set of code to be generated at compile-time. It makes little sense to generate the code for administrative tasks such as reading and writing data to file, special algorithms like adaptive mesh refinement, etc. Handling data files is conveniently and efficiently done in Python while a subject like adaptive mesh refinement calls for specific, very efficient compiled code, preferably in C++ where sophisticated data structures are well supported. In both cases, generic software components can be written and collected in reusable libraries.

Meta-programming as described above can be a powerful tool for combining generality and efficiency. However, constructing such an automating system is challenging, as the whole meta-programming paradigm is complex. Especially, debugging the generating machine is hard, since errors might only be visible in the final output of the automating system. Also, since the generating machine is a compiler, the compile-time and the memory usage can sometimes be difficult to control.

19.3.2   Just-in-time compilation of variational problems

In Section 19.2, we demonstrated the implementation of algorithms for the solution of PDEs in a high-level scripting language. We then assumed the existence of a component for the creation of matrices and vectors arising from finite-element discretizations of variational problems. Such "matrix factories" may provide functionality for computing a limited set of predefined matrices. By expressing the variational problem as a meta-program and implementing a just-in-time compiler for variational problems, one may instead generate the matrix factory for any given variational problem automatically at runtime. Below, we discuss two approaches to creating such compilers.

Assume that a physical problem is recast into a standard variational form: Find $U \in V_h$ such that

$$a(v, U) = L(v) \quad \forall v \in \hat{V}_h \tag{19.28}$$

where $a : \hat{V}_h \times V_h \to \mathbb{R}$ is a bilinear form, $L : \hat{V}_h \to \mathbb{R}$ is a linear form and $\hat{V}_h$ and V_h are discrete finite-element spaces, typically piecewise polynomials over a triangulation of a domain. From this variational problem, the finite-element method generates a linear system with coefficient matrix $\mathbf{A}$ and right-hand side vector $\mathbf{b}$.

Computing the matrix $\mathbf{A}$ and vector $\mathbf{b}$ is normally expensive. Furthermore, implementing a completely general-purpose routine for this computation is difficult since all variational problems look different. Thus, the computation of the linear system is typically implemented by a different routine for each variational problem. When using a traditional Fortran, C or C++ library for finite-element computation, the end user must code by hand the expressions that generate the matrix and vector corresponding to the bilinear and linear forms. The code that the user must provide may be heavily decorated by Fortran, C, or C++ details such that the correspondence between the code and the mathematical formulation becomes weak. This coding process can be time-consuming and prone to errors.

19.3.3   FFC

In [32, 31, 33], it is described how efficient code for the computation of the matrix and vector of the linear system may be generated automatically for a large class of variational problems. The algorithms described in these references have been implemented in the open-source form compiler FFC [37], which is a domain-specific compiler for variational problems. FFC has recently been embedded as a just-in-time compiler inside the DOLFIN [26]

```
element = FiniteElement("Vector Lagrange", "tetrahedron", 1)

v = TestFunction(element) u = TrialFunction(element)

def epsilon(w):
    return 0.5*(grad(w) + transp(grad(w)))

a = dot(epsilon(v), epsilon(u))*dx
```

FIGURE 19.8: The specification of the bilinear form, see Equation (19.32), in the FFC-form language.

Python/C++ problem-solving environment. This allows the specification of variational problems in Python from which FFC generates efficient code for the computation of the components of the associated linear system.

As a demonstration, consider the generation of code for the following set of bilinear forms:

$$a(v, u) = \int_\Omega v\, u \, dx \tag{19.29}$$

$$a(v, u) = \int_\Omega \nabla v \cdot \nabla u \, dx \tag{19.30}$$

$$a(v, u) = \int_\Omega \mathbf{v} \cdot (\mathbf{w} \cdot \nabla)\mathbf{u} \, dx \tag{19.31}$$

$$a(v, u) = \int_\Omega \varepsilon(\mathbf{v}) : \varepsilon(\mathbf{u}) \, dx \tag{19.32}$$

The bilinear form (19.29) generates a mass matrix, form (19.30) is the bilinear form for the Poisson equation, the bilinear form (19.31) results from a linearization of the nonlinear term in the incompressible Navier-Stokes equations, and the bilinear form in Equation (19.32) is the strain–strain term of linear elasticity where $\varepsilon(\mathbf{v}) = \frac{1}{2}(\nabla \mathbf{v} + (\nabla \mathbf{v})^T)$.

In Figure 19.8, we present the Python code specifying the bilinear form in Equation (19.32) in the FFC-form language (embedded in Python). We note that by suitable operator overloading in Python, the bilinear form may be expressed in a form that is close to the mathematical notation (19.32). We also note that by working on top of a general-purpose language like Python, the full power of the general-purpose language is in the hands of the user. In this case, the user can easily specify the operator ϵ by a standard Python function `epsilon`. An excerpt of the generated code (in total, 561 lines of C++ code are generated) is given in Figure 19.9.

Domain-specific knowledge allows the optimizing compiler FFC to generate very efficient code for test cases (19.29)–(19.32). We demonstrate this in Table 19.1, where we present the speedup of the generated code compared to

```
void tabulate_tensor(double* A, ...) {
  ...
  const double G0_0_0 = det*Jinv00*Jinv00 + det*Jinv00*Jinv00
                      + det*Jinv00*Jinv00 + det*Jinv00*Jinv00
                      + det*Jinv01*Jinv01 + det*Jinv02*Jinv02
                      + det*Jinv01*Jinv01 + det*Jinv02*Jinv02;
  const double G0_0_1 = det*Jinv00*Jinv10 + det*Jinv00*Jinv10
                      + det*Jinv00*Jinv10 + det*Jinv00*Jinv10
                      + det*Jinv01*Jinv11 + det*Jinv02*Jinv12
                      + det*Jinv01*Jinv11 + det*Jinv02*Jinv12;
  ...
  const double G8_2_1 = det*Jinv21*Jinv12 + det*Jinv21*Jinv12;
  const double G8_2_2 = det*Jinv21*Jinv22 + det*Jinv21*Jinv22;

  const real tmp0_13 = 4.166666666666660e-02*G0_0_0;
  const real tmp0_38 = 4.166666666666662e-02*G0_2_1;
  const real tmp0_1 = -tmp0_13 + -4.166666666666661e-02*G0_1_0
   - 4.166666666666660e-02*G0_2_0;
  const real tmp0_37 = 4.166666666666661e-02*G0_2_0;
  const real tmp0_25 = 4.166666666666661e-02*G0_1_0;
  const real tmp0_26 = 4.166666666666662e-02*G0_1_1;


  ...
  const real tmp8_139 = 4.166666666666662e-02*G8_2_2;
  const real tmp8_125 = 4.166666666666661e-02*G8_1_0;
  const real tmp8_100 = -tmp8_101 + 4.166666666666661e-02*G8_0_1
    + 4.166666666666661e-02*G8_0_2 + 4.166666666666662e-02*G8_1_1
    + 4.166666666666662e-02*G8_1_2 + 4.166666666666662e-02*G8_2_1
    + 4.166666666666662e-02*G8_2_2;
  const real tmp8_126 = 4.166666666666662e-02*G8_1_1;
  const real tmp8_113 = 4.166666666666660e-02*G8_0_0;
  const real tmp8_138 = 4.166666666666662e-02*G8_2_1;
  A[0] = tmp0_0;
  A[1] = tmp0_1;
  A[2] = tmp0_2;
  ...
  A[141] = tmp6_141;
  A[142] = tmp6_142;
  A[143] = tmp6_143;
}
```

FIGURE 19.9: Excerpt of the code generated by FFC for the computation of the 144 entries of the 12×12 "element stiffness matrix" corresponding to the input code of Figure 19.8.

TABLE 19.1: Speedups in two and three space dimensions as a function of the polynomial degree q for test cases (19.29)–(19.32): the mass matrix (M), Poisson's equation (P), the incompressible Navier–Stokes equations (NS) and linear elasticity (E).

Form	$q = 1$	$q = 2$	$q = 3$	$q = 4$	$q = 5$	$q = 6$	$q = 7$	$q = 8$
M2D	12	31	50	78	108	147	183	232
M3D	21	81	189	355	616	881	1442	1475
P2D	8	29	56	86	129	144	189	236
P3D	9	56	143	259	427	341	285	356
NS2D	32	33	53	37	—	—	—	—
NS3D	77	100	61	42	—	—	—	—
E2D	10	43	67	97	—	—	—	—
E3D	14	87	103	134	—	—	—	—

a standard quadrature-based implementation, i.e., an implementation where all integrals present in the definition of the variational problem are evaluated by quadrature over each local cell in the mesh at runtime. We also plot the number of lines of C++ code generated for each of the test cases as a function of the polynomial degree of the finite-element function spaces in Figure 19.10. We see from this figure that for basis functions of high degree, the routine for computing the element matrix and vector may have over 10,000 lines of optimized code. It is obvious that manual coding of such a routine would be a tedious and error-prone process.

FFC is able to generate very efficient code by computing a special tensor representation, see [33], of the variational problem at compile-time, factoring out all integrals from the definition of the variational problem. This makes it possible to precompute the integrals once on a reference element and then at compile-time map the precomputed quantities to each cell in the mesh during the computation of the matrix **A** or vector **b**.

As seen in Table 19.1, the speedup ranges between one and three orders of magnitude, with larger speedups for higher-degree elements. It should be noted that the speedup here refers to the speedup of an isolated portion (computing the "element stiffness matrix") of the overall solution process of a PDE. The overall global speedup depends also on the problem being solved, the efficiency of the interaction with the mesh, the data structure for the global sparse matrix, the linear solvers and preconditioners, etc.

19.3.4 SyFi

An approach similar to FFC, but which is based on symbolic computations prior to the code generation, is utilized in SyFi [2, 38, 39]. SyFi is an open source C++/Python library that supports a variety of different finite

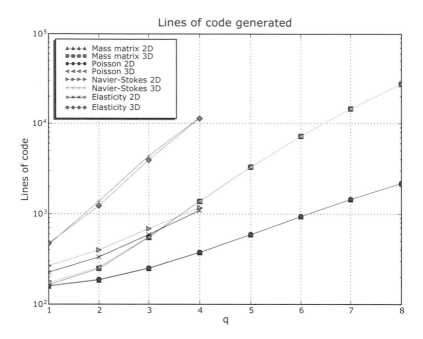

FIGURE 19.10: The number of lines of C++ code generated by FCC for test cases (19.29)–(19.32) as a function of the polynomial degree q.

elements[¶]. The basis functions of these elements are represented as symbolic polynomials which can be differentiated and integrated over polygonal domains. Hence, the computation of element matrices and vectors based on variational forms and finite elements is readily available analytically. SyFi relies on the computational engine GiNaC [23], which is a powerful and efficient C++ package for symbolic computation, and its Python interface Swiginac [51].

Earlier in Section 19.2 we showed an example of a fully implicit scheme for incompressible flow which gave rise to a system of nonlinear equations. We now describe how the linearized systems for the nonlinear convection part are computed symbolically. The formulas for the Jacobian matrix and the right-hand side are given in (19.25) and (19.26). The corresponding Python code for defining this matrix and vector is shown in Figure 19.11. Notice that this is code for $\mathbf{J}^c$ and $\mathbf{F}^c$ is completely independent of the element type and that

---

[¶]Supported elements include the Arnold-Falk-Winther element [3], the Crouzeix-Raviart element [19], the standard Lagrange elements, the Nédélec elements [42, 43], the Raviart-Thomas element [47], and the robust Darcy-Stokes element [40].

```
def F(Ni, U, G, Ginv):
    gradU = grad(U, Ginv)
    UxgradU = (U.transpose() * gradU)
    Fi_convection = inner(UxgradU.transpose(), Ni)
    return Fi_convection

def J(uj, Ni, U, G, Ginv):
    Fi = F(Ni, U, G, Ginv)
    Jji = diff(Fi, uj)
    return Jji
```

FIGURE 19.11: Jacobian element matrix and right-hand side for a nonlinear convection problem.

this code can be used to generate element matrices for arbitrary elements. Furthermore, the Jacobian matrix is computed directly by differentiating $\mathbf{F}^c$, i.e., there is no need for tedious calculations with pencil and paper.

In Figure 19.12 we display how the definitions of element matrices and element vectors from Figure 19.11 are used to generate the C++ code shown in Figure 19.13, which is then compiled into a shared library/Python extension module. This extension module is then used in the PyCC framework to assemble the global matrices and vectors.

We believe code generation with form compilers constitutes a generic approach for generating efficient code for finite-element methods. We have described two different form compilers, each with its strengths and weaknesses. One of the particular strong sides of FFC is that it produces very optimized code. The advantage of SyFi is the direct computation of the Jacobian in case

```
triangle = ReferenceTriangle() order = 2 fe =
VectorLagrange(triangle, order)

JC_form = NLMatrixForm(J, name="Jacobian", num_coefficients=1)
FC_form = VectorForm(F, name="Rhs", num_coefficients=1)

# generate and compile C++ code for the computation of the # element
matrix and vector, return a compiled form object Jacobian_form =
compile_form(JC_form, fe) rhs_form      = compile_form(FC_form, fe)

# assemble the linear system JC = gal.assemble_matrix(Jacobian_form,
u) FC = gal.assemble_vector(rhs_form, u)
```

FIGURE 19.12: Computing the Jacobian in a nonlinear convection-diffusion problem using SyFi.

```
void tabulate_tensor(double* A,
                     const double * const * w,
                     const ufc::cell& c) const
{
  // coordinates
  double x0=c.coordinates[0][0]; double y0=c.coordinates[0][1];
  double x1=c.coordinates[1][0]; double y1=c.coordinates[1][1];
  double x2=c.coordinates[2][0]; double y2=c.coordinates[2][1];

  // affine map
  double G00 = x1 - x0;
  double G01 = x2 - x0;

  double G10 = y1 - y0;
  double G11 = y2 - y0;

  double detG_tmp = G00*G11-G01*G10;
  double detG = fabs(detG_tmp);

  double Ginv00 =  G11 / detG_tmp;
  double Ginv01 = -G10 / detG_tmp;
  double Ginv10 = -G01 / detG_tmp;
  double Ginv11 =  G00 / detG_tmp;

  memset(A, 0, sizeof(double)*36);

  A[6*0 + 0] = detG*(-w[0][0]*Ginv01/12.0-Ginv11*w[0][3]/48.0
    - w[0][5]*Ginv11/48.0+w[0][4]*Ginv01/48.0
    + w[0][2]*Ginv00/48.0-w[0][3]*Ginv10/48.0
    - w[0][1]*Ginv10/24.0-Ginv00*w[0][0]/12.0
    - w[0][2]*Ginv01/48.0-w[0][5]*Ginv10/48.0
    - Ginv00*w[0][4]/48.0 -w[0][1]*Ginv11/24.0);
  A[6*0 + 1] = detG*(-w[0][1]*Ginv01/24.0-w[0][1]*Ginv00/24.0
    + Ginv00*w[0][3]/24.0+w[0][5]*Ginv01/24.0);
  A[6*0 + 2] = ...
  ...
  A[6*5 + 5] = detG*(w[0][0]*Ginv01/48.0+Ginv11*w[0][3]/48.0
    + w[0][5]*Ginv11/12.0+w[0][4]*Ginv01/24.0
    + w[0][3]*Ginv10/24.0-w[0][1]*Ginv10/24.0
    + w[0][2]*Ginv01/48.0-w[0][1]*Ginv11/48.0);
}
```

FIGURE 19.13: Generated C++ code for the Jacobian element matrix.

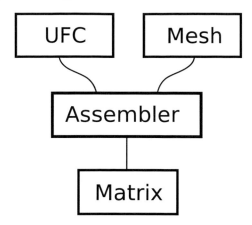

FIGURE 19.14: Main components in a finite-element assembly application.

of nonlinear PDEs. To enable users to experiment with both form compilers within the same framework, the systems agree on a common format for the generated C++ code, which is the topic of the next section.

19.4   A Unified Framework for Finite-Element Assembly

Large parts of a finite-element program are similar from problem to problem and can therefore be coded as a general, reusable library. Mesh handling, finite element assembly, linear and nonlinear solvers, preconditioners and other linear algebra tasks are examples of operations that are naturally coded in a problem-independent way and made available in reusable libraries [25, 50, 5, 18, 11, 6, 30, 44, 4, 52, 10]. However, some parts of a finite element program are difficult to code in a problem-independent way. In particular, this includes the code for evaluation of the so-called element matrix, that is, the evaluation of the local contribution from a local finite-element during the assembly of the global sparse matrix representing a discretized differential operator.

In this section, we present a unified framework for finite-element assembly, where the key is the specification of a fixed interface between the problem-independent and problem-specific parts of a finite element code. The interface is named Unified Form-assembly Code (UFC) [1], and is currently being developed as part of the FEniCS project [25, 20]. The two form compilers FFC and SyFi discussed in the previous section are both capable of generating code that conforms to the UFC specification. Thus, code generated by FFC and SyFi may be used interchangeably in finite-element applications.

19.4.1   Finite-element assembly

To present the UFC interface, we first outline the standard finite-element assembly algorithm used in many finite-element codes [54, 27, 34]. In Algorithm 19.2, a global sparse matrix A is assembled by adding the local contributions from each cell K of a finite element mesh $\mathcal{T} = \{K\}$. On each cell, the element matrix A^K is computed. This computation depends on which discrete operator (variational form) is being assembled and must thus be implemented in a problem-specific way. Furthermore, on each mesh the local-to-global mapping ι_K is computed, that is, the mapping from local degrees of freedom (the row and column indices in A^K) to global degrees of freedom (the row and column indices in A). The local-to-global mapping is used to add the entries of the element matrix A^K to the corresponding entries in the global matrix A.

Algorithm 19.2: Assembling a global matrix A over a mesh.

$A = 0$
for each $K \in \mathcal{T}$
    Tabulate the local-to-global mapping ι_K
    Compute the element matrix A^K
    Add A_i^K to $A_{\iota_K(i),\iota_K(j)}$ for all (i,j)

The algorithm assembly may be extended to assembly of tensors of arbitrary ranks (including matrices, vectors and scalars) and to assembly of contributions from boundaries or interior edges/faces.

19.4.2   The UFC interface

The UFC interface isolates the problem-specific parts of the standard finite-element assembly algorithm. Looking at Algorithm 19.2, we note that it relies on essentially four different components: (i) communication with the mesh, (ii) communication with the global sparse matrix, (iii) tabulation of the local-to-global mapping and (iv) tabulation of the element matrix. The first two components may be implemented by a mesh library and a linear algebra library, respectively, while components (iii) and (iv) are problem-specific and need to be implemented manually for any given variational problem or may be generated by a form compiler like FFC or SyFi.

What the UFC interface provides is a fixed interface for communication between the assembly algorithm and the problem-specific components (iii) and (iv). Thus, the assembly algorithm may itself be implemented as part of a reusable software library that calls the generated problem-specific code to tabulate the degrees of freedom and the element matrix.

A complete description of the UFC specification is beyond the scope of this chapter, but we briefly mention some key points of the interface. First, a general mathematical definition of the element tensor is presented, of which the element matrix A^K in Algorithm 19.2 is a special case. Next the components of this definition are related to the abstract classes in UFC.

The interface covers the assembly of arbitrary rank forms defined as integrals over the cells, boundary facets and interior facets (edges or faces) of a finite-element mesh. For $\{V_h^j\}_{j=1}^r$ a given set of finite element function spaces defined on a triangulation $\mathcal{T}$ of $\Omega \subset \mathbb{R}^d$, we consider a general form a defined on the product space $V_h^1 \times V_h^2 \times \cdots \times V_h^r$:

$$a : V_h^1 \times V_h^2 \times \cdots \times V_h^r \to \mathbb{R} \qquad (19.33)$$

For $j = 1, 2, \ldots, r$, we let $\{\phi_i^j\}_{i=1}^{|V_h^j|}$ denote the (nodal) basis of V_h^j and assemble the rank r tensor A given by

$$
\begin{aligned}
A_i &= a(\phi_{i_1}^1, \ldots, \phi_{i_r}^r; w_1, \ldots, w_n) \\
&= \sum_{K \in \mathcal{T}} \int_K C_K(\phi_{i_1}^1, \phi_{i_2}^2, \ldots, \phi_{i_r}^r; w_1, w_2, \ldots, w_n) \, \mathrm{d}x \qquad (19.34) \\
&+ \sum_{S \in \partial \mathcal{T}_e} \int_S E_S(\phi_{i_1}^1, \phi_{i_2}^2, \ldots, \phi_{i_r}^r; w_1, w_2, \ldots, w_n) \, \mathrm{d}s \qquad (19.35) \\
&+ \sum_{S \in \partial \mathcal{T}_i} \int_S I_S(\phi_{i_1}^1, \phi_{i_2}^2, \ldots, \phi_{i_r}^r; w_1, w_2, \ldots, w_n) \, \mathrm{d}s \qquad (19.36)
\end{aligned}
$$

$$\forall i = \{i_1, i_2, \ldots, i_r\} \in \Pi_{j=1}^r [1, |V_h^j|]$$

Here, $\mathcal{T}$, $\partial \mathcal{T}_e$ and $\partial \mathcal{T}_i$ denote the sets of cells, exterior facets (the facets on the boundary) and interior facets (the facets not on the boundary), respectively. Similarly, C_K, E_S and I_S denote the integrands on cells, exterior and interior facets, respectively. The integrands may optionally depend on a set of fixed functions $w_1, w_2, \ldots, w_n$, defined on an auxiliary set of finite-element function spaces $\{V_h^j\}_{j=r+1}^{r+n}$.

The general form (19.33) is specified by the class `ufc::form`. This class provides basic information about the compiled form, like the rank r of the element tensor and the number of coefficients n, and provides factory functions to construct concrete implementations of the other UFC classes. Through these factory functions, a `ufc::form` can construct a `ufc::finite_element` and `ufc::dof_map` for each function space $V^j, j = 1, 2, \ldots, r+n$. Figure 19.15 shows one of the function `tabulate_dofs` from `ufc::dof_map` for tabulation of the local-to-global mapping. The form object can also construct any number of integral objects (`ufc::cell_integral`, `ufc::exterior_facet_integral`, `ufc::interior_facet_integral`) representing (19.34-19.36). Each such integral object has a function `tabulate_tensor` that may be called to compute the contribution from the cells, exterior and interior facets, respectively. Each of these three functions has a slightly different signature, with one of them

```
namespace ufc {

  class dof_map
  {
    ...
    virtual void tabulate_dofs(unsigned int* dofs,
                               const mesh& m,
                               const cell& c) const = 0;

    ...
  };

  class cell_integral
  {
    ...
    virtual void tabulate_tensor(double* A,
                                 const double * const * w,
                                 const cell& c) const = 0;

    ...
  };

}
```

FIGURE 19.15: Part of the UFC interface for tabulation of the local-to-global mapping ι^K and the element tensor A^K.

shown in Figure 19.15. Examples of UFC code generated by FFC and SyFi were given in Figures 19.9 and 19.13.

In order not to introduce dependencies on particular mesh formats, two low-level data structures `ufc::mesh` and `ufc::cell` are used to pass information about the mesh to the UFC functions. The class `ufc::mesh` contains basic information about the size and type of mesh. In the `ufc::cell` structure, we include global coordinates of the cell vertices and global indices for each local mesh entity contained in the cell (vertices, edges and facets). The UFC data structures `ufc::mesh` and `ufc::cell` are not meant to be used to build a complete data structure for a whole finite-element mesh, but to provide a *view* (e.g., in terms of C/C++ pointers) of the necessary mesh and cell data for the UFC functions.

19.4.3   Implementing the UFC interface

A key design goal for the UFC interface is to allow complicated variational formulations in combination with most of the finite-element spaces in use today. It is imperative that the interface is flexible enough for complex problems, and that it allows efficient implementation of finite-element assembly. Since the interface is in relatively low-level C/C++ code, efficient implementations are straightforward to implement. It uses virtual functions for flexibility, and raw pointers for efficient argument passing. Since the UFC interface consists of a single header file with no external dependencies, this gives a high degree of decoupling between the software packages that implement and use it.

It should be noted that the UFC interface is not tied to form compilers or code generation; the various functions in the interface may well be implemented manually. We also hope that the UFC interface may be adopted by other PDE software packages in the future. Code that complies with the UFC specification may then be used interchangeably in different finite-element software.

19.5 Conclusions

We have presented a new way of designing large-scale finite element codes. One design goal is to offer the user a simple way to define the problem and the basic steps in the overall solution algorithm. Another design goal is to ensure that the numerically intensive parts of the code are highly optimized and well suited for supercomputers. Reusing mature and well-tested general libraries is a third design goal. A key idea is to use high-level Python scripts as the user's interface to define the problem and the basic steps of the solution algorithm. The problem-dependent parts of the code are automatically generated by form compilers, of which FFC and SyFi are two different choices. Both compilers write out UFC-compliant code such that generic libraries can reach all information about finite elements and element matrix/vector computations through a unified interface. We believe that UFC is an important step for achieving a higher degree of interoperability between various finite-element packages.

In a complete system for the evaluation and assembly of variational forms, formalizing the concept of a form compiler might be useful. In such a system UFC defines the output format of the back end, and SyFi and FFC constitute concrete compiler implementations. A common language for the front ends is, however, missing. We plan to define such a language, Unified Form Language (UFL), to be used by both SyFi and FFC, such that the end user can have full flexibility in his or her choice of form compiler.

Parallel computing has not yet been explicitly discussed. The simplest way of parallelization of the proposed type of finite-element codes is to provide support for parallel assembly in the general library. With parallel, "global" matrices and vectors, linear algebra libraries like PETSc, Hypre and Trilinos can be used to solve the linear and nonlinear systems. For many problems, an efficient parallelization technique is to apply overlapping or nonoverlapping domain decomposition [16, 49] at the PDE level, either as a direct solution algorithm or as a preconditioner. Essentially, this domain decomposition algorithm can be implemented as a new framework "on top" of serial solvers (following the ideas of [13], for instance). Each subproblem over a subdomain is then treated by the PyCC code snippets shown in the second section. Again, the general library must support parallel assembly and parallel matrices and vectors. The FFC and SyFi parts of the code will not be affected by either of these parallelization techniques. Also the PyCC code can be kept very much intact in a parallel context, but the technical demonstration of this assertion will be the topic of a future paper.

Along with petascale computing, scientists will get multicore desktop machines. This new architecture constitues great challenges to code writers. Our suggested way of splitting a finite-element program into different parts also has significant potential for multicore computers as the high-level problem definition remains the same, but the generic libraries must support this architecture. Whether multicore architectures affect the form compilers is an open question, but for large element tensors (higher-order elements, multiphysics models), concurrent computing of a single element tensor at the multicore level might be beneficial. The associated low-level code will be generated by the form compiler anyway.

The important idea is that a form compiler and a generic library can be written "once and for all," while the really human-resource-consuming part of the problem-solving process lies in the scientists' application programs. These programs have a much longer life than the hardware on which they are running. We believe that the suggested software design in this chapter helps to keep the numerically intensive core of the applications up-to-date for decades by occasionally using future form compilers to generate new code on new architectures and by linking with future generic libraries tailored to future architectures.

19.6   Acknowledgments

The authors thank Joachim Berdal Haga and Johan Hake for constructive feedback on an earlier version of the manuscript.

References

[1] M. S. Alnæs, H. P. Langtangen, A. Logg, K.-A. Mardal, and O. Skavhaug. UFC, 2007. http://www.fenics.org/ufc/.

[2] M. S. Alnæs and K.-A. Mardal. SyFi, 2007. http://www.fenics.org/syfi/.

[3] D. N. Arnold, R. S. Falk, and R. Winther. Mixed finite element methods for linear elasticity with weakly imposed symmetry. *Math. Comp.*, 76:1699–1723, 2007.

[4] S. Balay, K. Buschelman, W. D. Gropp, D. Kaushik, M. Knepley, L. C. McInnes, B. F. Smith, and H. Zhang. PETSc Users Manual. Technical Report ANL-95/11 - Revision 2.1.5, Mathematics and Computer Science Division, Argonne National Laboratory, 2003.

[5] W. Bangerth, R. Hartmann, and G. Kanschat. deal.II Differential Equations Analysis Library, 2006. http://www.dealii.org/.

[6] P. Bastian, K. Birken, S. Lang, K. Johannsen, N. Neuss, H. Rentz-Reichert, and C. Wieners. UG - a flexible software toolbox for solving partial differential equations. *Computing and Visualization in Science*, 1:27–40, 1997.

[7] J. Bey and G. Wittum. Downwind numbering: A robust multigrid method for convection diffusion problems on unstructured grids. *Applied Numerical Mathematics*, 23(2):177–192, 1997.

[8] S. C. Brenner and L. R. Scott. *The Mathematical Theory of Finite Element Methods*. Springer-Verlag, 1994.

[9] D. L. Brown, R. Cortez, and M. L. Minion. Accurate projection methods for the incompressible Navier–Stokes equations. *J. Comp. Physics*, 168:464–499, 2001.

[10] A. M. Bruaset, H. P. Langtangen, et al. *Diffpack*, 2006. http://www.diffpack.com/.

[11] Cactus Code Server. http://www.cactuscode.org.

[12] J. Cahouet and J. P. Chabard. Some fast 3D finite element solvers for the generalized Stokes problem. *International Journal for Numerical Methods in Fluids*, 8:869–895, 1988.

[13] X. Cai. Domain decomposition. In H. P. Langtangen and A. Tveito, editors, *Advanced Topics in Computational Partial Differential Equations – Numerical Methods and Diffpack Programming*. Springer, 2003.

[14] X. Cai and H. P. Langtangen. Parallelizing PDE solvers using the Python programming language. In A. M. Bruaset and A. Tveito, editors, *Numerical Solution of Partial Differential Equations on Parallel Computers*, volume 51 of *Lecture Notes in Computational Science and Engineering*, pages 295–325. Springer, 2006.

[15] X. Cai, H. P. Langtangen, and H. Moe. On the performance of the Python programming language for serial and parallel scientific computations. *Scientific Programming*, 13(1):31–56, 2005.

[16] T. F. Chan and T. P. Mathew. Domain decomposition algorithms. In *Acta Numerica 1994*, pages 61–143. Cambridge University Press, 1994.

[17] P. G. Ciarlet. *Numerical Analysis of the Finite Element Method*. Les Presses de l'Universite de Montreal, 1976.

[18] COMSOL Multiphysics. `http://www.comsol.com`.

[19] M. Crouzeix and P.A. Raviart. Conforming and non-conforming finite element methods for solving the stationary Stokes equations. *RAIRO Anal. Numér.*, 7:33–76, 1973.

[20] T. Dupont, J. Hoffman, C. Johnson, R. C. Kirby, M. G. Larson, A. Logg, and L. R. Scott. The FEniCS Project. Technical Report 2003–21, Chalmers Finite Element Center Preprint Series, 2003.

[21] Fastflo. `http://www.cmis.csiro.au/fastflo`.

[22] FreeFEM. `http://www.freefem.org`.

[23] GiNaC, 2006. http://www.ginac.de.

[24] W. Hackbusch. On the feedback vertex set problem for a planar graph. *Computing*, 58(2):129–155, 1997.

[25] J. Hoffman, J. Jansson, C. Johnson, M. G. Knepley, R. C. Kirby, A. Logg, L. R. Scott, and G. N. Wells. *FEniCS*, 2006. `http://www.fenics.org/`.

[26] J. Hoffman, J. Jansson, A. Logg, and G. N. Wells. *DOLFIN*, 2006. `http://www.fenics.org/dolfin/`.

[27] T. J. R. Hughes. *The Finite Element Method: Linear Static and Dynamic Finite Element Analysis*. Prentice-Hall, 1987.

[28] Hypre. `http://acts.nersc.gov/hypre/`.

[29] E. Jones, T. Oliphant, P. Peterson, et al. SciPy: Open source scientific tools for Python, 2001–. `http://www.scipy.org/`.

[30] Kaskade. `http://www.zib.de/Numerik/numsoft/kaskade/index.de.html`.

[31] R. C. Kirby and A.Logg. A compiler for variational forms. *ACM Transactions on Mathematical Software*, 32(3):417–444, 2006.

[32] R. C. Kirby, M. G. Knepley, A. Logg, and L. R. Scott. Optimizing the evaluation of finite element matrices. *SIAM J. Sci. Comput.*, 27(3):741–758, 2005.

[33] R. C. Kirby and A. Logg. Efficient compilation of a class of variational forms. *ACM Transactions on Mathematical Software*, 32(3, Article No. 17), 2007.

[34] H. P. Langtangen. *Computational Partial Differential Equations - Numerical Methods and Diffpack Programming*, volume 1 of *Texts in Computational Science and Engineering*. Springer, 2nd edition, 2003.

[35] H. P. Langtangen. *Python Scripting for Computational Science*. Texts in Computational Science and Engineering, vol 3. Springer, second edition, 2006.

[36] H. P. Langtangen, K.-A. Mardal, and R. Winther. Numerical methods for incompressible viscous flow. *Advances in Water Resources*, 25:1125–1146, 2002.

[37] A. Logg. *FFC*, 2006. http://www.fenics.org/ffc/.

[38] K.-A. Mardal. SyFi - An element matrix factory. In *Proc. Workshop on State-of-the-Art in Scientific and Parallel Computing (PARA'06)*, Umeå, Sweden, June 2006.

[39] K.-A. Mardal, O. Skavhaug, G. T. Lines, G. A. Staff, and Å. Ødegård. Using Python to solve partial differential equations. *Computing in Science & Engineering*, 9(3):48–51, 2007.

[40] K.-A. Mardal, X.-C. Tai, and R. Winther. A robust finite element method for Darcy–Stokes flow. *SIAM J. Numer. Anal.*, 40:1605–1631, 2002.

[41] K.-A. Mardal and R. Winther. Uniform preconditioners for the time dependent Stokes problem. *Numerische Mathematik*, 98(2):305–327, 2004.

[42] J.-C. Nédélec. Mixed finite elements in R^3. *Numer. Math.*, 35(3):315–341, October 1980.

[43] J.-C. Nédélec. A new family of mixed finite elements in R^3. *Numer. Math.*, 50(1):57–81, November 1986.

[44] Overture. http://www.llnl.gov/casc/Overture.

[45] PyCC, 2007. Software framework under development. http://www.simula.no/pycc.

[46] *Python.* http://www.python.org/.

[47] P. A. Raviart and J. M. Thomas. A mixed finite element method for 2-order elliptic problems. In *Mathematical Aspects of Finite Element Methods*, Lecture Notes in Mathematics, No. 606, pages 295–315. Springer Verlag, 1977.

[48] J. R. Rice and R. F. Boisvert. *Solving Elliptic Problems Using ELL-PACK*. Springer, 1984.

[49] B. Smith, P. Bjørstad, and W. Gropp. *Domain Decomposition – Parallel Multilevel Methods for Elliptic Partial Differential Equations*. Cambridge University Press, 1996.

[50] Sundance. `http://software.sandia.gov/sundance`.

[51] Swiginac, 2006. http://swiginac.berlios.de/.

[52] Trilinos. `http://software.sandia.gov/trilinos`.

[53] S. Turek. *Efficient Solvers for Incompressible Flow Problem*. Springer Verlag, 1999.

[54] O. C. Zienkiewicz, R. L. Taylor, and J. Z. Zhu. *The Finite Element Method — Its Basis and Fundamentals, 6th edition*. Elsevier, 2005, first published in 1967.

Chapter 20

Programming Petascale Applications with Charm++

Laxmikant V. Kalé, Eric Bohm, Celso L. Mendes, Terry Wilmarth, Gengbin Zheng

Department of Computer Science, University of Illinois at Urbana-Champaign

20.1   Motivation

The increasing size and complexity of parallel machines, along with the increasing sophistication of parallel applications, makes development of parallel applications a challenging task for petascale machines. The National Science Foundation has planned the deployment of a sustained petaflops (PFLOPS) machine by 2010, which is likely to have several hundred thousand processor cores. The Blue Gene/L already has 128K cores, and some future designs with low-power processors may have over a million cores. Machines in the intervening years are slated to perform at over a PFLOPS peak performance. Further, as portended by the Roadrunner project at Los Alamos National Laboratory (LANL), some of the large machines will have accelerators along with commodity processors.

Meanwhile, parallel applications are growing in sophistication: it is understood that "increasing resolution everywhere" is not the best way to utilize higher compute power. Instead, applications are increasingly using adaptive refinement to match the geometric variation over space, and dynamic refinement and coarsening to deal with variations in the phenomenon being simulated as it evolves over time. Modern algorithms, such as Barnes-Hut for gravity calculations, reduce operation counts to make large simulations feasible, but in the process make parallelization challenging. Applications often include independently developed parallel modules that must be effectively composed. Many applications are now multiphysics in nature, making such multi-module composition even more important.

MPI is the dominant programming methodology today, and there is some hope that it will continue to be used for some petascale applications. However, there is clearly a need for raising the level of abstraction beyond MPI to increase programmer productivity, especially for the sophisticated next-generation applications mentioned above.

Charm++ is a parallel programming system developed over the past 15 years or more, aimed at enhancing productivity in parallel programming while enhancing scalable parallel performance. A guiding principle behind the design of Charm++ is to automate what the "system" can do best, while leaving to the programmers what they can do best. In particular, we believe that the programmer can specify *what* to do in parallel relatively easily, while the system can best decide which processors own which data units and execute which work units. This approach requires an *intelligent runtime system*, which Charm++ provides.

Thus, Charm++ employs the idea of "overdecomposition" or "processor virtualization" based on *migratable objects*. This idea leads to programs that automatically respect locality, in part because objects provide a natural encapsulation mechanism. At the same time, it empowers the runtime system to automate resource management. The combination of features in Charm++ and associated languages makes them suitable for the expression of parallelism for a range of architectures, from desktops to PFLOPS-scale parallel machines.

In this chapter, we first present an overview of the Charm++ programming model. Charm++ is a mature system with libraries and sophisticated performance analysis tools that are being adapted to petascale. We describe these as well as other basic capabilities enabled by its adaptive runtime system. Adaptive MPI, a full MPI implementation built on top of Charm++, ensures that the benefits of Charm++ are available to the broad class of applications written using MPI. We then describe three full-scale applications that have scaled well to over 16,000 processors. These applications are in active use by scientists. They illustrate the potential of Charm++ in addressing petascale programming issues. BigSim, described in Section 20.4, is a system designed to analyze the performance of Charm++ and MPI applications on petascale machines using emulations and simulations on a relatively smaller number of processors. In section 20.5, we review a few new parallel programming abstractions on top of Charm++ that we believe substantially enhance programmer productivity.

20.2   Charm++ and AMPI: Programming Model

Charm++ programs are C++ programs, along with a few small interface description files. A computation (running program) consists of C++ objects that are organized into indexed collections. The indexing allows for sparse or large multidimensional collections, as well as other arbitrary indexing mechanisms such as strings or bit-vectors. Objects communicate via asynchronous method invocations. The objects that accept such (remote) method invocations are called *chares*. The collections are called *chare arrays* by analogy with data arrays; it is important to remember that each element of a chare array is NOT a base-type, but a chare, which can be migrated across processors by the runtime system, and typically holds a coarse grain of data and computation. Typical applications may have tens of chares on each processor, while for some applications there may be tens of thousands of them.

The execution of a program begins with the creation of a specially designated *main chare*, which typically creates multiple chare arrays. The control on each processor resides with a scheduler which works with a prioritized queue of messages, each holding (in the simple case of Charm++ programs) a method invocation for a chare on that processor. In the base model, all methods are nonblocking; thus, the scheduler repeatedly selects method invocations from its queue and executes them without interruption. The methods invoked on chares asynchronously are called *entry methods* and are specified in the interface description files mentioned above.

Charm++ also supports "threaded" and "sync" methods. A user level, lightweight thread is associated with each threaded method, which allows the thread to block and return control to the Charm++ scheduler. One particular reason to block is provided by the "sync" methods: unlike the asynchronous entry methods, which have no return value, these methods can return a value; more importantly, the caller object is blocked until the method is executed and returns, which also provides an additional synchronization mechanism.

Chares can migrate across processors; messages are delivered to chares correctly, with an efficient caching and forwarding mechanism. This empowers the runtime system to optimize program execution in myriad ways, as discussed later.

The message-driven split-phase style of Charm++ programming is not alien to developers of event-driven programs on the desktop, which is one reason why we believe it will be a good language to handle the desktop parallelism unleashed by multicore chips. However, it is more challenging for the computational science and engineering (CSE) community, where developers are accustomed to the single-control-thread model of MPI. For this reason, Charm++ supports a notation called structured dagger [13], which allows a clean expression of the life cycle of a single object. This notation is translated into normal Charm++ programs without resorting to the use of threaded methods, by the

Charm++ translator which also handles the interface files.

Charm++'s main strength is its adaptive runtime system (RTS). Its use of C++, and asynchronous method invocation are orthogonal to this strength. With this realization, and to further widen the applicability of this RTS, we developed adaptive MPI (AMPI), which is an implementation of MPI on top of Charm++: each MPI "process" is implemented as a user-level thread embedded in a Charm++ object.

20.2.1 Dynamic load balancing

Load balancing is an important and increasing need for large-scale simulations, as adaptive methods are increasingly being used, and simulations often deal with evolving physical models. For example, finite element method (FEM) simulations involve dynamic geometry, and use adaptive techniques to solve highly irregular problems.

Charm++ provides an application-independent automatic dynamic load-balancing capability. It is based on migratable objects in Charm++ and migratable threads in AMPI [33]. By migrating existing tasks among processors, the Charm++ runtime system distributes computation uniformly across all processors, taking the object load into account, while minimizing the communication between them. The most up-to-date application and background load and the communication structure are automatically instrumented and collected by the runtime system without requiring *a priori* application knowledge. This load and communication information can be used to predict the application's behavior in the near future due to the *principle of persistence* [12]. The principle of persistence is an empirical observation that the object computation times and communication patterns (number and bytes of messages exchanged between each communicating pair of objects) *tend to* persist over time in most scientific and engineering applications.

Charm++ implements a range of object-based load-balancing strategies, with approaches ranging from centralized to distributed. The centralized load balancers have been shown to work effectively on up to a few thousand processors for various applications, but beyond that they start becoming a bottleneck due to high memory and CPU overheads; fully distributed load balancers avoid the bottleneck, but lead to poor quality in the balance they achieve. We have demonstrated hierarchical techniques that combine aspects of centralized and fully distributed balancers, in principle, on several Charm++ benchmarks that can scale to a very large number of processors.

For machines with very large numbers of processors, it becomes important to take the topology of the machine into consideration due to the possible network contention caused by suboptimal object-to-processor mapping. Charm++ provides several topology-sensitive load-balancing techniques [1] for typical topologies like the torus of Blue Gene/L and Cray XT3. These load-balancing strategies minimize the impact of topology by heuristically minimizing the total number of *hop-bytes* (the total size of inter-processor communication in bytes weighted by the distance between the respective end-processors) communicated.

20.2.2 Projections

Projections is a sophisticated performance analysis framework developed to support the discovery of performance problems in large-scale Charm++ and AMPI applications. It consists of a highly flexible performance instrumentation and tracing infrastructure built into the Charm++ runtime system as well as a Java visualization toolkit that reads trace logs generated by the instrumentation component to present the performance information in a manner that is easily understood by application developers seeking to debug performance issues.

Performance analysis is human-centric. As illustrated in Figure 20.1, starting from visual distillations of overall application performance characteristics, the analyst employs a mixture of application domain knowledge and experience with visual cues expressed through Projections in order to identify general areas (e.g., over a set of processors and time intervals) of potential performance problems. The analyst then zooms in for more detail and/or seeks additional perspectives through the aggregation of information across data dimensions (e.g., processors). The same process is repeated, usually with higher degrees of detail, as the analyst homes in on a problem or zooms into another area to correlate problems. The richness of information coupled with the tool's ability to provide relevant visual cues contribute greatly to the effectiveness of this analysis process.

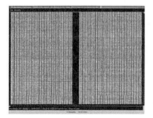 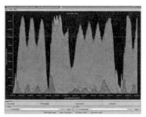 

(a) Overview of 512-processor run of NAMD over several seconds.

(b) A time profile of 512 processors over a 70ms range of interest.

(c) Detailed timeline of sample processors over the same 70ms range.

FIGURE 20.1: Projections performance analysis, zooming to different levels of detail and viewing different data perspectives.

In the drive to scale applications to petascale systems, Projections has been enhanced to scalably handle the increase in the volume of performance trace data generated on systems with many thousands of processors. We take a two-pronged approach to the problem. First, Projections has to present performance information that is pertinent and avoid overwhelming the analyst. For instance, it is very tedious to manually identify a small number of processor execution profiles that appear to behave poorly or differently from thousands of other processors. We employ various heuristics and algorithms to aid the identification of the set of extrema and representative processors for the analyst. Second, Projections must control the total volume of data generated. We do this by doing automatic parallel analysis at the end of a parallel run so that only the trace buffers of processors that are deemed "interesting" are written. For processors that are "uninteresting," a much smaller time-aggregated data format is written in its place.

20.2.3   Summary of other features

Since Charm++ programs typically consist of many more chares than processors, each processor houses many chares. As a result, the RTS must have a scheduler on each processor that decides which object executes next. Note that this is a user-level scheduler; for the operating system (OS), there is only one process running, and the OS scheduler is not involved in scheduling chares. The Charm++ scheduler works with a queue of generalized messages: each message in a simple Charm++ program corresponds to an asynchronous method invocation sent to one of the objects on that processor. It selects a message (by default in first-in, first-out (FIFO) order, but in general in order of application-specified priorities) from this queue and invokes a method on an object, both indicated by the envelope of the message. The scheduler allows the method to run to completion, and then repeats the process by selecting the next message in the queue. There is no preemption. For more complex Charm++ programs, as well as for AMPI programs, the object may return control to the scheduler when it is blocked waiting for some data. In this case, the method invoked must be a "threaded" method, and it is supported by a user-level threads package in the RTS.

This message-driven execution renders Charm++ programs automatically tolerant of communication latencies. If one object is waiting for data, another object which has its message can continue execution without any user code to orchestrate such switching. This automatic and adaptive overlap of communication also extends across parallel modules: two concurrently executing parallel modules may overlap on processors, and yet automatically interleave their execution adaptively. Such concurrent composition capability is essential for modularity in parallel programs: without this (as in MPI), a programmer will be tempted to break abstraction boundaries between two modules (in MPI, inserting wild-card receives in each module to transfer control to the other module, for example) to avoid performance penalties resulting from the inability to overlap idle time in one module with useful computation in the other.

Some of the impressive capabilities of Charm++ follow directly from the migratability of its objects. Since objects can migrate across processors, automatic checkpointing is supported by "migrating" a copy of each object to disk. Charm++ also supports much faster checkpointing in another processor's memory for applications which have a small footprint in between time steps (i.e., at checkpoint times). Charm++ programs gain this capability without additional user code development, and it is available in the standard Charm++ distribution. A more ambitious scheme being developed for the petascale context implements a sender-side message-logging protocol specialized to migratable objects. With this, when a fault occurs, all processors are *not* rolled back to their checkpoint; rather only the objects on the failed processors are resurrected on multiple other processors, where they recover concurrently, while the rest of the computation continues to make progress to the extent possible. Since recovery is parallelized, the system can make progress even when the MTBF (mean time between failures) is smaller than the checkpoint period. This capability is currently available only in a research version of Charm++.

The sets of processors allocated to a Charm++ job can be changed at runtime. This shrink/expand capability has been used to design efficient cluster schedulers. When a new job arrives, for example, a running job can be "shrunk" to make the minimum number of processors needed by the new job available to it. On a "parallel computer" consisting of idle desktop workstations, this same capability can be used to migrate objects away from a workstation, when its owner desires exclusive use of it.

For applications that require much more memory than a parallel machine has available, Charm++ can support automatic out-of-core execution by selectively moving chares to and from the disk. This is facilitated by its ability to "peek" into the scheduler's queue to decide which objects are needed in the near future and to prefetch them. This double-buffering capability is also exploited in the ongoing implementation of Charm++ on the IBM Cell Broadband Engine (Cell BE) processor. We expect that future machines that use accelerators of any form (GPGPUs, Cell BEs, etc.) will benefit from this capability.

Since the Charm++ RTS mediates all communication between chares, it can observe the patterns of communication and optimize communication; for example, it can replace one all-to-all algorithm by another at runtime [16].

The Charm++ LiveViz module allows applications to display images while the simulation is running, on machines where such external communication is allowed. The client-server library in the RTS used for this purpose is also leveraged by an online debugger [10], which supports such features as freeze-and-inspect, memory-visualization and analysis, and record-and-replay for catching nondeterministic bugs.

20.3   Charm++ Applications

20.3.1   NAMD

NAMD is a parallel program for simulation of biomolecular assemblies consisting of proteins, DNA molecules, cell membranes, water molecules, etc. NAMD is one of the earliest CSE applications developed using Charm++. As a result, several features in Charm++ have coevolved with NAMD, including its dynamic load-balancing framework. NAMD exemplifies our research group's objective of "application-oriented but computer science-centered research" [11], i.e., we aim to develop enabling technologies motivated by and in the context of a "real" application, while making sure those technologies are broadly applicable by honing them with multiple applications.

Parallelization of NAMD is described in earlier in this book (see Chapter 9, [23]). Here, we only note that adaptive overlap of communication and computation across modules as well as dynamic load balancing are two of the Charm++ features especially useful for effectively parallelizing NAMD. Features such as the topology-aware and hierarchical load balancers (see Section 20.2) being developed in Charm++ will be instrumental in scaling NAMD to petascale machines.

20.3.2   LeanCP

The Car-Parrinello *ab initio* molecular dynamics (CPAIMD) method [4, 21] can model complex systems with nontrivial bonding and events such as chemical bond forming and breaking, and thus has enormous potential to impact science and technology. Kalé, Martyna, and Tuckerman have collaborated as part of an interdisciplinary team funded under the U.S. National Science Foundation's Information Technology Research program (NSF-ITR, Grant 0121357) in order to (1) improve the methodology and sampling algorithms to permit larger scale simulations of higher accuracy; (2) employ next generation software engineering tools to design and implement parallel algorithms for CPAIMD which scale to thousands of processors and to instantiate them in a package to be referred to as *LeanCP* [27]; (3) use the new techniques and LeanCP software for applications in technologically important areas; (4) transfer the knowledge thus gained to the community through diversity-conscious outreach and education programs ranging from the secondary to graduate levels as well as through international academic and industrial collaborations. These goals have been largely realized on IBM's Blue Gene/L (see Figure 20.2). Like all Charm++ codes, it is portable to most clusters and supercomputers, and the specialized adaptations for Blue Gene/L are being enhanced to perform with similar efficiency on other torus architectures, such as the Cray XT4.

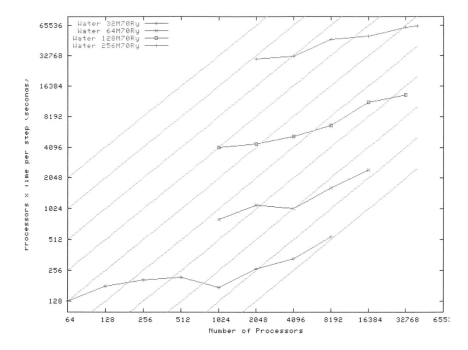

FIGURE 20.2: Parallel performance of LeanCP on IBM's Blue Gene/L (horizontal line is perfect scaling)

The fine-grained decomposition using Charm++ has freed the CPAIMD method from its traditional limitation of only scaling effectively to number of processors equal to the number of electronic states in the system being studied. Due to the $O(N^2)$ and $O(N^3)$ nature of the computation methods, where N is the number of states, systems with more than a few hundred states are considered very large and require a great deal of computational power to resolve them to convergence for the nanoseconds of simulated time required to capture interesting events.

Through the use of Charm++'s fine-grained decomposition, adaptive overlap of communication with computation, topology aware placement to minimize communication costs, and new methods development, LeanCP has scaled the CPAIMD method to over 30 or even 60 times the number of processors as there are electronic states (consuming all 40k processors of the IBM TJ Watson Blue Gene/L). This extreme scaling is necessary if we are to reduce the time to solution for interesting problems to within manageable limits by applying ever larger supercomputers.

Special attention must be given to managing communication in order to scale efficiently to large torus networks. Unlike traditional fat-tree networks, the cost of communication in bandwidth and time increases substantially in

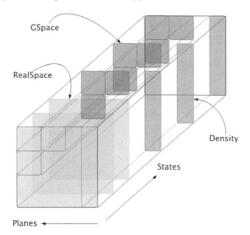

FIGURE 20.3: Mapping LeanCP onto IBM's Blue Gene/L 3D torus

proportion to the distance in hops from source to destination. If multi-cast and many-to-many communication operations are naively mapped onto the torus, they can easily degrade into spanning the entire torus. This will severely limit scaling for communication-bound algorithms. For complex applications with multiple interlocking communication patterns, it is critical that the mapping of tasks to processors be flexible. Charm++ decouples the computational task from its processor placement so that these issues can be treated separately.

In the case of LeanCP, there are seventeen chare object arrays which must be placed such that the computation of each phase (represented by one or more arrays) is well parallelized, while minimizing the communication cost from one phase to the next. This is accomplished by partitioning the interacting chares within interlocking sub-tori as shown in Figure 20.3. LeanCP decomposes the computation along electronic states and along plane-wise slices of each electronic state. In Figure 20.3, the two sides of the plane wave in "real" space before a 3D FFT and in "g" space after a 3D FFT are mapped in orthogonal sub-tori. These are chosen to minimize the communication in the three phases shown in the following manner: the state-wise communication in the FFT phase operates within sub-tori along the torus Z axis, the plane-wise 3D ortho-normalization matrix multiplication in "g" space operates within sub-tori prisms in the torus X and Y axes, and the "real" space plane-wise density computation operates within sub-tori slabs along the torus X-axis. Charm++ allows us to experiment with intricate mapping schemes independent from the computation and parallel driver code and to adjust them for different systems and torus sizes as needed.

Going forward, there are many ways in which petascale machines could be used to improve scientific studies using LeanCP. The key idea in this area is to study molecular systems large enough to contain all relevant effects for a sufficient duration to capture interesting events, such as bonds forming or breaking, with enough accuracy to be confident in the predictive power of the result. Planned modifications to LeanCP include several ways to improve scalability, to improve accuracy, and to widen the variety of systems suitable for study.

There are some important algorithmic extensions that parallelize well, or even trivially, but will yield important scientific advantages. Of particular interest, we will discuss path integral molecular dynamics [25, 22, 26, 20] which allows the quantum mechanical treatment of light nuclei, in particular, hydrogen. In this way, important chemical reactions and physical processes involving quantum tunneling (isotope effects) can be properly examined. In the Feynman path integral technique, the classical nuclei are replaced by ring polymers that interact with each other in an orderly fashion, with all beads of index i interacting with all other beads of index i on different atoms, via the external potential. Beads associated with the same atom interact with nearest-neighbor beads $(i, i + 1)$ via harmonic potentials. The computational cost of the method is P, the number of beads, times the cost of the standard computation. The method is easy to parallelize because one simply spawns P electronic structure computations, one for each bead. The electronic structure computations do not require communication between them. The interaction is transmitted by the intra-chain interactions between beads on the same atom. For a relatively simple system such as water, one would typically choose $P = 64$ beads, which would result in 64 times the work of that same system. Thus, to keep the time to solution for all 64 beads of the system about the same as that with non-path integral studies, 64 times the number of processors must be used. Since this additional scalability from path-integral studies comes with negligible overhead, LeanCP benchmark systems which efficiently scale non-path integral experiments to 8K processor cores (i.e. systems with 256+ states) will scale similar path-integral studies up to $P * 8k$ cores (*e.g.*, $64 * 8k = 512k$ cores for water), when such petascale machines become available for use.

20.3.3 ChaNGa

Cosmological simulators are becoming increasingly important in the study of the formation of galaxies and large-scale structures. The more general study of the evolution of interacting particles under the effects of Newtonian gravitational forces, also known as the N-body problem, has been extensively reported in the literature. A popular method to simulate such problems was proposed by Barnes and Hut [3]. That method associates particles to a hierarchical structure comprising a tree, and reduces the complexity of the problem from $O(N^2)$ to $O(N \log N)$, where N is the number of particles.

Hierarchical methods for N-body simulations have been adopted for quite some time by astronomers [17, 28]. Two of the most widely used codes in this area are PKDGRAV [6] and GADGET [24]. Both codes contain good

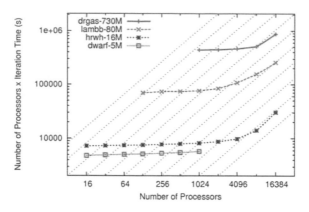

FIGURE 20.4: Parallel performance of ChaNGa on IBM's Blue Gene/L.

physical-modeling capabilities, and have been scaled to over a hundred processors, but they do not scale effectively on modern machines with thousands of processors. The main reason for the poor scalability of these codes is the intrinsic complexities associated with the problem which causes load imbalances and non-negligible communication overheads to arise when the machine size grows. Thus, it becomes very challenging to use the available processors efficiently.

To address these problems, we developed a new cosmological simulator in collaboration with Thomas Quinn from the University of Washington. This simulator, named ChaNGa (Charm++ N-body Gravity solver), was first publicly released in February 2007. ChaNGa was built as a Charm++ application that employs a tree data structure to represent the simulation space. This tree is segmented into elements named *TreePieces*, and constructed globally over all particles in the simulation. The various TreePieces are distributed by the Charm++ runtime system to the available processors for parallel computation of the gravitational forces. Each TreePiece is implemented in ChaNGa as a Charm++ *chare* and can migrate between processors (e.g., during load balancing) as the simulation evolves.

Because various TreePieces may reside on the same processor, ChaNGa employs a software-caching mechanism that accelerates repeated accesses to the same remote particle data. Each processor contains a cache, implemented as a Charm++ *group*, to store remotely fetched particles and tree nodes. The code contains various optimizations that effectively utilize this cache and allow a significant overlap between computation and communication. This overlap is essential to achieve good parallel scalability. All these optimizations are controlled by parameters that can be defined by the user when launching a simulation. Another important feature in ChaNGa is the use of a multi-step integration scheme, which on the smallest time step updates only those particles with the highest accelerations, and only infrequently calculates the forces on all the particles. This feature allows large efficiency gains in simulations of interest, because such simulations may span a large dynamic range of timescales.

Like other Charm++ applications, ChaNGa is portable to a variety of platforms, from commodity clusters to large parallel systems. As an example, Figure 20.4 shows the gravity calculation performance of ChaNGa on IBM's Blue Gene/L with different cosmological data sets (the size of each data set, in millions of particles, is represented in each data set name). In this diagram, horizontal plot lines represent perfect scalability, while diagonal lines correspond to reduced scalability. This figure shows that the code scales well to over 8,000 processors. The good scalability in these plots comes mainly from the optimizations that overlap computation and communication. Although ChaNGa is already in production use by astronomers, we continue to work on enhancements and further extensions for it. In particular, we are working on a new class of load balancers that can deal with the multiple phases of the computation separately, which is a requirement arising from the multistep characteristic of the code. In addition, we intend to add hydrodynamics as a complement to the current gravitational capability.

20.3.4 Other applications

Several other applications (or their components) have been implemented using Charm++, including a structural dynamics code *Fractography* [31] designed for studying crack propagation via cohesive elements, Rocstar[9] for simulation of solid-motor rockets, a space–time-meshing code based on the discontinuous Galerkin method, etc.

We believe that significant classes of future petascale applications can benefit from Charm++: the multicomponent, multiphysics applications discussed by [2] are one example. Also, codes similar to the community climate models can utilize AMPI to overlap computations of modules across processors. AMPI currently requires a single executable on all processors, but facilitates integration of independently developed MPI modules via cross communicators. We think that the combination of features in Charm++/AMPI, including automatic resource management and fault tolerance, makes it a compelling platform for petascale applications.

ParFUM [18] leverages the Charm++ runtime system to provide a framework for the parallelization of unstructured meshes and the applications that use them. ParFUM inherits the many capabilities provided by Charm++ such as processor virtualization, communication and computation overlap, dynamic load balancing and portability to many platforms. The base ParFUM implementation is in AMPI, allowing for rapid porting of existing MPI simulations to ParFUM. ParFUM provides serial and parallel mesh partitioning via METIS [15] and ParMETIS [14], access to remote entities or *ghosts* along partition boundaries, updates of ghost data and reductions on shared nodes along partition boundaries, topological adjacencies and adaptive mesh refinement and coarsening.

ParFUM also benefits from Charm++'s load-balancing capabilities [31]. Each partition of a ParFUM mesh is associated with a single VP, and the domain is typically decomposed into many more partitions than physical processors. These partitions can then be migrated to improve load balance during a simulation. In highly dynamic simulations [19], this ability is key to obtaining good performance scaling. Simulations may use adaptive mesh refinement and coarsening, involve rapidly evolving geometries, have variation in material states and thereby computation times, all of which may apply to confined regions of the mesh and vary dramatically over time. ParFUM is excellently suited to such applications.

POSE [29] is a general-purpose optimistically synchronized parallel discrete event simulation (PDES) environment built with Charm++. It was specifically designed to handle problematic simulation models with fine computation granularity and a low degree of parallelism. Simulation entities in POSE are known as *posers* and perform sequential computations in response to events generated by other posers. Posers are designed to be smaller and lighter weight than the traditional logical processors (LPs) of PDES, and can be migrated to maintain load balance. POSE was the first system to introduce speculative synchronization strategies that perform potentially out-of-order work on processors to improve the cache performance and reduce the simulation overhead. Even though there is a risk that such work may be rolled back, we have demonstrated that the overall performance improves significantly.

20.4   Simulation of Large Systems

The development of PFLOPS-class computers, both in progress and planned for the future, points to a need for specialized development tools to prepare major applications for such architectures before they become operational. The BigSim project [32, 35, 34] is aimed at creating such tools that allow one to develop, and debug applications, and tune, scale and predict their performance before these machines are available. In addition, BigSim allows easier off-line experimentation with parallel performance-tuning strategies — without using the full parallel computer. To machine architects, BigSim provides a method for modeling the impact of architectural choices (including the communication network) on actual, full-scale applications. The BigSim system consists of an emulator and a simulator, capable of modeling a broad class of machines.

The BigSim emulator can take any Charm++ or AMPI program and "execute" it on a specified number of simulated processors, P, using the number of simulating processors, Q, available to the emulator. For example, one can run an MPI program meant for P=100,000 processors using only Q=2,000 available processors. If the memory requirements of the application exceed the available memory on the Q processors, the emulator employs a built-in out-of-core execution scheme that uses the file system to store the processor's data when not in use. The emulator can be used to test and debug an application, especially for scaling bugs (such as a data structure of size $P * P$). One can, for example, (a) monitor memory usage, data values and output, (b) debug for correctness, and (c) address algorithmic-scaling issues such as convergence of numerical schemes, and scaling of operation counts with problem size, all at full scale. The emulator can be used to generate traces that are used for coarse-timing predictions and for identification of performance bottlenecks, when used together with the BigSim simulator that is described next.

The BigSim simulator is a trace-driven parallel discrete event simulator built with POSE, a parallel discrete event simulation environment developed using Charm++. With reasonable detail, it simulates an integrated model for computation (processors) and communication (interconnection networks). It models architectural parameters of the target machine, including (optionally) a detailed model of the communication network. It can be used to identify potential performance bottlenecks for the simulated application such as load imbalances, communication contention and long critical paths. The simulator also generates performance traces just as a real program running on the target machine would, allowing one to carry out normal performance visualization and analysis. To predict performance of sequential code segments, the simulator allows a variable-resolution model, ranging from simple scale factors to interpolation based on performance counters (and possibly cycle-accurate simulators). To analyze performance of communication networks, one can plug in either a very simple latency model, or a detailed model of the entire communication fabric. The simulator is parallel, which potentially allows it to run very large networks.

The BigSim emulator captures traces in log files for a collection of sequential execution blocks (SEBs) on a number of processors. For each SEB, the traces store their dependencies and relative timings of messages sent by them, with source and destinations of those messages, along with additional parameters that characterize their execution time. The logs are read by the BigSim simulator which simulates the execution of the original SEBs by elapsing time, satisfying dependencies, and spawning additional messages that trigger other SEBs. Messages may be passed through a detailed network contention model called BigNetSim [30, 5]. This generates corrected times for each event which can be used to analyze its performance on the target machine.

The network contention component of the BigSim simulator, BigNetSim, has a modular design: new topologies and routing algorithms can be easily plugged into the system. We typically use virtual cut-through packet switching with a credit-based flow control to keep track of packets in the network. The system supports virtual topologies for virtual channel routing which is essential for deadlock-free routing algorithms on most topologies. Topologies already implemented include N-dimensional meshes and tori, N-dimensional hypercubes and K-ary N-trees and hybrid topologies. All topologies have physical and virtual channel-routing algorithms. Some routing algorithms are adaptive. To support adaptivity based on the network load, we developed a contention model and a load model for the interconnection network. Each port of a switch has information which is dynamically updated and fed to the routing engine to make informed decisions to minimize contention. The load model maintains load information on each of the neighbors while the contention model maintains information about the number of packets contending for a particular output port of a switch.

20.5   New Parallel Languages

The message driven runtime in Charm++ makes efficient concurrent composition of multiple modules possible, as argued earlier. Further, these modules do not have to be written in Charm++: as long as they are virtualized using the same run-time system (RTS), each module can use a different coordination mechanism to communicate and synchronize among its entities. The common RTS is called Converse, which underlies Charm++ and supports a common interface to the machine's communication capabilities in addition to a user-level threads package. Sometimes, as we saw in the case of AMPI, languages/paradigms are easier to implement on top of Charm++, instead of the lower Converse level. We expect this *multi-paradigm interoperability* to play a significant role in enhancing productivity in the future, by allowing adoption of an appropriate parallel programming paradigm for each module. This will also lead to the reuse of libraries developed in one paradigm in a broad variety of applications, independent of the paradigm used by them.

On our part, we have developed two higher level mini-languages:

1. Multiphase shared arrays (MSA) allows AMPI (or Charm++) threads to communicate via shared data arrays. However, each array is assumed to be in only one mode at a time: either read-only, exclusive-write, accumulate, or "owner-computes." At user-demarcated synchronization points, each array may change its mode. This model provides disciplined, race-free access to shared data.

2. Charisma[7] enriches Charm++ by allowing a clear expression of the global view of control, while separating parallel and sequential code. It provides object-level parallelism and follows a producer-consumer model

(or macro dataflow model) in which data is sent out as soon as it becomes available. Charisma constructs allow easy expression of various communication patterns, such as point-to-point, broadcasts, multicasts, and reductions. The Charisma compiler generates an equivalent Charm++ program.

Each language is expressive enough to capture a significant subset of applications. However, these languages are not complete by themselves: For each language, there are classes of programs that it cannot express well or at all. Yet together they are very useful because they can be used in conjunction with each other and with other paradigms, due to the interoperability mentioned above.

In the same spirit, we developed an implementation of ARMCI on top of Charm++ [8]. We invite implementors of other paradigms to create virtualized implementations on top of Charm++, to orthogonally take advantage of its intelligent runtime system and to allow efficient interoperability across a broad spectrum of languages.

20.6   Summary

In this chapter, we presented features of Charm++ and adaptive MPI, programming methodologies that we believe are well-suited for programming petascale machines. They provide a programming model that is independent of the number of processors, automate resource management, and support concurrent composition. Several full-scale applications developed using Charm++ were shown to have scaled to tens of thousands of processors, in spite of their irregular and dynamic nature. The adaptive runtime system at the heart of Charm++ can be used for supporting multiple programming models, as well as enabling interoperability among modules developed using them.

Charm++ has attained a degree of success less common among academically developed programming models: e.g., 15-20% of the CPU cycles at two national centers (National Computational Science Alliance (NCSA) and Pittsburgh Supercomputing Center (PSC)) during a one-year period were spent on Charm++ applications, mainly NAMD. It is a mature, stable programming system, with mature performance analysis tools. Thus, it should not be dismissed as just an experimental system, but rather be accepted as a mainstream model that can have an impact on the state of the art in parallel computing. We hope that the next generation of applications being developed using Charm++ will further cement its place as one of the significant models for parallel programming.

20.7 Acknowledgments

The authors wish to thank many members of the Parallel Programming Laboratory (see http://charm.cs.uiuc.edu) for their contributions to the Charm++ system. In addition, Abhinav Bhatele, Chao Huang, Chee Wai Lee, Sayantan Chakravorty, Jim Philips and a few others contributed directly to this chapter. Our collaborators, including Professors Thomas Quinn (University of Washington), Glenn Martyna (IBM), Mark Tuckerman (NYU), Klaus Schulten, Michael Heath, Philippe Geubelle, and Bob Haber, (University of Illinois at Urbana-Champaign), motivated our work, and are responsible for the success of the application codes discussed in the paper.

References

[1] T. Agarwal, A. Sharma, and L. V. Kalé. Topology-aware task mapping for reducing communication contention on large parallel machines. In *Proceedings of IEEE International Parallel and Distributed Processing Symposium*, April 2006.

[2] S. F. Ashby and J. M. May. Multiphysics simulations and petascale computing. In D. A. Bader, editor, *Petascale Computing: Algorithms and Applications*. Chapman & Hall/CRC Press, 2007.

[3] J. Barnes and P. Hut. A hierarchical $O(N \log N)$ force-calculation algorithm. *Nature*, 324:446–449, December 1986.

[4] R. Car and M. Parrinello. Unified approach for molecular dynamics and density functional theory. *Phys. Rev. Lett.*, 55:2471, 1985.

[5] N. Choudhury, Y. Mehta, T. L. Wilmarth, E. J. Bohm, and L. V. Kalé. Scaling an optimistic parallel simulation of large-scale interconnection networks. In *Proceedings of the Winter Simulation Conference*, 2005.

[6] M. D. Dikaiakos and J. Stadel. A performance study of cosmological simulations on message-passing and shared-memory multiprocessors. In *Proceedings of the International Conference on Supercomputing - ICS'96*, pages 94–101, Philadelphia, PA, December 1996.

[7] C. Huang and L. V. Kalé. Charisma: Orchestrating migratable parallel objects. In *Proceedings of 16th IEEE International Symposium on High Performance Distributed Computing (HPDC)*, pages 75–84, Monterey, CA, July 2007.

[8] C. Huang, C. W. Lee, and L. V. Kalé. Support for adaptivity in armci using migratable objects. In *Proceedings of the Workshop on Performance Optimization for High-Level Languages and Libraries*, Rhodes Island, Greece, 2006.

[9] X. Jiao, G. Zheng, P. A. Alexander, M. T. Campbell, O. S. Lawlor, J. Norris, A. Haselbacher, and M. T. Heath. A system integration framework for coupled multiphysics simulations. *Engineering with Computers*, 22(3):293–309, 2006. Special infrastructure issue.

[10] R. Jyothi, O. S. Lawlor, and L. V. Kale. Debugging support for Charm++. In *PADTAD Workshop for IPDPS 2004*, page 294. IEEE Press, 2004.

[11] L. V. Kale. Application oriented and computer science centered HPCC research, 1994.

[12] L. V. Kalé. The virtualization model of parallel programming: Runtime optimizations and the state of art. In *LACSI*. Albuquerque, NM, October 2002.

[13] L. V. Kale and M. Bhandarkar. Structured dagger: A coordination language for message-driven programming. In *Proceedings of Second International Euro-Par Conference*, volume 1123-1124 of *Lecture Notes in Computer Science*, pages 646–653, September 1996.

[14] G. Karypis and V. Kumar. A coarse-grain parallel formulation of multilevel k-way graph partitioning algorithm. *Proc. of the 8th SIAM Conference on Parallel Processing for Scientific Computing*, 1997.

[15] G. Karypis and V. Kumar. A fast and high quality multilevel scheme for partitioning irregular graphs. *SIAM J. Sci. Comput.*, 20(1):359–392, 1998.

[16] S. Kumar. *Optimizing Communication for Massively Parallel Processing*. Ph.d. thesis, University of Illinois at Urbana-Champaign, May 2005.

[17] G. Lake, N. Katz, and T. Quinn. Cosmological N-body simulation. In *Proceedings of the Seventh SIAM Conference on Parallel Processing for Scientific Computing*, pages 307–312, Philadelphia, PA, February 1995.

[18] O. Lawlor, S. Chakravorty, T. Wilmarth, N. Choudhury, I. Dooley, G. Zheng, and L. Kale. ParFUM: A parallel framework for unstructured meshes for scalable dynamic physics applications. *Engineering with Computers*, 2005.

[19] S. Mangala, T. Wilmarth, S. Chakravorty, N. Choudhury, L. V. Kale, and P. H. Geubelle. Parallel adaptive simulations of dynamic fracture events. Technical Report 06-15, Parallel Programming Laboratory, Department of Computer Science, University of Illinois, Urbana-Champaign, 2006.

[20] G.J. Martyna, A. Hughes, and M.E. Tuckerman. Molecular dynamics algorithms for path integrals at constant pressure. *J. Chem. Phys.*, 110:3275, 1999.

[21] D. Marx and J. Hutter. Ab initio molecular dynamics: Theory and implementation. *Modern Methods and Algorithms for Quantum Chemistry*, 1:301, 2000.

[22] D. Marx and M. Parrinello. Ab initio path integral molecular dynamics: basic ideas. *J. Chem. Phys.*, 104:4077, 1996.

[23] K. Schulten, J. C. Phillips, L. V. Kalé, and A. Bhatele. Biomolecular modeling in the era of petascale computing. In D. A. Bader, editor, *Petascale Computing: Algorithms and Applications*. Chapman & Hall/CRC Press, 2007.

[24] V. Springel. The cosmological simulation code GADGET-2. *Monthly Notices of the Royal Astronomical Society*, 364:1105–1134, 2005.

[25] M. Tuckerman, G. Martyna, M. Klein, and B. Berne. Efficient molecular dynamics and hybrid Monte Carlo algorithms for path integrals. *J. Chem. Phys.*, 99:2796–2808, 1993.

[26] M. E. Tuckerman, D. Marx, M. L. Klein, and M. Parrinello. Efficient and general algorithms for path integral Car-Parrinello molecular dynamics. *J. Chem. Phys.*, 104:5579, 1996.

[27] R. V. Vadali, Y. Shi, S. Kumar, L. V. Kale, M. E. Tuckerman, and G. J. Martyna. Scalable fine-grained parallelization of plane-wave-based ab initio molecular dynamics for large supercomputers. *Journal of Computational Chemistry*, 25:2006–2022, October 2004.

[28] M. S. Warren and J. K. Salmon. Astrophysical N-body simulations using hierarchical tree data structures. In *Proceedings of Supercomputing 92*, November 1992.

[29] T. Wilmarth and L. V. Kalé. Pose: Getting over grainsize in parallel discrete event simulation. In *International Conference on Parallel Processing*, pages 12–19, August 2004.

[30] T. L. Wilmarth, G. Zheng, E. J. Bohm, Y. Mehta, N. Choudhury, P. Jagadishprasad, and L. V. Kale. Performance prediction using simulation of large-scale interconnection networks in pose. In *Proceedings of the Workshop on Principles of Advanced and Distributed Simulation*, pages 109–118, 2005.

[31] G. Zheng, M. S. Breitenfeld, H. Govind, P. Geubelle, and L. V. Kale. Automatic dynamic load balancing for a crack propagation application. Technical Report Tech. Rep. 06-08, Parallel Programming Laboratory, Department of Computer Science, University of Illinois at Urbana-Champaign, June 2006.

[32] G. Zheng, G. Kakulapati, and L. V. Kalé. BigSim: A parallel simulator for performance prediction of extremely large parallel machines. In *18th International Parallel and Distributed Processing Symposium (IPDPS)*, page 78, Santa Fe, NM, April 2004.

[33] G. Zheng, O. S. Lawlor, and L. V. Kalé. *Multiple Flows of Control in Migratable Parallel Programs*. IEEE Computer Society, Columbus, OH, August 2006.

[34] G. Zheng, T. Wilmarth, P. Jagadishprasad, and L. V. Kalé. Simulation-based performance prediction for large parallel machines. *International Journal of Parallel Programming*, 33:183–207, 2005.

[35] G. Zheng, T. Wilmarth, O. S. Lawlor, L. V. Kalé, S. Adve, D. Padua, and P. Geubelle. Performance modeling and programming environments for petaflops computers and the Blue Gene machine. In *NSF Next Generation Systems Program Workshop, 18th International Parallel and Distributed Processing Symposium (IPDPS)*, page 197, Santa Fe, NM, April 2004. IEEE Press.

Chapter 21

Annotations for Productivity and Performance Portability

Boyana Norris

Mathematics and Computer Science Division, Argonne National Laboratory, 9700 S. Cass Ave., Argonne, IL 60439, norris@mcs.anl.gov

Albert Hartono

Department of Computer Science and Engineering, Ohio State University, 2015 Neil Ave., Columbus, OH 43210, hartonoa@cse.ohio-state.edu

William D. Gropp

Mathematics and Computer Science Division, Argonne National Laboratory, 9700 S. Cass Ave., Argonne, IL 60439, gropp@mcs.anl.gov

21.1 Introduction

In many scientific applications, a significant amount of time is spent in tuning codes for a particular high-performance architecture. There are multiple approaches to such tuning, ranging from the relatively nonintrusive (e.g., by using compiler options) to extensive code modifications that attempt to exploit specific architecture features. In most cases, the more intrusive code tuning is not easily reversible and thus can result in inferior performance on a different architecture or, in the worst case, in wholly non-portable code. Readability is also greatly reduced in such highly optimized codes, resulting in lowered productivity during code maintenance.

We introduce an extensible annotation system that aims to improve both performance and productivity by enabling software developers to insert annotations into their source codes that trigger a number of low-level performance optimizations on a specified code fragment. The annotations are special structured comments inside the source code and are processed by a pre-compiler to produce optimized code in a general-purpose language, such as C, C++, or Fortran.

21.2   Implementation

In this section, we describe the current implementation of our annotation software system. The annotations language is designed to be embeddable in general-purpose languages, such as C/C++ and Fortran. Our design goal is to construct an annotation system that is general, flexible, and easily extensible with new annotation syntax and corresponding code optimizations. In the following subsections, we describe the overall design of the system, followed by an overview of the annotation language syntax and code generation modules implemented to date.

21.2.1   Overall design

Our proposed approach is to exploit semantic comments, henceforth referred to as *annotations*, which are inserted into application source code. Annotations allow programmers to simultaneously describe the computation and specify various performance-tuning directives. Annotations are treated as regular comments by the compiler, but recognized by the annotation system as syntactical structures that have particular meaning.

Figure 21.1 depicts at a high level the structure and operation of the annotation system. The system begins with scanning the application source code that contains inserted annotations, and then breaking up the code into different annotated code regions. Each annotated code region is then passed to the corresponding code generator for potential optimizations. As a final point, target language code with various applied optimizations is generated for the annotated regions.

The annotation system consists of one or more code generators, each implemented as a module. Modules can be added to the system at any time without requiring modifications to the existing infrastructure. Each code-generation module can define new syntax or extend the syntax of an existing annotation definition. Using the information supplied in the annotated region, each module performs a distinct optimization transformation prior to generating the optimized code. These optimizations can span different types of code transformations that are not provided by compilers in some cases, such as memory alignment, loop optimizations, various architecture-specific optimizations, high-level algorithmic optimizations, and distributed data management.

21.2.2 Annotation language syntax

Annotations are specified by programmers in the form of comments and do not affect the correctness of the program written. We denote annotations using stylized C/C++ comments that start with /*@ and end with @*/. Both of these markers are called opening and closing *annotation delimiters*, respectively. As an example, the annotation /*@ end @*/ is used syntactically to indicate the end of an annotated code region.

Table 21.1 shows the simple grammar of the annotation language syntax. The structure of an *annotated code region* fundamentally comprises three main parts: a *leader annotation*, an *annotation body block*, and a *trailer annotation*. An annotation body block can either be simply empty or contain C/C++ source code that may include other annotated regions. A leader annotation records the *name of code-generation module* that will be loaded dynamically

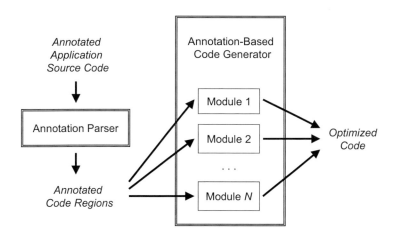

FIGURE 21.1: Overview of the annotation system.

TABLE 21.1: Annotation language grammar excerpt.

annotated-code-region	::=	*leader-annotation*
		annotation-body-block
		trailer-annotation
leader-annotation	::=	/*@ begin *module-name*
		(*module-body-block*)
		@*/
annotation-body-block	::=	
	|	*non-annotation-code annotation-body-block*
	|	*annotated-code-region annotation-body-block*
trailer-annotation	::=	/*@ end @*/

by the annotation system to optimize and generate the annotated application code. Moreover, a high-level description of the computation and several performance hints are specified in the *module body block* inside the leader annotation, and will be used as parametric input information during the optimization and code-generation phases. A trailer annotation is utilized to close an annotated code region, and it is uniformly defined as /*@ end @*/.

An example of annotated application code can be seen in Figure 21.2, where lines 2–6 make up the annotated code region with line 2 and line 6 as the leader and trailer annotations, respectively, and lines 3–5 make up the annotation body block. The annotation code-generation module identified in this example has the `Variable` name, and acquires the "x[],y[]" coding text as one of its input parameters.

```
1.  void axpy_1(int n, double *y, double a, double *x) {
2.    /*@ begin Variable (x[],y[]) @*/
3.    int i;
4.    for (i=0; i < n; i++)
5.      y[i] = y[i] + a * x[i];
6.    /*@ end @*/
7.  }
```

FIGURE 21.2: Example of annotated application source code.

21.2.3   System extensibility

As we have seen in Section 21.2.2, provided with the module name specified in the leader annotation, the annotation system dynamically seeks the corresponding code-generation module and then loads and utilizes it to transform and generate code. If the pertinent module cannot be found in the system, an error message will be reported to the users and then the annotation system process is suspended. In this fashion, the annotation system becomes flexible and can be extended easily without evolving the annotation software system that is currently available.

21.2.4   Code-generation module

Figure 21.3 portrays the general structure of the annotation-based code-generation module. In order to generate an optimized code, each module takes two classes of input parameters: code texts that are specified in the *module-body-block* and *annotation-body-block* fields. The module body normally includes information that is essential for performing code optimization and generation, such as multidimensional array variables, loop structures, loop-blocking factors, and so forth. In order to extract such information, new

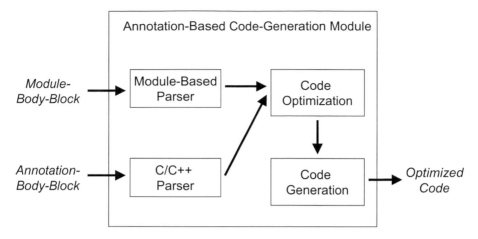

FIGURE 21.3: Structure of code-generation module.

language syntax and a corresponding parser component are therefore crucially required to be implemented in each code-generation module. On the other hand, extracting information from the annotation body segment may necessitate a challenging task to build a full-blown C/C++ parser. Fortunately, many optimization cases can still be performed properly without parsing the entire C/C++ source; thus, such a C/C++ parser is *optional*.

It is to be noted that a nested annotation code is possible when the annotation body block contains other annotated regions. Hence, the optimization and code generation must be carried out recursively by the annotation system to handle nested annotations.

Some details on several code-generation modules that have been developed and incorporated in the annotation systems are given as follows.

21.2.4.1   Memory-alignment module

The objective of this code-generation module is to exploit memory alignment optimizations on the Blue Gene/L architecture. The dual floating-point unit (FPU) of the Blue Gene/L's PowerPC 440d processor provides special instructions for parallelizing floating-point computations. The IBM XL compiler attempts to pair contiguous data values on which it can operate in parallel. Therefore, computations performance can be improved by specifying floating-point data objects that reside in contiguous memory blocks and are correctly aligned. In order to facilitate such parallelization, the compiler requires additional directives aimed *to remove possibilities of aliasing* and *to check for data alignment*.

We illustrate the implementation of this module using a simple example previously shown in Figure 21.2. As we can see at the specification in the

```
1.   void axpy_1(int n, double *y, double a, double *x) {
2.     /*@ begin Variable (x[],y[]) @*/
3.     #pragma disjoint (*x, *y)
4.     if ((((int)(x)|(int)(y)) & 0xF) == 0) {
5.       __alignx(16,x);
6.       __alignx(16,y);
7.       int i;
8.       for (i=0; i < n; i++)
9.         y[i] = y[i] + a * x[i];
10.    } else {
11.      int i;
12.      for (i=0; i < n; i++)
13.        y[i] = y[i] + a * x[i];
14.    }
15.    /*@ end @*/
16. }
```

FIGURE 21.4: Optimized version of annotated code shown in Figure 21.2.

leader-annotation segment, this module is named `Variable`. The module body comprises a list of array variables. In this example, `x[]` and `y[]` are the array variables to be parallelized.

The resulting optimized version that corresponds to the example given above can be seen in Figure 21.4. A `#pragma disjoint` directive (line 3) is injected into the optimized code to inform the compiler that none of the listed identifiers shares the same storage location within the scope of their use. Knowing such information will enable the compiler to evade the overhead cost of reloading data values from memory each time they are referenced, and to operate on values already resident in registers. Note that this directive demands that the two identifiers must be really disjoint. If the identifiers in fact share the same memory address, the computations will function incorrectly.

Furthermore, the Blue Gene/L architecture requires the addresses of the two data values, which are loaded in parallel in a single cycle, to be aligned such that the loaded values do not cross a cache-line boundary. If they cross this boundary, a severe performance penalty will be imposed due to the alignment trap generated by the hardware. So, testing for data alignment is important. In the optimized code example, checking for data alignment is executed in line 4. The following two lines (lines 5 and 6) show calls to the `__alignx` intrinsic functions. These function calls are used to notify the compiler that the arriving data is correctly aligned, so the compiler can generate more efficient loads and stores.

The complete grammar for the new language syntax of memory alignment module is shown in Table 21.2. In addition to one-dimensional arrays, multidimensional arrays can also be specified in the module-body block. One example is `a[i][]`, which will refer the compiler to the starting address location of the ith row of the two-dimensional array `a`. The empty bracket is basically used to refer to the starting address where a sequence of adjacent data to be computed is stored in the memory. Another legitimate example is `b[i][j][]`.

TABLE 21.2: New language grammar of the memory alignment module.

align-module-body-block	::=	*array-variable-list*
array-variable-list	::=	*array-variable*
	|	*array-variable* , *array-variable-list*
array-variable	::=	*array-variable-name dimension-list*
dimension-list	::=	*dimension*
	|	*dimension dimension-list*
dimension	::=	[]
	|	[*variable-name*]
	|	[*constant*]

However, the `c[][i][j]` specification is invalid because the empty bracket position contradicts with the fact that C/C++ stores its array elements in a row-major order. The `c[][i][j]` array variable will only become valid when used to annotate Fortran source, since Fortran employs a column-major arrangement rule. Such a data arrangement rule can be enforced easily by this module using a simple semantic analysis.

It is to be noted that the statements in the annotation-body block (lines 3–5 in Figure 21.2) are used directly by this module with no parsing mechanism. Thus, a complete C/C++ parser component is not needed, simplifying the implementation of this module.

21.2.4.2 Loop-optimization module

The primary goal of this code-generation module is to provide extensible high-level abstractions for expressing generic loop structures in conjunction with a variety of potential low-level optimization techniques, such as loop unrolling, skewing, and blocking for cache, including other architecture-specific optimizations. Two optimization strategies that have been constructed and integrated into the annotation system are *loop unrolling* and *automated simdization*.

An overview of the new language syntax introduced by the code-generation module can be found in Table 21.3. Essentially, *a subset of C statements* and a newly defined *transformation statement* constitute the language grammar of this module. For compactness, further details on each of the C statement clauses are not given in this grammar. Many of the C-language features, such as declarations, variable pointers, switch statements, enumeration constants, cast expressions, and so forth, are excluded from the language grammar selection due to the purpose of reducing the implementation complexity of this module.

A new transformation statement clause is added into the grammar to achieve the flexibility of extending the loop optimization module with new *transformation submodules*. Using the provided submodule name, the loop optimization

TABLE 21.3:   Overview of language structure of the loop optimization module.

loop-opt-module-body-block	::=	*statement-list*
statement-list	::=	*statement*
	|	*statement statement-list*
statement	::=	*labeled-statement*
	|	*expression-statement*
	|	*compound-statement*
	|	*selection-statement*
	|	*iteration-statement*
	|	*jump-statement*
	|	*transformation-statement*
transformation-statement	::=	**transform** *submodule-name*
		(*keyword-argument-list*) *statement*
keyword-argument-list	::=	*keyword-argument*
	|	*keyword-argument* **,**
		keyword-argument-list
keyword-argument	::=	*keyword-name* **=** *expression*

module dynamically searches the corresponding submodule and then utilizes it to transform the transformation statement body. Additional data specified in the keyword argument list are obtained by transformation submodule to perform its code transformation procedure.

The example in Figure 21.5 demonstrates how to annotate application code with a simple *loop unrolling* optimization that aims to increase the cache-hit rate and to reduce branching instructions by combining instructions that are executed in multiple loop iterations into a single iteration. The keyword used to identify the loop-optimization module is `LoopOpt`. The `Loop` name denotes the transformation submodule, of which the most basic function is to represent general loop structures. There are four fundamental parameters used to create a loop structure: the index variable name (`index`), the index's lower-bound value (`lower_bound`), the index's upper-bound value (`upper_bound`), and the iteration-step size (`step`). For instance, the simple loop structure shown below

```
for (i = 0; i <= n-1; i++)
   x[i] = x[i] + 1;
```

can be represented using the following transformation statement.

```
transform Loop(index=i, lower_bound=0, upper_bound=n-1, step=1)
   x[i] = x[i] + 1;
```

Annotating a loop structure with loop-unrolling optimization is straightforward: add another keyword argument of the form "`unroll = n`," where `n` signifies how many times the loop body will be unrolled/replicated in the generated code. In the Figure 21.5 example, the loop body is unrolled four times, resulting in the unrolled-loop structure shown in lines 15–21. The following loop (lines 22–23) is generated for the remaining iterations that are not executed in the unrolled loop. Additionally, the generated code also includes the original loop (lines 12–13) that can be executed through setting the `ORIGLOOP` (line 11) preprocessor variable accordingly.

```
1.  void ten_reciprocal_roots(double* x,        1.  void ten_reciprocal_roots(double* x,
2.                           double* f)          2.                           double* f)
3.  {                                            3.  {
4.      int i;                                   4.      int i;
5.      /*@ begin LoopOpt(                       5.      /*@ begin LoopOpt(
6.          transform Loop(unroll=4,             6.          transform Loop(unroll=4,
7.                  index=i, lower_bound=0,       7.                  index=i, lower_bound=0,
8.                  upper_bound=10, step=1)       8.                  upper_bound=10, step=1)
9.              f[i] = 1.0 / sqrt(x[i]);          9.              f[i] = 1.0 / sqrt(x[i]);
10.     ) @*/                                    10.     ) @*/
11.     for (i = 0; i < 10; i++)                 11.     #if ORIGLOOP
12.         f[i] = 1.0 / sqrt(x[i]);             12.     for (i = 0; i < 10; i++)
13.     /*@ end @*/                              13.         f[i] = 1.0 / sqrt(x[i]);
14. }                                            14.     #else
                                                 15.     for (i = 0; i <= 10 - 3; i += 4)
                                                 16.     {
                                                 17.         f[i] = 1.0 / sqrt(x[i]);
                                                 18.         f[i + 1] = 1.0 / sqrt(x[i + 1]);
                                                 19.         f[i + 2] = 1.0 / sqrt(x[i + 2]);
                                                 20.         f[i + 3] = 1.0 / sqrt(x[i + 3]);
                                                 21.     }
                                                 22.     for (; i <= 10; i += 1)
                                                 23.         f[i] = 1.0 / sqrt(x[i]);
                                                 24.     #endif
                                                 25.     /*@ end @*/
                                                 26. }
```

FIGURE 21.5: Cache-optimization annotation example: annotated code (left) and resulting generated code with an unrolled loop body (right) are shown.

As mentioned earlier in Section 21.2.4.1, on Blue Gene/L architecture, the IBM's XL C/C++ and XL Fortran compilers enable us to speed up computations by exploiting the PowerPC 440d's Double Hummer dual FPU to execute two floating-point operations in parallel. The XL compilers include a set of highly-optimized *built-in functions* (also called *intrinsic procedures*) that have an almost one-to-one correspondence with the Double Hummer instruction set. These functions are designed to efficiently manipulate complex-type variables, and also include a function that converts noncomplex data to complex types. Hence, programmers can manually parallelize their code by using these intrinsic functions.

When simple arithmetic operations (addition, subtraction, and multiplication) involve complex data types, the compiler automatically parallelizes the computations by using parallel add, subtract, and multiply instructions.

These parallel arithmetic operations are referred to as single-instruction, multiple-data (SIMD) operations. Since none of the available intrinsic functions performs complex arithmetic calculations, the compiler is incapable of automatically generating parallel instructions for complex arithmetic operations, such as the assignment statement below:

```
z[0] = a[0] + b[0] + 8.5 * c[0];
z[1] = a[1] + b[1] + 8.5 * c[1];
```

One resolution to parallelize the above expression is first to divide the complex expression into a sequence of simple arithmetic expressions, and then to translate each simple operation to its corresponding intrinsic functions. We refer to this process as *automated simdization*. For the above expression example, we can transform it using an intermediate variable i to perform the following two-step computation:

```
i[0] = b[0] + 8.5 * c[0];
i[1] = b[1] + 8.5 * c[1];
z[0] = a[0] + i[0];
z[1] = a[1] + i[1];
```

which can be mechanically simdized into the parallel code fragment below:

```
double _Complex i, _i_1, _i_2, _i_3, _i_4;
_i_1 = __lfpd(&b[0]) ;
_i_2 = __lfpd(&c[0]) ;
i = __fxcpmadd(_i_1, _i_2, 8.5) ;
_i_3 = __lfpd(&a[0]) ;
_i_4 = __fpadd(i, _i_3) ;
__stfpd(&z[0], _i_4) ;
```

We have developed a simdization transformation module as an extension of the loop-optimization module. An example of an automated simdization annotation can be observed in Figure 21.6. The simdization submodule is denoted with BGLSimd. This annotated code example shows the case when the statement to be simdized occurs inside the body of the loop that will be unrolled. Therefore, simdization and unrolling transformations are applied simultaneously. In this coupled-transformation process, each simdized statement must be associated to a particular unrolled loop. To create this association, a keyword argument that has a loop_id keyword identifier (lines 6 and 10) must be included. Loop identification is especially necessary when the statement to be simdized is contained within multiple-nested unrolled loops.

We note that automated simdization requires that the associated loop to be unrolled must have unit-stride access (i.e., step_size = 1). Another important semantic constraint in this case is that, given the fact that the number of parallel floating-point units of Blue Gene/L is two, the associated loop-unrolling factor must be divisible by two.

```
1.  void vector_sub(double* x, double* a,      1.  void vector_sub(double* x, double* a,
2.                   double* b, int n)           2.                   double* b, int n)
3.  {                                            3.  {
4.      int i;                                   4.      int i;
5.      /*@ begin LoopOpt(                       5.      /*@ begin LoopOpt(
6.          transform Loop(loop_id=lp1,          6.          transform Loop(loop_id=lp1,
7.              unroll=4, index=i,               7.              unroll=4, index=i,
8.              lower_bound=0,                   8.              lower_bound=0,
9.              upper_bound=n-1, step=1)         9.              upper_bound=n-1, step=1)
10.         transform BGLSimd(loop_id=lp1)       10.         transform BGLSimd(loop_id=lp1)
11.             x[i] = a[i] - b[i];              11.             x[i] = a[i] - b[i];
12.     ) @*/                                    12.     ) @*/
13.     for (i = 0; i < n; i++)                  13.     #if ORIGLOOP
14.         x[i] = a[i] - b[i];                  14.     for (i = 0; i < n; i++)
15.     /*@ end @*/                              15.         x[i] = a[i] - b[i];
16. }                                            16.     #else
                                                 17.     for (i = 0; i <= n - 1 - 3; i += 4)
                                                 18.     {
                                                 19.         {
                                                 20.         double _Complex _i_1, _i_2, _i_3;
                                                 21.         _i_1 = __lfpd(&a[i]);
                                                 22.         _i_2 = __lfpd(&b[i]);
                                                 23.         _i_3 = __fpsub(_i_1, _i_2);
                                                 24.         __stfpd(&x[i], _i_3);
                                                 25.         }
                                                 26.         {
                                                 27.         double _Complex _i_1, _i_2, _i_3;
                                                 28.         _i_1 = __lfpd(&a[i + 2]);
                                                 29.         _i_2 = __lfpd(&b[i + 2]);
                                                 30.         _i_3 = __fpsub(_i_1, _i_2);
                                                 31.         __stfpd(&x[i + 2], _i_3);
                                                 32.         }
                                                 33.     }
                                                 34.     for (; i <= n - 1; i += 1)
                                                 35.         x[i] = a[i] - b[i];
                                                 36.     #endif
                                                 37.     /*@ end @*/
                                                 38. }
```

FIGURE 21.6: Annotation example of automatic simdization for Blue Gene/L architecture: annotated code (left) and resulting generated code with simdized and unrolled loop body (right) are shown.

We can further speed up the simdized code by exploiting *common subexpression elimination* (CSE), a classical compiler-optimization approach used to reduce the number of operations, where intermediates are identified that can be computed once and stored for use multiple times later. Identification of effective common subexpressions is employed during the subdivision of the complex arithmetic expression into a sequence of simple expressions. We have developed an exhaustive CSE algorithm that is guaranteed to find optimal solutions. However, the exponential growth of its search time makes an exhaustive search approach prohibitively expensive for solving complex arithmetic equations. Therefore, one of our future work goals is to develop a heuristic CSE algorithm that is able to find a near-optimal solution in polynomial time.

TABLE 21.4:   Memory bandwidth of
$a = b + ss * c$ on the Blue Gene/L, where a, b, and c
are arrays of size n, and ss is a scalar.

Array Size n	No Annotations (MB/s)	Annotations (MB/s)
10	1920.00	2424.24
100	3037.97	6299.21
1,000	3341.22	8275.86
10,000	1290.81	3717.88
50,000	1291.52	3725.48
100,000	1291.77	3727.21
500,000	1291.81	1830.89
1,000,000	1282.12	1442.17
2,000,000	1282.92	1415.52
5,000,000	1290.81	1446.48

21.3   Performance Studies

In this section we present some performance results for performance annotations applied to operations for which tuned-library implementations do not exist or perform inadequately.

21.3.1   STREAM benchmark

Preliminary results from employing simple annotations for uniprocessor optimizations are given in Table 21.4. These data describe the performance of an example array operation from the STREAM benchmark [13]. This computation is similar to some in accelerator-modeling codes, such as VORPAL's particle push methods [14]. The achieved memory bandwidth of the compiler-optimized version is significantly lower than that of the annotated version. The latter includes annotations specifying that the array variables are disjoint and should be aligned in memory, if possible, and that the loop should be unrolled. The same compiler options were used for both the original and the annotated versions. Given the annotated code as input, the annotation tool generates many tuned versions of the same operation, using different optimization parameters. This annotation-driven empirical optimization must be redone only when the semantics or the optimization parameters of the annotations are changed or the code is ported to a new platform.

Figure 21.7 shows a simple annotation example for the Blue Gene/L that targets memory-alignment optimizations. Here, the annotations are shown

```
void axpy_1(int n, double *y,              void axpy_1(int n, double *y,
            double a, double *x)                       double a, double *x)
{ /*@ begin Variable (x[],y[])  @*/        { /*@ begin Variable (x[],y[])  @*/
  int i;                                     #pragma disjoint (*x, *y)
  for (i=0; i < n; i++)                       if ((((int)(x)|(int)(y)) & 0xF) == 0) {
    y[i] = y[i] + a * x[i];                     __alignx(16,x);
  /*@ end @*/ }                                 __alignx(16,y);
                                                int i;
                                                for (i=0; i < n; i++)
                                                  y[i] = y[i] + a * x[i];
                                              } else {
                                                int i;
                                                for (i=0; i < n; i++)
                                                  y[i] = y[i] + a * x[i];
                                              }
                                              /*@ end @*/ }
```

FIGURE 21.7: Memory-related annotation example: annotated code (left) and resulting generated code with optimized Blue Gene/L pragmas and alignment intrinsic calls (right) are shown.

as C comments starting with /*@. The Variable annotation directive results in the generation of architecture-specific preprocessor directives, such as pragmas, and calls to memory-alignment intrinsics, including a check for alignment. Even these simple optimizations can lead to potentially significant performance improvements. Table 21.4 shows gains of up to 60% in memory bandwidth with annotations.

What makes annotations especially powerful is that they are not limited to certain operations and can be applied to complex computations involving many variables and assignments containing long expressions. Thus, annotations can be used for arbitrary operations, exploiting the developer's understanding of the application to perform low-level code optimizations. Such optimizations may not be produced by general-purpose compilers because of the necessarily conservative nature of program analysis for languages such as Fortran and C/C++. These optimizations include low-level tuning for deep-memory hierarchies, through loop blocking, tiling, and unrolling, as well as composing linear algebra operations and invoking specialized algorithms for key computations. A simple unrolling optimization example for computations involving one-dimensional arrays is shown in Figure 21.5. More advanced optimizations on higher-dimensional arrays or other data structures, such as matrices, would present even greater opportunities for cache optimizations. Our aim is to use existing tools for performing such code optimization transformations where possible; the examples here merely illustrate the sorts of transformations that are sometimes necessary for performance and, because they are both ugly and system specific, are rarely performed in application codes.

21.3.2   AXPY operations

We consider generalized *AXPY* operations of the form $y = y + a_1x_1 + \cdots + a_nx_n$, where $a_1, \ldots, a_n$ are scalars and $y, x_1, \ldots, x_n$ are one-dimensional arrays. These operations are more general forms of the triad operation discussed in the previous section. Figure 21.8 shows the performance of this computation for various array sizes when $n = 4$ on the Blue Gene/L at Argonne National Laboratory. Included are timing and memory bandwidth results for five versions of the code: a simple loop implementation without any library calls (labeled "Original"), two BLAS-based implementations that use the Goto BLAS library [5, 6] and the ESSL [4], respectively, and two annotated versions. The first annotated version contains only variable alignment and loop unrolling annotations, while the second additionally contains a BGLSimd annotation similar to the one illustrated in Fig. 21.6. For our earliest experiments, the ESSL was the only BLAS library available; Goto BLAS was added more recently. All versions were compiled with the same aggressive compiler optimization options. The performance improvement of the annotated version over the simple loop (original) version is between 78% and 488% (peaking for array size 100). SIMD operations were significantly effective only for certain array sizes, resulting in a factor of 6 improvement over the simple loop version. ESSL exhibited very poor performance compared to Goto BLAS. Both annotated versions outperformed the Goto BLAS version by 33% to 317% depending on the array sizes. Improvement over BLAS can be typically expected in most cases where several consecutive interdependent calls to BLAS subroutines are made. The AXPY and similar computations dominate certain types of codes, such as some automatically generated Jacobian computations, but tuned library implementations do not support such operations directly; hence, annotation-driven optimization can have significant positive impact on performance. Implementations that rely on calls to multiple tuned library subroutines suffer from loss of both spatial and temporal locality, resulting in inferior memory performance.

21.4   Related Work

In this section we present a brief overview of other approaches to performance optimization through raising the level of abstraction. We have divided related work into several categories corresponding to the main characteristics of each approach.

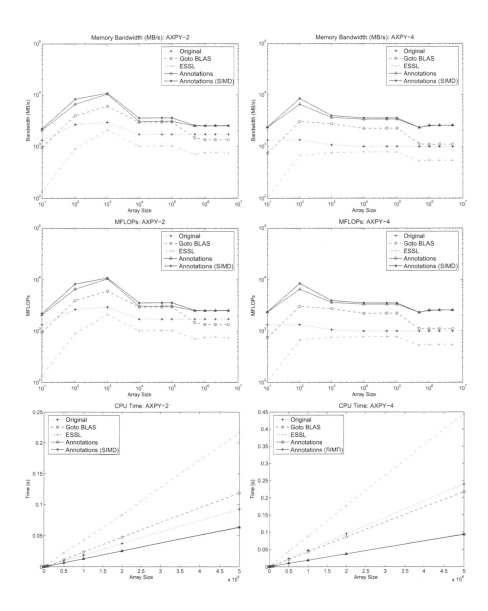

FIGURE 21.8: Performance on the Blue Gene/L for AXPY-2 (left) and AXPY-4 (right) operations is shown.

21.4.1   Self-tuning libraries and code

Active libraries. Active libraries [17, 2] such as ATLAS [20], unlike traditional libraries, are geared toward the generation and optimization of executable code. Some active libraries, such as the Blitz++ library [16], rely on specific language features and exploit the compiler to generate optimized code from high-level abstractions. The Broadway [12] compiler can be viewed as a specific instance of a system for supporting active libraries. Broadway gives domain-specific compiler optimizations based on user-specified annotation files expressing domain expertise.

Meta-programming techniques. *Expression templates* furnish a C++ meta-programming technique for passing expressions as function arguments [15]. The Blitz++ library [16] employs expression templates to generate customized evaluation code for array expressions. This approach remedies performance problems due to the noncomposability of operations when using traditional libraries such as the BLAS or language features such as operator overloading.

Programmable syntax macros [19] deliver a portable mechanism for extending a general-purpose compiler; they enable the person writing the macro to act as a compiler writer. The macro language is C, extended with abstract syntax tree (AST) types and operations on ASTs. While programmable syntax macros are general and powerful, the software developer must have significant compiler-writing expertise in order to implement desired language extensions.

A *meta-object protocol* (MOP) [11, 1] is an object-oriented interface for programmers enabling them to customize the behavior and implementation of programming languages. MOPs supply control over the compilation of programs. For example, a MOP for C++ can provide control over class definition, member access, virtual function invocation, and object creation.

Our annotations approach differs from these meta-programming techniques in that it is meant to be easily extensible without requiring a developer to have compiler expertise. Because of its generality and extensibility, it is not specific to a particular library, domain, or programming language.

21.4.2   Compiler approaches

Domain-specific languages and compilers. Domain-specific languages (DSLs) provide specialized syntax that raises the level of abstraction for a particular problem domain. Examples of DSLs include YACC for parsing and compilers, GraphViz for defining directed graphs, and Mathematica for numerical and symbolic computation. DSLs can be either stand-alone and used with an interpreter or compiler or they can also be embedded in general-purpose languages (e.g., as macros) and preprocessed into the general-purpose language prior to compilation. Our annotations approach includes the use of an embedded language, but it is a more general, extensible language, not a domain-specific one.

Telescoping languages. The Telescoping languages project [10, 9, 8] defines a strategy for generating high-performance compilers for scientific domain languages. To date, these efforts have focused on extensions to Matlab as defined by domain-specific toolboxes. The compiler generation process consists of several steps. First, a library-preprocessing phase is applied to the domain-specific code to extensively analyze and optimize collections of libraries that define an extended language. Results of this analysis are collected into annotated libraries and used to generate a library-aware optimizer. That optimizer uses the knowledge gathered during preprocessing to directly carry out fast and effective optimization of high-level scripts.

Aspect-oriented programming. Aspect-oriented programming [3] provides a way to describe and manage crosscutting parts of a program. The classic example is support for logging, which is defined in one place (the logging module) but used throughout the code. Aspect-oriented programming provides a way to make such crosscutting items part of the language, providing full language support for the aspects. Aspect-oriented programming has been applied to scientific computing such as sparse-matrix programming [7]. Our annotations approach trades the advantages of such full integration with the language with the flexibility and speed to quickly adapt to the needs of applications.

Unlike compiler approaches, we do not implement a full-blown compiler or compiler generator; rather, we define a pre-compiler that parses the language-independent annotations and includes code generation for multiple general-purpose languages, such as C and Fortran.

21.4.3 Performance-related user annotations

User annotations are used for other performance-related purposes not directly related to code optimization. One example is *performance assertions* [18], which are user annotations for explicitly declaring performance expectations in application source code. A runtime system gathers performance data based on the user's assertion and verifies this expectation at runtime. Unlike our annotations system, this approach does not guide or perform any code modifications; rather, it automates the testing of performance properties of specific portions of complex software systems.

The Broadway [12] compiler mentioned earlier also employs annotations to guide the generation of library calls. Thus, annotation files are associated with a particular *library*, and each library specifies its own analysis problems and code transformations. The annotation language is used to define dependence information about the library interface, as well as domain-specific analysis problems, which are used in the compiler's program analyses of the application. The annotations are also used to describe domain-specific optimizations, which are expressed as code transformations contingent on the analysis results. The annotations language has extensive syntax in order to allow the expression of dependencies, analyses, and transformations. Significant compiler expertise is needed in order to create an annotation file for a given library. By contrast, the performance annotations we describe in this chapter are more general, with a simpler syntax, and are meant to be associated with particular, usually small, code fragments within arbitrary applications. No compiler expertise is required of the program developer in order to use performance annotations to specify code optimization hints.

Our annotations system differs from the approaches described above in that it is general purpose and embedded in a general-purpose language. In addition to code generation, it includes a performance-evaluation runtime system that enables the selection of the best-performing version of generated code. Finally, it is designed to be easily extensible, allowing any developer to add both general and architecture-specific optimizations without requiring compiler expertise.

21.5   Summary and Future Directions

We have described the initial implementation of an annotation-based performance tuning system that is aimed at improving both performance and productivity in scientific software development. The annotations language is extensible and embeddable in general-purpose languages. We have demonstrated performance improvements in several computational kernels.

In addition to code optimizations targeting single-processor performance, we plan to expand our annotation language with syntax for distributed operations and data structures commonly used in scientific computing, such as parallel grid updates for problems discretized on a regular grid. In that case, the user annotation will describe the grid at a very high level, using global dimensions and the type and width of the stencil used. Then, we will define high-level annotations for initialization and point update using global grid coordinates (i.e., basically using sequential code). The job of the annotation processor would be to take the annotated source code and generate efficient parallel implementation of the distributed operations expressed in the annotations and global-coordinate code. An advantage of annotations over other language-based approaches is that the data structure support can be customized to the application. For example, support for staggered grids or C-grids (semi-regular grids with special properties, particularly at the boundaries) can be added quickly with an annotations-based approach.

21.6 Acknowledgments

This work was supported by the Mathematical, Information, and Computational Sciences Division subprogram of the Office of Advanced Scientific Computing Research, Office of Science, U.S. Department of Energy, under contract DE-AC02-06CH11357.

References

[1] S. Chiba. A metaobject protocol for C++. In *ACM SIGPLAN Conference on Object-Oriented Programming Systems, Languages and Applications*, pages 285–299, October 1995.

[2] K. Czarnecki, U. Eisenecker, R. Glück, D. Vandevoorde, and T. Veldhuizen. Generative programming and active libraries (extended abstract). In M. Jazayeri, D. Musser, and R. Loos, editors, *Proceedings of Generic Programming*, volume 1766 of *Lecture Notes in Computer Science*, pages 25–39. Springer-Verlag, 2000.

[3] T. Elrad, R. E. Filman, and A. Bader. Aspect-oriented programming: Introduction. *Commun. ACM*, 44(10):29–32, 2001.

[4] Engineering scientific subroutine library (ESSL) and parallel ESSL. http://www-03.ibm.com/systems/p/software/essl.html, 2006.

[5] K. Goto. High-performance BLAS by Kazushige Goto, 2007. http://www.tacc.utexas.edu/~kgoto/.

[6] K. Goto and R. van de Geijn. High-performance implementation of the Level-3 BLAS. Technical Report TR-2006-23, The University of Texas at Austin, Department of Computer Sciences, 2006.

[7] J. Irwin, J. Loingtier, J. Gilbert, G. Kiczales, J. Lamping, A. Mendhekar, and T. Shpeisman. Aspect-oriented programming OS sparse matrix code. In *Proceedings of the Scientific Computing in Object-Oriented Parallel Environments First International Conference*, pages 249–256. Springer-Verlag, 1997.

[8] K. Kennedy. Telescoping languages: A compiler strategy for implementation of high-level domain-specific programming systems. In *Proceedings of IPDPS 2000*, May 2000.

[9] K. Kennedy, B. Broom, A. Chauhan, R. Fowler, J. Garvin, C. Koelbel, C. McCosh, and J. Mellor-Crummey. Telescoping languages: A system for automatic generation of domain languages. *Proceedings of the IEEE*, 93(3):387–408, 2005. This provides a current overview of the entire Telescoping Languages Project.

[10] K. Kennedy and *et al.* Telescoping languages project description. `http://telescoping.rice.edu`, 2006.

[11] G. Kiczales, J. des Rivieres, and D. G. Bobrow. *The Art of the Meta-Object Protocol*. MIT Press, Cambridge, MA, 1991.

[12] C. Lin and S. Z. Guyer. Broadway: A compiler for exploiting the domain-specific semantics of software libraries. *Proceedings of the IEEE*, 93(2):342–357, July 2005.

[13] J. McCalpin. STREAM: Sustainable Memory Bandwidth in High Performance Computers. `http://www.cs.virginia.edu/stream/`, 2006.

[14] P. Messmer and D. L. Bruhwiler. A parallel electrostatic solver for the VORPAL code. *Comp. Phys. Comm.*, 164:118, 2004.

[15] T. Veldhuizen. Expression templates. *C++ Report*, 7(5):26–31, June 1995.

[16] T. L. Veldhuizen. Blitz++: The library that thinks it is a compiler. In E. Arge, A. M. Bruaset, and H. P. Langtangen, editors, *Modern Software Tools for Scientific Computing*. Birkhauser (Springer-Verlag), Boston, 1997.

[17] T. L. Veldhuizen. *Active Libraries and Universal Languages*. PhD thesis, Indiana University, Computer Science Department, May 2004.

[18] J. Vetter and P. Worley. Asserting performance expectations. In *Proceedings of the SC2002*, 2002.

[19] D. Weise and R. Crew. Programmable syntax macros. In *SIGPLAN Conference on Programming Language Design and Implementation*, pages 156–165, 1993.

[20] R. C. Whaley and J. Dongarra. Automatically tuned linear algebra software. `http://www.supercomp.org/sc98/TechPapers/sc98_FullAbstracts/Whaley814/INDEX.HTM`, 1998. Winner, best paper in the systems category, SC98: High Performance Networking and Computing.

Chapter 22

Locality Awareness in a High-Productivity Programming Language

Roxana E. Diaconescu

Yahoo! Inc., Burbank, CA

Hans P. Zima

Jet Propulsion Laboratory, California Institute of Technology, Pasadena, CA
Institute of Computational Science, University of Vienna, Austria

22.1 Introduction

Efficient management of locality is a key requirement for today's high performance computing (HPC) systems, most of which have a physically distributed memory. The standard programming paradigm for these systems has been based for more than a decade on the extension of sequential programming languages with message-passing libraries, in a processor-centric model for programming and execution. It is commonly understood that this approach leads to complex and error-prone programs, due to the way in which algorithms and communication are inextricably interwoven.

Some programming languages, such as High Performance Fortran (HPF) [15], provide high-level support for controlling locality by associating distributions with arrays, focusing on a set of built-in distribution classes such as *block*, *cyclic*, and *indirect*. However, such languages have the disadvantages of being constrained by the semantics of their base language, of providing just a single level of data parallelism, and of supporting only a limited range of distributions.

We are working on a new language called *Chapel* [9, 5] which is being designed to improve productivity for programmers of parallel machines. Chapel supports general parallel computation via a global-view, locality-aware, multithreaded programming model. It strives to narrow the gap between mainstream sequential languages and current parallel languages by supporting object-oriented programming, generic programming, and type and value safety. In Chapel, data locality is expressed via first-class objects called *distributions*. Distributions apply to collections of indices represented by *domains*, which determine how *arrays* associated with a domain are to be mapped and allocated across abstract units of uniform-memory access called *locales*. Chapel offers an *open* concept of distributions, supported by a set of classes which establish the interface between the programmer and the compiler. Components of distributions are overridable by the user, at different levels of abstraction, with varying degrees of difficulty. Well-known standard distributions can be specified along with arbitrary irregular distributions using the same uniform framework. The vision is that Chapel will be an open-source programming language, with an open-distribution interface that allows experts and non-experts the design of new distribution classes and the construction of distribution libraries that can be reused, extended, and optimized. Data-parallel computations are expressed in Chapel via `forall` loops, which concurrently iterate over domains.

Our design is governed by the following goals:

- **Orthogonality between data mapping and algorithms:** One of our goals is to separate the specification of data mapping, or distribution, from the algorithm and thus allow programmers the formulation of data-parallel programs in a sequential-looking fashion. We approach this goal through a flexible design, in which algorithms operating on dense, sparse, or irregular structures can switch between different distributions without needing to change the core computations on data aggregates.

- **Increased productivity of data parallel programming:** Another goal is to increase the productivity for writing data-parallel applications. We achieve this goal in two ways, first, through reuse and composition of distributions, and secondly, by concealing synchronization, communication, and thread management from the programmer. Once the programmer has specified the distribution aspects of the problem such as data mapping and layout, the compiler transparently handles the aspects of thread management, synchronization, and communication.

- **Increased efficiency of the resulting target programs:** The real end goal of parallel computations is to speed up an application by taking full advantage of the underlying architecture. We achieve this goal by giving programmers explicit control of data mapping, layout, and iteration based on their knowledge of the problem. We expect the resulting target code efficiency to be similar to that for fully manually parallelized programs using the Message Passing Interface (MPI) [14] library.

The rest of this chapter describes in detail the object-oriented design of the distribution interfaces. Section 22.2 describes the main Chapel abstractions that play a central role in the data-parallel model and thus, in the distribution framework. Section 22.3 describes in detail the distribution interface, showing how programmers can incrementally specialize distributions to fine-tune the efficiency of their parallel algorithms. Section 22.4 provides examples that show how various complex problems and algorithms can be expressed using our design and the benefits of this approach. Section 22.5 describes our current implementation, and outlines efficiency considerations. Section 22.6 contrasts our approach to related work. Finally, Section 22.7 concludes the chapter and outlines future research directions.

22.2   Basic Chapel Concepts Related to Data Parallelism

Chapel's data-parallel-programming model relies on the concepts of *domains* and *arrays*. This section provides a brief introduction to these language features.

22.2.1   Domains

The primary component of a domain is its *index set* — a set of index values that can be distributed across multiple locales, used to allocate data aggregates (arrays), and iterated over to specify serial or parallel computation. Domains are first-class entities, generalizing the *region* concept introduced in ZPL [6]. An example of a simple domain declaration in Chapel is as follows:

var D: **domain**(1) = [1..n];

This declaration creates a one-dimensional (1D) arithmetic domain, D, and initializes it to represent the set of indices $\{1, 2, \ldots, n\}$. Domains can be used to declare arrays, which represent mappings from the domain's index set to a set of variables of a given type. A, as declared below, is an array that contains a floating-point variable for each index in domain D:

var A: [D] **float**;

A domain's index set can consist of tuples of integers as in Fortran 90 arrays, but it can be much more general using arbitrary values and object references as in modern scripting languages. Chapel introduces a special *index type* which is parameterized by a domain and constrains values of that type to be members of the domain's index set. For example, given the declarations above, a variable, *lo*, storing an index of D could be declared as follows:

```
var  lo:  index(D)  =  D.first();
```

Index types aid readability, since they provide a context for an index variable's legal values. In addition, they also often allow the compiler to eliminate runtime bounds checks, since accessing arrays with index variables of their defining domains is guaranteed to be a safe operation. As an example, consider the following loop:

```
forall  i  in  D  do
   A(i)  =  ...
```

Chapel's *for* and *forall* loop constructs automatically declare the iterator variable (*i* in this example) to be of the index type of the domain over which the iteration is occurring. Thus, *i* here is of the type *index(D)*. Since A is declared in terms of domain D, the access of A using i is guaranteed to be in bounds and no runtime bounds check is required.

For a given domain, subdomains associated with a subset of the domain's index set can be defined. Chapel provides general mechanisms for the creation of subdomains, allowing the construction of arbitrary sparse data structures. The index type of a subdomain is considered a subtype of the parent domain's index type. A simple example of a subdomain declaration is given here:

```
var  InteriorD:  subdomain(D)  =  [2..n−1];
```

All domains can be queried for their *extent*, or number of elements. Associated iterators specify sequential and parallel pointwise iteration over their index set.

Arithmetic domains

Arithmetic domains are characterized by index sets that are Cartesian products of arithmetic sequences. The *rank* of an arithmetic domain specifies the number of its dimensions; it must be a compile-time constant. The indices in an arithmetic domain are linearly ordered based on lexicographic ordering.

An arithmetic domain can be declared and optionally initialized as follows:

```
var  aD:  domain(3)  =  [1..m,  0..n,  −1..p];
```

The shape of the domain corresponds to the cross product of the arithmetic sequences which define the bounds of the domain in each dimension.

Chapel also supports strided and sparse arithmetic domains in which subsets of a given Cartesian space can be represented efficiently. By using domains to specify index sets and iteration, Chapel expresses computations over arrays in a manner independent of whether they are sparse or dense. This is attractive due to the fact that sparse arrays are merely a space-efficient representation of a conceptually dense data aggregate.

Indefinite domains

Indefinite domains can have indices of any legal Chapel value type. Such indices can be values of primitive types or class references as well as tuples and records of such. In contrast to arithmetic domains, the index sets of indefinite domains are inherently unbounded. An indefinite domain of object references would be declared and initialized as follows:

```
class C {...}
var iD: domain(C);
for ... {
  var myC = C();  // construct instance
  iD.add(myC);    // add to domain
}
```

As shown here, indefinite domains can use an explicit **add** method to add a new index value. Similarly, indices can be removed from an indefinite domain by calling the **remove** method. Calling these methods on indefinite domains causes the arrays defined on them to be reallocated appropriately. In contrast to arithmetic domains, no order is defined for the indices in an indefinite domain.

Indefinite domains are an important concept in the language as they are a key to expressing irregular problems and supporting associative arrays that can grow and shrink.

22.2.2 Arrays

Arrays are defined on domains and map domain indices to variables of a common type. The rank, shape, and order of arrays are the same as for the domain on which they are defined. Arrays are classified as either *arithmetic* or *associative*, depending on their domain:

Arithmetic arrays are defined on arithmetic domains. For example:

```
var aA: [aD] float;
var aB: [1..n] float;
```

The arithmetic array **aA** is defined on the arithmetic domain **aD** and its elements are of type **float**. The arithmetic array **aB** is defined with the index set **[1..n]** of **float** elements. The compiler automatically inserts an *anonymous domain* for the array in this case.

Associative arrays are defined on indefinite domains. Every time an index is added to or removed from the indefinite domain, a new array element corresponding to the index is defined or removed. Associative arrays are a powerful abstraction for applications of a highly dynamic structure; they allow the user to avoid less efficient dynamic structures (such as lists) to express certain problems.

The declarations of arithmetic and associative arrays are virtually identical since the differences in their structures are factored into their domain declarations. This allows code to be written that iterates over and indexes into arrays independently of their implementation details.

22.3   Data Distributions

Data distributions are means for the programmer to exploit locality. The distribution of a domain specifies a mapping from the domain's index set to a collection of *locales*; in addition the arrangement of data in a locale can be controlled.

22.3.1   Basic approach

Chapel provides a predefined data type called a *locale*. The number of locales for a program execution is determined at the beginning of that execution, and remains invariant thereafter. This is achieved by using a predefined configuration variable, `num_locales`. The following code fragment illustrates the predefined declaration of the rank-1 array variable `Locale`, the elements of which represent the locales accessible to the execution of the program. This is called the *execution locale set*:

```
config const num_locales: integer;
const Locales: [1..num_locales] locale;
```

Every variable is associated with a locale, which can be queried in the form `<v>.locale`. Likewise, every computation is associated with a locale, which can be determined by calling the function `this_locale`. This can be used to reason in a program about the locality of accesses to data structures. An optional "on" clause that can be used before any statement allows the programmer to control where a computation occurs and where variables are allocated.

The following code excerpt shows the association of a distribution with an arithmetic domain:

```
class Block: Distribution {
    ...
    function map(i: index(source)): index(target);
    ...
}
```

```
read(m, n, p);
var D: domain(3) distributed(Block()) on Locales
             = [1..m, 0..n by 2, 1..p by −1];
var a: [D] float;
```

```
forall ijk in D { a(ijk) = ...}
```

The class `Block` is supplied either directly by the programmer, or it is imported from a library. Class `Distribution` is the root of the distribution hierarchy and all user-defined or library-provided distributions must be derived from it. The programmer must at least specify the mapping function. We will discuss other interface functions that can be overriden in more detail in the next section. Every distribution has a source domain, which in this case is D and a target domain, which is a subset of the locales available to the program. The index set of this domain is defined as the Cartesian product of linearly ordered sets $I \times J \times K$, where $I = \{1, \ldots, m\}$, $J = \{j \mid 0 \leq j \leq n, mod(j, 2) = 0\}$, and $K = \{p, p - 1, \ldots, 1\}$.

Array a is defined over the domain D and assigned in the `forall` loop.

The following code excerpt exemplifies the association of a distribution with an indefinite domain:

```
class GraphPart: Distribution {
  function map(i: index(source)): index(target);
  ...
}
```

```
class Vertex {...}
read(size);
var D: domain(Vertex) distributed(GraphPart()) on Locales;
var a: [D] float;
for i in 1..size D.add(Vertex(i));
```

```
forall i in D {a(i) = ...}
```

In this example the domain is a set of vertices, which are of a user-defined type `Vertex`. The `GraphPart` distribution class is also user or library defined and may specify a graph-partitioning strategy. The graph has `size` nodes and the domain D is initialized to `size` vertices. The `forall` loop iterates over the domain and assigns values into array a.

Our uniform treatment of regular and irregular data structures goes a step beyond existing approaches to data mapping and support for distributed execution. This feature is crucial for the relevance of the language. While many applications deal with linear data structures and regular distributions (e.g., block, cyclic), applications that deal with complex data structures and that require more complicated decompositions abound.

Note that even though the domains and distributions in the two examples are very different, this does not overtly complicate the code that the Chapel programmer must write. Although very simple, the two examples would look dramatically different if, for instance, they were written in C or Fortran and MPI.

The next section explains the distribution interface and how it can be incrementally refined by programmers for efficient parallel execution.

22.3.2   The distribution interface

The distribution interface in Figure 22.1 shows the methods bound to the abstract distribution class which are visible to the programmer.

```
class Distribution {
  var source: domain;  // source domain to distribute
  var target: domain;  // target domain to map to

  function getSource(): Domain;
  function getTargetDomain(): Domain;
  function getTargetLocales(): [target] locale;

  function map(i: index(source)): locale;
  iterator DistSegIterator(loc: index(target))
                              : index(source);
  function GetDistributionSegment(loc:index(target)):Domain;
```

FIGURE 22.1: The published distribution interface.

The source and target domains, as well as the subset of target locales of the distribution are transparently set up upon encountering a domain declaration containing a distribution specification. Thus, the programmer can query their values.

For each index in the source domain, the `map` function specifies a locale in the target locales. The set of all indices associated with a locale is called its *distribution segment.*

The iterator `DistSegIterator` and the function `GetDistributionSegment` specify the inverse computation of the map, in two variants. The former produces the elements in the distribution segment associated with a locale, as a sequence of source domain indices. The latter, applied to a locale, defines the domain associated with its corresponding distribution segment. Although the compiler by default generates code which inverts the `map` function, there are situations when the programmer expresses a more efficient specification of the inversion. The default version uses an exhaustive search based on the `map` function and is described in Figure 22.2. The default implementation of the `GetDistributionSegment` function uses a similar mechanism.

```
iterator Distribution.DistSegIterator(loc:locale):
index(source){
    forall i in source on loc do
        if (map(i) == loc) then yield(i);
}
```

FIGURE 22.2: The default distribution segment iterator.

Figure 22.3 illustrates the user-defined iterator for a `Cyclic` distribution. It is expected that the user-defined version is more efficient than the default one, since it is a direct computation, based on the symbolic formula corresponding to the inverse of the mapping function.

```
class Cyclic: Distribution {
    function map(i: index(source)): locale {
        return Locales((( i-1 mod num_locales) + 1);
    }

    iterator DistSegIterator(loc: locale) {
        for i in locale2integer(loc)..source.extent()
                by num_locales { yield(i); }
    }
}
```

FIGURE 22.3: Programmer-defined distribution segment iterator for a cyclic distribution.

```
class LocalSegment: Domain {
    function getLocale(): locale;
    function layout(i: index(source)): index(getLocalDomain())
    function setLocalDomain(ld: Domain);
    function getLocalDomain() : Domain;
}
```

FIGURE 22.4: Local-segment interface for controlling on-locale array storage.

22.3.3 The on-locale allocation policy

One key feature of our approach is the possibility to control the on-locale allocation policy for data associated with a distribution segment.

For this purpose, the user has access to an interface called `LocalSegment`. Logically, this class plays the role of the domain for the local portion of data within the corresponding arrays. Therefore, it subclasses the `Domain` class. The user can extend the `LocalSegment` interface and override its published functionality with problem-specific behavior. The user can also affect state. In turn, the compiler automatically sets up a corresponding `LocalArraySegment` class which actually allocates the array data (there is one such instance for each portion of the arrays associated with the domain). The interface for the `LocalSegment` is shown in Figure 22.4.

A `locale` value is set for each `LocalSegment` based on the mapping function. The programmer can query this value in order to reason about locality.

The system uses the `layout` function in conjunction with the `map` function to uniquely identify the location of a data item in a locale. This function contains the translation of *global indices* into corresponding *local indices* as dictated by the mapping and allocation policies specified by the user or decided by the system.

The user can set and query a `LocalSegment`'s *local domain*. The local domain is the local index domain for all the arrays associated with the global domain. The system transparently uses this variable to generate the corresponding local arrays. By default, this domain corresponds to the indices in the distribution segment and has the same type as the parent domain (the global domain for which it is defined).

This is an advanced level of difficulty at which the programmer operates. However, the design aims at making data representation orthogonal to algorithm and thus eases the pressure on the specification of the algorithm. The vision is that libraries of various layouts, e.g., compressed sparse row (CSR), will be written and most programmers can simply use these libraries. This will be illustrated in the following section.

22.4 Examples and Discussion

22.4.1 A load-balanced block distribution

This section presents a concrete example for the explicit specification of a mapping from source domain to target locales. The arrangement of locale-internal structures and translation between global and local domain indices are left to the system. The programmer specializes the `Distribution` class overriding the method `map` and the iterator `DistSegIterator`.

Figure 22.5 illustrates a programmer-provided load-balanced multidimensional block distribution. The programmer assigns each index to a locale in the map function. Then, in the iterator, the programmer computes for a given locale the subset of global indices which belong to it. Client code for the distribution might appear as follows:

```
param n : integer = ...;
var N₁, ... Nₙ: integer;
read(N₁, ..., Nₙ);
var P₁ .. Pₙ: integer;
read(P₁, ..., Pₙ);
var locDom: domain(n) = [1..P₁, ..., 1..Pₙ];
// reshape the locales as an n-dimensional topology:
var locales_nD: [locDom] - reshape(Locales);
var D: domain(n)
distributed(LoadBalancedBlock(n))
        on locales_nD = [1..N₁, ..., 1..Nₙ];

// distributed forall loop:
forall i in D do ...

class LoadBalancedBlock: Distribution {
param n: integer =...; // compile-time constant
const tl:[target] locale=getTargetLocales();
const ft: index(target) = target.first();

// All variables below are n-tuples of integers:
const N: n*integer = source.extent();
const P: n*integer = target.extent();
const q: n*integer = floor(N/P);
const u: n*integer = ceil(N/P);
const r: n*integer = mod(N,P);

function map(i: index(source)): locale {
  const f: index(source)= source.first();
  const tx: index(target);
  forall d in 1..n {
    tx(d)=if (i(d) <= (q(d) + 1) * r(d) + fs(d) - 1)
          then ceil((i(d)-fs(d)+1)/(q(d)+1))+ft(d)-1;
          else ceil((i(d)-fs(d)+1-r(d))/q(d))+ft(d)-1;
  }
  return tl(tx);
}
```

```
iterator DistSegIterator(loc:locale):index(source){
  const k: index(target) = locale_index(loc);
  var cdom: domain(n);
  var firstIndexInLoc: index(source);

  forall d in 1..n {
    if (k(d)<=r(d)) { //first r(d) locales: blocksize u(d)
      firstIndexInLoc(d)=(k(d)-ft(d))*u(d)+source.first(d);
      cdom(d)=firstIndexInLoc(d).. firstIndexInLoc(d)+ u(d)-1;
    }
    else {//remaining P(d)-r(d) locales: blocksize q(d)
      firstIndexInLoc(d)=r(d)+(k(d)-ft(d))*q(d)+fs(d);
      cdom(d)=firstIndexInLoc(d).. firstIndexInLoc(d)+q(d)-1;
    }
  }
  for c in cdom do yield(c);
  }
}
```

FIGURE 22.5: Specification of a regular load-balanced multi-dimensional block distribution.

In the code fragment above, an **n**-dimensional domain is distributed on an **n**-dimensional set of locales, with each dimension of the source domain being distributed over the corresponding dimension in the target domain.

22.4.2   A sparse data distribution

In terms of building the distribution, the generation of a distributed sparse structure differs from that of a dense domain in at least the following points:

- It is necessary to deal with two domains and their interrelationship: the algorithm writer formulates the program based on the original dense domain, i.e., indexing data collections in the same way as if they were dense. In contrast, the actual representation of the data and the implementation of the algorithm are based on the sparse subdomain of the dense domain.

- In many approaches used in practice, the distribution is determined in two steps:

  1. First, the dense domain is distributed, i.e., a mapping is defined for *all* indices of that domain, including the ones associated with zeroes. In general, this will result in an irregular partition, reflecting the sparsity pattern and communication considerations.

  2. Secondly, the resulting local segments are represented using a sparse format, such as CRS (compressed row storage).

The approach for user-defined distributions in Chapel is powerful enough to deal with this problem. We illustrate this with the example below, which assumes the sparse structure to be invariant.

Distributed CRS representation for sparse data structures

In the example code of Figure 22.6 we assume the sparsity pattern to be predefined. This is an approach used in many applications, where the pattern is derived from an irregular mesh. In the program, this assumption is reflected by the (unspecified) assignment to DD, which represents the sparse domain.

```
type eltType;
const n: integer = ...;
const m: integer = ...;
const D: domain(2) = [1..n, 1..m]; // dense data domain

// myBRD initialized with instance of distribution
class BRD: var myBRD: BRD = BRD(...);

// declaration of sparse subdomain DD of D and its layout.
// The unspecified assignment initializes DD based on the
// pre-determined sparse structure:
const DD: sparse domain(D) distributed(myBRD, CRS()) = ...;
var A: [DD] eltType;
var x: [1..n] eltType;
var y: [1..m] eltType;
...
forall (i,j) in DD {
    y(i) = sum reduce(dim=2) A(i,j)*x(j);...
}
```

FIGURE 22.6: Sparse-matrix vector multiplication.

The original dense domain, D, is an arithmetic domain of rank 2 with index set [1..n,1..m]. It is partitioned into a set of rectangular "boxes," based on the sparsity pattern, load-balancing, and communication considerations. In general, this partition is irregular; the boxes represent the distribution segments. The global mapping establishes a one-to-one map between boxes and the target locales. Two partitioning approaches that have been used in practice include *Binary Recursive Decomposition (BRD)* [4] and *Multiple Recursive Decomposition (MRD)* [22]. The upper half of Figure 22.7 illustrates such a distribution for $D = [1..10, 1..8]$ and the target index domain 1..4. Zero elements are represented by empty fields; nonzeros are explicitly specified and numbered in row major order (for simplicity, we have chosen as the value its

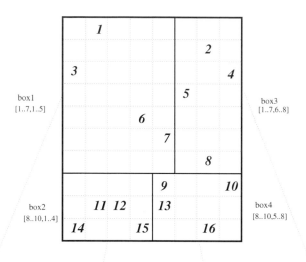

LocalSegment(1)

Local Domain	data	cx
1	1	2
2	3	1
3	6	4
4	7	5

r	ro(r)
1	1
2	2
3	2
4	3
5	3
6	4
7	5
8	5

LocalSegment(2)

Local Domain	data	cx
1	11	2
2	12	3
3	14	1
4	15	4

r	ro(r)
8	1
9	1
10	3
11	5

LocalSegment(4)

Local Domain	data	cx
1	9	5
2	10	8
3	13	5
4	16	7

r	ro(r)
8	1
9	3
10	4
11	5

LocalSegment(3)

Local Domain	data	cx
1	2	7
2	4	8
3	5	6
4	8	7

r	ro(r)
1	1
2	1
3	2
4	3
5	4
6	4
7	4
8	5

FIGURE 22.7: Sparse-matrix distribution.

position in this numbering scheme).

The four boxes defining the subdomains of D generated by the partition in the example are given as: $box^1 = [1..7, 1..5]$, $box^2 = [8..10, 1..4]$, $box^3 = [1..7, 6..8]$, and $box^4 = [8..10, 5..8]$. The global map is defined as $map^D(i,j) = k$ for all $(i,j) \in box^k$.

The sparse subdomain of D, as shown in the example, is given as:

$$DD = \{(1,2), (2,7), (3,1), (3,8), (4,6), (5,4), (6,5), (7,7), (8,5),$$
$$(8,8), (9,2), (9,3), (9,5), (10,1), (10,4), (10,7)\}$$

Given the partition of D and the global mapping, a corresponding partition of DD and its mapping can be immediately inferred. The inferred mapping is the restriction of the original map to indices associated with nonzero elements:

$$map^{DD} = map^D \mid DD$$

We turn now to the specification of the layout, for which we choose the *compressed row storage (CRS)* distribution format in this example. This means that in each locale, k:

- The local data domain, `LocalDomain`, is established as a one-dimensional arithmetic domain, representing the nonzero elements in the associated box, box^k, in the lexicographic order of their indices.

- The *column index vector* maps the local index of each nonzero element (i.e., its index in the local-data domain) to the second component of its index in the global-dense domain, D.

- The *row vector* determines for each row in D that is part of box^k the local index of the first nonzero element in that row (if such an element exists).

The lower half of Figure 22.7 illustrates the CRS representation for the four boxes in the example. Figure 22.8 outlines the definition of the layout (class CRS).

```
class BRD: Distribution
  {
   ...
   function map( i : index ( source )): locale { ... }
   // mapping dense domain
   // this yields the box associated with loc:
   function GetDistributionSegment ( loc : locale ): Domain { ... }
   ...
  }
  class CRS: LocalSegment {
  const loc:  locale = this . getLocale ();
  //dense distribution segment for loc:

  const locD:  domain ( 2 );
  // sparse distribution segment for loc:
  const locDD : sparse domain( locD )= GetDistributionSegment ( loc );
  //  number of elements in locDD:
  const nnz : integer= locDD . extent ();
  // row numbers of first and last index in locDD:
  const l1 :  integer= locDD ( 1 ). first ();
  const u1 :  integer = locDD ( 1 ). last ();
  // local data domain and extension:
  const LocalDomain :  domain ( 1 )= 1 .. nnz ;
  const xLocalDomain :  domain ( 1 )= 1 .. nnz +1;

  // persistent data structures in the local segment for
  // all arrays associated with the sparse domain:
```

```
var  cx : [ LocalDomain ]  index ( locD ( 2 ) );  //column  index  vector
var  ro :  [ l1 .. u1 +1]  index ( xLocalDomain );  //row  vector



// auxiliary functions:
  // mapping local index z to its global index:
  function nz2x ( z : index ( LocalDomain ):  index ( locDD ))  { ... };
  // mapping global index i to its local index:
  function x2nz ( i : index ( locDD )):  index ( LocalDomain ) { ... };
  // the following function , applied to row r, yields
  // true iff the local sparse subdomain contains an
  // element (r, c) for some c. Then the index of the
  // first such element is returned via argument firstz :
  function exists_sparse_index ( r :  index ( locD ( 1 )) ,
                          out firstz : LocalDomain ):  bool  { ... };

  function define_columnVector ( )  do
     [ z in LocalDomain ]  cx ( z )= nz2x ( z ) ( 2 );

  function define_rowVector ( )  {
     ro ( u1 +1)= nnz +1;
     for  r in  1 .. u1 by −1 do
        ro ( r ) = if exists_sparse_index ( r , firstz )
                  then firstz else ro ( r +1);
  }

  function layout ( i : index ( D )):  index ( LocalDomain )
                                       return ( x2nz ( i ));

  constructor LocalSegment ( )  {
     define_column_vector ( );  define_row_vector ( );
  }
}
```

FIGURE 22.8: BRD distribution with CRS layout.

22.5 Implementation

This section discusses the status of the Chapel compiler implementation and the strategy for implementing distributions. This is a work in progress.

22.5.1 Compiler implementation status

Since Chapel is a new language, much effort has gone into the implementation of the (serial) *base language*. Our intent was to develop a highly productive parallel language from first principles, unconstrained by features irrelevant for that purpose. Specifically, the design goals have focused on the complete and consistent specification of concurrency, threading, and synchronization in the framework of a powerful base language.

The base language has been designed with the dual goals of providing generality and supporting analysis and optimization of programs. Its features include type and value safety, static typing, value and reference classes, and type parameterization. Also, Chapel syntax is designed for high productivity, and as a result, programs written in Chapel are more compact and easier to read and understand than their counterparts in languages such as Java and C++.

The current implementation covers the majority of language features described in the language document [9]. This includes arithmetic and indefinite domains as well as their respective arrays.

We decided to take a source-to-source compilation approach in order to support rapid development and prototyping of the compiler, and to be able to run the generated code on a variety of platforms. The Chapel compiler is written in C++. The compiler-generated code and runtime libraries are being developed in ISO C in order to closely match the target machine architectures, and to avoid relying on obtaining production-grade performance from C++ compilers on parallel architectures (since they have typically not received as much attention as C compilers). While it is tempting to use C++ as our target language in order to ease the burden of implementing Chapel's generic programming and object-oriented features, we believe that Chapel's features are sufficiently more aggressive than that of C++ so that the benefits would be minimal.

22.5.2 Distribution-implementation strategy

Arrays and domains are implemented as standard Chapel modules using the Chapel language itself. Arithmetic and associative arrays both specialize an abstract `Array` class, while arithmetic and indefinite domains both specialize an abstract `Domain` class.

Chapel supports both task and data parallelism. However, we focus on the latter model in this chapter.

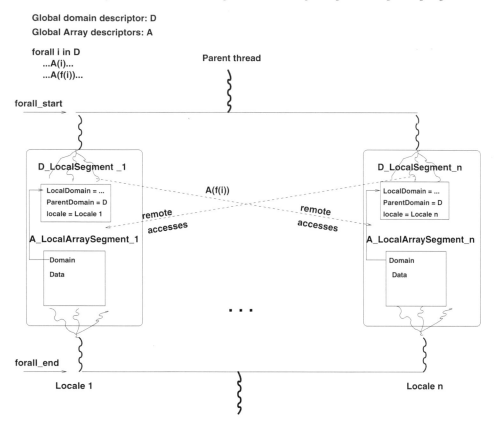

FIGURE 22.9: Data-parallel execution model.

In Figure 22.9 we depict a typical transformation of a `forall` loop. That is, one thread per locale involved in the computation is spawned upon encountering the `forall` statement. These are heavyweight threads, which account for coarse-grain data parallelism. Within each locale, a number of lightweight threads are spawned and scheduled depending on the locale internal configuration. Accesses to data are classified as either local or global: an access is considered *local* if the compiler can determine that it references data in the same locale as the executing thread, otherwise it is *global*. This reflects the fact that Chapel is a global-view language, in which a program can access data without taking into account the locality of access.* Although for regular codes compiler analyses can often distinguish between local and

---

*In certain contexts the language allows the programmer to assert locality with a special attribute, in order to enhance performance.

remote accesses, using this information for the optimization of communication [18, 1, 17], in general this decision can only be made at execution time. In such cases, for example when accessing irregular data structures, optimization of communication must be based on runtime analysis.

Most partitioned global address space (PGAS) languages, such as Co-Array Fortran [11] or X10 [12, 8], deal with this issue by making an explicit distinction between local and non-local accesses at the *source level*. This creates a burden for the programmer but makes it easier for the compiler to generate efficient code.

We have developed a general translation scheme for the data-parallel idioms in Chapel. Our scheme primarily addresses, (1) processing of domain declarations, including distribution and layout assertions, (2) array declarations over domains, (3) array references and subscript translation, (4) transformations of `forall` loops, (5) remote-access optimization, and (6) runtime support for distributed storage management and execution.

At present, we are developing an experimental framework that will allow a comparison of the performance of programs written in Chapel with manually parallelized codes based on a combination of a sequential language and MPI.

22.6 Related Work

This section does not provide an extensive review of general parallel and concurrent programming languages and systems. Such overviews are given elsewhere [21, 20]. Instead, we review efforts which are most closely related to the focus of this chapter.

Fortran D [13], Vienna Fortran [7], and Connection Machine Fortran [2], the major predecessors of High Performance Fortran [19], all offered facilities for combining multidimensional array declarations with the specification of a data distribution or alignment. These languages follow the data-parallel programming model and provide built-in data distributions and alignments. In addition, Vienna Fortran introduced a capability for user-defined mappings from Fortran arrays to a set of abstract processors, and for user-defined alignments between arrays in a Fortran 77 language framework.

ZPL [6] supports a concept of dimensional distributions organized into five types: block, cyclic, multi-block, non-dist, and irregular. These types give the compiler the information it needs to generate loop nests and communication, abstracting the details of the distribution from the compiler's knowledge. This strategy was detailed in [10].

Other language developments include the class of partitioned global address space (PGAS) languages, with Co-Array Fortran [11], Unified Parallel

C (UPC) [16], and Titanium [23] as their main representatives. These languages are based on the single-program, multiple-data (SPMD) paradigm, provide features for the partitioning of the address space, and support for one-sided communication. They represent a significant productivity improvement over communication libraries such as MPI, however, while still relying on a processor-centric programming model that requires users to manually decompose their data structures and control flow into per-processor chunks.[†]

Chapel is more general than these languages for a number of reasons. First, its model of concurrency covers data as well as task parallelism. Secondly, Chapel's indefinite domains and generalized arrays constitute a more powerful base language. Third, Chapel provides a general object-oriented framework for the specification of user-defined distributions, which is the only way for introducing distributions: there are no built-in distribution classes as with the other languages. Furthermore, Chapel is the only programming language which offers facilities for specifying the on-locale data layout rather than relying on compiler-selected standard representations. As illustrated with the distributed sparse matrix example, this framework is general enough to deal with arbitrarily complex data structures.

Closer to the goals represented by Chapel are two languages developed along with Chapel in DARPA's High Productivity Computing Systems (HPCS) program: X10, designed in the PERCS project led by IBM [12, 8], and Fortress [3], developed at SUN Microsystems. X10 and Fortress both provide built-in distributions as well as the possibility to create new distributions by combining existing ones. However, they do not contain a framework for specifying user-defined distributions and layouts such as Chapel.

22.7   Conclusion and Future Work

Today's high-performance computing systems are characterized by massive parallelism and a physically distributed memory. The state-of-the-art approach for programming these systems has been based on the extension of sequential programming languages, such as Fortran, C, and C++, with the MPI message-passing library. We believe that in view of emerging systems with tens of thousands of processors and the growing complexity of applications, a higher level of programming such architectures should be adopted. Specifically, we consider it necessary to provide the user with an inherently parallel language that allows the management of concurrency and locality at

---

[†]The one exception to this is UPC which has support for 1D block-cyclic distributions of 1D arrays over a 1D set of processors (*threads*) and a stylized *upc_forall* loop that supports an affinity expression to map iterations to threads.

a high level of abstraction. This is the main motivation for the design of Chapel, a general-purpose parallel-programming language with support for object-oriented programming, type and value safety, data and task parallelism, and explicit locality control.

After outlining the powerful features of the Chapel base language — in particular the unified treatment of regular and irregular data structures under generalized concepts of domains and arrays — this chapter focused on the discussion of an object-oriented framework for the specification of user-defined distributions, orthogonal to the specification of algorithms. The generation of communication, synchronization, and thread management is relegated to the compiler and runtime system; however, a sophisticated user may control many low-level details using the distribution framework.

Future language work will focus on the refinement of the distribution framework and on extending the semantics of replication and alignment, in particular, with a view on performance and data consistency. A major goal for our work on the compiler and runtime system is the validation of our assertion that the distribution framework will lead to higher productivity, compared to the message-passing approach as well as to other high-level languages. First, we will show that our reliance on the distribution framework for standard distributions does not lead to a loss of performance compared to languages with built-in distributions. Secondly, we will demonstrate that exploiting the full generality of the distribution framework will not only allow a high level of abstraction for the specification of complex data structures such as those occurring in sparse-matrix algorithms, but also guarantee performance comparable to that of manually parallelized programs.

22.8   Acknowledgments

This work was funded in part by the Defense Advanced Research Projects Agency under contract no. NBCH3039003. We thank Bradford Chamberlain and Steve Deitz from Cray Inc. for interesting discussions and their valuable contributions to this chapter.

References

[1] G. Agrawal. Interprocedural communication and optimizations for message passing architectures. In *Proceedings of Frontiers 99*, pages 123–130,

February 1999.

[2] E. Albert, K. Knobe, J. D. Lukas, and G. L. Steele Jr. Compiling Fortran 8x Array features for the connection machine computer system. In *PPEALS '88: Proceedings of the ACM/SIGPLAN Conference on Parallel Programming: Experience with Applications, Languages and Systems*, pages 42–56. ACM Press, 1988.

[3] E. Allen, D. Chase, V. Luchangco, J.-W. Maessen, S. Ryu, G. Steele Jr., and S. Tobon-Hochstadt. The Fortress language specification version 0.707. Technical Report, Sun Microsystems, Inc., July 2005.

[4] M. J. Berger and S. H. Bokhari. A partitioning strategy for nonuniform problems on multiprocessors. *IEEE Trans. Computing*, 36(5):570–580, 1987.

[5] D. Callahan, B. L. Chamberlain, and H. P. Zima. The Cascade High Productivity Language. In *9th International Workshop on High-Level Parallel Programming Models and Supportive Environments (HIPS'04)*, pages 52–60, April 2004.

[6] B. L. Chamberlain. *The Design and Implementation of a Region-Based Parallel Programming Language*. Ph.d. thesis, Department of Computer Science and Engineering, University of Washington, 2001.

[7] Barbara M. Chapman, Piyush Mehrotra, and Hans P. Zima. Programming in Vienna Fortran. *Scientific Programming*, 1(1):31–50, 1992.

[8] P. Charles, C. Grothoff, V. Saraswat, C. Donawa, A. Kielstra, K. Ebcioglu, C. von Praun, and V. Sarkar. X10: An object-oriented approach to non-uniform cluster computing. In *Conference on Object-Oriented Programming Systems, Languages and Applications*, pages 519–538, 2005.

[9] Cray Inc. *Chapel Specification 4.0*, February 2005. http://chapel.cs.washington.edu/specification.pdf.

[10] S. J. Deitz. *High-Level Programming Language Abstractions for Advanced and Dynamic Parallel Computations*. Ph.d. thesis, University of Washington, 2004.

[11] Y. Dotsenko, C. Coarfa, and J. Mellor-Crummey. A multi-platform Co-Array Fortran compiler. In *PACT '04: Proceedings of the 13th International Conference on Parallel Architectures and Compilation Techniques*, pages 29–40, Washington, DC, 2004. IEEE Computer Society.

[12] K. Ebcioglu, V. Saraswat, and V. Sarkar. X10: Programming for hierarchical parallelism and non-uniform data access. In *3rd International Workshop on Language Runtimes, ACM OOPSLA 2004*, Vancouver, BC, October 2004.

[13] G. Fox, S. Hiranandani, K. Kennedy, C.s Koelbel, U. Kremer, C.-W. Tseng, and M.-Y. Wu. Fortran D language specification. Technical Report CRPC-TR90079, Rice University, Center for Research on Parallel Computation, Houston, TX, December 1990.

[14] W. Gropp, E. Lusk, and A. Skjellum. *Using MPI: Portable Parallel Programming with the Message Passing Interface.* The MIT Press, Cambridge, MA, second edition, 1999.

[15] High Performance Fortran Forum. High Performance Fortran language specification, version 2.0. Technical report, Rice University, Center for Research on Parallel Computation, January 1997.

[16] P. Husbands, C. Iancu, and K. Yelick. A performance analysis of the Berkeley UPC compiler. In *ICS '03: Proceedings of the 17th Annual International Conference on Supercomputing*, pages 63 73, New York, NY, 2003. ACM Press.

[17] M. Kandemir, P. Banerjee, A. Choudhary, J. Ramanujam, and N. Shenoy. A global communication optimization technique based on data-flow analysis and linear algebra. *ACM Transactions on Programming Languages and Systems*, 21(6):1251–1297, 1999.

[18] K. Kennedy and N. Nedeljković. Combining dependence and data-flow analyses to optimize communication. In *Proceedings of the 9th International Parallel Processing Symposium*, Santa Barbara, CA, 1995.

[19] C. Koelbel, D. Loveman, R. Schreiber, G. Steele Jr., and M. Zosel. *The High Performance Fortran Handbook.* The MIT Press, 1994.

[20] K. Pingali. *Parallel and Vector Programming Languages.* Wiley Encyclopedia of Electrical and Electronics Engineering. John Wiley & Sons, 1999.

[21] David B. Skillicorn and Domenico Talia. Models and languages for parallel computation. *ACM Computing Surveys*, 30(2):123,169, June 1998.

[22] M. Ujaldon, E. L. Zapata, B. M. Chapman, and H. P. Zima. Vienna Fortran/HPF extensions for sparse and irregular problems and their compilation. *IEEE Trans. Parallel and Distributed Systems*, 8(11), 1997.

[23] K. Yelick, L. Semenzato, G. Pike, C. Miyamoto, B. Liblit, A. Krishnamurthy, P. Hilfinger, S. Graham, D. Gay, P. Colella, and A. Aiken. Titanium: A high-performance Java dialect. In ACM, editor, *ACM 1998 Workshop on Java for High-Performance Network Computing*, New York, 1998. ACM Press.

Chapter 23

Architectural and Programming Issues for Sustained Petaflop Performance

Uwe Küster

High Performance Computing Center Stuttgart (HLRS), University of Stuttgart, Germany

Michael Resch

High Performance Computing Center Stuttgart (HLRS), University of Stuttgart, Germany

23.1 Introduction

The development of computer technology brought an unprecedented performance increase during the last five decades. Coming from a few hundreds of floating-point operations per second we will reach at the end of this decade more than a petaflop per second (PFLOPS). In the same time frame numerical research got a big impetus. We will show that the appearance of a new computer technology also changed numerical algorithms. This also will be true for the large machines of the petascale class. It may be that the petascale class of machines will force even larger changes in numerical paradigms. In the following we will give a short synopsis of the development of modern computer systems for numerical purposes together with mathematical methods which where developed and heavily used at the same time. This should show how architectures and algorithms correspond over time. Discussing technical issues of recent and future computer architectures we will try to anticipate the implications of petascale computers on algorithms. Further, we will make some remarks on the economical impact of petascale computing.

23.2   A Short History of Numerical Computing and Computers

We first want to give a short overview of the history of supercomputing in order to better understand the challenges ahead and to show how algorithms have changed over time depending on available hardware architectures.

23.2.1   The Sixties

The most important and interesting machine in the 1960s was Cray's CDC 6600, first delivered in 1964. The cycle time of this computer was 100 nsec. A floating point multiply needed 10 cycles. The machine had a memory access time of 0.2 μsec or 2 cycles. The bandwidth was 32 MWords/sec for a peak performance of 3 MFLOPS. Different from the current development, neither memory access time nor bandwidth were an issue for performance. The main bottleneck was the floating-point performance. Today's processors may have hundreds of cycles of memory latencies and deliver 1 byte per potential floating-point operation. The CDC 6600 was already a pipelined system. It had a second level memory (ECS) and in-parallel operating peripheral processors for I/O. These machines were able to solve discretized partial differential equations on meshes with small cell numbers. To save memory, orthogonal meshes with a small number of points on simple geometries were used. But the development of FEM packages still in use today (e.g., NASTRAN at NASA) began. The involved linear systems were solved by direct methods based on sparse matrix representations. Amdahl gave the pessimistic prediction that parallel computers would never be successful [2]. Fortran became the standard language for numerical computing. Computers were an expensive resource and all computers were supercomputers.

23.2.2   The Seventies

Computers still were an expensive resource but were accessible for all technical universities. The user interface was the batch queue. Memory sizes were still less than 256 KWords. The FEM packages for the solution of static and dynamic engineering problems developed which are still widely used in a lot of different areas. These codes were written using out-of-core methods for program instructions (overlays) and data (paging techniques) — a technique reused for small local memories. Direct sparse solvers developed for the increasing size of FEM problems. Much work was dedicated to decrease the bandwidth of these matrices and the computational effort in solving the respective linear equations. First libraries for dense matrix problems were developed.

Linpack and Eispack assembled modern algorithms for the solution of linear and eigenvalue systems. The computing kernels of these libraries were collected in the Basic Linear Algebra Subprograms (BLAS) library and optimized for different architectures. The main difficulty was the optimization of floating-point operations (which since then describe system performance). Memory bandwidth played some role but not the memory latency. The Linpack benchmark was established as a test for system performance. This simple test seduced computer architects in the following decades to neglect memory bandwidth.

23.2.3 The Eighties

In the early 1980s computers like CRAY 1, Cyber 205, NEC SX-1, Fujitsu VP-200, IBM 3090 spread out for numerical computing. They all had vector architectures and provided a performance jump compared to previous systems in all disciplines: vector and scalar floating-point performance, memory bandwidth and I/O. The Convex minisupercomputer closed the gap to workstations. The parallel-programming model was directives similar to OpenMP.

The appearance of vector computers led to widespread use of Krylov space procedures for the solution of large, sparse linear systems which provide long vectors for suitable column-ordered data structures of the sparse matrices. Later on, Krylov space procedures for the solution of discretized partial differential equations turned out to be easily and efficiently parallelizable. Lapack was developed as a comprehensive library for dense linear and eigenvalue problems based on optimized BLAS-2 and -3 kernels. The reduction of memory traffic per floating-point operation in these kernels was an excellent anticipation of the needs of hierarchical memory architectures in the following era.

The late 1980s brought the first parallel architectures. The German Suprenum machine was one of the early parallel machines and combined scalar and vector processors. Suprenum was built with multigrid algorithms in mind. Systems like the Connection machine and Maspar were massively parallel computers which enforced a parallel-programming paradigm that became the "array syntax" part of Fortran 90. Their SIMD concept is seen again today in the ClearSpeed card. Ncube was a massively parallel machine comparable to the IBM Blue Gene machine. All these approaches suffered from the lack of parallelized software.

23.2.4 The Nineties

During the early 1990s massively parallel systems began to influence scientific programming — like the Thinking Machines CM-5, CRAY T3E, SGI Origin, and Intel Paragon. Further, workstation architectures were developed into shared-memory parallel systems such as SGI Power Challenge, Sun Enterprise, the IBM SP systems, and DEC Alpha-based systems. Indispensable for the further development of parallel architectures was the standardization of MPI and OpenMP as parallel-programming methods.

The peak performance of processors was continuously growing mainly due to the frequency improvement but also because of architectural changes like better pipelining, support of out-of-order execution, higher associativity of caches, and better branch prediction. The memory and memory bus systems could not and still cannot keep up with this. Especially for shared memory systems, this often implied that parallelized programs were not essentially faster than the serial original.

Later in the 1990s, computing as a whole was influenced by the pervasive usage of PCs, replacing workstations, triggering the development of Linux as a replacement for UNIX. Based on these PCs, first Beowulf clusters appeared on the scene. Visionaries declared the end of traditional supercomputers. Vendors started to move away from the concept.

In the field of application development, parallel computers had a clear impact. With increasing size of overall main memory, adaptive complex methods become popular. As domain decomposition became the algorithmic paradigm load balancing and adaptation were implemented. Beyond mesh-based methods, particle methods became interesting on large-scale systems. For both approaches pointers became essential. As FORTRAN did not support unstructured approaches with pointers it was partially replaced by C.

The close connection to the hardware base was broken. Performance became identified more with scaling of code on a parallel machine rather than single-processor performance. The gap between peak and sustained performance began to widen.

On the other hand, the pessimistic Amdahl assumption of the impossibility of parallelizing a fixed workload across a growing number of processors to reduce compute time — later called "strong scaling" — was replaced by the more optimistic assumption of Gustafson of growing the amount of work for a fixed calculation time on a growing number of processors — later called "weak scaling."

23.2.5   2000 and beyond

Recent years were marked by the propagation of PC clusters leading to inexpensive solutions in high performance computing (HPC) for a wide community. Standardization in processor technology, network technology and operating systems made these clusters extremely successful and interesting for ISVs. Nevertheless special machines are not out of business as is shown by systems such as the Cray XT3, the IBM Blue Gene, and the SGI Origin/Altix.

The main hard physical limit for all these systems is latency while bandwidth can somehow be managed. In effect these relatively high latencies reduce the effective bandwidth for small-sized objects. At the processor level, latencies theoretically can be hidden by prefetching methods. However, little support for this is available in programming languages and compilers as of today. For inter-processor communication, MPI latencies have become a bottleneck. They are not decreasing at the same speed that node performance is increasing. A work-around is the reduction of messages by collecting small messages into a single larger one. This is supported by special engines (Quadrics, Infiniband).

At the algorithmic level the increasing size of main memory allows much large simulations but leads to runtimes in the range of weeks or even months due to time step limitations. This is true for both structured and unstructured meshes. As structured meshes lead to higher cache reuse, software developers partially at least return to structured mesh approaches. This "archaic" approach makes parallelization, load balancing, vectorization, and compiler optimization simpler and more efficient.

23.3 Architectures

In this chapter some aspects of recent processor development are described. We specifically refer to the power consumption problem and to parallel programming paradigms.

23.3.1 Processor development

For 35 years, clock frequency has increased continuously with decreased feature size following Moore's law. However, frequency for a given technology is also directly attached to electrical power consumption:

$$power = capacitance * voltage^2 * frequency \qquad (23.1)$$

An increased frequency implies a linear increase of voltage and an increase in power consumption of 2.7 to 3. This is one of the reasons for the current frequency stagnation. A second one is the signal delay on long on-chip wires due to the decreasing size of the technology and the increase of its per-length capacities. A third reason is the decrease of the number of gates which can be crossed during the shortened clock cycle. A fourth is the increase of the leakage power losses [3] towards more than 50%. A fifth reason is the increase of the memory latency in terms of processor clock cycles because of the complicated memory systems of shared-memory processors and their need for cache-coherency protocols. Optical interconnections might help in accessing the memory with higher bandwidth and less electrical losses and the number of lines [12] but it may take a while before such optical technology can be found in standard processors.

So today we are faced with the challenge of getting performance boosts mainly by an increase in the number of active cores or parallel units in a processor. Decreasing the frequency by a factor of two today allows for the implementation of 8 identical cores within the same power budget. Switching to a smaller feature size there will be an even larger gain and first samples of chips with 80 caches have been presented recently by Intel.

As the number of cores increases, programs unaware of the underlying architecture may experience high losses in performance. Some algorithms may benefit. Dense matrix algebra for larger matrices may provide enough operations for hiding memory accesses efficiently. Another good example is molecular dynamics as long as interactions for every pair of atoms or molecules have to be calculated in an $n \times n$ operation. These operations today are the basis for special accelerators like MD-Grape. But large memory bandwidth is the key to performance for a lot of other numerical algorithms, specifically in computational fluid dynamics and sparse-matrix algebra arising from finite element (FE), finite volume (FV) or finite difference (FD). MPI-based domain decomposition approaches will fail because of a lack of spatial and/or temporal data locality and memory reuse.

Multicore architectures rely on local caches to minimize concurrency of memory requests. The efficient use of these caches implies the usage of hardware and software prefetching mechanisms to limit the memory latency effects. Any parallel-programming model has to allow controlling the locality of the core for a specified process or thread. A way to do this might be an enhanced OpenMP model. This would have to support core locality and hierarchical caches. For larger numbers, hierarchical filters or directory-based protocols (SGI Origin/Altix) are needed to decrease the cache-coherence traffic.

Present languages do not provide portable means to separate data fetch and store on one side and the calculations on the other side. Today's compilers enable the use of prefetching hints. These can be directive-based but are not portable. Data for local memories instead of caches have to be loaded and stored explicitly. They will complicate programs considerably. But they give the chance of a perfect data locality over a long time and are not disturbing the data access of concurrent processes.

We see this kind of separation of data movement and calculation as a basic principle of the Cell Broadband Engine processor (see next section). Transactional memory is able to mark memory cells forcing the requesting processor to repeat the memory operation if a concurrent access has changed its state. In that way it is possible to use the same data concurrently by different processors or threads allowing for reduction of the overhead of critical sections. This is useful for the case of rare concurrent updates of the same data by different processors.

A further challenge might be that cores are not identical but have varying architectures and tasks. We might see floating-point accelerators, MPI accelerators, TCP/IP off-load engines, or graphic accelerators on a single die. The chips will support different architectures and instruction sets on the same die. The compiling system will have to reflect the different purposes. The general executables must include the special purpose executables. All these arrangements could cut latencies and remove obstacles for sustained high performance.

23.3.2 Cell Broadband Engine processor

The Cell Broadband Engine processor [10] was initially developed to power the Sony PlayStation 3 and intended to be used also for high definition television (HDTV). Given the mass market for game consoles it is a relatively cheap and at the same time relatively fast computing system — hence very attractive for HPC.

The machine has an uncommon architecture. On a single chip a PowerPC processor serves as a host for 8 synergistic processor elements (SPE). It runs the operating system, does the I/O and controls the different tasks. The PowerPC allows for two hardware threads and operates in order. This simplifies the architecture and allows for higher clock frequencies. However, the burden for the compiler becomes heavier.

The 512 KB L2-cache of the Cell BE processor is on die and directly accessible by the SPEs and the host processor. The SPEs are like independent vector processors. Each has its own local memory of 256 KB. The total single precision peak performance of all SPEs of a 3.2 GHz processor is around 200 GFLOPS while the IEEE double precision performance is 26 GFLOPS. The double precision performance is expected to increase substantially for future versions.

A special direct memory access (DMA) engine in each SPE maps parts of the main memory to the local memory and initiates and performs block-data transfers. This engine allows for the separation of computation and memory access which is crucial to achieve high sustained performance. The block-data transfers may also be performed between different SPEs' of different Cell BEs which are connected by the fast FLEXIO interface of aggregated 76 GB/s bandwidth. The negative consequences of relatively low bandwidth should be attenuated by the SPEs local stores. They enable the programmers to optimize the codes by decoupling heavily used data from the main memory.

The high peak performance of the Cell BE processor makes it attractive for technical computing. The architectural restrictions will generate new algorithmic techniques. Even though it is not a shared-memory system, OpenMP looks like a promising programming model for this system. Vectorization of codes will become important again as the SPEs can be programmed like vector coprocessors. Most important is the ability of the Cell BE processor also to boost relatively small numerical problems to high performance. Potential applications may be gene sequencing while sparse-matrix applications may well fail to harness the power of the vector-like SPEs.

23.3.3　ClearSpeed card

The ClearSpeed card offers a very different paradigm. It is an independent card installed in a standard bus. The low frequency (250 MHz) of the processors guarantees very small heat dissipation. Its 96 functional units lead to a peak performance of 50 GFLOPS. Each processing element is equipped with a small local memory (6KB). The SIMD approach is similar to the Connection and Maspar machines (see above). The programming style is similar to a vector model, using vectors of methods on data instead of data vectors. The parallel-programming model is based on a C dialect with additional syntax elements for declaration of variables.

The decisive bottleneck of the machine is the memory bandwidth to the onboard memory. The CSX 600 model consists of two onboard processors with their own memory on a PCI-X card. Loading the data from the host computer to the accelerator's memory may be time consuming, at least for solving sets of small problems. Special-data movement instructions allow for distributing data to the processing elements. But we notice an insufficient card memory bandwidth limiting the sustained performance for problems with modest data intensity. Because of the bandwidth problems, there are only a few examples showing the performance potential of the two processor cards. Dense matrix operations, fast Fourier transforms and some techniques of molecular dynamics may benefit from the card.

As the concept shows high aggregated performance and low power consumption, ClearSpeed cards have been part of a recent large installation at the Tokyo Institute of Technology (TITech). For the future we expect such SIMD-like parallel-processing elements to be part of standard processors.

23.3.4　Vector-like architectures

With increasing frequencies deeper pipelining seemed to be unnecessary and counterproductive. It was in conflict with the object-oriented programming paradigm which is trying to hide even regular data layout from the programmer and from the compiler. A vector architecture, however, is no obstacle to object-oriented programming. Instead of using sets of atomic small objects, large objects of sets have to be used. The loops in their methods should not work on the atomic objects but directly on buffers. This approach saves the full potential of vectorization.

The situation has changed since clock-frequency limits became apparent and the power consumption became a topic. Hence, there is a growing interest in vectorization and SIMD techniques. Special purpose processors (GPU, Cell BE processor, ClearSpeed) and special processor instructions (SSE), and further new approaches (IRAM [9]) complement the vector architectures of CRAY and NEC.

Whereas traditional technologies use pipelined vector registers, the other approaches focus on data-parallel mechanisms on register files of a moderate size. Both approaches may boost the performance by factors from 2 to 100 depending on the number of functional units in the system. Sustainable bandwidth to support the operations is essential. The bandwidth may be provided by the central or global memory with a large number of expensive banks. However, this is getting more and more difficult. Larger (vector) caches are the most elegant solution for the programmer but may suffer from system overhead (coherency requirements and administration of cache lines). Nearby local memories allow for deterministic data access and exact blocking but impose penalties for the programmer and for context switches. They may, however, be a good solution if they are not too small.

Traditional vector machines are expected to keep up with the multicore progress. They can get the same cost advantages as the microprocessor development by the decreased feature size. They can handle a higher power dissipation per chip because the number of chips to get the same sustained performance will be smaller and cooling is more effective. The vector machines will be part of hybrid concepts taking advantage of microprocessor multicore development as well as the special vector design. They will also in future offer a relatively high memory bandwidth which still is one key to success for sustained performance.

23.3.5 Power consumption and cost aspects

Aggregating a large number of parts results in very high total power consumption. Petaflops computers are currently expected to require in excess of 10 MW. Cooling power in the same range has to be added. With this increase in power consumption comes an increase in costs. Power is becoming a major factor in the total cost-of-ownership calculation. Furthermore, it is increasingly difficult for HPC centers to handle the infrastructure requirements.

As a consequence, HPC simulation may become so expensive it may be cheaper to do experiments. This runs against the purpose of simulation and may well hamper further progress in the field. Hence, performance/watt will become a key figure for evaluation of future hardware architectures.

Furthermore, we see rising costs for HPC systems. While in 1995 a center could compete in the top 10 with a budget in the range of $20 Million, today a factor of at least 5 (i.e. $100 Million) is required. Even though large-scale clusters still come with a price/performance bonus, the gap between large-scale systems and average HPC installations is widening dramatically. This may lead to a decoupling of user communities. The damage done to HPC may be substantial.

23.3.6 Communication network

In a very large computer with millions of nodes the interconnecting network plays a dominant role. From the technical point of view it would be desirable to have multiple networks with a small number of hubs or routers between any pair of nodes. This would lead to well-known solutions like hypercubes (but only for 2^n machines) or 3D tori or fat trees to maintain the total bandwidth between any pair of nodes. On the other hand, the network can be expected to be very large because of the high number of nodes. Therefore a large part of the total investment will go into the network. This may exclude excellent but expensive interconnects.

The machines will have large physical sizes. This implies varying latencies through the whole machine. An important feature will be that a running application program will able to differentiate latency and bandwidth of the different lines. It would also be helpful to recognize hierarchical structures in the network for numerical purposes. Definitely communication protocols on the networks have to support pipelining.

23.3.7   Communication protocols and parallel paradigms

Currently, the most important parallel paradigm is the distributed memory based on MPI. Within the last ten years a lot of codes have been parallelized in that way and showed good speedup even on a large number of processors. But MPI has a deep calling stack and is not designed to pipeline multiple messages. As MPI is independent of the programming language there is hardly a chance for optimization by the compiler. As a consequence, MPI latencies are decreasing only slowly. There is some hope for hardware support of MPI as we see it with the InfiniPath/Pathscale chip. On the other hand, special cores could be reserved for special MPI accelerators on future multicore systems. As shared memory nodes will have an increasing number of cores in the future, OpenMP will play a more important role.

In addition to these parallel-programming paradigms, we see Partitioned Global Address Space (PGAS) languages as a supplement for the complete system but also for the parallelization on multicore chips. UPC and Co-Array Fortran are of this type. Co-Array Fortran is proposed for the next Fortran standard. The role of new languages like Chapel, X-10 and Fortress is not yet fully clear. The portability of these languages to other platforms will be important.

23.4   Algorithms for Very Large Computers

23.4.1   Large-scale machines

Considering the question of algorithms for a petaflop system, we have to look at potential systems able to provide that level of performance at least theoretically. What we find is the following:

- Any future large-scale system will be composed of a large number of components. This will also be true for vector systems even though the number may be an order of magnitude smaller. Any algorithm will have to consider that it truly has to be massively parallel but also that optimum usage of memory and of networks may require a re-design.

- Any future large-scale system will be hybrid by nature. Vendors will assemble the best parts in order to achieve maximum performance. Combinations like the ones discussed in the Roadrunner project or the Japanese next generation HPC project will become standard. A first flavor of this mix of components can be found in the most recent TITech installation.

For the algorithms that want to achieve a sustained level of performance in the petaflops range the consequences are:

- Algorithms will have to be developed that are able to feed hundreds of thousands of components. This will require decoupling as much as possible. Decoupling of algorithmic building blocks will be as essential as well as decoupling of computation, communication and I/O.

- Given current limitations of algorithms for very large problem sizes we have to find ways to solve small problems on large-scale systems. This is going to be much more important than solving even larger problems. Approaches in the right direction are coming from multiphysics and multi-scale. In both cases additional compute power is used to bring in additional quality rather than quantity.

- As a consequence of memory and interconnect deficiencies we will have to return to structured data handling. Such regular data structures can be flexible. However, they have to give compilers a chance to fill functional units as much as possible considering memory hierarchies and communication network limitations.

23.4.2 Linpack will show limits

Since the 1980s the Linpack benchmark serves as a well-accepted criterion of all types of computers. For larger problem sizes it measures the performance of a matrix multiply (DGEMM) which is dominated by floating-point operations. Lack of memory bandwidth can be hidden by clever programming. The new High Performance Linpack (HPL) test moves from a fixed size measurement to maximum performance achieved for an unrestricted size of system, ignoring runtime.

To see whether Linpack also is suited to analyze system performance in the future, we analyze the runtime estimations for the "increasing-ring (modified) variant" in the "long variant" given in [8] by the formula

$$T_{hpl} \geq \frac{2}{3}\gamma_3 n^3 \frac{1}{PQ} + \beta n^2 \frac{3P+Q}{2PQ} + \alpha n \left(\left(1 + \frac{1}{nb}\right) \log(P) + \frac{1}{nb}P \right) \quad (23.2)$$

with

- α time to set up a message between two different processors,

- $\beta = 1/bandwidth$ time to transfer an 8-byte word

- γ_3 time for one floating-point operation in a matrix multiply subprogram

- nb block size of matrix blocks used to get single processor performance. nb must not be too small

- z number of matrix blocks per processor

- P, Q parameters defining a rectangular processor array, $M_{proc} = P Q$ is the number of processors

- The memory consumption per processor $Mem = \frac{n^2}{M_{proc}}$ in 8-byte words is assumed to be constant (weak scaling)

We neglect the $\log(P)$ part and replace the total matrix size by $n = (Mem \, P \, Q)^{\frac{1}{2}}$. After minimizing the time T_{hpl} with respect to P, Q we get the expression

$$T_{hpl} \geq \sqrt{M_{proc}}\sqrt{Mem}\left(\frac{2}{3}\gamma_3 Mem + \sqrt{6\beta^2 Mem + 2\alpha\beta z^{\frac{1}{2}} M_{proc}^{\frac{1}{2}}}\right)(23.3)$$

If we neglect the term $6\beta^2 Mem$ we receive

$$T_{hpl} \geq \sqrt{M_{proc}}\sqrt{Mem}\left(\frac{2}{3}\gamma_3 Mem + \sqrt{2\alpha\beta}z^{\frac{1}{4}} M_{proc}^{\frac{1}{4}}\right) \quad (23.4)$$

To make the implications clearer we compare this with the time, T_1, for the local problem size

$$\frac{T_{hpl}}{T_1} \geq M_{proc}^{\frac{1}{2}}\left(1 + \frac{\sqrt{2\alpha\beta}}{\frac{2}{3}\gamma_3 \, Mem}z^{\frac{1}{4}} M_{proc}^{\frac{1}{4}}\right) \quad (23.5)$$

Both parts are strongly increasing with the number of processors M_{proc}, the second much stronger. This indicates that the benchmark needs more and more time if the processor number increases. That cannot be the intention of a benchmark.

A detailed analysis shows that the second term is connected to the communication parameters. Millions of processors are needed to make this term dominant. On the other hand, the slower increasing first term shows that the benchmark time increases as the machine size increases. The growth of the

relative execution time with the square root of the processor number is un-
avoidable because the process size cannot be decreased under a certain limit
to maintain good per-processor performance.

This example shows the behavior of all algorithms for which the computing
time needed is increasing faster than their size. Their runtime will grow with
processor counts if the problem size for a single processor is not allowed to
shrink. Doing this would, however, reduce processor-level performance. The
reason for this is that the immanent processor latencies become more dominant for small sizes. To show the consequence of this we analyze Figures 23.1
and 23.2.

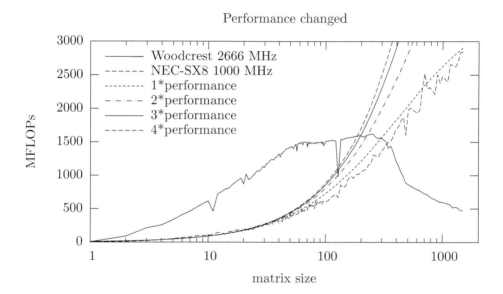

FIGURE 23.1: Changed peak performance for fit to the NEC SX-8 performance curve.

They show a comparison of the Eispack routine SVD for the Intel Woodcrest
processor with a frequency of 2666 MHz and the vector machine NEC SX-8
with a frequency of 1000 MHz which has a larger inherent parallelism. The
results are limited to relatively small matrices and show that, for smaller cases,
the PC processor is significantly faster. In the figures we add performance
curves for

$$performance = \frac{1}{latency/size + \Delta t} \tag{23.6}$$

where $\frac{1}{\Delta t}$ is the peak performance of the code segment; *latency* assembles
all setup times as loop overhead and memory latencies. This is a very simple

approximation of the performance curve but accurate enough for our purposes.

We fitted the parameters and then varied *latency* as well as Δt. In Figure 23.1, the Δt for the vector system is changed. In Figures 23.2, the latency is reduced. Apparently the peak performance plays an insignificant role for the behavior of the curve in contrast to the latency. It is interesting to note that an eight-times smaller latency for the NEC SX-8 fits the first part of the performance curve of the Woodcrest processor.

Figure 23.3 shows a similar problem for an OpenMP-parallelized version of a sparse-matrix vector multiplication. Here latency is to be understood as the losses in the startup and the ending phases of the parallel-code segment. We see that the absolute performance may only be increased for quite large cases. This is again caused by the startup latency which has been extracted from a fit to the initial part of the curves. The latency (=startup) curve is the upper limit of the parallel performance of that machine for an infinite number of processors. Additional curves show the effect of smaller startups. We see the consequences of Amdahl's Law.

Decreasing the problem size without larger penalties is possible only by decreasing parasitoid times like startup, loop overhead, memory latencies or overlap of latencies with productive times. With a fixed frequency the architecture must be improved to achieve higher performance.

It is definitely not possible to use MPI for that purpose because of the inherent overhead. Communication between nearby processors has to be done directly by special machine instructions. Only in this case it can be assured

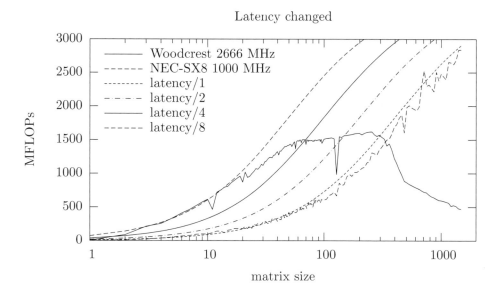

FIGURE 23.2: Changed latency for fit to the NEC SX-8 performance curve.

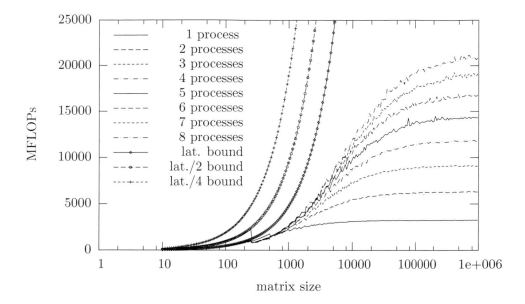

FIGURE 23.3: Sparse-matrix vector performance independent of the number
of processors with absolute limits for assumed latencies.

that latencies become as small as clock-cycle times. For the purpose of local
synchronization the on-chip-cores architecture should provide special wires
and registers for fast barriers and fast reductions.

23.4.3 Narrowing the path to algorithms

The Linpack benchmark as shown above is a good example for a scalable
benchmark. For tightly coupled simulations, results of an analysis would
be even worse because the amount and the frequency of communication are
larger. The grid density of a Lattice-Boltzmann experiment cannot further be
refined if results have to be simulated in reasonable time. Real-time predictions of turbulence will hence remain utopian with current technology unless
the processor speed increases dramatically and communication parameters
improve. Reduction of cell sizes in weather prediction will only be possible as
long as the calculation time does not exceed the time frame predicted.

Changing numerics in a way that more computational efforts are done for
less entities (nodes, elements, cells) will reduce the negative impact of the
enforced weak scalability. Examples are higher order finite elements such
as Discontinuous Galerkin or spectral element techniques or usage of locally
known solutions.

On the other hand, only algorithms with linear numerical complexity are well suited to scale for a fixed time frame. One important candidate of algorithms fulfilling this condition on multigrid ones. The disadvantage is the reduction of points at different levels. The best we can get is a logarithmic dependence of the computing time on the problem size. This seems to be acceptable. But the multigrid has a relatively high need of memory bandwidth per flop. For the simplest case of a 3D Poisson equation on a rectangular grid, more than 6 bytes/flop are necessary. Suppose a chip with 64 cores is running at a moderate frequency of 1 GHz. Each core should have a peak performance of 2 operations per clock. Aggregated memory bandwidth would have to be 6 bytes/operation × 2 operations × 64 cores × 1 GHz = 768 GB/sec. This is beyond the speed even of current vector systems.

H-matrices [7] also provide algorithms of low numerical complexity. This may fit into the computer science approach of so called hierarchical tiled matrices (HTA) [1, 13] which are proposed to have direct compiler support for the localization of data. But even if the localization of data is increased there is still a high demand for sustained memory bandwidth. This can be provided by optical links and forces the access to a memory system consisting of large and expensive numbers of independent banks.

Examples for effectively working projects may be the composition of known parameterized solutions to new larger interacting phenomena. The simplest solution will be to solve trivially parallel numerical problems. Parameterized investigations are needed in most engineering disciplines for design and optimization. These investigations need capacity machines. Tightly coupled parallel high-performance machines are only needed for the single (smaller) job. Based on strategies like genetic algorithms, gradient-based search, and solution of the dual equations, many (parallel) solutions of the underlying problem are required and a lot of them can be done at the same time. A lot of life sciences and molecular dynamics problems are of this type.

Meta-programming languages are necessary for that purpose, which allow for a simple formulation of the optimization workflow on top of the parallel simulation experiment. The optimization strategy would be formulated in the meta language independent of the specific task. Examples for these meta languages for workflow environments are NIMROD, ILAB [6, 14], SEGL [4, 5]. The workflow environment has to administrate the large amount of large data sets, has to be aware of the scheduling policy of the supercomputer and has to integrate batch mechanisms and interactive access to the running experiment. The program itself could have some message-passing-like interfaces for communication with other programs. Multidisciplinary design is part of this approach. Unless diverse equations define a stiff combined equation that must be solved by an integrated algorithm, they may be solved in different programs running at the same time on the same machine, but communicating from time to time. This requires a different communication layer which is not only simply transporting data but also changing their semantics. The grids for an electromagnetic calculation on the same domain as a connected FEM calculation may be very different. An appropriate interpolation must take place. MPCCI [11] has been developed for that purpose.

Multi-scale technologies can be combined such that different scales are calculated at the same time on the machine in different parts and exchanging parameter data iteratively. An example might be Steve Ashby's/LLNL technique of using different models for a stressed material. Most of the elements will conform to the continuum hypothesis. For the elements not conforming a molecular dynamics code will be called by "remote procedure invocation" which handles the local problem by a part of the parallel machine in a dynamic way. Load balancing is difficult and assumes large resources of the machine.

Programming will be more difficult and assumes also programming of the meta level combining the different solution steps in the machine. The user program needs closer connection to the machine scheduler. The program must be able to anticipate the machine status for some future time. Machine resources such as sets of nodes, hardware accelerators, file space, communication connections have to be prefetched to be ready at the right time. The allocation of resources should be signaled early enough to the machine scheduler.

23.5   Impact on Other Technologies

What will be the impact of the large-scale systems that we will se in the next five years? First there will be an increased level of sustained performance for a number of applications. How many applications will actually benefit remains to be seen. We have pointed out some of them in this chapter. It is most likely that over the next years the number of applications to reach extreme levels of performance will be quite limited. On the other hand, much higher levels of performance will allow tackling new problems. Some of these problems can be found in genomics and bioinformatics.

In general, however, there is a widening gap between standard applications — as we have seen them for years in computational fluid dynamics (CFD) and other fields — and applications being able to exploit petaflops systems. This widening gap is due to the fact that HPC is deviating from mainstream computing. It goes without saying that HPC for a long time has been different from workstations and PCs. But during the 1970s and 1980s, HPC was a development tool for new standard systems and at the same time the only means to do realistic simulations. The current situation is different in that users have more options. The use of a medium-sized cluster will allow for achieving reasonable performance to do reasonably large simulations.

The simulation community may become split in purpose. Large-scale systems may only be used for theoretical research looking at basic problems in physics, astrophysics and so forth. At the same time, applied research may turn away from sophisticated HPC architectures and focus on mainstream architectures. Whether specialized hardware like the Cell BE processor or Blue Gene will ever become part of standard hardware remains to be seen. If this is not the case then it will become increasingly more difficult for HPC centers to justify the expense for a large-scale system.

On the other hand, the increased speed of multicore systems will open up entirely new fields. To mention only one, medical applications will benefit. We can expect to see embedded multicore processors in medical devices allowing embedding of simulation in the process of diagnostic procedures. This may lead to a new market for specialized hardware further driving the development of architectures like the Cell Broadband Engine processor.

Summarizing, it is extremely difficult to achieve sustained petaflops on those systems we expect to see in computer rooms over the next five years. We expect to see a separation between basic research that may be able to exploit such architectures and standard technologies like CFD or weather forecasting which are hooked to the classical single processor model and have only gradually been adapted to massively parallel systems. This gap between basic and applied research may be widened by independent software vendors that already today have begun to ignore nonstandard architectures. As a consequence, these independent software packages lose their ability to bring new users into the HPC arena. Risks and chances are associated with the introduction of petaflops systems. Our role as leaders in HPC is to build on the chances and find ways to avoid the risks.

References

[1] G. Almasi, L. De Rose, J. Moreira, and D. Padua. Programming for locality and parallelism with hierarchically tiled arrays. In *Proc. 16th International Workshop on Languages and Compilers for Parallel Computing (LCPC)*, College Station, TX, October 2003.

[2] G. Amdahl. Validity of the single processor approach to achieving large-scale computing capabilities. *AFIPS Conference Proceedings*, 30:483–485, 1967.

[3] R. W. Brodersen, M. A. Horowitz, D. Markovic, B. Nikolic, and V. Stojanovic. Methods for true power minimization. In *Proc. IEEE/ACM Int'l. Conference on Computer-Aided Design (ICCAD)*, pages 35–42, San Jose, CA, November 2002.

[4] N. Currle-Linde, F. Bös, and M. Resch. GriCoL: A language for scientific

grids. In *Proc. 2nd IEEE International Conference on e-Science and Grid Computing*, Amsterdams, December 2006.

[5] N. Currle-Linde, U. Küster, M. Resch, and B. Risio. Experimental grid laboratory (SEGL) dynamical parameter study in distributed systems. In *Proc. International Conference on Parallel Computing (ParCo)*, pages 49–56, Malaga, Spain, September 2005.

[6] A. de Vivo, M. Yarrow, and K. McCann. A comparison of parameter study creation and job submission tools. Technical Report NAS-01002, NASA Ames Research Center, Moffet Field, CA, 2000.

[7] HLib. http://www.hlib.org/.

[8] HPL. http://www.netlib.org/benchmark/hpl/scalability.html.

[9] IRAM. http://iram.cs.berkeley.edu.

[10] J. A Kahle, M. N. Day, H.P. Hofstee, C. R. Johns, T. R. Maeurer, and D. Shippy. Introduction to the Cell multiprocessor. *IBM J. Res. & Dev.*, 49(4/5), July/September 2005.

[11] MPCCI. http://www.scai.fraunhofer.de/mpcci.html.

[12] A. Shacham, K. Bergman, and L. P. Carloni. Maximizing GFLOPS-per-watt: High-bandwidth, low power photonic on-chip networks. In *Proc. $P = ac^2$ Conference*, pages 12–21, IBM T.J. Watson Research Center, Yorktown Heights, NY, October 2006.

[13] M. Snir. Software for high performance computing: Requirements & research directions. In *Insightful Understanding of China's Higher Education and Research in Computer Science and Information Technology: A U.S. Delegation Visit to China*, May 2006. http://dimacs.rutgers.edu/Workshops/China/Presentations/snir.pdf.

[14] J. Yu and R. Buyya. A taxonomy of workflow management systems for grid computing. *Journal of Grid Computing*, 3(3-4):171–200, 2005.

Chapter 24

Cactus Framework: Black Holes to Gamma Ray Bursts

Erik Schnetter

Center for Computation & Technology and Department of Physics and Astronomy, Louisiana State University

Christian D. Ott

Steward Observatory and Department of Astronomy, The University of Arizona

Gabrielle Allen

Center for Computation & Technology and Department of Computer Science, Louisiana State University

Peter Diener

Center for Computation & Technology and Department of Physics and Astronomy, Louisiana State University

Tom Goodale

Center for Computation & Technology, Louisiana State University, and School of Computer Science, Cardiff University, UK

Thomas Radke

Max-Planck-Institut für Gravitationsphysik, Albert-Einstein-Institut, Germany

Edward Seidel

Center for Computation & Technology and Department of Physics, Louisiana State University

John Shalf

CRD/NERSC, Lawrence Berkeley National Laboratory

24.1 Current Challenges In Relativistic Astrophysics and the Gamma-Ray Burst Problem

Ninety years after Einstein first proposed his General Theory of Relativity (GR), astrophysicists more than ever and in greater detail are probing into regions of the universe where gravity is very strong and where, according to GR's geometric description, the curvature of space-time is large.

The realm of strong curvature is notoriously difficult to investigate with conventional observational astronomy, and some phenomena might bear no observable electro magnetic signature at all and may only be visible in neutrinos (if sufficiently close to Earth) or in gravitational waves — ripples of space-time itself which are predicted by Einstein's GR. Gravitational waves have not been observed directly to date, but gravitational-wave detectors (e.g., LIGO [21], GEO [14], VIRGO [36]) are in the process of reaching sensitivities sufficiently high to observe interesting astrophysical phenomena.

Until gravitational-wave astronomy becomes a reality, astrophysicists must rely on computationally and conceptually challenging large-scale numerical simulations in order to grasp the details of the energetic processes occurring in regions of strong curvature that are shrouded from direct observation in the electromagnetic spectrum by intervening matter, or that have little or no electromagnetic signature at all. Such astrophysical systems and phenomena include the birth of neutron stars (NSs) or black holes (BHs) in collapsing evolved massive stars, coalescence of compact* binary systems, gamma-ray bursts (GRBs), active galactic nuclei harboring super-massive black holes, pulsars, and quasi-periodically oscillating NSs (QPOs). In Figure 24.1 we present example visualizations of binary BH and stellar-collapse calculations carried out by our groups.

From these, GRBs, intense narrowly beamed flashes of γ-rays of cosmological origin, are among the most scientifically interesting and the riddle concerning their central engines and emission mechanisms is one of the most complex and challenging problems of astrophysics today.

GRBs last between 0.5–1000 secs, with a bimodal distribution of duration [23], indicating two distinct classes of mechanisms and central engines. The short-hard (duration $\lesssim 2$ secs) group of GRBs (hard, because their γ-ray spectra peak at a shorter wavelength) predominantly occurs in elliptical galaxies with old stellar populations at moderate astronomical distances [38, 23].

---

*The term "compact" refers to the compact stellar nature of the binary members in such systems: white dwarfs, neutron stars, black holes.

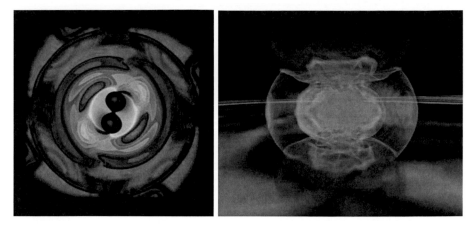

FIGURE 24.1: (See color insert following page 18.) *Left:* Gravitational waves and horizons are shown in a binary black hole in spiral simulation. Simulation is by AEI/CCT collaboration, image by W. Benger (CCT/AEI/ZIB). *Right:* A rotationally deformed proto-neutron star formed in the iron core collapse of an evolved massive star is pictured. Shown are a volume rendering of the rest-mass density and a 2D rendition of outgoing gravitational waves. Simulation is by [28], image by R. Kähler.

The energy released in a short-hard GRB and its duration suggest [38, 23] a black hole with a ~0.1 solar-mass ($M_\odot$) accretion disk as the central engine. Such a BH–accretion-disk system is likely to be formed by the coalescence of NS–NS or NS–BH systems (e.g., [33]).

Long-soft (duration ~2–1000 secs) GRBs, on the other hand, seem to occur exclusively in the star-forming regions of spiral or irregular galaxies with young stellar populations and low metallicity.[†] Observations that have recently become available (see [23] for reviews) indicate features in the x-ray and optical afterglow spectra and luminosity evolutions of long-soft GRBs that show similarities with spectra and light curves obtained from Type-Ib/c core-collapse supernovae, whose progenitors are evolved massive stars ($M \gtrsim 25\,M_\odot$) that have lost their extended hydrogen envelopes, and probably also a fair fraction of their helium shells. These observations support the collapsar model [38] of long-soft GRBs that envisions a stellar-mass black hole formed in the aftermath of a stellar core-collapse event with a massive ~1 $M_\odot$ rotationally supported accretion disk as the central engine, powering the GRB jet that punches through the compact and rotationally evacuated polar stellar envelope reaching ultra-relativistic velocities [23].

[†]The metallicity of an astrophysical object is the mass fraction in chemical elements other than hydrogen and helium. In big-bang nucleosynthesis, only hydrogen and helium were formed. All other elements are ashes of nuclear-burning processes in stars.

Although observations are aiding our theoretical understanding, much that is said about the GRB central engine will remain speculation until it is possible to self-consistently model (i) the processes that lead to the formation of the GRB central engine, and (ii) the way the central engine utilizes gravitational (accretion) and rotational energy to launch the GRB jet via magnetic stresses and/or polar neutrino pair-annihilation processes. The physics necessary in such a model includes general relativity, relativistic magneto-hydrodynamics, nuclear physics (describing nuclear reactions and the equation of state of dense matter), neutrino physics (weak interactions), and neutrino and photon radiation transport. In addition, it is necessary to adequately resolve physical processes with characteristic scales from ∼100 meters near the central engine to ∼5–10 million kilometers, the approximate radius of the collapsar progenitor star.

24.1.1   GRBs and petascale computing

The complexity of the GRB central engine and its environs requires a multi-physics, multi-length-scale approach that cannot be fully realized on present-day computers. Computing at multiple sustained petaflops (PFLOPS) of performance will allow us to tackle the full GRB problem and provide complete numerical models whose output can be compared with observations.

In this chapter we outline our petascale approach to the GRB problem and discuss the computational toolkits and numerical codes that are currently in use, and that will be scaled up to run on emerging PFLOPS-scale computing platforms in the near future.

Any comprehensive approach to GRBs must naturally draw techniques and tools from both numerical relativity and core-collapse supernova and neutron star theory. Hence, much of the work presented and suggested here builds upon the dramatic progress that has been made in these fields in the past decade. In numerical relativity, immense improvements in the long-term stability of 3D GR vacuum and hydrodynamics evolutions (e.g., [1, 30]) allow for the first time long-term stable binary black hole merger, binary neutron star merger, neutron star and evolved massive star collapse calculations. Supernova theory, on the other hand, has made giant leaps from spherically symmetric (1D) models with approximate neutrino radiation transport in the early 1990s, to Newtonian or approximate GR to 2D and the first 3D [13] calculations, including detailed neutrino and nuclear physics and energy-dependent multispecies Boltzmann neutrino transport [6] or neutrino flux-limited diffusion [8] and magneto-hydrodynamics [7].

As we shall discuss, our present suite of terascale codes, comprised of the space-time evolution code `Ccatie` and the GR hydrodynamics code `Whisky`, can be and has already been applied to the realistic modeling of the in-spiral and merger phase of NS–NS and NS-BH binaries, to the collapse of polytropic (cold) supermassive NSs, and to the collapse and early post-bounce phase of a core-collapse supernova or a collapsar. As the codes will be upgraded and readied for petascale, the remaining physics modules will be developed and integrated. In particular, energy-dependent neutrino transport and magneto-hydrodynamics, both likely to be crucial for the GRB central engine, will be given high priority.

To estimate roughly the petaflopage and petabyteage required for a full collapsar-type GRB calculation, we assume a Berger-Oliger-type [5] adaptive-mesh refinement setup with 16 refinement levels, resolving features with resolutions of 10,000 km down to 100 m across a domain of 5 million cubic km. To simplify things, we assume that each level of refinement has twice the resolution as the previous level and covers approximately half the domain. Taking a base grid size of 1024^3 and 512 3D grid functions, storing the curvature and radiation-hydrodynamics data on each level, we estimate a total memory consumption of ~ 0.0625 petabytes (64 terabytes). To obtain an estimate of the required sustained petaflopage, we first compute the number of time steps that are necessary to evolve for 100 s in physical time. Assuming a time step that is half the light-crossing time of each grid cell on each individual level, the base grid has to be evolved for ~ 6000 time steps, while the finest grid will have to be evolved for $2^{16-1} \times 6000$ individual time steps. `Ccatie` plus `Whisky` require approximately 10k flops per grid-point per time step. When we assume that additional physics (neutrino and photon radiation transport, magnetic fields; some of which may be evolved with different and varying time-step sizes) requires on average an additional 22k flops, one time step of one refinement level requires 10^{-5} PFLOPS. Summing up over all levels and time steps, we arrive at a total petaflopage of ~ 13 million. On a machine with 2 PFLOPS sustained, the runtime of the simulation would come to ~ 75 days. GRBs pose a true petascale problem.

24.2 The Cactus Framework

To reduce the development time for creating simulation codes and encourage code reuse, researchers have created computational frameworks such as the Cactus framework [16, 9]. Such modular component-based frameworks allow scientists and engineers to develop their own application modules and use them in conjunction with existing modules to solve computational problems. Cactus provides tools ranging from basic computational building blocks to complete toolkits that can be used to solve a range of application problems. Cactus runs on a wide range of hardware ranging from desktop PCs, large supercomputers, to "grid" environments. The Cactus framework and core toolkits are distributed with an open-source license from the Cactus Website [9], are fully documented, and are maintained by active developer and user communities.

The Cactus framework consists of a central part ("flesh") and components ("thorns"). The flesh has minimal computational functionality and serves as a module manager, coordinating the flow of data between the different components to perform specific tasks. The components or "thorns" perform tasks ranging from setting up a computational grid, decomposing the grid for parallel processing, setting up coordinate systems, boundary and initial conditions, communication of data from one processor to another, solving partial differential equations, to input and output and streaming of visualization data. One standard set of thorns is distributed as the Cactus Computational Toolkit to provide basic functionality for computational science.

Cactus was originally designed for scientists and engineers to collaboratively develop large-scale, parallel scientific codes which would be run on laptops and workstations (for development) and large supercomputers (for production runs). The Cactus thorns are organized in a manner that provides a clear separation of the roles and responsibilities between the "expert computer scientists" who implement complex parallel abstractions (typically in C or C++), and the "expert mathematicians and physicists" who program thorns that look like serial blocks of code (typically, in F77, F90, or C), implementing complex numerical algorithms. Cactus provides the basic parallel framework supporting several different codes in the numerical relativity community used for modeling black holes, neutron and boson stars and gravitational waves. This has led to over 150 scientific publications in numerical relativity which have used Cactus and the establishment of a Cactus Einstein toolkit of shared community thorns. Other fields of science and engineering are also using the Cactus framework, including quantum gravity, computation fluid dynamics, computational biology, coastal modeling, applied mathematics, etc., and in some of these areas community toolkits and shared domain-specific tools and interfaces are emerging.

Cactus provides a range of advanced development and runtime tools including an HTTPD thorn that incorporates a simplified web server into the simulation allowing for real-time monitoring and steering through any web interface; a general timer infrastructure for users to easily profile and optimize their codes; visualization readers and writers for scientific visualization packages; and interfaces to grid application toolkits for developing new scientific scenarios taking advantage of distributed computing resources.

Cactus is highly portable. Its build system detects differences in machine architecture and compiler features, using automatic detection where possible and a database of known information where auto-detection is impractical — e.g., which libraries are necessary to link Fortran and C code together. Cactus runs on all variants of the Unix and on Windows operating systems. Codes written using Cactus have been run on some of the fastest computers in the world, such as the Japanese Earth Simulator and the IBM Blue Gene/L, Cray X1e, and the Cray XT3/XT4 systems [25, 26, 24, 11, 19, 34] This enables Cactus developers to write and test code on their laptop computers, and then deploy the very same code on the full scale systems with very little effort.

24.3 Space-time and Hydrodynamics Codes

24.3.1 `Ccatie`: Space-time evolution

In strong gravitational fields, such as in the presence of neutron stars or black holes, it is necessary to solve the full Einstein equations. Our code employs a 3 + 1 decomposition [3, 39], which renders the four-dimensional spacetime equations into hyperbolic time-evolution equations in three dimensions, plus a set of constraint equations which have to be satisfied by the initial condition. The equations are discretized using high-order finite differences with adaptive mesh refinement and using Runge–Kutta time integrators, as described below.

The time evolution equations are formulated using a variant of the BSSN formulation described in [2] and coordinate conditions described in [1] and [35]. These are a set of 25 coupled partial differential equations which are first order in time and second order in space. One central variable describing the geometry is the three-metric γ_{ij}, which is a symmetric positive definite tensor defined everywhere in space, defining a scalar product which defines distances and angles.

`Ccatie` contains the formulation and discretization of the right-hand sides of the time-evolution equations. Initial data and many analysis tools, as well as time integration and parallelization, are handled by other thorns. The current state of the time evolution, i.e., the three-metric γ_{ij} and related variables, is communicated into and out of `Ccatie` using a standard set of variables (which is different from `Ccatie`'s evolved variables), which makes it possible to combine unrelated initial data solvers and analysis tools with `Ccatie`, or to replace `Ccatie` by other evolution methods, while reusing all other thorns.

We have a variety of initial conditions available, ranging from simple test cases, analytically known and perturbative solutions to binary systems containing neutron stars and black holes.

The numerical kernel of `Ccatie` has been hand-coded and extensively optimized for floating-point unit (FPU) performance where the greatest part of the computation lies. (Some analysis methods can be similarly expensive and have been similarly optimized, e.g., the calculation of gravitational-wave quantities.) The Cactus/Ccatie combination has won various awards for performance.[‡]

24.3.2 `Whisky`: General relativistic hydrodynamics

The `Whisky` code [4, 37] is a GR hydrodynamics code originally developed under the auspices of the European Union research training network "Sources

---

[‡]See http://www.cactuscode.org/About/Prizes

of Gravitational Waves" [12].

While `Ccatie` in combination with `Cactus`'s time-integration methods provides the time evolution of the curvature part of space-time, `Whisky` evolves the "right-hand side," the matter part, of the Einstein equations. The coupling of curvature with matter is handled by `Cactus` via a transparent and generic interface, providing for modularity and interchangeability of curvature and matter-evolution methods. `Whisky` is also fully integrated with the `Carpet` mesh refinement driver discussed in Section 24.4.2.

`Whisky` implements the equations of GR hydrodynamics in a semi-discrete fashion, discretizing only in space and leaving the explicit time integration to `Cactus`. The update terms for the hydrodynamic variables are computed via flux-conservative finite-volume methods exploiting the characteristic structure of the equations of GR hydrodynamics. Multiple dimensions are handled via directional splitting. Fluxes are computed via piecewise-parabolic cell-interface reconstruction and approximate Riemann solvers to provide right-hand side data that are accurate to (locally) third-order in space and first-order in time. High-temporal accuracy is obtained via Runge-Kutta-type time-integration cycles handled by `Cactus`.

`Whisky` has found extensive use in the study of neutron-star collapse to a black hole, incorporates matter excision techniques for stable non-vacuum BH evolutions, has been applied to BH–NS systems, and NS rotational instabilities.

In a recent upgrade, `Whisky` has been endowed with the capability to handle realistic finite-temperature nuclear equations of state and to approximate the effects of electron capture on free protons and heavy nuclei in the collapse phase of core-collapse supernovae and/or collapsars [28]. In addition, a magneto-hydrodynamics version of `Whisky`, `WhiskyMHD`, is approaching the production stage [15].

24.4 Parallel Implementation and Mesh Refinement

Cactus has generally been used to date for calculations based upon explicit finite-difference methods. Each simulation routine is called with a block of data — e.g., in a 3-dimensional simulation the routine is passed as a cuboid, in a 2-dimensional simulation as a rectangle — and integrates the data in this block forward in time. In a single-processor simulation the block would consist of all the data for the whole of the simulation domain, and in a multiprocessor simulation the domain is decomposed into smaller sub-domains, where each processor computes the block of data from its sub-domain. In a finite-difference calculation, the main form of communication between these sub-domains is on the boundaries, and this is done by *ghost-zone exchange* whereby each

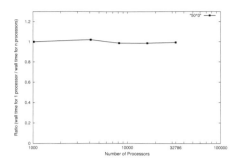

FIGURE 24.2: PUGH scaling result from Blue Gene/L. (See also Figure 1.6 in Chapter 1 of this book [27] for similar Cactus/PUGH benchmarks.)

sub-domain's data-block is enlarged by the nearest boundary data from neighboring blocks. The data from these ghost zones is then exchanged once per iteration of the simulation.

In Cactus, infrastructure components providing storage handling, parallelization, mesh refinement, and I/O methods are implemented by thorns in the same manner that numerical components provide boundary conditions or physics components provide initial data. The component which defines the order in which the time evolution is orchestrated is called the *driver*.

There are several drivers available for Cactus. In this chapter we present results using the unigrid PUGH driver and using the adaptive mesh-refinement (AMR) driver Carpet.

24.4.1 PUGH

The Parallel UniGrid Hierarchy (PUGH) driver was the first parallel driver available in Cactus. PUGH decomposes the problem domain into one block per processor using the Message Passing Interface (MPI) for the ghost-zone exchange described above. PUGH has been successfully used on many architectures and has proven scaling up to many thousands of processors [25]. Figure 24.2 shows benchmark results from Blue Gene/L. (See also Figure 1.6 in the Chapter 1 of this book [27].)

24.4.2 Adaptive mesh refinement with Carpet

Carpet [32, 10] is a driver which implements Berger–Oliger mesh refinement [5]. Carpet refines parts of the simulation domain in space and/or time by factors of two. Each refined region is block-structured, which allows for efficient representations, e.g., as Fortran arrays.

In addition to mesh refinement, Carpet also provides parallelism by distributing grid functions onto processors, corresponding I/O methods for ASCII and HDF5 output and for checkpointing and restart, interpolation of values to arbitrary points, and reduction operations such as norms and maxima.

Fine-grid boundary conditions require interpolation in space. This is currently implemented up to the seventh order, and the fifth order is commonly used. When refining in time, finer grids take smaller time steps to satisfy the local CFL criterion. In this case, Carpet may need to interpolate coarse-grid data in time to provide boundary conditions for fine grids. Similarly, time interpolation may also be required for interpolation or reduction operations at times when no coarse-grid data exist. Such time interpolation is currently implemented with up to a fourth order accuracy, although only the second order is commonly used.

In order to achieve convergence at mesh-refinement boundaries when second spatial derivatives appear in the time-evolution equations, we do not apply boundary conditions during the substeps of a Runge–Kutta time integrator. Instead we extend the refined region by a certain number of grid-points before each time step (called *buffer zones*) and only interpolate after each complete time step. We find that this is not necessary when no second derivatives are present, such as in discretizations of the Euler equations without the Einstein equations. The details of this algorithm are laid out in [32].

Carpet is currently mostly used in situations where compact objects need to be resolved in a large domain, and Carpet's mesh-refinement adaptivity is tailored for these applications. One can define several *centers of interest*, and space will be refined around these. This is ideal, e.g., for simulating binary systems, but is not well suited for resolving shock fronts as they, e.g., appear in collapse scenarios. We plan to address this soon in future versions of Carpet.

24.4.3   I/O

Our I/O methods are based on the HDF5 [17] library, which provides a platform-independent high-performance file parallel format. Data sets are annotated with Cactus-specific descriptive attributes providing meta-data, e.g., for the coordinate systems, tensor types, or the mesh-refinement hierarchy.

Some file systems can only achieve high performance when each processor or node writes its own file. However, writing one file per processor on massively concurrent systems will be bottlenecked by the meta-data server, which often limits file-creation rates to only hundreds of files per second. Furthermore, creation of many thousands of files can create considerable meta-data management headaches. Therefore the most efficient output setup is typically to designate every nth processor as an output processor, which collects data from $n - 1$ other processors and writes the data into a file.

Input performance is also important when restarting from a checkpoint. We currently use an algorithm where each processor reads data only for itself, trying to minimize the number of input files which it examines.

Cactus also supports novel server-directed I/O methods such as PANDA, and other I/O file formats such as SDF and FlexIO. All of these I/O methods

TABLE 24.1: Our benchmarks and their characteristics. The
`PUGH` and `Carpet_1lev` benchmarks evolve the vacuum Einstein
equations without mesh refinement, with identical setups but
using different communication strategies. `Carpet_8lev` features 8
fixed levels with the same number of grid-points on each level.
`Whisky_8lev` evolves the relativistic Euler equations in addition to
the Einstein equations. `BenchIO_HDF5_801` writes several large
files to disk using the Cactus checkpointing facility.

Name	type	complexity	physics
Bench_Ccatie_PUGH	compute	unigrid	vacuum
Bench_Ccatie_Carpet_1lev	compute	unigrid	vacuum
Bench_Ccatie_Carpet_8lev	compute	AMR	vacuum
Bench_Ccatie_Whisky_Carpet	compute	AMR	hydro
BenchIO_HDF5_801	I/O	unigrid	vacuum

are interchangeable modules that can be selected by the user at compile time
or runtime depending on the needs and local performance characteristics of
the cluster file systems. This makes it very easy to provide head-to-head comparisons between different I/O subsystem implementations that are nominally
writing out exactly the same data.

24.5   Scaling on Current Machines

We have evaluated the performance of the kernels of our applications on
various contemporary machines to establish the current state of the code. This
is part of an ongoing effort to continually adapt the code to new architectures.
These benchmark kernels include the time-evolution methods of `Ccatie` with
and without `Whisky`, using as a driver either `PUGH` or `Carpet`. We present
some recent benchmarking results below; we list the benchmarks in Table
24.1 and the machines and their characteristics in Table 24.2.

Since the performance of the computational kernel does not depend on
the data which are evolved, we choose trivial initial data for our space-time
simulations, namely the Minkowski space-time (i.e., vacuum). We perform no
analysis on the results and perform no output. We choose our resolution such
that approximately 800 MByte of main memory is used per process, since
we presume that this makes efficient use of a node's memory without leading
to swapping. We run the simulation for several time steps requiring several
wall-time minutes. We increase the number of grid-points with the number
of processes. We note that the typical usage model for this code favors weak
scaling as the resolution of the computational mesh is a major limiting factor
for accurate modeling of the most demanding problems.

TABLE 24.2:   The machines used for the benchmarks in this section. Note that we speak of "processes" as defined in the MPI standard; these are implicitly mapped onto the hardware "cores" or "CPUs."

Name	Host	CPU	ISA	Interconnect
Abe	NCSA	Clovertown	x86-64	InfiniBand
Damiana	AEI	Woodcrest	x86-64	InfiniBand
Eric	LONI	Woodcrest	x86-64	InfiniBand
Pelican	LSU	Power5+		Federation
Peyote	AEI	Xeon	x86-32	GigaBit

Name	# proc	cores/ node	cores/ socket	memory/ proc	CPU freq.
Abe	9600	8	4	1 GByte	2.30 GHz
Damiana	672	4	2	2 GByte	3.00 GHz
Eric	512	4	2	1 GByte	2.33 GHz
Pelican	128	16	2	2 GByte	1.90 GHz
Peyote	256	2	1	1 GByte	2.80 GHz

Our benchmark source code, configurations, parameter files, and detailed benchmarking instruction are available from the Cactus Web site [9].§ There we also present results for other benchmarks and machines.

24.5.1   Floating-point performance

Figure 24.3 compares the scaling performance of our code on different machines. The graphs show scaling results calculated from the wall time for the whole simulation, but excluding startup and shutdown. The ideal result in all graphs is a straight horizontal line, and larger numbers are better. Values near zero indicate lack of scaling.

Since the benchmarks `Bench_Ccatie_PUGH` and `Bench_Ccatie_Carpet_1lev` use identical setups, they should ideally also exhibit the same scaling behavior. However, as the graphs show, PUGH scales, e.g., up to 1024 processors on Abe, while Carpet scales only up to 128 processors on the same machine. The differences are caused by the different communication strategies used by PUGH and Carpet, and likely also by the different internal bookkeeping mechanisms. Since Carpet is a mesh-refinement driver, there is some internal (but no communication) overhead even in unigrid simulations.

PUGH exchanges ghost-zone information in three sequential steps, first in the x, then in the y, and then in the z directions. In each step, each processor communicates with two of its neighbors. Carpet exchanges ghost-zone information in a single step, where each processor has to communicate with all 26

---

§See http://www.cactuscode.org/Benchmarks/

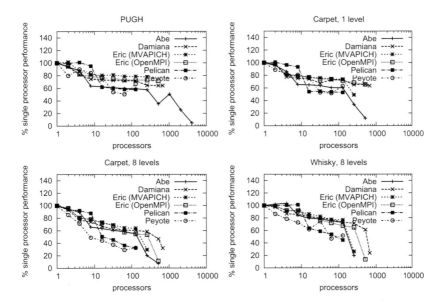

FIGURE 24.3: Weak-scaling tests using our four benchmarks on various machines are shown. The graphs show the fraction of single processor performance which is achieved on multiple processors. The `PUGH` and `1lev` benchmarks have identical setups, but use different drivers with different communication strategies. On Eric, we also compare two different MPI implementations.

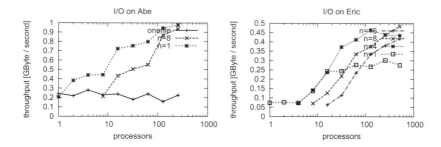

FIGURE 24.4: I/O scaling on various machines. This benchmark was only run on Abe and Eric. For each processor, 369 MByte of data are written to disk.

neighbors. In principle, Carpet's algorithm should be more efficient, since it does not contain several sequential steps; in practice, many MPI implementations seem to have trouble since there are more messages sent. Both PUGH and Carpet use `MPI_Irecv MPI_Isend`, and `MPI_Wait` to exchange data.

The falloff at small processor numbers coincides with the node boundaries. For example, Abe drops off at the 8-processor boundary, and Pelican drops off at the 16-processor boundary. We assume that this is because a fully loaded node provides less memory bandwidth to each process.

The scaling drop-off at large processor numbers is likely caused by using "too much" memory on each node. This drop-off is different for different MPI implementations, as is, e.g., evident on Eric, where OpenMPI scales further than MVAPICH. This problem could be caused by translation lookaside buffer (TLB) thrashing: a TLB miss causes potentially up to a 3-4k cycle stall, so it may only take a 1% miss rate to have a dramatic effect on performance.

24.5.2   I/O performance

Figure 24.4 compares the scaling of different I/O methods: collecting the output onto a single processor, having every nth processor perform output, and outputting on every processor. Each output processor creates one file. In these benchmarks, for each compute processor, 369 MBytes need to be written to disk. There is a small additional overhead per generated file which is less than 1 MByte.

The throughput graphs of Abe and Eric show that the maximum single-node throughput is limited, but the overall throughput can be increased if multiple nodes write simultaneously. While Abe has sufficient bandwidth to support every CPU writing at the same time, it is more efficient on Eric to collect the data first onto fewer processors.

24.6 Developing for Petascale

Petascale computing requires additional ingredients over conventional parallelism. The large number of available processors can only be fully exploited with mechanisms for significantly more fine-grained parallelism, in particular more fine grained than that offered by the MPI standard's collective communication calls. In order to ensure correctness in the presence of more complex codes and of increasingly likely hardware failures, simulations should be divided into *packets* describing a certain amount of computation. These packets can then be autonomously and dynamically distributed over the available CPU power, and can be rescheduled in response to hardware failure, thus achieving both load balancing and end-to-end correctness, akin to the way the IP protocol is used to deliver reliability over unreliable components. Furthermore, the current static-simulation schedule — often described by a sequence of subroutine calls — needs to become dynamic. The latter can be achieved, e.g., by listing pre- and postconditions of each substep, and determining an efficient execution order only at runtime, akin to the way UNIX `make` functions.

Clearly, these are only some measures, and a reliable methodology for petascale computing requires a comprehensive set of concepts which enable and encourage the corresponding algorithms and programming styles. It will very likely be necessary to develop new algorithms for petascale machines. These algorithms may not necessarily run efficiently on single processors or today's popular small workstation clusters, in the same way in which vectorised codes do not work well on non-vector machines, or in which MPI codes may not work well on single-processor machines.

One example of a non-petascale parallelization algorithm which is currently widely employed is the use of ghost zones to distribute arrays onto multiple processors. As the number of processors increases, each processor receives a correspondingly smaller part of the original array, while the *overhead* (the ratio between the number of ghost zones to the number of owned points) increases. Some typical numbers illustrate this. Our fourth-order stencils require a ghost-zone layer which is 3 points wide. If a processor owns 20^3 grid-points of a 3D array, the overhead is $(26^3 - 20^3)/20^3 \approx 20\%$. If a processor owns only 10^3 grid-points, then the overhead increases to $(16^3 - 10^3)/10^3 \approx 310\%$: the number of ghost zones is more than three times the number of owned points. Clearly this cannot scale in its current form.

Another problem for petascale simulations is time integration, i.e., the solution of hyperbolic equations. Since each time step depends on the result of the previous time step, time integration is inherently serial. If a million time steps are required, the total time to solution is at least a million times the time for a single time step.

The Cactus framework with its clear division between physical calculations and computational infrastructure already provides a partial pathway to peta-scale computations. By improving or exchanging the driver, existing physics modules can be used in novel ways, e.g., invoking them on smaller parts of the domain, or scheduling several at the same time to overlap computation and communication. Such improvement would be very difficult without a computational framework. However, achieving full petascale performance will also require improvements to the framework itself, i.e., to the way in which a physics calculation describes itself and its interface to the framework.

24.6.1   Physics: Radiation transport

The most conceptually and technically challenging part of the petascale development will be related to the implementation of neutrino and photon radiation transport. While photons play an important role in the jet propagation and, naturally, in the gamma-ray emission of the GRB, neutrinos are of paramount importance in the genesis and evolution of the GRB engine, in particular in the collapsar context. During the initial gravitational collapse to a proto-neutron star and in the latter's short cooling period before black-hole formation, neutrinos carry away $\sim$99% of the liberated gravitational binding energy. After black-hole formation, neutrino cooling of the accretion disk and polar neutrino-antineutrino-pair annihilation are likely to be key ingredients for the GRB mechanism.

Ideally, the neutrino radiation field should be described and evolved via the Boltzmann equation, making the transport problem 7-dimensional and requiring the inversion of a gigantic semi-sparse matrix at each time step. This matrix inversion will be at the heart of the difficulties associated with the parallel scaling of radiation transport algorithms and it is likely that even highly integrated low-latency petascale systems will not suffice for full Boltzmann radiation transport, and sensible approximations will have to be worked out and implemented.

One such approximation may be neutrino, multi-energy-group, flux-limited diffusion (MGFLD) along individual radial rays that are not coupled with each other and whose ensemble covers the entire sphere/grid with reasonable resolution of a few degrees. When, in addition, energy-bin coupling (downscattering of neutrinos, a less-than-10% effect) and neutrino flavor changes are neglected, each ray for each energy group can be treated as a mono-energetic spherically symmetric calculation. Each of these rays can then be domain decomposed, and the entire ensemble of rays and energy groups can be updated in massively parallel and entirely scalable petascale fashion.

A clear downside of MGFLD is that all local angular dependence of the radiation field is neglected, making it (for example) fundamentally difficult to consistently estimate the energy deposited by radiation-momentum angle-dependent neutrino-antineutrino annihilation. An alternative to the MGFLD approximation that can provide an exact solution to the Boltzmann transport equation, while maintaining scalability, is the statistical Monte Carlo approach that follows a significant set of sample particle random walks. Monte Carlo may also be the method of choice for photon transport.

24.6.2 Scalability

In its current incarnation, the Cactus framework and its drivers PUGH and Carpet are not yet fully ready for petascale applications. Typical petascale machines will have many more processing units than today's machines, with more processing units per node, but likely with less memory-per-processing unit.

One immediate consequence is that it will be impossible to replicate meta-data across all processors. Not only the simulation data themselves, but also all secondary data structures will need to be distributed, and will have to allow asynchronous remote access. The driver thorns will have to be adapted to function without global knowledge of the simulation state, communicating only between neighboring processors, making global load balancing a challenge.

The larger number of cores per node will make hybrid communication schemes feasible and necessary, where intra-node communication uses shared memory while inter-node communication remains message-based. Explicit or implicit multithreading within a node will reduce the memory requirements, since data structures are kept only once per node, but will require a more complex overall orchestration to keep all threads well fed. Multithreading will have direct consequences for and may require alterations to almost all existing components. Multithreading will require a change of programming paradigm, as programmers will have to avoid race conditions, and debugging will be much more difficult than for a single-threaded code.

With the increasing number of nodes, globally synchronous operations will become prohibitively expensive, and thus the notion of "the current simulation time" will need to be abolished. Instead of explicitly stepping a simulation through time, different parts will advance at different speeds. Currently global operations, such as reduction or interpolation operations, will need to be broken up into pieces which are potentially executed at different times, where the framework has to ensure that these operations find the corresponding data, and that the results of such operations are collected back where they were requested.

Current Cactus thorns often combine the physics equations which are to be solved and the discretization methods used to implement them, e.g. finite differences on block-structured grids. Petascale computing may require different discretization techniques, and it will thus be necessary to isolate equations from their discretization so that one can be changed while keeping the other. Automated code generation tools such as, e.g., Kranc [18, 20] can automatically generate physics modules from given equations, achieving this independence.

As a framework, Cactus will not implement solutions to these problems directly, but will instead provide abstractions which can be implemented by a variety of codes. As external AMR drivers mature and prove themselves, they can be connected into Cactus. We are currently engaged in projects to incorporate the PARAMESH [29, 22] and SAMRAI [31] AMR infrastructures into Cactus drivers in the projects *Parca* (funded by NASA) and *Taka*, respectively.

Instead of continuing to base drivers on MPI, we plan to investigate existing new languages such as, e.g., co-array Fortran, Titanium, and UPC to provide parallelism, while we will also examine novel parallel-computing paradigms such as ParalleX¶. Cactus will be able to abstract most of the differences between these, so that the same physics and/or numerics components can be used with different drivers.

It should be noted that the Cactus framework does not prescribe a particular computing model. For example, after introducing AMR capabilities into Cactus, most existing unigrid thorns could be AMR-ified with relatively little work; the existing abstraction that each routine works on a block-structured set of grid-point was sufficient to enable the introduction of mesh refinement. We expect that the existing abstractions will need to be updated for petascale computing, containing, e.g., information about simultaneous action of different threads and removing the notion of a global-simulation time, but we also expect that this same abstraction will then cover a multitude of possible petascale-computing models.

24.6.3   Tools

Petascale computing provides not only new challenges for programming methodologies, but will also require new tools to enable programmers to cope with the new hardware and software infrastructure. Debugging on $100,000+$ processors will be a formidable challenge, and only good and meaningful profiling information for the new dynamic algorithms will make petascale performance possible. These issues are being addressed in three U.S. National Science Foundation (NSF) funded projects.

In the *ALPACA* (Application Level Profiling And Correctness Analysis) project, we are developing interactive tools to debug and profile scientific applications at the petascale level. When a code has left the stage where it has segmentation faults, it is still very far from giving correct physical answers. We plan to provide debuggers and profilers which will not examine the program from the outside, but will run together with the program, being coupled to the program via the framework, so that it has first-class information about the current state of the simulation. We envision a smooth transition between debugging and watching a production run progressing, or between running a

---

¶See   http://www.cs.sandia.gov/CSRI/Workshops/2006/HPC_WPL_workshop/
Presentations/22-Sterling-ParalleX.pdf.

special benchmark and collecting profiling information from production runs on the side.

The high cost of petascale simulations will make it necessary to treat simulation results similar to data gathered in expensive experiments. Our *XiRel* (CyberInfrastructure for Numerical Relativity) project seeks to provide scalable adaptive mesh refinement on the one hand, but also to provide means to describe and annotate simulation results, so that these results can later be interpreted and analyzed unambiguously so that the provenance of these data remains clear. With ever-increasing hardware and software complexity, ensuring reproducibility of numerical results is becoming an important part of scientific integrity.

The *DynaCode* project is providing capabilities and tools to adapt and respond to changes in the computational environment, such as, e.g., hardware failures, changes in the memory requirement of the simulation, or user-generated requests for additional analysis or visualization. This will include interoperability with other frameworks.

24.7   Acknowledgments

We acknowledge the many contributions of our colleagues in the Cactus, Carpet, and Whisky groups, and the research groups at the CCT, AEI, and at the University of Arizona. We thank especially Luca Baiotti and Ian Hawke for sharing their expertise on the Whisky code, and Maciej Brodowicz and Denis Pollney for their help with the preparation of this manuscript. This work has been supported by the Center for Computation & Technology at Louisiana State University (LSU), the Max-Planck-Gesellschaft, European Union (EU) Network HPRN-CT-2000-00137, NASA CAN NCCS5-153, NSF PHY 9979985, NSF 0540374 and by the Joint Institute for Nuclear Astrophysics sub-award no. 61-5292UA of NFS award no. 86-6004791.

We acknowledge the use of compute resources and help from the system administrators at the AEI (Damiana, Peyote), LONI (Eric), LSU (Pelican), and NCSA (Abe).

References

[1] M. Alcubierre, B. Brügmann, P. Diener, M. Koppitz, D. Pollney, E. Seidel, and R.f Takahashi. Gauge conditions for long-term numerical black

hole evolutions without excision. *Phys. Rev. D*, 67:084023, 2003.

[2] M. Alcubierre, B. Brügmann, T. Dramlitsch, J. A. Font, P. Papadopoulos, E. Seidel, N. Stergioulas, and R. Takahashi. Towards a stable numerical evolution of strongly gravitating systems in general relativity: The conformal treatments. *Phys. Rev. D*, 62:044034, 2000.

[3] R. Arnowitt, S. Deser, and C. W. Misner. The dynamics of general relativity. In L. Witten, editor, *Gravitation: An Introduction to Current Research*, pages 227–265. John Wiley, New York, 1962.

[4] L. Baiotti, I. Hawke, P. J. Montero, F. Löffler, L. Rezzolla, N. Stergioulas, J. A. Font, and E. Seidel. Three-dimensional relativistic simulations of rotating neutron star collapse to a Kerr black hole. *Phys. Rev. D*, 71:024035, 2005.

[5] Marsha J. Berger and Joseph Oliger. Adaptive mesh refinement for hyperbolic partial differential equations. *J. Comput. Phys.*, 53:484–512, 1984.

[6] R. Buras, H.-T. Janka, M. Rampp, and K. Kifonidis. Two-dimensional hydrodynamic core-collapse supernova simulations with spectral neutrino transport. II. Models for different progenitor stars. *Astron. Astrophys.*, 457:281, October 2006.

[7] A. Burrows, L. Dessart, E. Livne, C.D. Ott, and J. Murphy. Simulations of magnetically-driven supernova and hypernova explosions in the context of rapid rotation. *Astrophys. J.*, 664(1):416–434, 2007.

[8] A. Burrows, E. Livne, L. Dessart, C.D. Ott, and J. Murphy. Features of the acoustic mechanism of core-collapse supernova explosions. *Astrophys. J.*, 655:416, January 2007.

[9] Cactus computational toolkit. http://www.cactuscode.org/.

[10] Mesh refinement with Carpet. http://www.carpetcode.org/.

[11] J. Carter, L. Oliker, and J. Shalf. Performance evaluation of scientific applications on modern parallel vector systems. In *VECPAR: High Performance Computing for Computational Science*, Rio de Janeiro, Brazil, July 2006.

[12] EU astrophysics network. http://www.eu-network.org.

[13] C.L. Fryer and M.S. Warren. The collapse of rotating massive stars in three dimensions. *Astrophys. J.*, 601:391, January 2004.

[14] GEO 600. http://www.geo600.uni-hannover.de/.

[15] B. Giacomazzo and L. Rezzolla. WhiskyMHD: a new numerical code for general relativistic magnetohydrodynamics. *Class. Quantum Grav.*, 24:S235–S258, 2007.

[16] T. Goodale, G. Allen, G. Lanfermann, J. Massó, T. Radke, E. Seidel, and J. Shalf. The Cactus framework and toolkit: Design and applications. In *Vector and Parallel Processing – VECPAR'2002, 5th International Conference, Lecture Notes in Computer Science*, Berlin, 2003. Springer.

[17] HDF 5: Hierarchical Data Format version 5. `http://hdf.ncsa.uiuc.edu/HDF5/`.

[18] S. Husa, I. Hinder, and C. Lechner. Kranc: a Mathematica application to generate numerical codes for tensorial evolution equations. *Comput. Phys. Comm.*, 174:983–1004, 2006.

[19] S. Kamil, L. Oliker, J. Shalf, and D. Skinner. Understanding ultrascale application communication requirements. In *IEEE International Symposium on Workload Characterization*, 2005.

[20] Kranc: Automatic code generation. `http://numrel.aei.mpg.de/Research/Kranc/`.

[21] LIGO: Laser Interferometer Gravitational Wave Observatory. `http://www.ligo.caltech.edu/`.

[22] P. MacNeice, K. M. Olson, C.k Mobarry, R. de Fainchtein, and C. Packer. Paramesh: A parallel adaptive mesh refinement community toolkit. *Computer Physics Communications*, 126(3):330–354, 11 April 2000.

[23] P. Mészáros. Gamma-ray bursts. *Reports of Progress in Physics*, 69:2259, 2006.

[24] L. Oliker, A. Canning, J. Carter, J. Shalf, and S. Ethier. Scientific computations on modern parallel vector systems. In *Proc. SC04: International Conference for High Performance Computing, Networking, Storage and Analysis*, Pittsburgh, PA, November 2004.

[25] L. Oliker, A. Canning, J. Carter, J. Shalf, S. Ethier, T. Goodale, et al. Scientific application performance on candidate petascale platforms. In *Proc. IEEE International Parallel and Distributed Processing Symposium (IPDPS)*, Long Beach, CA, March 2007.

[26] L. Oliker, J. Carter, J. Shalf, D. Skinner, S. Ethier, R. Biswas, J. Djomehri, and R. Van der Wijngaart. Evaluation of cache-based superscalar and cacheless vector architectures for scientific computations. In *Proc. SC03: International Conference for High Performance Computing, Networking, Storage and Analysis*, November 2003.

[27] L. Oliker, J. Shalf, J. Carter, A. Canning, S. Kamil, and M. Lijewski. Performance characteristics of potential petascale scientific applications. In D. A. Bader, editor, *Petascale Computing: Algorithms and Applications*. Chapman & Hall/CRC Press, 2007.

[28] C.D. Ott, H. Dimmelmeier, A. Marek, H.-T. Janka, I. Hawke, B. Zink, and E. Schnetter. 3D collapse of rotating stellar iron cores in general relativity including deleptonization and a nuclear equation of state. *Phys. Rev. Lett.*, 98(261101), 2007.

[29] PARAMESH: Parallel adaptive mesh refinement. `http://www.physics.drexel.edu/~olson/paramesh-doc/Users_manual/amr.html`.

[30] F. Pretorius. Evolution of binary black hole spacetimes. *Phys. Rev. Lett.*, 95:121101, 2005.

[31] SAMRAI: Structured adaptive mesh refinement application infrastructure. `http://www.llnl.gov/CASC/SAMRAI/`.

[32] E. Schnetter, S. H. Hawley, and I. Hawke. Evolutions in 3D numerical relativity using fixed mesh refinement. *Class. Quantum Grav.*, 21(6):1465–1488, March 2004.

[33] S. Setiawan, M. Ruffert, and H.-T. Janka. Three-dimensional simulations of non-stationary accretion by remnant black holes of compact object mergers. *Astron. Astrophys.*, 458:553–567, November 2006.

[34] J. Shalf, S. Kamil, L. Oliker, and D. Skinner. Analyzing ultra-scale application communication requirements for a reconfigurable hybrid interconnect. In *Proc. SC05: International Conference for High Performance Computing, Networking, Storage and Analysis*, Seattle, WA, November 2005.

[35] J. van Meter, J. G. Baker, M. Koppitz, and D. Choi. How to move a black hole without excision: Gauge conditions for the numerical evolution of a moving puncture. *Phys. Rev. D*, 73:124011, 2006.

[36] VIRGO. `http://www.virgo.infn.it/`.

[37] Whisky, EU Network GR hydrodynamics code. `http://www.whiskycode.org/`.

[38] S.E. Woosley and J.S. Bloom. The supernova gamma-ray burst connection. *Ann. Rev. Astron. Astrophys.*, 44:507, September 2006.

[39] J. W. York. Kinematics and dynamics of general relativity. In L. L. Smarr, editor, *Sources of Gravitational Radiation*, pages 83–126. Cambridge University Press, Cambridge, U.K., 1979.

Index

Printed and bound by CPI Group (UK) Ltd, Croydon, CR0 4YY

23/10/2024

01777708-0011